Wind Loading of Structures

Third Edition

Wind Loading of Structures

Third Edition

John D. Holmes

JDH Consulting, Australia

CRC Press
Taylor & Francis Group
Boca Raton London New York

CRC Press is an imprint of the
Taylor & Francis Group, an **informa** business
A SPON PRESS BOOK

Reprint 2019

CRC Press
Taylor & Francis Group
6000 Broken Sound Parkway NW, Suite 300
Boca Raton, FL 33487-2742

First issued in paperback 2017

ISBN: 978-0-367-24180-3 (hbk)

Digitally Printed at Replika Press Pvt. Ltd.

Library of Congress Cataloging-in-Publication Data

Holmes, John D., 1942-
Wind loading of structures / John D. Holmes. -- Third edition.
pages cm
Includes bibliographical references and index.
ISBN 978-1-4822-2919-6 (hardback)
1. Wind-pressure. 2. Structural dynamics. 3. Buildings--Aerodynamics. I. Title.

TA654.5.H65 2015
624.1'75--dc23
2015006883

Visit the Taylor & Francis Web site at
http://www.taylorandfrancis.com

and the CRC Press Web site at
http://www.crcpress.com

For sale in India, Pakistan, Nepal, Bhutan, Bangladesh and Sri Lanka only.

Contents

Preface to the first edition

The wind loading of structures has had significant research effort in many countries during the last 30–35 years. Several thousand research papers have been published in journals and conference proceedings in all aspects of the subject. In many countries, wind loading governs the design of many structures; yet, even there, a good knowledge and understanding of wind loading among practising engineers is not widespread, despite the wealth of material available. Why is this the case? There are probably several reasons. The multi-disciplinary nature of the subject – involving probability and statistics, meteorology, the fluid mechanics of bluff bodies and structural dynamics – undoubtedly is a deterrent to structural engineers whose expertise is in the analysis and design of structures under nominally static loads. The subject is usually not taught in University and College courses, except as final-year undergraduate electives, or at post-graduate level, although exposure to wind-loading codes of practice or standards often occurs in design courses. Like many subjects, the jargon used by specialists and researchers in wind loading can be a deterrent to many non-specialists.

This book has been written with the practising structural engineer in mind, based on many years of experience working with clients in this profession. I hope it may also find use in advanced university courses. Although there are several other books on the subject, in this one, I have attempted to fill gaps in a number of areas:

- An overview of wind loading on structures of all types is given (not just buildings).
- The method of effective static wind load distributions is covered in some detail (mainly in Chapter 5). I have found this approach to be fluctuating and dynamic wind loading to have good acceptance among structural engineers, raised on a diet of static load analysis.
- Internal pressures are discussed in some detail (Chapter 6).
- An attempt has been made (Appendix D) to give an overview of extreme and design wind speeds for the whole world. This is probably a first anywhere, but it is an important step, and one that needs to be expanded in the future, as design projects are now routinely carried out by structural engineers in countries other than their own. The need for such information will become more important in the future as the expansion in world trade (including engineering services) continues.

I have tried to minimise the amount of mathematics, and concentrate on the physical principles involved. In some chapters (e.g. Chapter 5), I have found it necessary to include a significant amount of mathematics, but, hopefully, not at the expense of the physical principles. These sections could be omitted in a first reading.

I have been influenced by the work of many outstanding researchers and colleagues in this field over a period of 30 years. They are too many to list but most of their names will be found in the reference lists attached to each chapter. However, a number of people have assisted with the production of this book: Professor K.C.S. Kwok for contributing most of Section 15.9; Dr. John Ginger, Michael Syme, Dr. Ignatius Calderone and Dr. Jannette Frandsen for reading parts of the manuscript; Heather Fordham, Paul Bowditch, Maryjeanne Watt and Harry Fricke for the drafting of figures; Shoba Narayan for typing permission letters and Elizabeth Gray for assisting with indexing. I am most grateful for their assistance. I would also like to thank the staff of E.F. and N. Spon for their patience in waiting for the delivery of the manuscript.

I would be most happy to receive constructive comments and suggestions from readers.

J.D. Holmes
Mentone, Australia

Preface to the second edition

More than 5 years have elapsed since the first edition of 'Wind Loading of Structures' was published. At that time, the need for a good design of structures against the effects of wind has become even more obvious, with the great increase in destructive wind storms that have affected many parts of the world. In particular, this was emphasised by the unprecedented two successive hurricane seasons of 2004 and 2005 on the southern coastline of the United States.

Although the original chapter topics of the first edition of the book have remained unchanged, there have been changes to nearly every chapter and appendix. The importance of strong winds generated by thunderstorms has become more recognised by wind-engineering researchers in the last 5 years, and Chapters 1 and 3 have been expanded with new material on the structure of tornadoes and downbursts. Chapter 5 also includes a new section on the transient dynamic response to winds of this type. In Chapter 7, the laboratory simulation of tornadoes is discussed; although pioneer work in this area was carried out in the 1960s and 1970s, this was not included in the first edition. Chapter 7 has also been expanded with a new section on the simulation of internal pressures in a wind tunnel.

There are changes and additions to Chapters 8, 9, 11, 12 and 14, but none in Chapters 10 and 13. However, Chapter 15 on 'Wind loading codes and standards' has been completely re-written. This has been necessary since four out of the six major codes and standards reviewed in the first edition have been revised extensively during the last 5 years (twice in the case of ASCE 7!). However, clearly, this chapter will become out of date again quite quickly.

Appendix D has also been greatly extended, with the basic extreme wind information given for an additional 16 countries or regions. I would like to thank many people who have provided me with additional information for that part of the book.

I would also like to thank many people who have provided useful comments on the first edition of the book, the several University staff who have adopted the book for teaching post-graduate courses, Dr. M. Matsui (Tokyo Polytechnic University) for providing Figure 11.6 and my daughter Lucy Holmes for the drafting of most of the new figures in the second edition. Finally, thanks are due to Taylor & Francis (Spon Press) for supporting this book into a second edition, and acceding to most of my requests.

John D. Holmes
Mentone, Australia

Preface to the third edition

Since the publication of the second edition of *Wind Loading of Structures*, strong windstorms have continued to occur at frequent intervals across the world: Hurricane 'Sandy' (2012) on the east coast of the United States, and Typhoon 'Haiyan' in the Philippines (2014) are just two recent damaging examples. Whether there is an increase in the number and/or intensity of these events as a result of global warming is not yet established, but it is certain that the number of structures exposed to the loads imparted by strong winds *has* escalated, and there is a strong demand by structural engineers worldwide for guidance on practical methods for the design of wind actions, with the appropriate balance between safety and economy.

In the same period, the largest structures have become even more wind sensitive. For example, the current world's tallest building (*Burj Khalifa*) is now nearly twice the height of the tallest building in 2007 (*Taipei 101*), requiring a large amount of wind engineering in the design process.

In the last 7 years, there has also been an explosion in the number of research papers on the topic of wind loading of structures through reports, conference presentations and journal papers, so that it is now virtually impossible for even a specialist to keep up with the output. However, I have endeavoured to expand parts of the book with new information, when it is of practical value to practising engineers, or if it provides new understanding of the basic phenomena involved.

In the third edition, there have been changes and additions to every chapter and to two appendices. In particular, Chapter 15 has been re-written in a more general way, focussing on the general format and methodologies of the major codes and standards, with less emphasis on particular editions of these documents, which tend to change at regular intervals. Appendix D has been expanded to now cover 84 countries and jurisdictions. However, the general layout of the book and the chapter headings seem to have been successful and have been retained.

Thanks are due to all the purchasers of this book – around 2000 copies for the first and second editions, and to those who have made constructive comments – in particular Drs. Leighton Cochran, John Ginger and Seifu Bekele. I have attempted to incorporate their comments into the third edition. I thank my daughter Lucy Holmes for the cover design for the third edition. I also am grateful to Taylor & Francis/CRC Press for continuing to support the book, and their patience for my attempts to meet their deadlines.

Finally, I would like to honour the memory of two pioneers in wind engineering from whom I learnt a lot, and who both passed away in the period since 2007 – Alan Davenport (1932–2009) and Jack Cermak (1922–2012). Their contributions to the field, including groundbreaking experimental and theoretical developments, were immense and will live on after their death.

John D. Holmes
Mentone, Australia

Author

Dr. John D. Holmes is the director of JDH Consulting, Mentone, Victoria, Australia. He has a BSc (Eng.) from the University of Southampton, UK, and a PhD from Monash University, Australia, is Fellow of Engineers Australia and a life member of the Australasian Wind Engineering Society. He is the author of more than 400 research papers and consulting reports. Among several awards, he received the Senior Award (A.G. Davenport Medal) from the International Association for Wind Engineering in 2011. He has been involved in the determination of design wind loads for many major (and minor) structures and industrial facilities, has been actively involved in the writing of several Australian Standards and is currently the Chair of the Wind Loads Subcommittee for Australia and New Zealand.

Chapter 1

The nature of wind storms and wind-induced damage

1.1 INTRODUCTION

Wind loading competes with seismic loading as the dominant environmental loading for structures. They have produced roughly equal amounts of damage over a long time period, although large damaging earthquakes occur less often than severe wind storms. On almost every day of the year, a severe wind storm is happening somewhere on Earth – although many storms are small and localised. In the tropical oceans, the most severe of all wind events – tropical cyclones (including hurricanes and typhoons) – are generated. When these storms make landfall on populated coastlines, their effects can be devastating.

In this introductory chapter, the meteorology of severe wind storms – *gales* produced by large extra-tropical depressions – *tropical cyclones* and *downbursts*, squall lines and *tornadoes* associated with thunderstorms, is explained, including the expected horizontal variation in wind speed during these events. The history of damaging wind events, particularly those of the last 30 years, is discussed, focussing on the lessons learnt from them by the structural engineering profession. The behaviour of flying debris, a major source of damage in severe wind storms, is outlined. Insurance aspects are discussed, including the recent development of loss models, based on historical data on the occurrences of large severe storms, the spatial characteristics for the wind speeds within them and assumed relationships between building damage and wind speed.

1.2 METEOROLOGICAL ASPECTS

Wind is air movement relative to the Earth, driven by several different forces, especially pressure differences in the atmosphere, which are themselves produced by differential solar heating of different parts of the Earth's surface, and forces generated by the rotation of the Earth. The differences in solar radiation between the poles and the Equator produce temperature and pressure differences. These together with the effects of the Earth's rotation set up large-scale circulation systems in the atmosphere, with both horizontal and vertical orientations. The result of these circulations is that the prevailing wind directions in the tropics, and near the poles, tend to be easterly. Westerly winds dominate in the temperate latitudes.

Local severe winds may also originate from local convective effects (*thunderstorms*), or from the uplift of air masses produced by mountain ranges (*downslope winds*). Severe tropical cyclones, known in some parts of the world as *hurricanes* and *typhoons*, generate extremely strong winds over some parts of the tropical oceans and coastal regions, in latitudes from 10 to about 30 degrees, both north and south of the Equator.

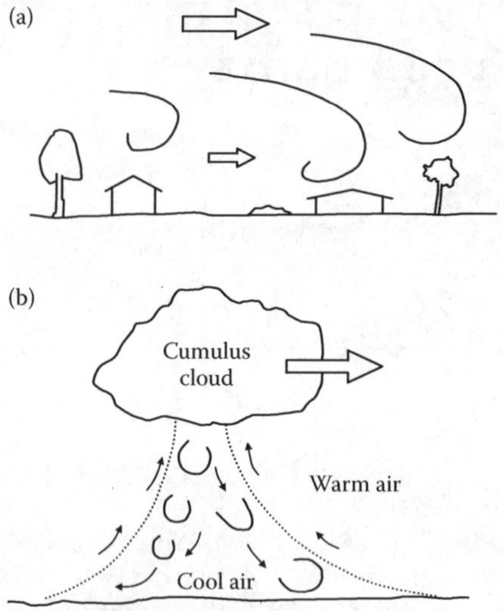

Figure 1.1 The generation of turbulence in (a) boundary-layer winds and, (b) thunderstorm downdrafts.

For all types of severe storms, the wind is highly turbulent or gusty. The turbulence or gustiness is produced by eddies or vortices within the airflow which are generated by frictional interaction at ground level, or by shearing action between air moving in opposite directions at a high altitude. These processes are illustrated in Figure 1.1 for downdrafts generated by thunderstorms, and for larger storms such as gales or tropical cyclones that are of the 'boundary-layer' type.

1.2.1 Pressure gradient

The two most important forces acting on the upper-level air in the 'free atmosphere', which are above the frictional effects of the Earth's boundary layer, are the pressure gradient force and the Coriolis force.

It is shown in elementary texts on fluid mechanics that, at a point in a fluid where there is a pressure gradient, $\partial p/\partial x$, in a given direction, x, in a Cartesian coordinate system, there is a resulting force per unit mass given by Equation 1.1.

$$\text{Pressure gradient force per unit mass} = -\left(\frac{1}{\rho_a}\right)\frac{\partial p}{\partial x} \tag{1.1}$$

where

ρ_a is the density of air

This force acts from a high-pressure region to a low-pressure region.

1.2.2 Coriolis force

The *Coriolis* force is an apparent force due to the rotation of the Earth. It acts to the right in motion direction in the Northern hemisphere, and to the left of the velocity vector, in

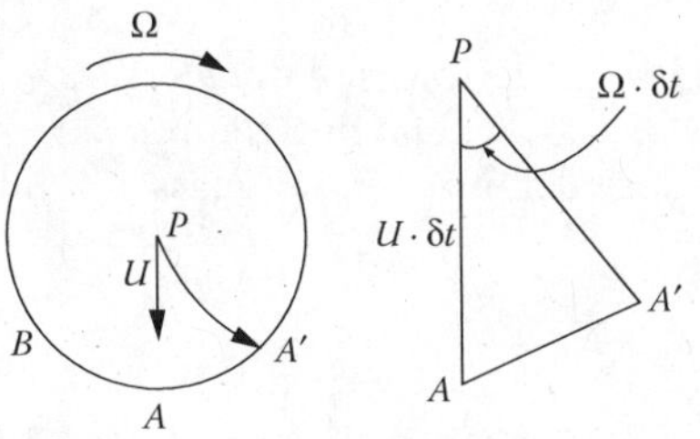

Figure 1.2 The apparent (Coriolis) force due to the Earth's rotation (Southern hemisphere).

the case of the Southern hemisphere; at the Equator, the Coriolis force is zero. Figure 1.2 gives a simple explanation of the Coriolis force by observing the motion of a particle of air northwards from the South Pole.

Consider a parcel of air moving horizontally away from the South Pole, P, with a velocity U, in the direction of point A (Figure 1.2, left). Since the Earth is rotating clockwise with angular velocity, Ω, the point originally at A, will have moved to B, and a point originally at A', will have moved to A, as the air parcel arrives. *Relative to the Earth's surface*, the particle will have appeared to follow the path PA', that is to have undergone a continuous deflection to the left. At the North Pole, the deflection is to the right. These deflections can be associated with an apparent acceleration acting at right angles to the velocity of the parcel – the Coriolis acceleration.

Consider a small time interval, δt, (Figure 1.2, right); AA' is then small compared with PA. In this case,

$$AA' = \Omega\, U\, (\delta t)^2 \tag{1.2}$$

Let the Coriolis acceleration be denoted by a. Since AA' is the distance travelled under this acceleration, then it can also be expressed by

$$AA' = (1/2)\, a(\delta t)^2 \tag{1.3}$$

Equating the two expressions for AA', Equations (1.2) and (1.3),

$$a = 2U\, \Omega \tag{1.4}$$

This gives the Coriolis acceleration, or force per unit mass, at the poles.

At other points on the Earth's surface, the angular velocity is reduced to $\Omega \sin \lambda$, where λ is the latitude. Then the Coriolis acceleration is equal to $2U\, \Omega \sin \lambda$. The term $2\, \Omega \sin \lambda$ is a constant for a given latitude, and is called the 'Coriolis parameter', often denoted by the symbol, f. The Coriolis acceleration is then equal to fU.

Thus, the Coriolis force is an apparent, or effective, force acting to the right of the direction of air motion in the Northern hemisphere, and to the left of the air motion in the Southern hemisphere. At the Equator, the Coriolis force is zero, and in the Equatorial region, within about five degrees of the Equator, it is negligible in magnitude. The latter explains why tropical cyclones (Section 1.3.2), or other cyclonic systems, will not form in the Equatorial regions.

1.2.3 Geostrophic wind

Steady flow under equal and opposite values of the *pressure gradient* and the *Coriolis* force is called 'balanced geostrophic flow'. Equating the pressure gradient force per unit mass from Equation 1.1 and the Coriolis force per unit mass, given by fU, we obtain:

$$U = -\left(\frac{1}{\rho_a f}\right)\frac{\partial p}{\partial x} \tag{1.5}$$

This is the equation for the *geostrophic wind speed*, which is proportional to the magnitude of the pressure gradient, $(\partial p/\partial x)$.

The directions of the pressure gradient and Coriolis forces, and of the flow velocity are shown in Figure 1.3, for both Northern and Southern hemispheres. It may be seen that the flow direction is parallel to the isobars (lines of constant pressure), in both hemispheres. In the Northern hemisphere, the high pressure is to the right of an observer facing the flow direction; in the Southern hemisphere, the high pressure is on the left. This results in anti-clockwise rotation of winds around a low-pressure centre in the Northern hemisphere, and a clockwise rotation in the Southern hemisphere. In both hemispheres, rotation about a low-pressure centre (which usually produces strong winds) is known as a 'cyclone' to meteorologists. Conversely, rotation about a high-pressure centre is known as an 'anti-cyclone'.

1.2.4 Gradient wind

If the isobars have significant curvature (as e.g. near the centre of a tropical cyclone), then the *centrifugal* force acting on the air particles cannot be neglected. The value of the centrifugal force per unit mass is (U^2/r), where U is the resultant wind velocity, and r is the radius of curvature of the isobars.

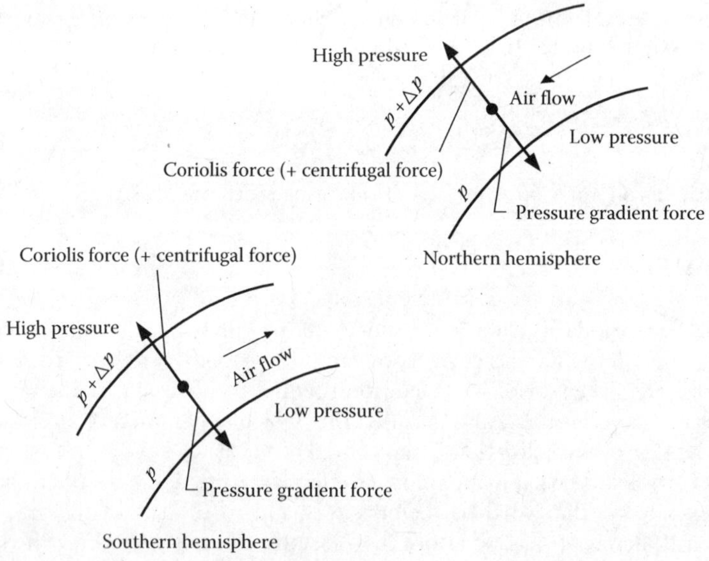

Figure 1.3 Balanced geostrophic flow in Northern and Southern hemispheres.

The direction of the force is away from the centre of curvature of the isobars. If the path of the air is around a high-pressure centre (anti-cyclone), the centrifugal force acts in the same direction as the pressure gradient force, and in the opposite direction to the Coriolis force. For flow around a low-pressure centre (cyclone), the centrifugal force acts in the same direction as the Coriolis force, and opposite to the pressure gradient force.

The equation of motion for a unit mass of air moving at a constant velocity, U, is then Equation 1.6 for an anti-cyclone, and Equation 1.7 for a cyclone:

$$\frac{U^2}{r} - |f|U + \frac{1}{\rho_a}\left|\frac{\partial p}{\partial r}\right| = 0 \tag{1.6}$$

$$\frac{U^2}{r} + |f|U - \frac{1}{\rho_a}\left|\frac{\partial p}{\partial r}\right| = 0 \tag{1.7}$$

Equations 1.6 and 1.7 apply to both hemispheres. Note that the pressure gradient $\frac{\partial p}{\partial r}$ is negative in an anti-cyclone and that f is negative in the Southern hemisphere. These equations are quadratic equations for the *gradient wind speed, U*. In each case, there are two theoretical solutions, but if the pressure gradient is zero, then U must be zero, so that the solutions become:

$$U = \frac{|f|r}{2} - \sqrt{\frac{f^2r^2}{4} - \frac{r}{\rho_a}\left|\frac{\partial p}{\partial r}\right|} \quad \text{for an anti-cyclone} \tag{1.8}$$

$$U = -\frac{|f|r}{2} + \sqrt{\frac{f^2r^2}{4} + \frac{r}{\rho_a}\left|\frac{\partial p}{\partial r}\right|} \quad \text{for a cyclone.} \tag{1.9}$$

Examining Equation 1.8, it can be seen that a maximum value of U occurs when the term under the square root sign is zero. This value is $|f|r/2$, which occurs when $|\partial p/\partial r|$ is equal to $\rho_a f^2 r/4$. Thus, in an anti-cyclone, there is an upper limit to the gradient wind; *anti-cyclones are normally associated with low wind speeds.*

Now, considering Equation 1.9, it is clear that the term under the square root sign is always positive. The wind speed in a cyclone is therefore only limited by the pressure gradient; *cyclones are therefore associated with strong winds.*

1.2.5 Frictional effects

As the Earth's surface is approached, *frictional* forces, transmitted through shear between layers of air in the atmospheric boundary layer, gradually play a larger role. This force acts in a direction opposite to that of the flow direction, which, in order to achieve a vector balance, is not parallel to the isobars, but directed towards the low-pressure region. Figure 1.4 shows the new balance of forces in the boundary layer.

Thus, as the ground surface is approached from above, the wind vector gradually turns towards the low-pressure centre, as the height reduces. This effect is known as the *Ekman*

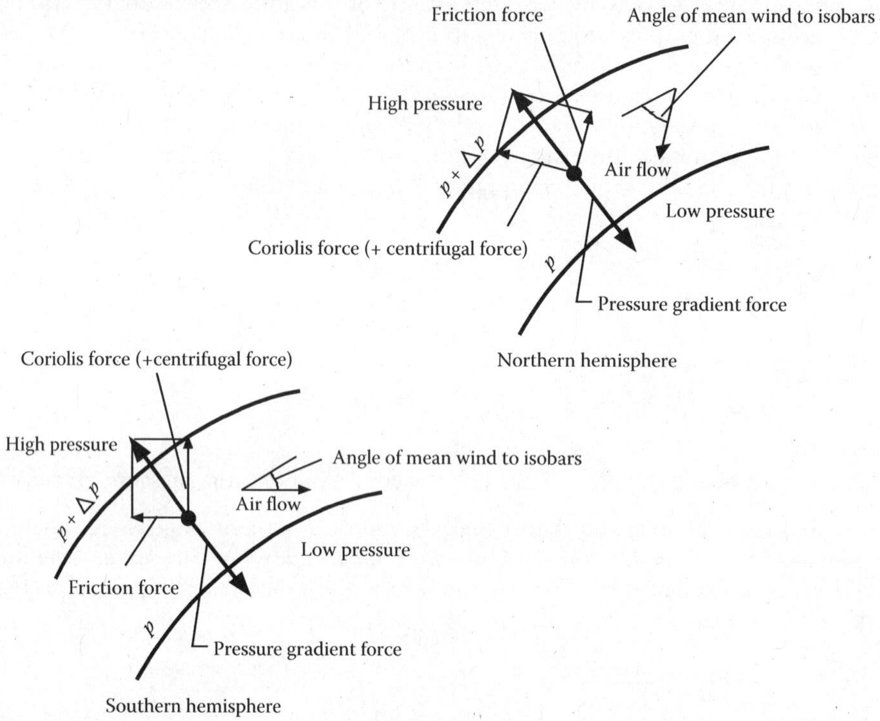

Figure 1.4 Force balance in the atmospheric boundary layer.

spiral. The total angular change between gradient height and the surface is about thirty degrees. However, the angular change over the height of most tall structures is quite small.

1.3 TYPES OF WIND STORMS

1.3.1 Gales from large depressions

In the mid-latitudes from about 40 to 60 degrees, the strongest winds are gales generated by large and deep depressions or (extra-tropical) cyclones, of synoptic scale. They can also be significant contributors to winds in lower latitudes. Navigators, particularly in sailing ships, are familiar with the strong westerly winds of the 'roaring forties', of which those of the North Atlantic, and at Cape Horn are perhaps the most notorious. As shown in Section 1.4.1, severe building damage has been caused by winter gales in northwest Europe.

These systems are usually large in horizontal dimension – they can extend for more than 1000 km; so, they can influence large areas of land during their passage – several countries in the case of Europe. They may take several days to pass, although winds may not blow continuously at their maximum intensity during this period. The winds tend to be quite turbulent near the ground, as the flow has adjusted to the frictional effects of the Earth's surface over hundreds of kilometres. The direction of the winds remains quite constant over many hours. These features are illustrated in a typical anemograph (wind speed and direction vs. time) from this type of event reproduced in Figure 1.5.

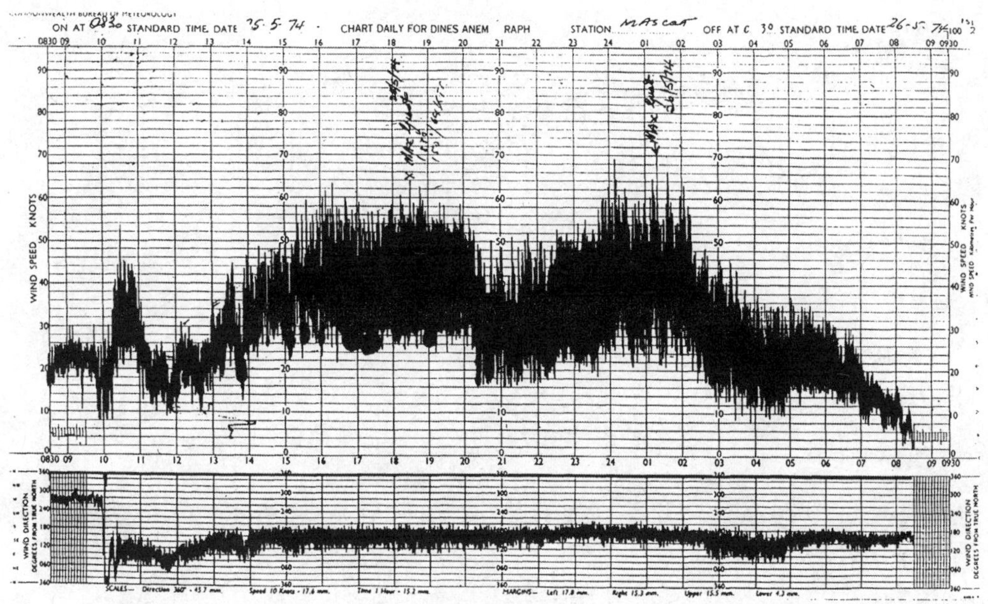

Figure 1.5 Anemograph for a large extra-tropical depression. Upper graph is wind speed in knots; lower graph is wind direction in degrees. Abcissa of both graphs is in hours.

1.3.2 Tropical cyclones

Tropical cyclones are intense cyclonic storms that occur over the tropical oceans, mainly in late summer and autumn. They are driven by the latent heat of the oceans, and require a minimum sea temperature of about 26°C to sustain them; they rapidly degenerate when they move over land, or into cooler waters. They will not form within about five degrees of the Equator, and do not reach full strength until they reach at least ten degrees latitude. They are usually at full strength when they are located between 20 and 30 degrees latitude, but can travel to higher latitudes if there are warm ocean currents to sustain them.

The strongest tropical cyclones have occurred in the Caribbean, where they are known as *hurricanes*, in the South China Sea, where they are called *typhoons*, and off the northwest coast of Australia. Areas of medium tropical cyclone activity are the eastern Pacific Ocean off the coast of Mexico, the southern Indian Ocean, the Bay of Bengal, the South Pacific, southern Japan, the Coral Sea (off Eastern Australia) and the southeast Atlantic Ocean. Regions of lesser activity or weaker storms are the Arabian Sea, the Gulf of Thailand and the north coast of Australia (including the Gulf of Carpentaria).

A developed tropical cyclone has a three-dimensional vortex structure, which is schematically shown in Figure 1.6. The horizontal dimensions of these storms are less than the extra-tropical cyclones, or depressions, discussed earlier, but their effects can extend for several hundred kilometres. The circulation flows with a radial component towards the 'eye', outside of which is a region of intense thermal convection with air currents spiralling upwards. Inside the eye is a region of relative calm with slowly sinking air; the diameter of the eye can range between 8 and 80 km. Often, clear skies have been observed in this region. The strongest winds occur just outside the eye wall.

Figure 1.7 gives an example of an anemograph measured at a height of 10 m above the ground for a tropical cyclone. This example shows a fortuitous situation when the eye of

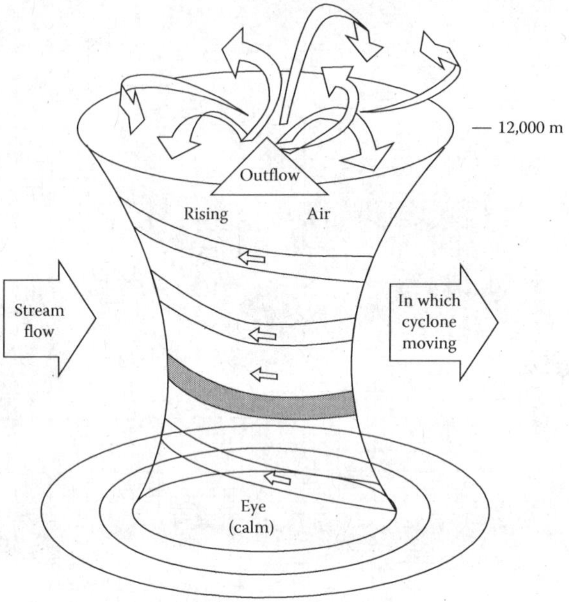

Figure 1.6 Three-dimensional structure for a developed tropical cyclone.

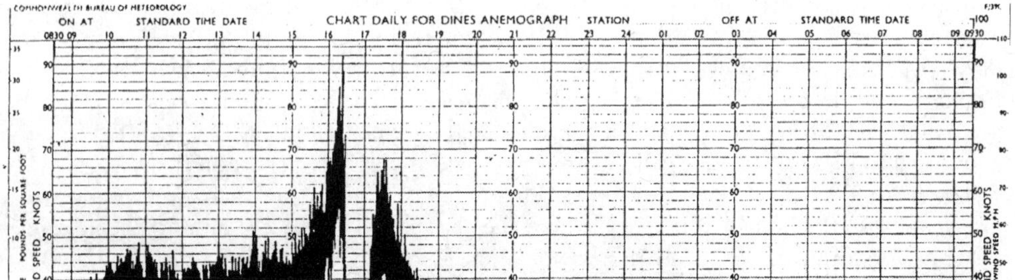

Figure 1.7 Anemograph at 10-m height for a tropical cyclone.

the storm passed nearly directly over the recording station, resulting in a period of about an hour of very low winds. The direction changed nearly 180 degrees during the passage of the vortex over the measuring station.

Outside of the eye of a tropical cyclone, the wind speed at upper levels decays with the radial distance from the storm centre. The gradient wind equation (Equation 1.9) can be used to determine this wind speed:

$$U = -\frac{|f|r}{2} + \sqrt{\frac{f^2r^2}{4} + \frac{r}{\rho_a}\left|\frac{\partial p}{\partial r}\right|} \tag{1.9}$$

where

f is the Coriolis parameter (=2 Ω sin λ)
r is the radius from the storm centre
ρ_a is the density of air
p is the atmospheric pressure

To apply Equation 1.9, it is necessary to establish a suitable function for the pressure gradient. A commonly assumed expression is the following (Holland, 1980):

$$\frac{p - p_o}{p_n - p_o} = \exp\left(\frac{-A}{r^B}\right) \tag{1.10}$$

where

p_o is the central pressure of the tropical cyclone
p_n is the atmospheric pressure at the edge of the storm
A and B are scaling parameters

The pressure difference ($p_n - p_o$) can be written as Δp, and is an indication of the strength of the storm.

Differentiating Equation 1.10 and substituting in Equation 1.9, we have:

$$U = -\frac{|f|r}{2} + \sqrt{\frac{f^2r^2}{4} + \frac{\Delta p\, AB}{\rho_a\, r^B}\exp\left(-\frac{A}{r^B}\right)} \tag{1.11}$$

This is an equation for the mean wind field at upper levels in a tropical cyclone as a function of radius from the storm centre, r, the characteristic parameters, A and B, the pressure drop across the cyclone, Δp and the Coriolis parameter, f.

Near the centre of a tropical cyclone, the Coriolis forces, that is the first two terms in Equations 1.9 and 1.11, are small, and it can be shown by differentiating the remaining term that the maximum value of U occurs when r equals $A^{1/B}$. Thus, $A^{1/B}$ is to a good approximation, the radius of maximum winds in the cyclone. The exponent B is found to be in the range 1.0–2.5, and to reduce with increasing central pressure, p_o (Holland, 1980).

Figure 1.8 shows the profiles of pressure and gradient wind speed with radial distance from the centre of the storm calculated from Equations 1.10 and 1.11, for Cyclone 'Tracy' which severely damaged Darwin, Australia, in 1974. The parameters A and B were taken as 23 and 1.5, (where r is measured in kilometres), following Holland (1980). The gradient

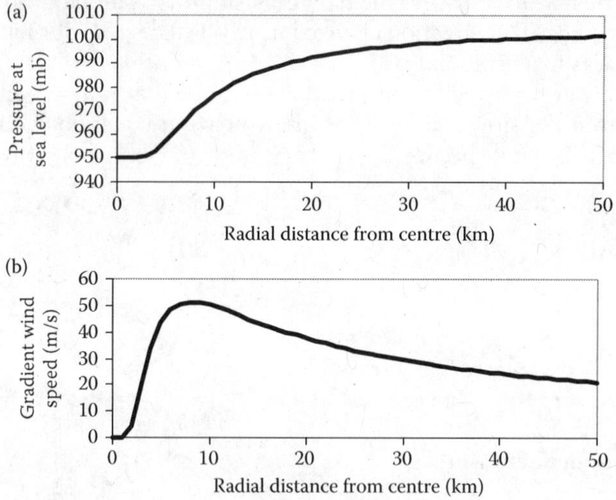

Figure 1.8 Pressure and gradient wind speeds for Cyclone 'Tracy', 1974. (a) Sea-level pressure. (b) Gradient wind speed.

Table 1.1 Saffir–Simpson intensity scale for hurricanes

Category	*Central pressure (millibars)*	*Wind speed range (3-s gust, m/s)*
I	>980	42–54
II	965–979	55–62
III	945–964	63–74
IV	920–944	75–88
V	<920	>88

wind speed in Figure 1.8b is approximately equal to the gust wind speed near the ground level. The radius of maximum winds, in this case about 8 km, approximately coincides with the maximum pressure gradient.

The forward motion of the moving storm adds an additional vector component to the wind speed given by Equation 1.11, which gives the wind speed *relative* to the moving storm.

An intensity scale for North Atlantic and Caribbean hurricanes has been proposed by Saffir and Simpson. This is reproduced in Table 1.1.

This scale is widely used for forecasting and emergency management purposes. However, the wind speed ranges given in Table 1.1 should be used with caution, as the estimated wind speeds in hurricanes are usually obtained from upper-level aircraft readings. A similar, but not identical, scale is used in the Australian region.

1.3.3 Thunderstorms

Thunderstorms, both isolated storms and those associated with advancing cold fronts, are small disturbances in the horizontal extent, compared with extra-tropical depressions and tropical cyclones, but they are capable of generating severe winds, through tornadoes and downbursts. They significantly contribute to the strongest gusts recorded in many countries, including the United States, Australia and South Africa. They are also the main source of high winds in the Equatorial regions (within about ten degrees of the Equator), although their strength is not high in these regions.

Thunderstorms also derive their energy from heat. Warm moist air is convected upwards to mix with the drier upper air. With evaporation, rapid cooling occurs and the air mass loses its buoyancy, and starts to sink. Condensation then produces heavy rain or hail that falls, dragging cold air with it. A strong downdraft reaches the ground, and produces a strong wind for a short period of time – perhaps 5–10 min. The strongest winds produced by this mechanism are known as *downbursts*, which are further sub-divided into *microbursts* and *macrobursts*, depending on their size. The strongest winds produced by these events have a large component of wind speed due to the forward motion of the convection cell.

The conditions for generation of severe thunderstorms are:

- Water vapour in the atmosphere at low levels, that is high humidity
- Instability in the atmosphere, that is a negative temperature gradient with height that is greater than the adiabatic rate of the neutral atmosphere
- A lifting mechanism that promotes the initial rapid convection – this may be provided by, for example, a mountain range, or a cold front

1.3.4 Tornadoes

The strongest convection cells, that often generate tornadoes, are known as *supercells*. They are larger and last longer than 'ordinary' convection cells. The tornado, a vertical funnel-shaped vortex created in thunderclouds, is the most destructive of wind storms. Fortunately, they are quite small in their horizontal extent – of the order of 100 m – but they can travel for quite long distances up to 50 km before dissipating, producing a long narrow path of destruction. They occur mainly in large continental plains in countries such as the United States, Argentina, Russia and South Africa.

Periodically, atmospheric conditions in the central United States are such that severe outbreaks with many damaging tornadoes can occur in a short period. For example, this has occurred in April 1974, May 2003 and April 2011. In 1974, 335 fatalities, and destruction of about 7500 dwellings resulted from the 'super-outbreak' of 148 tornadoes within a 2-day period (3–4 April, 1974) with 13 states affected. In 2003, a total of 393 tornadoes were reported in 19 states of the United States in a period of about a week. Of these, 15 resulted in 41 fatalities. In 2011, 325 deaths occurred in the outbreak of 25–28 April in the southeastern states. The year 2011 was also notable for the Joplin, Missouri, tornado of 22 May, 2011 which killed 158 persons. A detailed survey of tornadoes in South Africa has been given by Goliger et al. (1997). They occur in that country at the rate of about four per year, with a concentration in the Gauteng Province in the north of the country, with an occurrence rate of 1×10^{-4} per square kilometre per year. This compares with a rate of about 2×10^{-4} per square kilometre per year in the midwest United States.

Tornadoes are sometimes confused with downbursts (described in the following section); however, tornadoes can be identified by the appearance of the characteristic funnel vortex, a long narrow damage 'footprint', and the evidence of varying wind directions.

The wind speed in a tornado can be related to the radial pressure gradient by neglecting the Coriolis term in the equation of motion. Hence, from either Equation 1.7 or 1.9:

$$U = \sqrt{\frac{r}{\rho_a}\left|\frac{\partial p}{\partial r}\right|} \tag{1.12}$$

Table 1.2 Fujita intensity scale for tornadoes

Category	*Original Fujita wind speed range (m/s)*	*Enhanced Fujita wind speed range (m/s)*
F0	20–35	29–38
F1	35–52	39–49
F2	53–72	50–60
F3	72–93	61–74
F4	94–117	75–89
F5	117–142	Over 90

This is known as the *cyclostrophic* wind speed. Assuming that the pressure is constant along the edge of a tornado funnel (actually a line of condensed water vapour), Equation 1.12 has been used to estimate wind speeds in tornadoes.

Measurement of wind speeds in tornadoes is very difficult. Because of their small size, they seldom pass over a weather-recording station. If one does, the anemometer is quite likely to be destroyed. For many years, photogrammetric analyses of movie film shots by eyewitnesses were used to obtain reasonable estimates (Fujita et al., 1976; Golden, 1976). The method involves the tracking of clouds, dust and solid debris from the film frames, and was first applied to the Dallas, Texas, tornado of 2 April, 1957 by Hoecker (1960). This method is subject to a number of errors – for example, distortion produced by the camera or projector lenses or tracked large objects not moving with the local wind speed. Also, the method is not able to detect velocities normal to the image plane.

However, the photogrammetric method has enabled several significant features of tornadoes, such as 'suction vortices' – smaller vortex systems rotating around the main vortex core, and the high vertical velocities. In the latter case, analysis of a tornado at Kankakee, Illinois in 1963 (Golden, 1976) indicated vertical velocities of 55–60 m/s, at a height above the ground of <200 m.

Analyses of failures of engineered buildings in tornadoes have generally indicated lower maximum wind speeds in tornadoes than those obtained by photogrammetric or other methods (e.g. Mehta, 1976). After considering all the available evidence at that time, Golden (1976) estimated the maximum wind speeds in tornadoes to be not more than 110 m/s.

In recent years, portable Doppler radars have been successfully used in the United States for more accurate determination of wind speeds in tornadoes.

An intensity scale for tornadoes was originally proposed in 1971 (Fujita, 1971). Several F-scale classifications are associated with wind speed ranges, although, in practice, classifications are applied based on observed damage to buildings and other structures. The original scaling was criticised by engineers for several reasons: for example, for failing to account for variations in the quality of construction, and that it has not been based on a proper correlation of damage descriptions and wind speeds. The wind speed ranges of the original Fujita Scale and the Enhanced Fujita Scale (McDonald and Mehta, 2004) for F0 to F5 categories are given in Table 1.2. Ranges are given for 3-s average gust speeds in metres per second. The Enhanced Fujita Scale is now used for operational purposes in the United States.

An engineering model of wind speed distributions in a tornado is discussed in Section 3.2.7.

1.3.5 Downbursts

Figure 1.9 shows an anemograph from a severe thunderstorm downburst, recorded at the Andrew's Air Force Base, near Washington, DC, USA in 1983, with a timescale in minutes. The short duration of the storm is quite apparent, and there is also a rapid change of wind

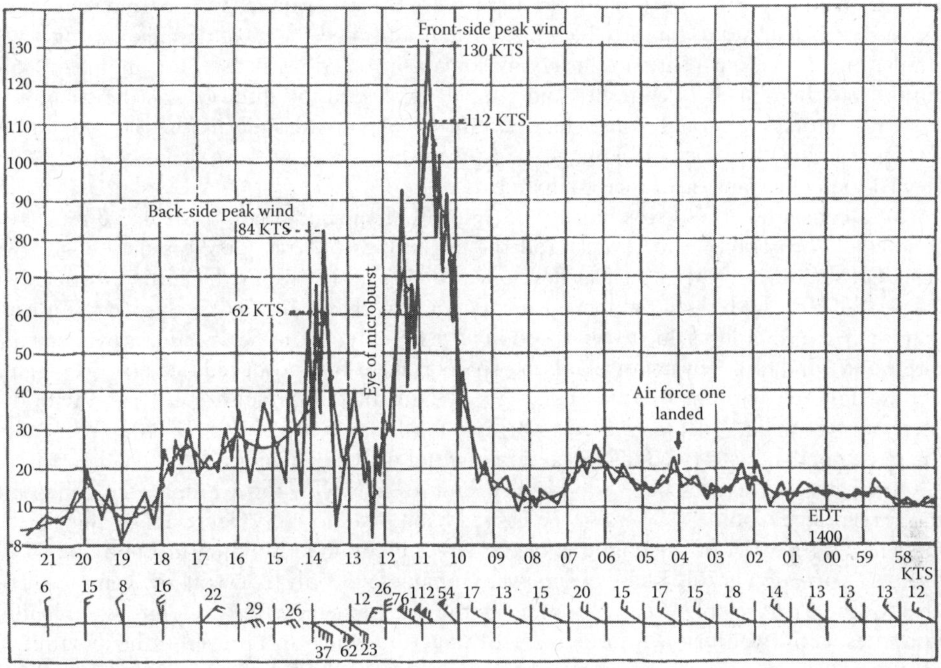

Figures 1.9 Anemograph for a severe downburst at Andrews Air Force Base, Maryland, USA, 1983. (From Fujita, T.T. 1985. *The Downburst.* Report on Projects NIMROD and JAWS. Published by the author at the University of Chicago.)

direction during its passage across the measurement station. Such events typically produce a damage footprint that is 2–3 km wide and 10–15 km long.

The horizontal wind speed in a thunderstorm downburst with respect to the moving storm is similar to that in a jet of fluid impinging on a plain surface. It varies approximately linearly from the centre of impact to a radius where the wind speed is maximum, and then decays with increasing radius. The forward velocity of the moving storm can be a significant component of the total wind speed produced at ground level, and must be added as a vector component to that produced by the jet.

1.3.6 Downslope winds

In certain regions such as those near the Rocky Mountains of the United States, Switzerland, Croatia, and the Southern Alps of New Zealand, extreme winds can be caused by thermal amplification of synoptic winds on the leeward slopes of mountains. The regions affected are usually quite small, but are often identified as 'special regions', in wind-loading codes and standards (see Appendix D).

1.4 WIND DAMAGE

Damage to buildings and other structures by wind storms has been a fact of life for human beings from the time they moved out of cave dwellings to the present day. Trial and error has played an important part in the development of construction techniques and roof shapes for small residential buildings, which have usually suffered the most damage during severe winds. In the past few centuries, heavy masonry construction, as used for important community buildings such as churches and temples, was seen, by intuition, as the solution to resist wind forces (although somewhat less effective against seismic action). For other types of construction, wind storm damage was generally seen as an 'Act of God', as it is still viewed today by some insurance companies.

The nineteenth century was important as it saw the introduction of steel and reinforced concrete as construction materials, and the beginnings of stress analysis methods for the design of structures. The latter was developed further in the twentieth century, especially in the second half, with the development of computer methods. During the last two centuries, major structural failures due to wind action have occurred periodically, and provoked much interest in wind forces by engineers. Long-span bridges often produced the most spectacular of these failures, with the Brighton Chain Pier, England (1836) (Figure 1.10), the Tay Bridge, Scotland (1879) and Tacoma Narrows Bridge, Washington State, USA (1940) being among the most notable, with the dynamic action of wind playing a major role.

Other large structures have experienced failures as well – for example, the collapse of the Ferrybridge Cooling Towers in the United Kingdom in 1965 (Figure 1.11), and the permanent deformation of the columns of the Great Plains Life Building in Lubbock, Texas, during a tornado (1970). These events were notable, not only as events in themselves, but also for the part they played as a stimulus to the development of research into wind loading in the respective countries. Another type of structure which has proved to be dynamically sensitive to wind is the guyed mast; it has also suffered a high failure rate – in one 10-year period (from the mid-1980s to the mid-1990s), there were 83 failures of this type of structure worldwide. In many cases of mast failures, a combination of wind and ice action was involved.

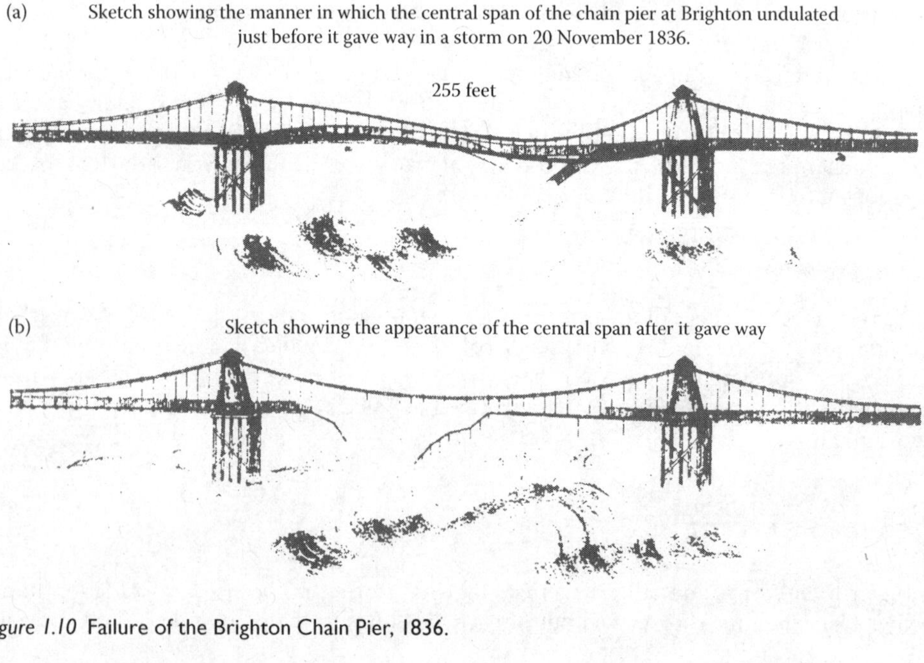

Figure 1.10 Failure of the Brighton Chain Pier, 1836.

Figure 1.11 Ferrybridge Cooling Tower failures, 1965 (photo).

Some major wind storms, which have caused large-scale damage to residential buildings, as well as some engineered structures, are also important for the part they have played in promoting research and understanding of wind loads on structures. The Yorkshire (UK) storms of 1962, Cyclone 'Tracy' in Darwin, Australia in 1974, and Hurricane 'Andrew' in Florida, USA, in 1992, can be mentioned as seminal events of this type. However, these extreme events occur intermittently, and it is unfortunate that the collective human memory after them is only about 10 years, and often, old lessons have to be re-learned by a new generation. However, an encouraging sign is the recent interest of some major insurance and re-insurance groups in natural hazards, in the estimation of the potential financial losses, and the beginnings of a realisation that any structure can be made wind resistant, with appropriate knowledge of the forces involved, and suitable design approaches.

1.4.1 Recent history of wind loss and damage

Figure 1.12 shows the annual insured losses in billions of $US from all major natural disasters, from 1970 to 2012. Wind storms account for about 70% of the total insured losses. Bearing in mind that property insurance is much less common in the less-developed

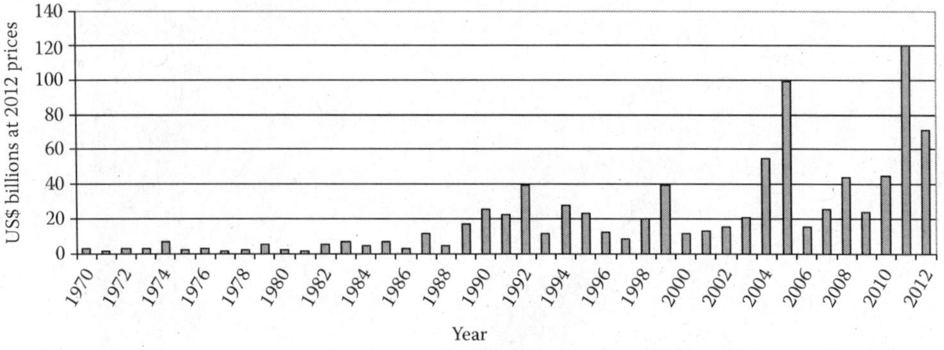

Figure 1.12 World insurance losses from major natural disasters 1970–2012. (From Swiss Reinsurance Company. With permission.)

economies, Figure 1.12 does not show the total property damage from natural events, and in fact is biased towards losses in Europe and North America. However, the graph does show that the level of insured losses from natural disasters increased dramatically after about 1987. The major contributor to the increase was wind storms, especially tropical cyclones such as Hurricanes 'Hugo' (1989) 'Andrew' (1992), 'Charley' (2004), 'Ivan' (2004), 'Katrina' (2005) and 'Sandy' (2012) in the United States, and winter gales in Europe in 1987, 1990 and 1999.

In 2012, there was an estimated $US71 billion of insured losses from natural disasters, of which the majority originated from hurricanes and typhoons.

Some notable wind storms and the losses resulting from them are listed in Table 1.3. All the events listed in Table 1.3 have had a major influence on the insurance industry, and structural engineering profession.

Table 1.3 does not include tornadoes. However, the aggregate damage from multiple events can be substantial. For example, in the 'super-outbreak' of 3–4 April, 1974, the total

Table 1.3 Some disastrous wind storms of the last 30 years

Year	*Name*	*Country or region*	*Approximate economic losses ($US mill)*	*Lives lost*
1974	Cyclone 'Tracy'	Australia	500	52
1987	Gales	Western Europe	3700	17
1989	Hurricane 'Hugo'	Caribbean, United States	9000	61
1990	Gales	Western Europe	15,000	230
1992	Hurricane 'Andrew'	United States	30,000	44
1999	Gales	France	10,000	140
2003	Typhoon 'Maemi'	Japan, Korea	6000	131
2004	Hurricane 'Ivan'	Caribbean, United States	11,000	124
2005	Hurricane 'Katrina'	Southern United States	150,000	1830
2011	Cyclone 'Yasi'	Australia	3500	1
2012	Hurricane 'Sandy'	United States	35,000	117

Source of data apart from Cyclone Tracy and Hurricane Katrina: Munich Reinsurance and Swiss Reinsurance.

Table 1.4 Annual count of tornadoes in the United States, and deaths caused by them

Year	*Number*	*Deaths*
2001	1219	40
2002	938	55
2003	1394	54
2004	1820	35
2005	1262	38
2006	1117	67
2007	1102	81
2008	1685	126
2009	1305	21
2010	1543	45
2011	1894	553
2012	1119	70

Source: National Oceanic and Atmospheric Administration (NOAA).

damage in the state of Ohio alone was estimated to be US$100 million. The estimated total damage from tornadoes in the United States, in the recent active year of 2011 was US$10 billion, with about US$3 billion of that coming from the Joplin, Missouri, tornado of 22 May, 2011.

Tornadoes are also notable for the loss of life they produce in the United States. Table 1.4 summarises the number of confirmed tornadoes in the United States and the loss of life for the period 2001–2012. The average number of tornadoes per annum is about 1370, with an average number of deaths per year produced by them of about 100.

1.5 WIND-GENERATED DEBRIS

As well as damage to buildings produced by direct wind forces – either overloads caused by overstressing under peak loads, or fatigue damage under fluctuating loads of a lower level, a major cause of damage in severe wind storms is flying debris. Penetration of the building envelope by flying missiles has a number of undesirable results: high internal pressures threatening the building structure, wind and rain penetration of the inside of the building, the generation of additional flying debris and the possibility of flying missiles inside the building endangering the occupants.

The area of a building most vulnerable to impact by missiles is the windward wall region, although impacts could also occur on the roof and side walls. As the air approaches the windward wall, its horizontal velocity reduces rapidly. Heavier objects in the flow with higher inertia will probably continue with their velocity little changed until they impact on the wall. Lighter and smaller objects may lose velocity in this region or even be swept around the building with the flow if they are not directed at the stagnation point (see Chapter 4).

1.5.1 Threshold of flight

Wills et al. (1998) carried out an analysis of debris flight conditions, and the resulting building damage in severe winds. They considered 'compact' objects, sheet objects and rods and

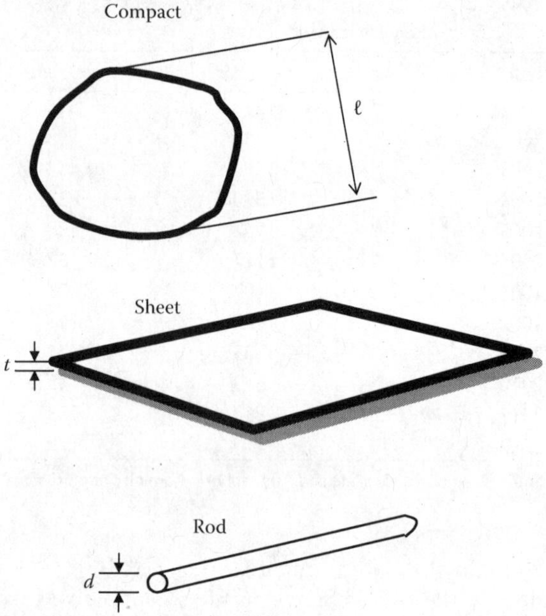

Figure 1.13 Three types of flying debris. (After Wills, J., Wyatt, T. and Lee, B.E. 1998. Warnings of high winds in densely populated areas. United Kingdom National Coordination Committee for the International Decade for Natural Disaster Reduction.)

poles (Figure 1.13), and established relationships between the body dimensions and the wind speed, U_f, at which flight occurs and the objects become missiles. For each of the three categories, these relationships are:

$$\ell = \frac{(1/2)\rho_a U_f^2 C_F}{I\rho_m g} \tag{1.13}$$

$$t = \frac{(1/2)\rho_a U_f^2 C_F}{I\rho_m g} \tag{1.14}$$

$$d = \frac{(2/\pi)\rho_a U_f^2 C_F}{I\rho_m g} \tag{1.15}$$

where

ℓ is a characteristic dimension for 'compact' objects
t is the thickness of sheet objects
d is the effective diameter of rod-type objects
ρ_a is the density of air
ρ_m is the density of the object material
C_F is an aerodynamic force coefficient (see Section 4.2.2)
U_f is the wind speed at which flight occurs

I is a fixing strength integrity parameter, that is the value of force required to dislodge the objects expressed as a multiple of their weight (for objects resting on the ground $I \cong 1$)
g is the gravitational constant

Equations 1.13 through 1.15 illustrate the important point that the larger the value of the characteristic dimension, ℓ, t or d, the higher the wind speed at which flight occurs. These equations also show that the higher the value of the density, ρ_m, the higher is the wind speed for lift off. Thus, as the wind speed in a cyclone builds up, the smaller lighter objects – for example, gravel, small loose objects in gardens and backyards – 'fly' first. At higher wind speeds, appurtenances on buildings are dislodged as the wind forces exceed their fixing resistance, and they also commence flight. At even higher wind speeds, substantial pieces of building structure, such as roof sheeting and purlins, may be removed, and become airborne.

As examples of the application of Equation 1.13, Wills et al. (1998) considered wooden compact objects ($\rho_m = 500$ kg/m^3) and stone objects ($\rho_m = 2700$ kg/m^3). Assuming $C_F = 1$, and $I = 1$, Equation 1.13 gives ℓ equal to 110 mm for the wooden missile, but only 20 mm for the stone missile, for a lift-off speed of 30 m/s.

For sheet objects, Equation 1.14 shows that the wind speed for flight depends on the thickness of the sheet, but not on the length and width. Wills et al. expressed Equation 1.14 in a slightly different form:

$$\rho_m t = \frac{(1/2)\rho_a U_f^2 C_F}{Ig} \tag{1.16}$$

The left-hand side of Equation 1.16 is the mass per unit area of the sheet. This indicates that the wind speed for the flight of a loose object depends essentially on its mass per unit area. Thus, a galvanised iron sheet of 1 mm thickness with mass per unit area of 7.5 kg/m^2 will fly at about 20 m/s ($C_F = 0.3$).

For 'rod'-like objects, which include timber members of a rectangular cross section, a similar formula to Equation 1.16 can be derived from Equation 1.15, with the 't' replaced by 'd', the equivalent rod diameter. Using this, Wills et al. calculated that a timber rod of 10 mm diameter will fly at about 11 m/s, and a 100-mm-by-50-mm timber member, with an equivalent diameter of 80 mm, will fly at about 32 m/s, assuming C_F is equal to 1.0.

1.5.2 Trajectories of compact objects

A missile, once airborne, will continue to accelerate until its flight speed approaches the wind speed, or until its flight is terminated by impact with the ground or with an object such as a building. The trajectories of compact objects are produced by drag forces (Section 4.2.2), acting in the direction of the relative wind with respect to the body.

Consider first the aerodynamic force on a compact object (such as a sphere) in a horizontal wind of speed, U. Neglecting the vertical air resistance initially, the aerodynamic force can be expressed as

$$\text{Accelerating force} = \frac{1}{2}\rho_a (U - v_m)^2 C_D A \tag{1.17}$$

where
v_m is the horizontal velocity of the missile with respect to the ground
A is the reference area for the drag coefficient, C_D (Section 4.2.2)

Table 1.5 Flight times and distances for a steel ball (neglecting vertical air resistance)

Object/speed	*Time taken (s)*	*Horizontal distance travelled (m)*
Steel ball to 20 m/s	5.4	71
Steel ball to 30 m/s	49	1270

Applying Newton's law, the instantaneous acceleration of the object (characteristic dimension, ℓ) is given by

$$\text{Acceleration} = \frac{dv_m}{dt} = \frac{(1/2)\rho_a(U - v_m)^2 C_D A}{\rho_m \ell^3} = \frac{(1/2)\rho_a(U - v_m)^2 C_D}{\rho_m \ell} \tag{1.18}$$

taking A equal to ℓ^2.

Equation 1.18 shows that heavier and larger objects have lower accelerations, and hence their flight speeds are likely to be lower than smaller or lighter objects. The equation also shows that the initial acceleration from rest ($v_m = 0$) is high, but the acceleration rapidly reduces as the difference between the missile speed and the wind speed reduces, so that the wind speed is approached very slowly. Of course, the missile speed cannot exceed the wind speed in steady winds.

Equation 1.18 can be integrated to obtain the time taken to accelerate to a given speed, v_m, and the distance travelled in this time. These equations are as follows:

Time taken to accelerate from 0 to v_m,

$$T = \frac{v_m}{kU(U - v_m)} \tag{1.19}$$

$$\text{Distance travelled} = U\left[T - \left(\frac{1}{kU}\right)\ln(1 + kUT)\right] \tag{1.20}$$

where $k = (\rho_a C_D)/(2\rho_m \ell)$ with units of (1/m).

Using Equation 1.20, the flight times and distance travelled by a steel ball of 8 mm diameter and 2-g mass have been calculated, for a wind speed, U, of 32 m/s, and are given in Table 1.5.

The calculations show that it takes nearly a minute and 1.27 km for the steel ball to reach 30 m/s – that is within 2 m/s of the wind speed. In reality, such a long flight time and distance would not occur since the object would strike a building, or the ground, and lose its kinetic energy.

A more accurate analysis of the trajectories of compact objects requires the vertical air resistance to be included, and the neglect of it results in underestimation of the missile speed and distance travelled in a given time (Holmes, 2004).

1.5.3 Trajectories of sheet and rod objects

Tachikawa (1983) carried out a fundamental study of the trajectories of missiles of the sheet type. Aerodynamic forces on auto-rotating plates were measured in a wind tunnel. These results were then used to calculate the trajectories of the plates released into a wind stream. Free-flight tests of model plates with various aspect ratios were made in a small wind tunnel, and compared with the calculated trajectories. A distinct change in the mode of motion and the trajectory, with an initial angle of attack of the plate, was observed. The calculated trajectories predicted the upper and lower limits of the observed trajectories, with reasonable accuracy. A later study by Tachikawa (1990) extended the experiments to small prismatic models as well as flat plates, and gave a method of estimating the position of a missile impact on a downstream building. The critical non-dimensional parameter for determination of trajectories is $K = \rho_a U^2 A/2$ mg, where

ρ_a is the density of air
U is the wind speed
A is the plan area of a plate
m is the mass of the missile
g is the gravitational constant

This parameter, now known as the 'Tachikawa Number' (Holmes et al., 2006a), represents the ratio of aerodynamic forces to gravity forces, and can also be expressed as the product of three other non-dimensional parameters:

$$K = \frac{1}{2}\frac{\rho_a}{\rho_m}\frac{U^2}{g\ell}\frac{\ell}{t} \tag{1.21}$$

where
ρ_m is the missile density
t is the plate thickness
ℓ is $\sqrt{A}$, that is a characteristic plan dimension

In Equation 1.21, ρ_a/ρ_m is a density ratio and $(U^2/g\ell)$ is a Froude Number, both of which are important non-dimensional quantities in aerodynamics (see also Section 7.4).

The equations of motion for horizontal, vertical and rotational motion of a flat plate moving in a vertical plane must be solved numerically. Good agreement has been obtained when such numerical solutions are compared with measurements of trajectories of many small plates in a wind tunnel (Lin et al., 2006; Holmes et al., 2006b).

1.5.4 Standardised missile-testing criteria

In regions subjected to hurricanes and tropical cyclones (Section 1.3.2), where the occurrence of damage to buildings by wind-generated missiles has been shown to be a major problem, standardised missile tests have been devised. These demonstrate the ability of wall claddings of various types to resist penetration by flying debris, or assist in the development of window protection screens.

When specifying appropriate test criteria for missile impact resistance, the following principles should be followed:

- The missiles should be representative of the actual objects available.
- The criteria should be physically realistic, that is if the flight threshold speed is greater than the expected wind speed in the storm, then the object should not be regarded as a potential missile.
- Realistic missile speeds should be specified for the expected separation distances between buildings.

Missile-testing criteria were included in the Darwin Area Building Manual, following Cyclone 'Tracy' in 1974, in Australia. This specified that windows and doors should withstand impact at any angle of a piece of 100-mm-by-50-mm timber weighing 4 kg, travelling at 20 m/s. A more severe test was specified for cyclone refuge shelters: 'end-on' impact of a piece of 100-mm-by-50-mm timber weighing 8 kg, travelling at 30 m/s. Later, the test requirement for windows and doors of buildings was modified to a piece of 100-mm-by-50-mm timber weighing 4 kg, travelling at 15 m/s. More recent debris speeds given in Australian Standards are linked to the regional gust speed used for wind-loading design, with a horizontal debris speed of 40% of that wind speed.

Wind-borne debris impact test standards in the United States were discussed by Minor (1994). Following investigations of glass breakage (mainly in high-rise buildings), during several U.S. hurricanes, Pantelides et al. (1992) proposed a test protocol involving impacts from small spherical missiles of 2 gr. This was taken up in South Florida following Hurricane Andrew in 1992. The Dade County and Broward County editions of the South Florida Building Code required windows, doors and wall coverings to withstand impacts from large and small debris. The large missile test, which is similar to the Australian one, is only applicable to buildings below 9 m in height. The small missile test is only applicable to windows, doors and wall coverings above 9 m, and differs between the two counties. The Dade County protocol uses 10 2-g pieces of roof gravel impacting simultaneously at 26 m/s, while the Broward County version uses 10 2-g steel balls impacting successively at 43 m/s.

1.6 WIND STORM DAMAGE AND LOSS PREDICTION

The trend towards increased losses from wind storms has provoked concern in the insurance and re-insurance industries, and many of these groups are requiring detailed assessments of the potential financial losses from the exposure of their portfolios of buildings to large-scale severe wind storms. Government bodies are also requiring predictions of damage and economic losses to aid in planning for disaster and emergency management.

The prediction of the average annual loss, or accumulated losses over an extended period, say 50 years, requires two major inputs: hazard models and vulnerability curves. The hazard model focusses on the wind storm hazard itself, and makes use of historical meteorological data and statistics to predict potential wind speeds at a site into the future. Vulnerability curves attempt to predict building (and sometimes contents) damage, given the occurrence of a particular wind speed.

1.6.1 Hazard models

The purpose of wind hazard models is to define the risk of occurrence of extreme wind speeds at the site of a single structure, on a system such as a transmission line, or on a complete city or region. The basis for these models is usually the historical record of wind speeds from anemometer stations, but often, larger-scale storm parameters such as central pressures for tropical cyclones, and atmospheric stability indices for thunderstorm occurrences, are studied. The methods of statistics and probability are extensively used in the development of hazard models in wind engineering.

The application of statistical methods to the prediction of extreme wind speeds is discussed in Chapter 2.

An understanding of the structure of the wind within a storm enables predictions of 'footprints' such as that shown in Figure 1.14 (Holmes and Oliver, 2000), which shows simulated contours of maximum wind speeds, occurring at some time during the passage

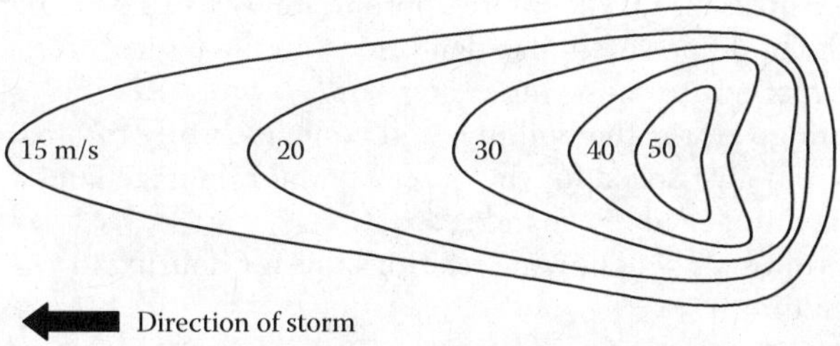

Figure 1.14 Wind speed threshold footprint during the passage of a downburst. (Reprinted from *Engineering Structures*, Holmes, J.D. and Oliver, S.E., An empirical model of a downburst. 22: 1167–72. Copyright 2000, with permission from Elsevier.)

of a downburst (Section 1.3.5). This information, in combination with knowledge of the strength or 'vulnerability' of structures, enables predictions of potential damage to be made.

1.6.2 Vulnerability curves

Insurance loss predictions are quite sensitive to the assumed variations of relative building and contents damage, as a function of the local wind speed – usually a gust speed (see Section 3.3.3). Such graphs are known as 'vulnerability curves'. Vulnerability curves can be derived in a number of ways. Leicester (1981) proposed the simplified form, with straight-line segments, for Australian houses, shown in Figure 1.15. The ordinate is a 'damage index' defined as follows for the building:

Damage index (D) = (Repair cost)/(Initial cost of building)

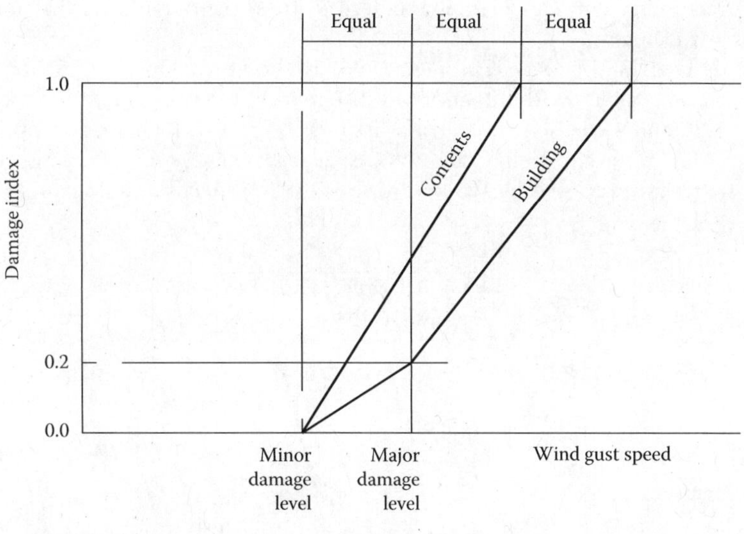

Figure 1.15 Form of a vulnerability curve proposed by Leicester, R.H. 1981. A risk model for cyclone damage to dwellings. *Proceedings, 3rd International Conference on Structural Safety and Reliability*, Trondheim, Norway.

For insurance purposes, it may be more appropriate to replace the denominator with the insured value of the building. A similar definition can be applied to the building contents, with 'replacement cost' in the numerator.

Separate lines are given for the building and contents. Only two parameters need to be specified – a threshold gust speed for the onset of minor damage, and a speed for the onset of a major building damage (damage index > 0.2).

Walker (1995) proposed the following relationships for housing in Queensland, Australia. For pre-1980 buildings:

$$D = 0.2\left(\frac{U-30}{30}\right)^2 + 0.5\left(\frac{U-30}{30}\right)^6 \tag{1.22}$$

For post-1980 buildings:

$$D = 0.2\left(\frac{U-37.5}{37.5}\right)^2 + 0.5\left(\frac{U-37.5}{37.5}\right)^6 \tag{1.23}$$

Clearly, in both cases, D is limited to the range 0–1.0.

The relationship of Equation 1.22 was also found to agree well with recorded damage and wind speed estimates in Hurricane 'Andrew' (see Table 1.1).

A simple form of a vulnerability curve for a fully engineered structure consisting of a large number of members or components with strengths of known probability distribution can be derived. The failure of each component is assumed to be independent of all the others, and they are all designed to resist the same wind load, or speed. Thus, the expected fractional damage to the complete structure, for a given wind speed, is the proportion of failed components expected at that wind speed. If all the components have the same probability distribution of strength, which would be true if they were all designed to the same codes, then the vulnerability curve can simply be derived from the cumulative distribution of the strength of any element.

A curve derived in this way (Holmes, 1996) is shown in Figure 1.16, for a structure composed of components with a lognormal distribution of strength, with a mean/nominal strength of 1.20 and a coefficient of variation of 0.13, values which are appropriate for steel components. The nominal design gust wind speed is taken as 65 m/s. This curve can be compared with that proposed by Walker, for post-1980 Queensland houses, in the tropical

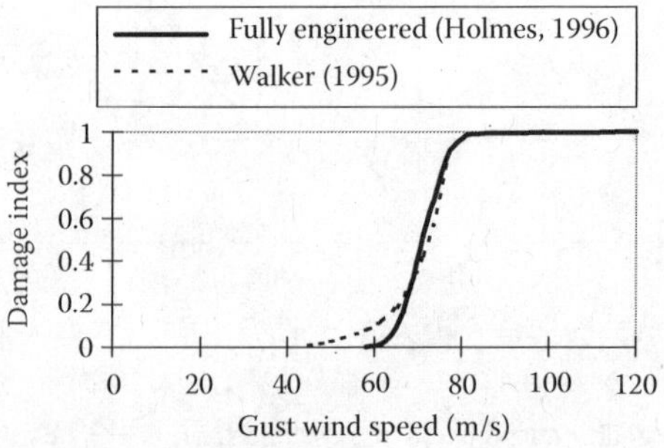

Figure 1.16 Theoretical and empirical vulnerability curves.

cyclone-affected coastal region (Equation 1.23). The theoretical curve, representing fully engineered structures, is steeper than the Walker curve, which has been derived empirically, and incorporates the greater variability in the components of housing structures.

Alternatively, vulnerability curves for individual building types can be derived through a probabilistic approach, based on assessing individual building component strengths, and on progressive failure simulation (e.g. Henderson and Ginger, 2007; Hamid et al., 2010). Following methods adopted for predicting damage due to earthquakes, this approach often introduces intermediate 'fragility curves' which plot the probability of exceeding a particular damage state against wind speed. Several of these are then combined, with a weighting based on the relative cost of repair of each state, to give a vulnerability curve.

1.6.3 Damage produced by flying debris

Wills et al. (1998) carried out an analysis of the damage potential of flying missiles, based on the assumption that the damage of a given missile is proportional to its kinetic energy in flight. A number of interesting conclusions arose from this work:

- For compact objects, lower-density objects have more damage potential.
- Sheet and rod objects have generally more damage potential than compact objects.
- Very little energy is required to break glass (e.g. 5-g steel ball travelling at 10 m/s is sufficient to break 6-mm-annealed glass).
- On the basis of an assumed distribution of the available missile dimensions, Wills et al. found that the total damage is proportional to U^n, where n is a power equal to about 5.

Probabilistic damage modelling has recently been extended to incorporate modelling of damage produced by flying debris, as well as that due to direct wind forces. The main features required for an engineering wind-borne debris damage model are summarised as follows (Twisdale et al., 1996):

- A *windfield model* – similar to those discussed in Section 1.6.1.
- A *debris generation model* – This is required to establish the source, numbers and generic shapes of debris items.
- A *debris trajectory model* – Once they become airborne, a debris trajectory model is required to predict consequent damage due to impact on downwind buildings.
- A *debris impact model* – This represents the magnitude of damage produced by an impacting wind-borne missile.

Twisdale et al. (1996) surveyed observed missiles after the 1995 Hurricanes 'Erin' and 'Opal' in the United States, and found that the clearly dominant contributors were roofing tiles, shingles, sheathing and structural members from roof trusses, with lesser contributions from wall cladding, miscellaneous house materials such as guttering, vents and yard items and accessories. This led to the development of a wind-borne debris damage model, with the generation part focussed on the generation of wind-borne roofing elements.

Twisdale et al. (1996) then described a debris generation model, which is essentially a wind load failure model for roofing elements of low-rise buildings. It is based on simulating wind loads on elements of a simple representative gable-roof building, and checking whether these exceed the resistance of roof sheeting and roof truss elements. The resistance is based on the pull-out strength of nails, and an assessment of errors in construction – that is an assessment of the number of nails attaching plywood roof sheathing to the underlying roof trusses.

Damage produced by wind-borne debris is primarily generated by the horizontal energy and momentum at impact. The trajectories of missiles of the generic 'compact', 'rod' and 'sheet' types were discussed in Sections 1.5.2 and 1.5.3.

The horizontal velocities attained by all three types are mainly determined by the mass of the debris object, and product of the average drag coefficient and the exposed area, during the trajectories.

It was shown, numerically and experimentally (Baker, 2007; Lin et al., 2007), that the *horizontal* velocity component of a wind-borne missile can be well represented by the following function:

$$\frac{u_m}{V_s} \cong 1 - \exp\left[-b\sqrt{x}\right] \tag{1.24}$$

where

u_m is the horizontal missile velocity
V_s is the local (gust) wind speed
x is the horizontal distance travelled (this can be related to the average building spacing)
b is a dimensional parameter depending on the shape of the missile and its drag coefficient, and its mass (Equation 1.25)

$$b = \sqrt{\frac{\rho_a C_{D,av} A}{m}} \tag{1.25}$$

In Equation 1.25, $C_{D,av}$ is an *average* drag coefficient, averaged over the rotations of the body with respect to the relative wind. Note that the right-hand side of Equation 1.24 does not include the wind speed, and is only a function of the horizontal distance travelled (or building spacing in the case of an impact), and the missile properties in Equation 1.25.

Following from Equation 1.14, the momentum and kinetic energy at impact can be represented by Equations 1.26 and 1.27, respectively.

$$m \cdot u_m \cong mV_s\left\{1 - \exp\left[-b\sqrt{x}\right]\right\} \tag{1.26}$$

$$E = \tfrac{1}{2} m \cdot u_m^2 \cong \tfrac{1}{2} mV_s^2\left\{1 - \exp\left[-b\sqrt{x}\right]\right\}^2 \tag{1.27}$$

where

m is the mass of the missile

The previous discussion has indicated that the horizontal velocities (and hence impact momenta and kinetic energy) of wind-borne debris are relatively simple functions of wind gust speeds and the distance travelled – and hence of building separation. The damage produced by an impacting missile on a building surface is dependent on the component of momentum normal to the surface and/or its kinetic energy at impact. The change of momentum at impact is direct related to the force applied to a surface – it is equal to the impulse applied – the integral of force with respect to time. A perfectly elastic surface (i.e. with a coefficient of restitution at impact of 1.0) would not absorb any of the kinetic energy of the missile; it would be retained as kinetic energy of the missile moving away from the surface.

The total energy at impact must be conserved, and for many building materials suffering plastic deformation, most of the kinetic energy of the debris item will be dissipated through deformation of the material.

1.7 HURRICANE-DAMAGE MODELLING

The prediction of losses resulting from hurricane impact on buildings and facilities has become a major activity that several companies have embraced for the service of the insurance and re-insurance industry. While the details of most of these are commercial in confidence, some useful discussion of the background methodologies is in the public domain (e.g. Vickery et al., 2000a).

A publicly available model has been funded by the State of Florida (Hamid et al., 2010), and will be used to illustrate the main components and features of these models in the following paragraphs. The Florida Public Hurricane Loss Model (FPHLM) consists of three components:

1. An 'atmospheric science component'
2. An 'engineering' component
3. An 'actuarial' component

The atmospheric science component is essentially a form of the 'hazard model' as introduced in Section 1.6.1. The first function of this component is to model the annual hurricane occurrences within the defined area chosen to encompass the historical origin points of hurricanes affecting Florida – a circle with a radius of 1000 km centred on a particular location just off the southwest coast of Florida. The Poisson Distribution (Section C3.5) was used to model the annual rate of hurricane occurrences within the defined area of interest. The points of origin of the simulated hurricanes were derived from the historical record of land-falling Atlantic tropical cyclones known as 'HURDAT' with small random perturbations. In this way, thousands of years of simulated hurricane tracks are generated. The intensity of each storm, represented by the difference in barometric pressure at sea level at the centre of the storm, and that at the periphery, was also varied at 24-h intervals by sampling from an appropriate probability distribution.

The *wind-field model* is implemented when the simulated hurricane is close to a coastline. In the FPHLM, the slab boundary-layer model of Shapiro (1983) is used; a similar model was used by Vickery et al. (2000b). The model is initialised by a vortex in gradient balance (see Section 1.2.4), with a radially symmetric pressure profile given by the Holland (1980) expression (Equation 1.10).

As part of the engineering component, vulnerability curves (Section 1.6.2) were generated by Monte Carlo simulations for 168 cases for every combination of structural type (timber frame or masonry), geographical location in Florida and roof type (gable, hip, tile, shingle, etc.). The strength of building components is determined as a function of gust wind speed through a detailed wind and structural engineering approach that includes an allowance for wind-borne debris damage. The latter is included empirically rather than through the detailed approach outlined in Section 1.6.3.

In the actuarial model, the expected annual insurance losses for the building structure, contents and additional living expenses using the vulnerability matrices are derived as discussed above. The probability distribution of gust wind speeds for each zip code is derived from the simulated set of hurricanes. These wind speeds are applied to the vulnerability curves, and using the insured values, the expected losses are estimated for each policy.

Another public-domain hurricane-damage model is the Hazards U.S. Multi-Hazard (HAZUS-MH) described by Vickery et al. (2006a,b). The hurricane hazard model is based on the methods described by Vickery et al. (2000a,b). HAZUS-MH includes a detailed mechanics-based model of wind-borne debris impacts, and the hazard component includes estimates of rainfall rates to enable the prediction of damage due to water ingress into buildings.

1.8 PREDICTED EFFECTS OF CLIMATE CHANGE

It has often been suggested that global warming is having significant effects on the numbers and strengths of wind storms – particularly on tropical cyclones (including hurricanes and typhoons). For example, there have been a number of studies of the effects of global warming on tropical cyclones in recent years (e.g. Emanuel, 2005; Webster et al., 2005; Klotzbach, 2006; Kossin et al., 2007).

As discussed in Section 1.3.2, a sea surface temperature of 26°C is required for tropical cyclone formation in the current climate; hence, it might be expected that there would be an increase in the number of tropical cyclones worldwide with increasing average sea temperatures. In fact, Webster et al. (2005) found there was no significant trend in global cyclones of all strengths. However, they did show a statistically significant increasing trend in Category 4–5 storms from the 1970s to the decade 1995–2004. These mainly seem to have occurred in the North Atlantic basin.

Klotzbach (2006) extended the analysis to all basins with tropical cyclone activity, and excluded the data before 1986 on the basis that, before the mid-1980s, only *visible* satellite information was available and hence night-time observations were excluded; also, the quality and resolution of satellite imagery had improved greatly by the later period. Klotzbach's analysis, using the more recent (and more reliable) data, found only a small increase in Category 4–5 hurricanes in the North Atlantic and Northwest Pacific during the 20-year study period.

The following summary statement in the U.N. International Panel on Climate Change Report of 2007 is relevant:

> There is observational evidence for an increase in intense tropical cyclone activity in the North Atlantic since about 1970, correlated with increases in sea surface temperatures. There are also suggestions of increased intense tropical cyclone activity in some other regions where concerns over data quality are greater. ... There is no clear trend in the annual numbers of tropical cyclones.

Although some have suggested that higher design wind speeds should be introduced in the United States, as a result of trends in hurricane occurrences (e.g. Mudd et al., 2013; Liu and Pang, 2013), this is premature until the biasses resulting from changes in observational practices have been eliminated. Clearly, there is a great deal of uncertainty in the predictions, and regulators may be reluctant to impose additional economic costs of higher design wind loads on the community without more confidence in the trends. A rational approach to design in these circumstances might suggest a small increase in load factor to cover these uncertainties.

1.9 SUMMARY

In this chapter, the physical mechanisms and meteorology of strong wind storms of all types have been described. The balance of forces in a large-scale synoptic system was established, and the gradient wind equation was derived. Smaller-scale storms – tornadoes and downbursts – were also introduced.

The history of significant damaging wind storms was discussed. The mechanics of wind-generated flying debris was considered, and vulnerability curves relating the fractional damage potential to wind speed, for insurance loss prediction, were derived. The modelling of wind damage from direct wind forces and from wind-borne debris, for disaster management and insurance purposes, has been outlined. Two hurricane loss models that are publicly available in the United States have been described.

The predicted effects of climate change on the frequencies and strengths of tropical cyclones have also been briefly discussed.

1.10 THE FOLLOWING CHAPTERS AND APPENDICES

Following this introductory chapter, Chapters 2 through 7 are directed towards fundamental aspects of wind loading, common to all or most structures – for example, atmospheric wind structure and turbulence (Chapter 3), bluff-body aerodynamics (Chapter 4), resonant dynamic response of structures (Chapter 5) and wind-tunnel techniques (Chapter 7). Chapters 8 through 14 deal with aspects of wind loading for particular types of structures: buildings, bridges, towers and so on. Finally, Chapter 15 discusses contemporary wind-loading codes and standards – the most common point of contact of practising structural engineers with wind loads.

Appendices A and B cover the terminology of wind engineering and the symbols used in this book, respectively. Appendix C describes probability distributions relevant to wind loading. Appendix D attempts to summarise the extreme wind climate of more than 80 countries, and Appendix E gives some approximate formulae for natural frequencies of structures. Appendix F gives a simple example of the calculation of effective static wind load distributions.

REFERENCES

Baker, C.J. 2007. The debris flight equations. *Journal of Wind Engineering and Industrial Aerodynamics*, 95: 329–53.

Emanuel, K.A. 2005. Increasing destructiveness of tropical cyclones over the past thirty years. *Nature*, 436: 686–8.

Fujita, T.T. 1971. Proposed characterization of tornadoes and hurricanes by area and intensity. Report SMRP No. 91, University of Chicago.

Fujita, T.T. 1985. *The Downburst.* Report on Projects NIMROD and JAWS. Published by the author at the University of Chicago.

Fujita, T.T., Pearson, A.D., Forbes, G.S., Umenhofer, T.A., Pearl, E.W. and Tecson, J.J. 1976. Photogrammetric analyses of tornados. *Symposium on Tornadoes: Assessment of Knowledge and Implications for Man*, Texas Tech University, Lubbock, TX, 22–24 June 1976, pp. 43–88.

Golden, J.H. 1976. An assessment of windspeeds in tornados. *Symposium on Tornadoes: Assessment of Knowledge and Implications for Man*, Texas Tech University, Lubbock, TX, 22–24 June 1976, pp. 5–42.

Goliger, A.M., Milford, R.V., Adam, B.F. and Edwards, M. 1997. *Inkanyamba: Tornadoes in South Africa.* CSIR Building Technology and S.A. Weather Bureau, Pretoria, South Africa.

Hamid, S., Kibria, B.M.G., Gulati, S., Powell, M., Annane, B., Cocke, S., Pinelli, J.-P., Gurley, K. and Chen, S.-C. 2010. Predicting losses of residential structures in the State of Florida by the public hurricane loss evaluation model. *Statistical Methodology*, 7: 552–73.

Henderson, D.J. and Ginger, J.D. 2007. Vulnerability model of an Australian high-set house subjected to cyclonic wind loading. *Wind and Structures*, 10: 269–85.

Hoecker, W.H. 1960. Wind speed and airflow patterns in the Dallas tornado of 2 April, 1957. *Monthly Weather Review*, 88: 167–80.

Holland, G.J. 1980. An analytic model of the wind and pressure profiles in a hurricane. *Monthly Weather Review*, 108: 1212–8.

Holmes, J.D. 1996. Vulnerability curves for buildings in tropical-cyclone regions for insurance loss assessment. *ASCE EMD/STD Seventh Specialty Conference on Probabilistic Mechanics and Structural Reliability*, Worcester, Massachusetts, USA, 7–9 August.

Holmes, J.D. 2004. Trajectories of spheres in strong winds with applications to wind-borne debris. *Journal of Wind Engineering and Industrial Aerodynamics*, 92: 9–22.

Holmes, J.D. and Oliver, S.E. 2000. An empirical model of a downburst. *Engineering Structures*, 22: 1167–72.

Holmes, J.D., Baker, C.J. and Tamura, Y. 2006a. Tachikawa number: A proposal. *Journal of Wind Engineering and Industrial Aerodynamics*, 94: 41–7.

Holmes, J.D., Letchford, C.W. and Lin, N. 2006b. Investigations of plate-type windborne debris. II. Computed trajectories. *Journal of Wind Engineering and Industrial Aerodynamics*, 94: 21–39.

Klotzbach, P.J. 2006. Trends in global tropical cyclone activity in the last twenty years (1986–2005). *Geophysical Research Letters*, 33: L10805.

Kossin, J.P., Knapp, K.R., Vimont, D.J., Murnane, R.J. and Harper, B.A. 2007. A globally consistent reanalysis of hurricane variability and trends. *Geophysical Research Letters*, 34: L04815.

Leicester, R.H. 1981. A risk model for cyclone damage to dwellings. *Proceedings, 3rd International Conference on Structural Safety and Reliability*, Trondheim, Norway.

Lin, N., Holmes, J.D. and Letchford, C.W. 2007. Trajectories of windborne debris in horizontal winds and applications to impact testing. *ASCE Journal of Structural Engineering*, 133: 274–82.

Lin, N., Letchford, C.W. and Holmes, J.D. 2006. Investigations of plate-type windborne debris. I. Experiments in full scale and wind tunnel. *Journal of Wind Engineering and Industrial Aerodynamics*, 94: 51–76.

Liu, F. and Pang, W. 2013. Projection of future U.S. design wind speeds due to changes in hurricane activity: Storm frequency and sea surface temperature. *Proceedings, 12th Americas Conference on Wind Engineering*, Seattle, Washington, United States, 16–20 June, 2013.

McDonald, J.R. and Mehta, K.C. 2004. *A Recommendation for an Enhanced Fujita Scale*. Wind Science and Engineering Research Center, Texas Tech University, Lubbock, Texas, USA.

Mehta, K.C. 1976. Windspeed estimates: Engineering analyses. *Symposium on Tornadoes: Assessment of Knowledge and Implications for Man*, Texas Tech University, Lubbock, TX, 22–24 June 1976, pp. 89–103.

Minor, J.E. 1994. Windborne debris and the building envelope. *Journal of Wind Engineering and Industrial Aerodynamics*, 53: 207–27.

Mudd, L., Wang, Y., Letchford, C.W. and Rosowsky, D. 2013. Assessing climate change impact on the U.S. East Coast hurricane hazard: Sea temperature, hurricane frequency and hurricane track. *Proceedings, 12th Americas Conference on Wind Engineering*, Seattle, Washington, United States, 16–20 June, 2013.

Pantelides, C.P., Horst, A.D. and Minor, J.E. 1992. Post-breakage behaviour of architectural glazing in wind storms. *Journal of Wind Engineering and Industrial Aerodynamics*, 41–44: 2425–35.

Shapiro, L. 1983. The asymmetric boundary layer flow under a translating hurricane. *Journal of Atmospheric Sciences*, 40: 1984–98.

Tachikawa, M. 1983. Trajectories of flat plates in uniform flow with application to wind-generated missiles. *Journal of Wind Engineering and Industrial Aerodynamics*, 14: 443–53.

Tachikawa, M. 1990. A method for estimating the distribution range of trajectories of windborne missiles. *Journal of Wind Engineering and Industrial Aerodynamics*, 29: 175–84.

Twisdale, L.A., Vickery, P.J. and Steckley, A.C. 1996. *Analysis of Hurricane Windborne Debris Impact Risk for Residential Structures*. Report prepared for State Farm Mutual. Applied Research Associates, Rayleigh, North Carolina, Report 5503, March 1996.

Vickery, P.J., Skerjl, P.F. and Twisdale, L.A. 2000a. Simulation of hurricane risk in the United States using an empirical storm track modeling technique. *Journal of Structural Engineering*, 126: 1222–37.

Vickery, P.J., Skerjl, P.F., Steckley, A.C. and Twisdale, L.A. 2000b. A hurricane wind field model for use in simulations. *Journal of Structural Engineering*, 126: 1203–22.

Vickery, P.J., Lin, J., Skerjl, P.F., Twisdale, L.A. and Huang, K. 2006a. HAZUS-MH hurricane model methodology. I: Hurricane hazard, terrain and wind load modeling. *Natural Hazards Review*, 7: 82–93.

Vickery, P.J., Skerjl, P.F., Lin, J., Twisdale, L.A., Young, M.A. and Lavelle, F.M. 2006b. HAZUS-MH hurricane model methodology. II: Damage and loss estimation. *Natural Hazards Review*, 7: 94–103.

Walker, G.R. 1995. *Wind Vulnerability Curves for Queensland Houses*. Alexander Howden Insurance Brokers (Australia) Ltd., Sydney, NSW, Australia.

Webster, P.J., Holland, G.J., Curry, J.A. and Chang, H.R. 2005. Changes in tropical cyclone number, duration and intensity in a warming environment. *Science*, 309: 1844–6.

Wills, J., Wyatt, T. and Lee, B.E. 1998. Warnings of high winds in densely populated areas. United Kingdom National Coordination Committee for the International Decade for Natural Disaster Reduction.

Chapter 2

Prediction of design wind speeds and structural safety

2.1 INTRODUCTION AND HISTORICAL BACKGROUND

The establishment of appropriate design wind speeds is a critical first step towards the calculation of design wind loads for structures. It is also usually the most uncertain part of the design process for wind loads, and requires the statistical analysis of historical data on recorded wind speeds.

In the 1930s, the use of the symmetrical bell-shaped *Gaussian* distribution (Appendix C3.1) to represent extreme wind speeds, for the prediction of long-term design wind speeds, was proposed. However, this failed to take note of the earlier theoretical work of Fisher and Tippett (1928), establishing limiting forms of the distribution of the largest (or smallest) value in a fixed sample, depending on the form of the tail of the parent distribution. The identification of the three types of *extreme value distribution* was of prime significance to the development of probabilistic approaches in engineering in general.

The use of extreme value analysis for design wind speeds lagged behind the application to flood analysis. Gumbel (1954) strongly promoted the use of the simpler Type I Extreme Value Distribution for such analyses. However, von Mises (1936) showed that the three asymptotic distributions of Fisher and Tippett could be represented as a single *Generalised Extreme Value (GEV) Distribution* – this is discussed in detail in the following section. In the 1950s and the early 1960s, several countries had applied extreme value analyses to predict design wind speeds. In the main, the Type I (by now also known as the 'Gumbel Distribution') was used for these analyses. The concept of *return period* also arose at this time.

The use of probability and statistics as the basis for the modern approach to wind loads was, to a large extent, a result of the work of A.G. Davenport in the 1960s, recorded in several papers (e.g. Davenport, 1961).

In the 1970s and 1980s, the enthusiasm for the then standard 'Gumbel analysis' was tempered by events such as Cyclone 'Tracy' in Darwin, Australia (1974), and severe gales in Europe (1987), when the previous design wind speeds determined by a Gumbel fitting procedure, were exceeded considerably. This highlighted the importance of

- Sampling errors inherent in the recorded data base, usually <50 years
- The separation of data originating from different storm types

The need to separate the recorded data by the storm type was recognised in the 1970s by Gomes and Vickery (1977a).

The development of probabilistic methods in structural design generally, developed in parallel with their use in wind engineering, followed the pioneering work by Freudenthal (1947, 1956) and Pugsley (1966). This area of research and development is known as 'structural reliability' theory. Limit states design, which is based on probabilistic concepts, was steadily introduced into the design practice from the 1970s onwards.

This chapter discusses modern approaches to the use of extreme value analysis for the prediction of extreme wind speeds for the design of structures. Related aspects of structural design and safety are discussed in Section 2.6.

2.2 PRINCIPLES OF EXTREME VALUE ANALYSIS

The theory of extreme value analysis of wind speeds, or other geophysical variables, such as flood heights, or earthquake accelerations, is based on the application of one or more of the three asymptotic extreme value distributions identified by Fisher and Tippett (1928), and discussed in the following section. They are asymptotic in the sense that they are the correct distributions for the largest of an *infinite* population of independent random variables of a known probability distribution. In practice, of course, there will be a finite number in a population, but in order to make predictions, the asymptotic extreme value distributions are still used as empirical fits to the extreme data. Which one of the three is theoretically 'correct' depends on the form of the tail of the underlying parent distribution. However, unfortunately, this form is not usually known with certainty due to lack of data. Physical reasoning has sometimes been used to justify the use of one or other of the asymptotic extreme value distributions.

Gumbel (1954, 1958) has covered the theory of extremes in detail. A useful review of the various methodologies available for the prediction of extreme wind speeds, including those discussed in this chapter, has been given by Palutikof et al. (1999).

2.2.1 The GEV Distribution

The *GEV Distribution* introduced by von Mises (1936), and later re-discovered by Jenkinson (1955), combines the three extreme value distributions into a single mathematical form:

$$F_U(U) = \exp\{-[1 - k\,(U - u)/a]^{1/k}\} \tag{2.1}$$

where $F_U(U)$ is the cumulative probability distribution function (see Appendix C) of the maximum wind speed in a defined period (e.g. 1 year).

In Equation 2.1, k is a shape factor, a is a scale factor and u is a location parameter. When $k < 0$, the GEV is known as the *Type II Extreme Value* (or *Frechet*) Distribution; when $k > 0$, it becomes a *Type III Extreme Value* Distribution (a form of the *Weibull* Distribution). As k tends to 0, Equation 2.1 becomes Equation 2.2 in the limit. Equation 2.2 is the *Type I Extreme Value* Distribution, or *Gumbel* Distribution.

$$F_U(U) = \exp\{-\exp[-(U - u)/a]\} \tag{2.2}$$

The GEV with k equal to −0.2, 0 and 0.2 is plotted in Figure 2.1, in a form that the Type I appears as a straight line. As can be seen, the Type III ($k = +0.2$) curves in a way to approach a limiting value – it is therefore appropriate for variables that are 'bounded' on the high side. It should be noted that Types I and II predict unlimited values – they are therefore suitable distributions for variables that are 'unbounded'. Since we would expect that there is an upper limit to the wind speed that the atmosphere can produce, the Type III Distribution may be more appropriate for wind speed.

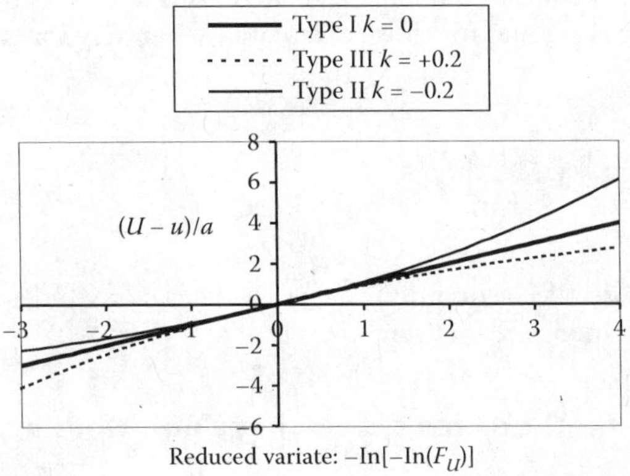

Figure 2.1 The GEV Distribution ($k = -0.2, 0, +0.2$).

A method of fitting the GEV Distribution to wind data is discussed in Section 2.4. An alternative method is the method of probability-weighted moments described by Hosking et al. (1985).

2.2.2 Return period

At this point, it is appropriate to introduce the term *return period, R*. It is simply the inverse of the complementary cumulative distribution of the extremes.

$$\text{i.e. Return period}, R = \frac{1}{\text{Probability of exceedence}} = \frac{1}{1 - F_U(U)}$$

Thus, if the annual maximum is being considered, then the return period is measured in years. Thus, a 50-year return period wind speed has a probability of exceedence of 0.02 (1/50) in any 1 year. It should not be interpreted as recurring regularly every 50 years. The probability of a wind speed, of a given return period, being exceeded in the lifetime of a structure is discussed in Section 2.6.3.

2.2.3 Separation by storm type

In Chapter 1, the various types of wind storm that are capable of generating winds strong enough to be important for the structural design were discussed. These different event types will have different probability distributions, and therefore should be statistically analysed separately; however, this is usually quite a difficult task as weather bureaus or meteorological offices do not normally record the necessary information. If anemograph records such as those shown in Figures 1.5 and 1.7 are available for older data, these can be used for identification purposes! Modern automatic weather stations (AWS) can generate wind speed and direction data at short intervals of as low as 1 min. These can be used to reconstruct time histories similar to those in Figures 1.5 and 1.7, and assist in identifying storm types.

The relationship between the combined return period, R_c, for a given extreme wind speed due to winds of either type, and for those calculated separately for storm types 1 and 2, (R_1 and R_2) is

$$\left(1 - \frac{1}{R_c}\right) = \left(1 - \frac{1}{R_1}\right)\left(1 - \frac{1}{R_2}\right) \tag{2.3}$$

Equation 2.3 relies on the assumption that exceedences of wind speeds from the two different storm types are independent events.

2.2.4 Simulation methods for tropical cyclone wind speeds

The winds produced by severe tropical cyclones, also known as 'hurricanes' and 'typhoons', are the most severe on Earth (apart from those produced by tornadoes which affect very small areas). However, their infrequent occurrence at particular locations often makes the historical record of recorded wind speeds an unreliable predictor for design wind speeds. An alternative approach, which gained popularity in the 1970s and early 1980s, was the simulation or 'Monte-Carlo' approach, introduced originally for offshore engineering by Russell (1971). In this procedure, satellite and other information on storm size, intensity and tracks are made use of to enable a computer-based simulation of wind speed (and in some cases direction) at particular sites. Usually, established probability distributions are used for parameters such as central pressure and radius to maximum winds. A recent use of these models is for damage prediction for insurance companies. The disadvantage of this approach is the subjective aspect resulting from the complexity of the problem. Significantly varying predictions could be obtained by adopting different assumptions. Clearly, whatever recorded data that are available should be used to calibrate these models.

2.2.5 Compositing data from several stations

No matter what type of probability distribution is used to fit historical extreme wind series, or what fitting method is used, extrapolations to high return periods for ultimate limit states design (either explicitly, or implicitly through the application of a wind load factor) are usually subject to significant sampling errors. This results from the limited record lengths usually available to the analyst. In attempts to reduce the sampling errors, a recent practice has been to combine records from several stations with perceived similar wind climates to increase the available records for extreme value analysis. Thus, 'superstations' with long records can be generated in this way.

For example, in Australia, stations in a huge region in the southern part of the country have been judged to have similar statistical behaviour, at least as far as the all-direction extreme wind speeds are concerned. A single set of design wind speeds has been specified for this region (Standards Australia, 1989, 2011; Holmes, 2002). A similar approach has been adopted in the United States (ASCE, 1998, 2010; Peterka and Shahid, 1998).

2.2.6 Correction for gust duration

Wind data are recorded and supplied by meteorological offices, or bureaus, after being averaged over certain time intervals. The most common averaging time is 10 min. Gust data obtained by modern AWS are typically averaged over 3 s. However, earlier data obtained from older anemometers and analogue-recording systems are usually of a shorter duration (e.g. Miller et al., 2013). A time series of gust data may need to be corrected to a common gust duration (see Section 3.3.3), before extreme value analysis is carried out. Correction factors can be derived using random process theory, with some assumptions on the form of the spectral density of the turbulence (Section 3.3.4), as a function of mean wind speed and turbulence intensity (e.g. Holmes and Ginger, 2012).

2.2.7 Wind direction effects and wind direction multipliers

Increased knowledge of the aerodynamics of buildings and other structures, through wind-tunnel and full-scale studies, has revealed the variation of structural response as a function of wind direction as well as speed. The approaches to probabilistic assessment of wind loads including direction can be divided into those based on the parent distribution of wind speed, and those based on extreme wind speeds. In many countries, the extreme winds are produced by rare severe storms such as thunderstorms and tropical cyclones, and there is no direct relationship between the parent population of regular everyday winds, and the extreme winds. For such locations (which would include most tropical and sub-tropical countries), the latter approach may be more appropriate. Where a separate analysis of extreme wind speeds by a direction sector has been carried out, the relationship between the return period, R_a, for exceedence of a specified wind speed from *all*-direction sectors, and the return periods for the same wind speed from direction sectors θ_1, θ_2 and so on is given in Equation 2.4.

$$\left(1-\frac{1}{R_a}\right)=\prod_{i=1}^{N}\left(1-\frac{1}{R_{\theta_i}}\right) \tag{2.4}$$

Equation 2.4 follows from an assumption that the wind speeds from each direction sector are statistically independent of each other, and is a statement of the following:

Probability that a wind speed U is *not* exceeded for all wind directions =
(probability that U is not exceeded from direction 1)
×(probability that U is not exceeded from direction 2)
×(probability that U is not exceeded from direction 3)
...and so on

Equation 2.4 is a similar relationship to Equation 2.3 for combining extreme wind speeds from different types of storms. A similar approach can be adopted for combining the extreme wind *pressure*, or a structural *response*, from contributions from different directions (e.g. Holmes, 1990, Section 9.11).

The use of Equation 2.4, or other full probabilistic methods of treating the directional effects of wind, adds a considerable degree of complexity when applied to the structural design for wind. To avoid this, simplified methods of deriving direction multipliers, or directional factors, have been developed (e.g. Cook, 1983; Melbourne, 1984; Cook and Miller, 1999). These factors, which have been incorporated into some codes and standards, allow climatic effects on wind direction to be incorporated into wind load calculations in an approximate way. Generally, the maximum (or minimum) value of a response variable, from any direction sector, is used for design purposes.

Three methods that have been suggested for deriving directional wind speeds, or direction multipliers, are:

a. The extreme wind speeds, given a direction sector, are fitted with an extreme value probability distribution, in the same way as the all-direction winds. Then direction multipliers are obtained by dividing the wind speed with a specified exceedence probability by the all-direction wind speed with the same exceedence probability. This is an *unconservative* approach when applied to structural response, as it ignores contributions to the combined probability of response from more than one direction sector.

b. The extreme wind speeds within a direction sector are fitted with an extreme value probability distribution in the same way as the all-direction winds. The target probability of exceedence of the all-direction wind is then divided by the number of direction sectors. The wind speed within each direction sector is then calculated for the reduced exceedence probability. These wind speeds are then divided by the all-direction wind speed with the original target probability to give wind direction multipliers. This approach (proposed by Cook, 1983) clearly gives higher direction multipliers than Method (a), and is generally regarded as being a conservative one, when applied to codes and standards for the design of buildings.
c. As an empirical approach between the unconservative Method Approach (a) and the conservative Method (b), Melbourne (1984) suggested what amounts to a modification to Method (a), to render it less unconservative, by specifying a probability of exceedence for the directional winds of the target probability of the all-direction wind speeds, divided by *one-quarter* of the number of direction sectors. In effect, Melbourne (1984) suggested that only two sectors from eight sectors in Equation 2.4, when $N = 8$, are effective, on average, in contributing to the combined probability of a structural response – that is a 90° sector from the total of 360°. Wind direction multipliers are then obtained by dividing the all-direction wind speed with the target probability of exceedence, as for Methods (a) and (b).

Kasperski (2007) derived direction wind speeds and multipliers, M_d, for Dusseldorf, Germany, using Methods (a) and (b). These values are shown in Table 2.1.

For this location, the southwesterly and westerly sectors (i.e. 210–270°) are the dominant ones for strong winds, due to the effect of Atlantic gales. It will be noted that Method (a) gives values of M_d that are all <1.0. The conservative nature of Method (b) is shown by the values of 1.02 for wind directions from 210° to 270°.

Table 2.1 Direction multipliers for Dusseldorf, Germany

Wind direction (deg)	*Md* *Method (a)*	*Md* *Method (b)*
0	0.70	0.86
30	0.75	0.91
60	0.75	0.91
90	0.73	0.89
120	0.73	0.89
150	0.83	0.97
180	0.90	1.00
210	0.97	1.02
240	0.97	1.02
270	0.95	1.02
300	0.84	0.98
330	0.83	0.96

Source: Re-calculated from data of Kasperski, M. 2007. *Journal of Wind Engineering and Industrial Aerodynamics*, 95: 1125–44.

Method (c), representing an empirical compromise between Methods (a) and (b), generally gives maximum values of M_d close to 1.0. This approach has been used to derive wind direction multipliers for Australia and New Zealand in the Standard AS/NZS 1170.2 (Standards Australia, 2011).

Besides the use of wind direction multipliers, or the direct application of Equation 2.4, there are several other methods of treating the varying effects of wind direction on the response of buildings. Several of these are discussed in Section 9.11.

2.3 EXTREME WIND ESTIMATION BY THE TYPE I DISTRIBUTION

2.3.1 Gumbel's method

Gumbel (1954) gave an easily usable methodology for fitting recorded annual maxima to the Type I Extreme Value Distribution. This distribution is a special case of the GEV Distribution discussed in Section 2.2.1. The Type I Distribution takes the form of Equation 2.2 for the cumulative distribution $F_U(U)$:

$$F_U(U) = \exp\{-\exp[-(U-u)/a]\}$$

where u is the mode of the distribution, and a is a scale factor.

The return period, R, is directly related to the cumulative probability distribution, $F_U(U)$, of the annual maximum wind speed at a site as follows:

$$R = \frac{1}{1-F_U(U)} \tag{2.5}$$

Substituting for $F_U(U)$ from Equation 2.5 in Equation 2.2, we obtain:

$$U_R = u + a\left\{-\log_e\left[-\log_e\left(1-\frac{1}{R}\right)\right]\right\} \tag{2.6}$$

For large values of return period, R, Equation 2.6 can be written as

$$U_R \cong u + a\log_e R \tag{2.7}$$

In Gumbel's original extreme value analysis method (applied to flood prediction as well as extreme wind speeds), the following procedure is adopted:

- The largest wind speed in each calendar year of the record is extracted.
- The series is ranked in order of smallest to largest: 1,2,...m.... to N.
- Each value is assigned a probability of non-exceedence, p, according to

$$p \approx m/(N+1) \tag{2.8}$$

- A reduced variate, y, is formed from:

$$y = -\log_e(-\log_e p) \tag{2.9}$$

- y is an estimate of the term in {} brackets in Equation 2.6.
- The wind speed, U, is plotted against y, and a line of 'best fit' is drawn, usually by means of linear regression.

As may be seen from Equation 2.7 and Figure 2.1, the Type I or Gumbel Distribution will predict unlimited values of U_R, as the return period, R, increases. That is as R becomes larger, U_R as predicted by Equation 2.6 or 2.7 will also increase without limit. As discussed in Section 2.2.1, this is not consistent with the physical argument that there are upper limits to the wind speeds that can be generated in the atmosphere in different types of storms.

2.3.2 Gringorten's method

The Gumbel procedure, as described in Section 2.3.1, has been used many times to analyse extreme wind speeds for many parts of the world.

Assuming that the Type I Extreme Value Distribution is in fact the correct one, the fitting method, due to Gumbel, is biased, that is Equation 2.8 gives distorted values for the probability of non-exceedence, especially for high values of p near 1. Several alternative fitting methods have been devised which attempt to remove this bias. However, most of these are more difficult to apply, especially if N is large, and some involve the use of computer programmes to implement. A simple modification to the Gumbel procedure, which gives nearly unbiased estimates for this probability distribution, is due to Gringorten (1963). Equation 2.8 is replaced by the following modified formula:

$$p \approx (m - 0.44)/(N + 1 - 0.88) = (m - 0.44)/(N + 0.12) \tag{2.10}$$

Fitting of a straight line to U versus the plotting parameter, p, then proceeds as for the Gumbel method.

2.3.3 Method of Moments

The simplest method of fitting the Type I Extreme Value Distribution to a set of data is known as the 'Method of Moments'. It is based on the following relationships between the mean and standard deviation of the distribution, and the mode and scale factor (or slope).

$$\text{Mean} = u + 0.5772a \tag{2.11}$$

$$\text{Standard deviation} = \left(\frac{\pi}{\sqrt{6}}\right)a \tag{2.12}$$

The method to estimate the parameters, u and a, of the distribution simply entails the calculation of the sample mean, μ, and standard deviation, σ, from the data, and then estimating u and a by the use of the inverse of Equations 2.11 and 2.12, that is

$$a \cong \left(\frac{\sqrt{6}}{\pi}\right)\sigma \tag{2.13}$$

$$u \cong \mu - 0.5772a \tag{2.14}$$

Once the parameters u and a have been determined, predictions of the extreme wind speed for a specified return period, R, are made using Equation 2.6 or 2.7.

Another procedure is the 'best linear unbiased estimators' proposed by Lieblein (1974), in which the annual maxima are ordered, and the parameters of the distribution are obtained by weighted sums of the extreme values.

2.3.4 Example of fitting the Type I Distribution to annual maxima

Wind gust data have been obtained from a military airfield at East Sale, Victoria, Australia, continuously since late 1951. The anemometer position has been constant throughout that period, and the height of the anemometer head has always been the standard meteorological value of 10 m. Thus, in this case, no corrections for height and terrain are required. Also, the largest gusts have almost entirely been produced by gales from large synoptic depressions (Section 1.3.1). However, the few gusts that were produced by thunderstorm downbursts were eliminated from the list, in order to produce a statistically consistent population (see Section 2.2.3).

The annual maxima for the 47 calendar years 1952–1998 are listed in Table 2.2. The values in Table 2.2 are sorted in order of increasing magnitude (Table 2.3) and assigned a probability, p, according to (i) the Gumbel formula (Equation 2.8), and (ii) the Gringorten formula (Equation 2.10). The reduced variate, $-\log_e(-\log_e p)$, according to Equation 2.9, is formed for both cases. These are tabulated in Table 2.3. The wind speed is plotted against the reduced variates, and straight lines are fitted by linear regression ('least squares' method). The results of this are shown in Figures 2.2 and 2.3, for the Gumbel and Gringorten methods, respectively. The intercept and slope of these lines give the mode, u, and slope, a, of the fitted Type I Extreme Value Distribution, according to Equation 2.1. u and a can also be estimated from the calculated mean and standard deviation (shown in Table 2.1) by the Method of Moments using Equations 2.13 and 2.14.

Predictions of extreme wind speeds for various return periods can then readily be obtained by application of either Equation 2.6 or 2.7. Table 2.4 lists these predictions based on the Gumbel and Gringorten fitting methods, and by the Method of Moments. For return periods up to 500 years, the predicted values by the three methods are within 1 m/s of each other. However, these small differences are swamped by sampling errors, that is the errors inherent in trying to make predictions for return periods of 100 years or more from <50 years of data. This problem is illustrated by the following exercise. The problem of high sampling errors can often be circumvented by compositing data, as discussed in Section 2.2.5.

EXERCISE

Re-analyse the annual maximum gust wind speeds for East Sale for the years 1952–1997, that is ignore the high value recorded in 1998. Compare the resulting predictions of design wind speeds for (a) 50-years return period, and (b) 1000-years return period, and comment.

2.3.5 General penultimate distribution

For extreme wind speeds that are derived from a Weibull parent distribution (see Section 2.5), Cook and Harris (2004) have proposed a 'General penultimate' Type I, or Gumbel Distribution. This takes the form of Equation 2.15.

$$F_U(U) = \exp\{-\exp[-(U^w - u^w)/a^w]\} \tag{2.15}$$

where w is the Weibull exponent of the underlying parent distribution (see Equation 2.22).

Table 2.2 Annual maximum gust speeds from East Sale, Australia 1952–1998

Year	*Maximum gust speed (m/s)*
1952	31.4
1953	33.4
1954	29.8
1955	30.3
1956	27.8
1957	30.3
1958	29.3
1959	36.5
1960	29.3
1961	27.3
1962	31.9
1963	28.8
1964	25.2
1965	27.3
1966	23.7
1967	27.8
1968	32.4
1969	27.8
1970	26.2
1971	30.9
1972	31.9
1973	27.3
1974	25.7
1975	32.9
1976	28.3
1977	27.3
1978	28.3
1979	28.3
1980	29.3
1981	27.8
1982	27.8
1983	30.9
1984	26.7
1985	30.3
1986	28.3
1987	30.3
1988	34.0
1989	28.8
1990	30.3
1991	27.3
1992	27.8
1993	28.8
1994	30.9
1995	26.2
1996	25.7
1997	24.7
1998	42.2
Mean	29.27
Standard deviation	3.196

Table 2.3 Processing of East Sale data

Rank	*Gust speed (m/s)*	*Reduced variate (Gumbel)*	*Reduced variate (Gringorten)*
1	23.7	−1.354	−1.489
2	24.7	−1.156	−1.226
3	25.2	−1.020	−1.069
4	25.7	−0.910	−0.949
5	25.7	−0.816	−0.848
6	26.2	−0.732	−0.759
7	26.2	−0.655	−0.679
8	26.7	−0.583	−0.604
9	27.3	−0.515	−0.534
10	27.3	−0.450	−0.467
11	27.3	−0.388	−0.403
12	27.3	−0.327	−0.340
13	27.3	−0.267	−0.279
14	27.8	−0.209	−0.220
15	27.8	−0.151	−0.161
16	27.8	−0.094	−0.103
17	27.8	−0.037	−0.045
18	27.8	0.019	0.013
19	27.8	0.076	0.071
20	28.3	0.133	0.129
21	28.3	0.190	0.187
22	28.3	0.248	0.246
23	28.3	0.307	0.306
24	28.8	0.367	0.367
25	28.8	0.427	0.428
26	28.8	0.489	0.492
27	29.3	0.553	0.556
28	29.3	0.618	0.623
29	29.3	0.685	0.692
30	29.8	0.755	0.763
31	30.3	0.827	0.837
32	30.3	0.903	0.914
33	30.3	0.982	0.995
34	30.3	1.065	1.081
35	30.3	1.152	1.171
36	30.9	1.246	1.268
37	30.9	1.346	1.371
38	30.9	1.454	1.484
39	31.4	1.572	1.607
40	31.9	1.702	1.744
41	31.9	1.848	1.898
42	32.4	2.013	2.075
43	32.9	2.207	2.285
44	33.4	2.442	2.544
45	34.0	2.740	2.885
46	36.5	3.157	3.391
47	42.2	3.861	4.427

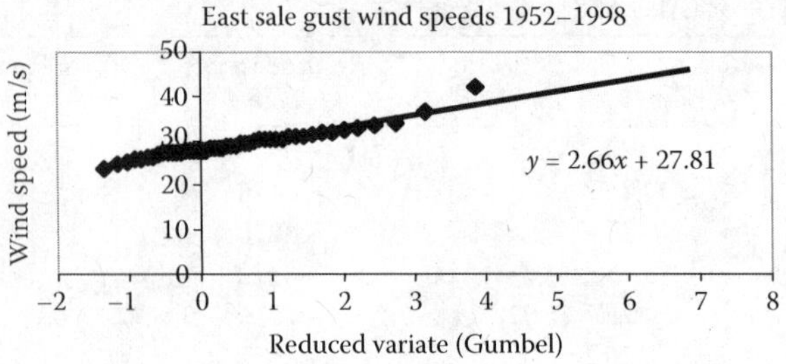

Figure 2.2 Analysis of the annual maximum wind gusts from East Sale using the Gumbel method.

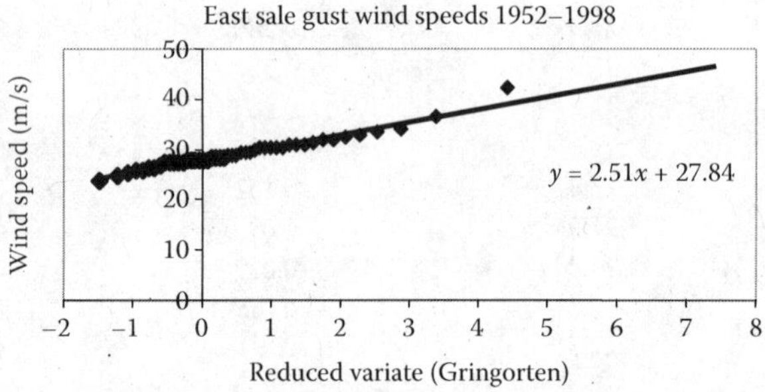

Figure 2.3 Analysis of East Sale data using the Gringorten fitting method.

Table 2.4 Prediction of extreme wind speeds for East Sale (synoptic winds)

Return period (years)	*Predicted gust speed (m/s) (Gumbel)*	*Predicted gust speed (m/s) (Gringorten)*	*Predicted gust speed (m/s) (method of moments)*
10	33.8	33.5	33.4
20	35.7	35.3	35.2
50	38.2	37.6	37.6
100	40.0	39.4	39.3
200	41.9	41.1	41.0
500	44.3	43.5	43.3
1000	46.2	45.2	45.0

Comparing Equation 2.15 with Equation 2.2, it can be seen that Equation 2.15 represents a Gumbel Distribution for a transformed variable, Z, equal to U^w.

If the parent wind speed data are available for a site, w can be obtained directly from fitting a Weibull Distribution to that site. Alternatively, the penultimate distribution of Equation 2.15 can be treated as a three-parameter (u, a and w) distribution and fitted directly to the extreme wind data, without knowing the parent distribution directly.

The Weibull exponent, w, is typically in the range of 1.3–2.0; in that case, when Equation 2.15 is plotted in the Gumbel form (Figure 2.2), the resulting line curves downwards, and is similar in shape to the Type III Extreme Value Distribution. The main difference is that the latter has a finite upper limit, whereas for the penultimate distribution, U^w, and hence U, is unlimited. However, for practical design situations, the two distributions give very similar predictions (Holmes and Moriarty, 2001).

2.4 THE PEAKS-OVER-THRESHOLD APPROACH

The approach of extracting a single maximum value of wind speed from each year of historical data obviously has limitations in that there may be many storms during any year, and only one value from all these storms is being used. A shorter reference period than a year could, of course, be used to increase the amount of data. However, it is important for extreme value analysis that the data values are statistically independent – this will not be the case if a period as short as 1 day is used. An alternative approach which makes use only of the data of relevance to extreme wind prediction is the *peaks*, or *excesses, over threshold* approach (e.g. Davison and Smith, 1990; Lechner et al., 1992; Holmes and Moriarty, 1999). The method is also known as the 'conditional mean exceedence' (CME) method.

A brief description of the method is given here. This is a method which makes use of all wind speeds from independent storms above a particular minimum threshold wind speed, u_o (say 20 m/s). There may be several of these events, or none, during a particular year. The basic procedure is as follows:

- Several threshold levels of wind speed are set: u_o, u_1, u_2 and so on (e.g. 20, 21, 22 ...m/s).
- The exceedences of the lowest level u_o by the maximum storm wind are identified, and the number of crossings of this level per year, λ, is calculated.
- The differences $(U - u_o)$ between each storm wind and the threshold level u_o are calculated and averaged (only positive excesses are counted).
- The previous step is repeated for each level, u_1, u_2 and so on, in turn.
- The mean excess is plotted against the threshold level.
- A scale factor, σ, and a shape factor, k, are determined from the following equations (Davison and Smith, 1990):

$$\text{Slope} = \frac{-k}{(1+k)} \qquad \text{Intercept} = \frac{\sigma}{(1+k)} \tag{2.16}$$

Prediction of the R-year return period wind speed, U_R, can then be calculated from:

$$U_R = u_o + \sigma[1 - (\lambda R)^{-k}]/k \tag{2.17}$$

In Equation 2.17, the shape factor, k, is normally found to be positive (usually around 0.1). As R increases to very large values, the upper limit to U_R of $u_o + (\sigma/k)$ is gradually approached.

When k is zero, it can be shown mathematically that Equation 2.17 reduces to Equation 2.18.

$$U_R = u_o + \sigma \log_e(\lambda R) \tag{2.18}$$

The similarity between Equations 2.7 and 2.18 should be noted.

The highest threshold level, u_n, should be set so that it is exceeded by at least 10 wind speeds. An example of this method is given in the following section.

2.4.1 Example of the use of the 'peaks over threshold' method

Daily wind gusts at several stations in the area of Melbourne, Australia, have been recorded since the 1940s. Those at the four airport locations of Essendon, Moorabbin, Melbourne Airport (Tullamarine), and Laverton are the most useful since the anemometers are located at positions most closely matching the ideal open-country conditions, and away from the direct influence of buildings. Table 2.5 summarises the data available from these four stations.

The two most common types of event-producing extreme wind in the Melbourne area are gales produced by the passage of large low pressure or frontal systems ('synoptic' winds – see Section 1.3.1), and severe thunderstorm 'downbursts' (Section 1.3.3). Downbursts are usually accompanied by thunder, but the occurrence of thunder does not necessarily mean that an extreme gust has been generated by a downburst. The occurrences of downbursts in the data from the four stations were identified by inspection of the charts stored by the Australian Bureau of Meteorology or the National Archives. Table 2.5 shows that the rate of occurrence of downbursts >21 m/s is quite low (around one per year at each station); however, as will be seen, they are significant contributors to the largest gusts.

The largest recorded gusts in the Melbourne area are listed in Table 2.6. Approximately half of these were generated by downbursts.

Extreme value analysis of the data was carried out in the following stages:

- Daily gusts over 21 m/s were retained for analysis.
- Gusts generated by downbursts were identified by inspection of anemometer charts, and separated from the synoptic gusts.
- The data from the four stations were composited into single data sets, for both downburst and synoptic gusts.
- The synoptic data were corrected to a uniform height (10 m), and approach terrain (open country), using correction factors according to the direction derived from wind-tunnel tests for each station.
- For both data sets, the 'excesses over threshold' analysis was used to derive relationships between wind speed and return period.

The last stage enabled a scale factor, σ, and a shape factor, k, to be determined in the relationship:

$$U_R = u_o + \sigma[1 - (\lambda R)^{-k}]/k$$

u_o is the lowest threshold – in this case 21 m/s, and λ is average annual rate of exceedence of u_o, for the combined data sets. For the current analysis, λ was 23.4 for the synoptic data, and 0.97 for the downburst data.

The results of the two analyses were expressed in the following forms for the Melbourne data:

Table 2.5 Summary of data for Melbourne stations

Station	*Station number*	*Years*	*Maximum recorded gust (m/s)*	*Rate/year (synoptic gusts > 21 m/s)*	*Rate/year (downburst gusts > 21 m/s)*
Essendon	86038	1940–1971	40.6	34.6	1.1
Moorabbin	86077	1972–1992	41.2	19.3	0.7
Tullamarine	86282	1970–1997	38.6	30.1	1.3
Laverton	87031	1946–1995	42.7	28.4	0.8

Note: 1953, 1954 and 1956 are missing from Laverton data.

Table 2.6 Largest recorded gusts in the Melbourne area 1940–1997

Date	*Station*	*Gust speed (knots)*	*Gust speed (m/s)*	*Type*
14/1/1985	Laverton	83	42.7	Synoptic
25/12/1978	Moorabbin	80	41.2	Downburst
6/9/1948	Essendon	79	40.6	Synoptic
15/11/1982	Tullamarine	75	38.6	Downburst
3/1/1981	Tullamarine	74	38.1	Downburst
26/10/1978	Laverton	71	36.5	Downburst
4/8/1947	Essendon	70	36.0	Synoptic
27/2/1973	Laverton	70	36.0	Downburst
8/11/1988	Tullamarine	70	36.0	Synoptic
1/7/1942	Essendon	67	34.5	Downburst
5/8/1959	Laverton	67	34.5	Synoptic
24/1/1982	Laverton	67	34.5	Downburst
10/8/1992	Tullamarine	67	34.5	Synoptic

For synoptic winds:

$$U_R = 68.3 - 39.3R_1^{-0.059} \tag{2.19}$$

For downburst winds:

$$U_R = 69.0 - 48.1R_2^{-0.108} \tag{2.20}$$

The combined probability of exceedence of a given gust speed from either type of wind is obtained by substituting in Equation 2.3:

$$\frac{1}{R_C} = 1 - \left[1 - \left(\frac{68.3 - U_R}{39.3}\right)^{\frac{1}{0.059}}\right]\left[1 - \left(\frac{69.0 - U_R}{48.1}\right)^{\frac{1}{0.108}}\right] \tag{2.21}$$

Equations 2.19 through 2.21 are plotted in Figure 2.4. The lines corresponding to Equations 2.19 and 2.20 cross at a return period of 30 years. It can also be seen that the combined wind speed return period relationship is asymptotic to the synoptic line at low return periods, and to the downburst line at high return periods.

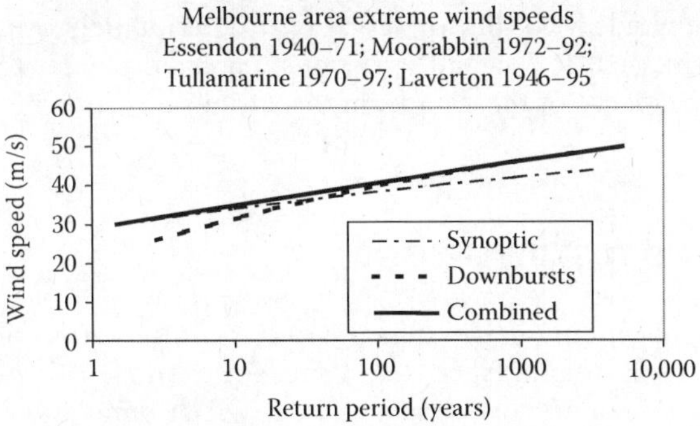

Figure 2.4 Gust wind speed versus return period for the Melbourne area.

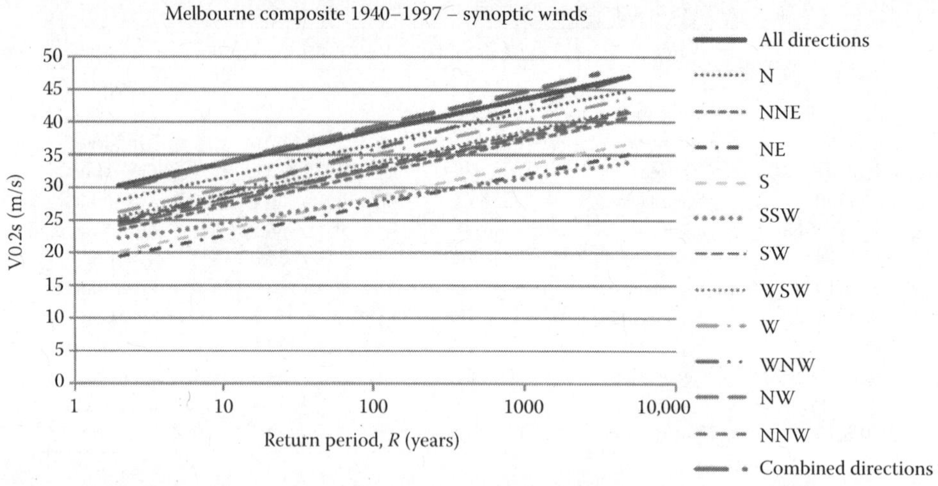

Figure 2.5 Gumbel probability distributions fitted to gust data for direction sectors from the Melbourne area.

2.4.2 Extreme winds by direction sector

The peaks-over-threshold approach can be applied to winds separated by a direction sector. This is done in Figure 2.5 which shows a wind speed versus return period for a number of cases using, again, combined gust data, from four stations in the Melbourne area, for synoptic wind events:

- Gust wind speeds separately analysed for 11 direction sectors of 22.5° width. For the remaining five sectors (ENE to SSE), there was insufficient data for a meaningful analysis to be carried out.
- Combined data from all-direction sectors, analysed as a single data set.
- A combined distribution obtained by combining distributions for directional sectors, according to Equation 2.4.

A peaks-over-threshold approach was used, but in this case, the shape factor, k, was 'forced' to be zero, and Equation 2.18 was used for the predictions for the 11 direction sectors and 'all directions' data. This effectively resulted in Gumbel Distributions, forming straight lines on the wind speed (linear) versus return period (logarithmic) graphs.

There is good agreement for the 'all-directions' line, and the 'combined directions' line, indicating the independence of the data from the various direction sectors, and the validity of Equation 2.4.

2.5 PARENT WIND DISTRIBUTIONS

For some design applications, it is necessary to have information on the distribution of the complete population of wind speeds at a site. An example is the estimation of fatigue damage for which an account must be taken of damage accumulation over a range of wind storms (see Section 5.6). The population of wind speeds produced by synoptic wind storms at a site is usually fitted with a distribution of the *Weibull* type:

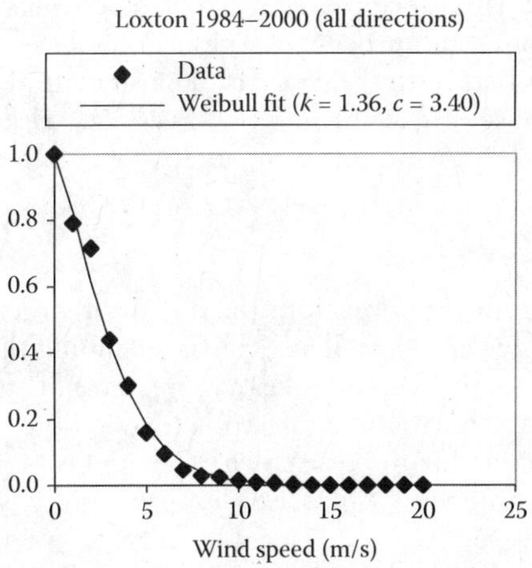

Figure 2.6 Example of a Weibull Distribution fit to the parent population of synoptic winds.

$$f_U(\bar{U}) = \frac{w\bar{U}^{w-1}}{c^w} \exp\left[-\left(\frac{\bar{U}}{c}\right)^w\right] \tag{2.22}$$

Equation 2.22 represents the probability density function for mean wind speeds produced by synoptic events. There are two parameters: a *scale factor*, c, which has units of wind speed, and a *shape factor*, w, which is dimensionless (see also Appendix C3.4). The probability of exceedence of any given wind speed is given by Equation 2.23

$$1 - F(\bar{U}) = \exp\left[-\left(\frac{\bar{U}}{c}\right)^w\right] \tag{2.23}$$

Typical values of c are 3–10 m/s, and w usually falls in the range 1.3–2.0. An example of a Weibull fit to recorded meteorological data is shown in Figure 2.6.

Several attempts have been made to predict extreme winds from knowledge of the parent distribution of wind speeds, and thus make predictions from quite short records of wind speed at a site (e.g. Gomes and Vickery, 1977b). The 'asymptotic' extreme value distribution for a Weibull parent distribution is the Type I, or Gumbel Distribution. However, for extremes drawn from a finite sample (e.g. annual maxima), the 'penultimate' Type I, as discussed in Section 2.3.2, is the more appropriate extreme value distribution.

However, it should be noted that both the Weibull Distribution and the Type I Extreme Value Distribution will give unlimited wind speeds with reducing probability of exceedence.

2.6 WIND LOADS AND STRUCTURAL SAFETY

The development of structural reliability concepts – that is the application of probabilistic methods to the structural design process – has accelerated the adoption of probabilistic methods into wind engineering since the 1970s. The assessment of wind loads is only one part of the total structural design process, which also includes the determination of other loads and the resistance of structural materials. The structural engineer must proportion the structure so that collapse or overturning has a very low risk of occurring, and defined serviceability limits on deflection, acceleration, and so on, are not exceeded very often.

2.6.1 Limit states design

Limit states design is a rational approach to the design of structures, which has gradually been accepted around the world. As well as explicitly defining the ultimate and serviceability limit states for the design, the method takes a more rational approach to structural safety by defining 'partial' load factors ('gamma' factors) for each type of loading, and a separate resistance factor ('phi' factor) for the resistance. The application of the limit states design method is not, in itself, a probabilistic process, but probability is usually used to derive the load and resistance factors.

A typical ultimate limit states design relationship involving wind loads is as follows:

$$\varphi R \geq \gamma_D D + \gamma_W W \tag{2.24}$$

where

φ is a resistance factor
R is the nominal structural resistance
γ_D is a dead load factor
D is the nominal dead load
γ_W is a wind load factor
W is the nominal wind load

In this relationship, the partial factors φ, γ_D and γ_W are adjusted separately to take account of the variability and uncertainty in the resistance, dead load and wind load. The values used also depend on what particular nominal values have been selected. Often, a final calibration of a proposed design formula is carried out by evaluating the safety, or reliability, index as discussed in the following section, for a range of design situations, for example, various combinations of nominal dead and wind loads.

2.6.2 Probability of failure and the safety index

A quantitative measure of the safety of structures, known as the *safety index*, or *reliability index*, is used in many countries as a method of calibrating the existing and future design methods for structures. As will be explained in this section, there is a one-to-one relationship between this index and a probability of failure, based on the exceedence of a design resistance by an applied load (but not including failures by human errors and other accidental causes).

The design process is shown in its simplest form in Figure 2.7. The design process consists of comparing a structural load effect, S, with the corresponding resistance, R. In the case of limit states associated with structural strength or collapse, the load effect could be an axial force in a member or a bending moment, or the corresponding stresses. In the case of serviceability limit states, S and R may be deflections, accelerations or crack widths.

The probability density functions $f_s(S)$ and $f_R(R)$ for a load effect, S, and the corresponding structural resistance, R, are shown in Figure 2.7. (Probability density is defined in Section C.2.1 in Appendix C). Clearly, S and R must have the same units. The dispersion or 'width' of the two distributions represents the uncertainty in S and R.

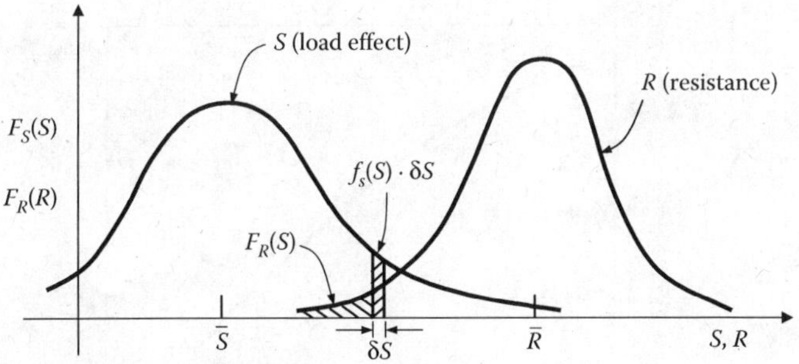

Figure 2.7 Probability densities for load effects and resistance.

Failure (or unserviceability) occurs when the resistance of the structure is less than the load effect. The probability of failure will now be determined assuming S and R are statistically independent.

The probability of failure occurring at a load effect between S and $S + \delta S$ = [probability of load effect lying between S and $S + \delta S$] × [probability of resistance, R, being less than S]

$$= [f_S(S)\delta S] \times F_R(S) \tag{2.25}$$

where $F_R(R)$ is the cumulative probability distribution of R, and

$$F_R(S) = \int_{-\infty}^{S} f_R(R)dR \tag{2.26}$$

The terms in the product in Equation 2.25 are the areas shown in Figure 2.7.

The total probability of failure is obtained by summing, or integrating, Equation 2.25 over all possible values of S (between $-\infty$ and $+\infty$):

$$p_f = \int_{-\infty}^{\infty} f_S(S) \cdot F_R(S)dS \tag{2.27}$$

Substituting for $F_R(S)$ from Equation 2.26 into 2.27

$$p_f = \int_{-\infty}^{\infty}\int_{-\infty}^{S} f_S(S) \cdot f_R(R) \cdot dR \cdot dS = \int_{-\infty}^{\infty}\int_{-\infty}^{S} f(S,R) \cdot dR \cdot dS \tag{2.28}$$

where $f(S, R)$ is the *joint* probability density of S, R.

The values of the probability of failure computed from Equation 2.28 are normally very small numbers, typically 1×10^{-2} to 1×10^{-5}.

The safety, or reliability, index is defined according to Equation 2.29, and normally takes values in the range 2–5.

$$\beta = -\Phi^{-1}(p_f) \tag{2.29}$$

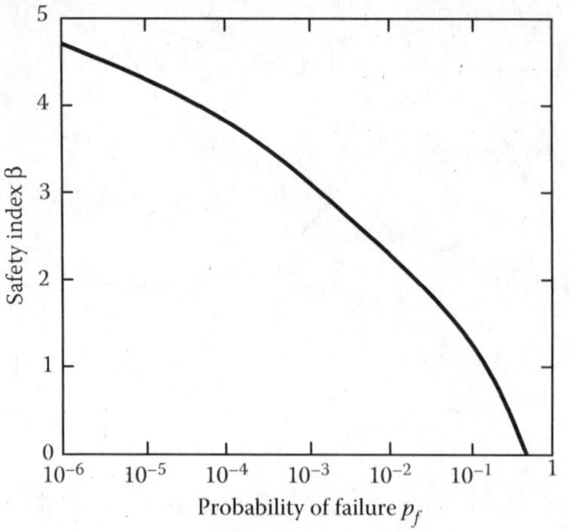

Figure 2.8 Relationship between the safety index and probability of failure.

where $\Phi^{-1}()$ is the inverse cumulative probability distribution of a unit normal (Gaussian) variate, that is a normal variate with a mean of zero and a standard deviation of one.

The relationship between the safety index, β, and the probability of failure, p_f, according to Equation 2.29 is shown and plotted in Figure 2.8.

Equations 2.28 and 2.29 can be evaluated exactly when S and R are assumed to have Gaussian (normal) or lognormal (Appendix C3.2) probability distributions. However, in other cases, (which includes those involving wind loading), numerical methods must be used. Numerical methods must also be used when, as is usually the case, the load effect, S, and resistance, R, are treated as combinations (sums and products) of separate random variables, with separate probabilistic characteristics.

Details of structural reliability theory and practice can be found in a number of texts on the subject (e.g. Blockley, 1980; Melchers, 1987; Ang and Tang, 1990).

2.6.3 Nominal return period for design wind speeds

The return periods (or annual risk of exceedence) for the nominal design wind speeds in various wind-loading codes and standards are discussed in Chapter 15. The most common choice is 50 years. There should be no confusion between the return period, R, and the expected lifetime of a structure, L. The return period is just an alternative statement of the annual risk of exceedence, for example, a wind speed with a 50-year return period is one with an expected risk of exceedence of 0.02 (1/50) in any 1 year. However the risk, r, of exceedence of a wind speed *over the lifetime,* can be determined by assuming that all years are statistically independent of each other.

Then,

$$r = 1 - \left[1 - \left(\frac{1}{R}\right)\right]^L \tag{2.30}$$

Equation 2.30 is very similar to Equation 2.4 in which the combined probability of exceedence of a wind speed occurring over a range of wind directions was determined.

Setting both R and L as 50 years in Equation 2.30, we arrive at a value of r of 0.636. Thus, there is a nearly 64% chance that the 50-year return period wind speed will be exceeded at least once during a 50-year lifetime – that is a better than even chance that it *will occur*. Wind loads derived from wind speeds with this level of risk must be factored up when used for the ultimate limit states design. Typical values of the wind load factor, γ_W, are in the range of 1.4–1.6. Different values may be required for regions with different wind speed/return period relationships, as discussed in Section 2.7.

The use of a return period for the nominal design wind speed, substantially higher than the traditional 50 years, avoids the need to have different wind load factors in different regions. This was an important consideration in the revision of the Australian Standard for Wind Loads in 1989 (Standards Australia, 1989), which, in previous editions, required the use of a special 'Cyclone Factor' in the regions of northern coastline affected by tropical cyclones. The reason for this factor was the greater rate of change of wind speed with return period in the cyclone regions. A similar 'hurricane importance factor' appeared in some editions of the American National Standard (ASCE, 1993), but was later incorporated into the specified basic wind speed (ASCE, 1998).

In AS1170.2-1989, the wind speeds for ultimate limit states design had a nominal probability of exceedence of 5% in a lifetime of 50 years (a return period of 1000 years, approximately).

However, a load factor of 1.0 was applied to the wind loads derived in this way – and this factor was the same in both cyclonic and non-cyclonic regions.

2.6.4 Uncertainties in wind load specifications

A reliability study of structural design involving wind loads requires an estimation of all the uncertainties involved in the specification of wind loads – wind speeds, multipliers for the terrain height and topography, pressure coefficients, local and area-averaging effects and so on. Some examples of this type of study for buildings and communication towers were given by Pham et al. (1983, 1992).

Table 2.7 shows estimates by Pham et al. (1983) of mean-to-nominal values of various parameters associated with wind-loading calculations for regions affected by tropical cyclones from the Australian Standard of that time. It can be seen from Table 2.7 that the greatest contributor to the variability and uncertainty in wind load estimation is the wind speed itself – particularly as it is raised to a power of 2 (or greater, when dynamic effects are important) when wind loads and effects are calculated. A secondary contributor is the uncertainty in the exposure parameter in Table 2.7, which is also squared, and includes uncertainties in the vertical profile of mean and gust speeds as discussed in the earlier sections of this chapter.

Table 2.7 Variability of wind-loading parameters

Parameter	*Mean/nominal*	*Coefficient of variation*	*Assumed distribution*
Wind speed (50-year maximum)	1.12	0.28	Gumbel
Directionality	0.9	0.05	Lognormal
Exposure	0.8	0.15	Lognormal
Pressure coefficient	0.8	0.15	Lognormal
Local and area reduction effects	0.85	0.10	Lognormal

Source: Reprinted from *Journal of Wind Engineering and Industrial Aerodynamics*, Pham, L., Holmes, J.D. and Leicester, R.H., Safety indices for wind loading in Australia, 14: 3–14, Copyright 1983, with permission from Elsevier.

2.7 WIND LOAD FACTORS

As discussed in the previous section, wind load factors, γ_w, in the range of 1.4–1.6 have traditionally been adopted for use with wind loads derived on the basis of a 50-year return period wind speeds. Wind load factors in this range have been derived on the basis of two assumptions:

1. Wind loads vary as the square of wind speeds. This is valid for small structures of high frequency with high natural frequencies. However, this assumption is not valid for tall buildings for which the effective wind loads vary to a higher power of wind speed, due to the effect of a resonant dynamic response. This power is up to 2.5 for a long-wind response, and up to 3 for cross-wind response (see Chapter 9).
2. The wind load factors have been derived for non-tropical-cyclone wind loads, that is climates for which the rate of change of wind speed with return period R is relatively low, compared with those in regions affected by tropical cyclones, typhoons or hurricanes.

For structures with a significant dynamic response to wind such as tall buildings, wind loads and effects vary with wind speed raised to a power somewhat greater than 2. Holmes and Pham (1993) considered the effect of the varying exponent on the safety index and showed that the use of the traditional approach of 'working stress' wind loads based on a 50-year return period wind speed, with a fixed wind load factor of 1.5, resulted in a significant reduction in safety as the exponent increased. However, a near-uniform safety (insensitive to the effects of dynamic response) was achieved by the use of a high return period nominal design wind speed together with a wind load factor of 1.0.

The need for higher load factors in locations affected by tropical cyclones is illustrated by Table 2.8. This shows the equivalent return period of wind speeds when a wind load factor of 1.4 is applied to the 50-year return period value for several different locations. For two locations in the United Kingdom, the application of the 1.4 factor is equivalent to calculating wind loads from wind speeds with return periods considerably greater than 1000 years. However, in Penang, Malaysia, where design wind speeds are governed by winds from relatively infrequent severe thunderstorms, the factor of 1.4 is equivalent to using a wind speed of about 500 years return period.

At Port Hedland (Western Australia), and Hong Kong, where tropical cyclones and typhoons are dominant, and extreme wind events are even rarer, the factored wind load based on V_{50} is equivalent to applying wind loads of only 150 and 220 years return period, respectively. In Australia, previous editions of AS1170.2 adjusted the effective return period by use of the 'Cyclone Factor' as discussed above. However, in AS/NZS1170:2010,

Table 2.8 Return periods of factored wind loads with a wind load factor of 1.4

Location	*Wind type*	V_{50} *(m/s)*	$\sqrt{(1.4\, V_{50}^2)}$ *(m/s)*	*Return period of factored wind speed (years)*
Cardington (UK)	Atlantic gales	42.9	50.7	1130
Jersey (UK)	Atlantic gales	45.6	53.9	1470
Penang (Malaysia)	Thunderstorms	27.6	32.7	510
Port Hedland (WA)	Tropical cyclones	60.5	71.6	150
Hong Kong	Typhoons	61.5	72.8	220

Note: (i) Values of wind speed shown are gust speeds at 10-m height over a flat open terrain, except for Hong Kong for which the gust speed is adjusted to 50-m height above the ocean. (ii) Probability distributions for Cardington and Jersey obtained from Cook, N.J. 1985. *The Designer's Guide to Wind Loading of Building Structures*. BRE-Butterworths, UK.

a nominal design wind speed with 500 years return period (80 m/s basic gust wind speed at Port Hedland) with a wind load factor of 1.0 is used for the design of most structures for ultimate limit states. An 'uncertainty' factor of 1.10 is also applied to the wind speeds in that part of Australia, further increasing the effective return period based on recorded data.

2.8 SUMMARY

In this chapter, the application of extreme value analysis to the prediction of design wind speeds has been discussed. In particular, the Gumbel and 'peaks over threshold' approaches were described in detail. The need to separate wind speeds caused by wind storms of different types was emphasised, and wind direction effects were considered.

The main principles of the application of probability to the structural design and safety under wind loads were also introduced.

REFERENCES

American Society of Civil Engineers. 1993. *Minimum Design Loads for Buildings and Other Structures.* ASCE Standard, ANSI/ASCE 7-93, American Society of Civil Engineers, New York.

American Society of Civil Engineers. 1998. *Minimum Design Loads for Buildings and Other Structures.* ASCE Standard, ANSI/ASCE 7-98, American Society of Civil Engineers, New York.

American Society of Civil Engineers. 2010. *Minimum Design Loads for Buildings and Other Structures.* ASCE/SEI 7-10, American Society of Civil Engineers, New York.

Ang, A.H. and Tang, W. 1990. *Probability Concepts in Engineering Planning and Design. Vol. II. Decision, Risk and Reliability*. Published by the authors.

Blockley, D. 1980. *The Nature of Structural Design and Safety.* Ellis Horwood, Chichester, England, UK.

Cook, N.J. 1983. Note on directional and seasonal assessment of extreme wind speeds for design. *Journal of Wind Engineering and Industrial Aerodynamics*, 12: 365–72.

Cook, N.J. 1985. *The Designer's Guide to Wind Loading of Building Structures*. BRE-Butterworths, UK.

Cook, N.J. and Harris, R.I. 2004. Exact and general FT1 penultimate distributions of extreme winds drawn from tail-equivalent Weibull parents. *Structural Safety*, 26: 391–420.

Cook, N.J. and Miller, C.A.M. 1999. Further note on directional assessment of extreme wind speeds for design. *Journal of Wind Engineering and Industrial Aerodynamics*, 79: 201–8.

Davenport, A.G. 1961. The application of statistical concepts to the wind loading of structures. *Proceedings Institution of Civil Engineers*, 19: 449–71.

Davison, A.C. and Smith, R.L. 1990. Models for exceedances over high thresholds. *Journal of the Royal Statistical Society, Series B*, 52: 339–442.

Fisher, R.A. and Tippett, L.H.C. 1928. Limiting forms of the frequency distribution of the largest or smallest member of a sample. *Proceedings, Cambridge Philosophical Society Part 2*, 24: 180–90.

Freudenthal, A.M. 1947. The safety of structures. *American Society of Civil Engineers Transactions*, 112: 125–59.

Freudenthal, A.M. 1956. Safety and the probability of structural failure. *American Society of Civil Engineers Transactions*, 121: 1337–97.

Gomes, L. and Vickery, B.J. 1977a. Extreme wind speeds in mixed wind climates. *Journal of Industrial Aerodynamics*, 2: 331–44.

Gomes, L. and Vickery, B.J. 1977b. On the prediction of extreme wind speeds from the parent distribution. *Journal of Industrial Aerodynamics*, 2: 21–36.

Gringorten, I.I. 1963. A plotting rule for extreme probability paper. *Journal of Geophysical Research*, 68: 813–4.

Gumbel, E.J. 1954. *Statistical Theory of Extreme Values and Some Practical Applications. Applied Math Series 33*, National Bureau of Standards, Washington, DC.

Gumbel, E.J. 1958. *Statistics of Extremes*. Columbia University Press, New York.
Holmes, J.D. 1990. Directional effects on extreme wind loads. *Civil Engineering Transactions, Institution of Engineers, Australia*, 32: 45–50.
Holmes, J.D. 2002. A re-analysis of recorded extreme wind speeds in Region A. *Australian Journal of Structural Engineering*, 4: 29–40.
Holmes, J.D. and Ginger, J.D. 2012. The gust wind speed duration in AS/NZS 1170.2. *Australian Journal of Structural Engineering* (I.E. Aust.), 13: 207–17.
Holmes, J.D. and Moriarty, W.W. 1999. Application of the generalized Pareto distribution to extreme value analysis in wind engineering. *Journal of Wind Engineering and Industrial Aerodynamics*, 83: 1–10.
Holmes, J.D. and Moriarty, W.W. 2001. Response to discussion by N.J. Cook and R.I. Harris of: Application of the generalized Pareto distribution to extreme value analysis in wind engineering. *Journal of Wind Engineering and Industrial Aerodynamics*, 89: 225–7.
Holmes, J.D. and Pham, L. 1993. Wind-induced dynamic response and the safety index. *Proceedings of the 6th International Conference on Structural Safety and Reliability*, Innsbruck, Austria, 9–13 August, pp. 1707–9, A.A. Balkema Publishers.
Hosking, J.R.M., Wallis, J.R. and Wood, E.F. 1985. Estimates of the generalized extreme value distribution by the method of probability-weighted moments. *Technometrics*, 27: 251–61.
Jenkinson, A.F. 1955. The frequency distribution of the annual maximum (or minimum) values of meteorological elements. *Quarterly Journal of the Royal Meteorological Society*, 81: 158–71.
Kasperski, M. 2007. Design wind speeds for a low-rise building taking into account directional effects. *Journal of Wind Engineering and Industrial Aerodynamics*, 95: 1125–44.
Lechner, J.A., Leigh, S.D. and Simiu, E. 1992. Recent approaches to extreme value estimation with application to wind speeds. Part 1: The Pickands method. *Journal of Wind Engineering and Industrial Aerodynamics*, 41: 509–19.
Lieblein, J. 1974. Efficient methods of extreme-value methodology. Report NBSIR 74-602, National Bureau of Standards, Washington.
Melbourne, W.H. 1984. Designing for directionality. *Workshop on Wind Engineering and Industrial Aerodynamics*, Highett, Victoria, Australia, July.
Melchers, R. 1987. *Structural Reliability – Analysis and Prediction*. Ellis Horwood, Chichester, England, UK.
Miller, C.A., Holmes, J.D., Henderson, D.J., Ginger, J.D. and Morrison, M. 2013. The response of the Dines anemometer to gusts and comparisons with cup anemometers. *Journal of Atmospheric and Oceanic Technology*, 30: 1320–36.
Palutikof, J.P., Brabson, B.B., Lister, D.H. and Adcock, S.T. 1999. A review of methods to calculate extreme wind speeds. *Meteorological Applications*, 6: 119–32.
Peterka, J.A. and Shahid, S. 1998. Design gust wind speeds in the United States. *Journal of Structural Engineering (ASCE)*, 124: 207–14.
Pham, L., Holmes, J.D. and Leicester, R.H. 1983. Safety indices for wind loading in Australia. *Journal of Wind Engineering and Industrial Aerodynamics*, 14: 3–14.
Pham, L., Holmes, J.D. and Yang, J. 1992. Reliability analysis of Australian communication lattice towers. *Journal of Constructional Steel Research*, 23: 255–72.
Pugsley, A.G. 1966. *The Safety of Structures*. Edward Arnold, London.
Russell, L.R. 1971. Probabilistic distributions for hurricane effects. *American Society of Civil Enginners Journal of Waterways, Harbours and Coastal Engineering*, 97: 139–54.
Standards Australia. 1989. *SAA Loading Code. Part 2: Wind Loads*. Standards Australia, North Sydney, New South Wales, Australia. Australian Standard AS1170.2-1989.
Standards Australia. 2011. *Structural Design Actions. Part 2: Wind Actions*. Standards Australia, Sydney, New South Wales, Australia. Australian/New Zealand Standard, AS/NZS1170.2:2011.
Von Mises, R. 1936. *La Distribution de la Plus Grande de n Valeurs. Reprinted in Selected Papers of Richard von Mises*, 2: 271–94, American Mathematical Society, Providence R.I., USA.

Chapter 3

Strong wind characteristics and turbulence

3.1 INTRODUCTION

As the earth's surface is approached, the frictional forces play an important role in the balance of forces on the moving air. For larger storms such as extra-tropical depressions, this zone extends up to 500–1000-m height. For thunderstorms, the boundary layer is much smaller – probably around 100 m (see Section 3.2.5). The region of frictional influence is called the 'atmospheric boundary layer' and is similar in many respects to the turbulent boundary layer on a flat plate or airfoil at high wind speeds.

Figure 3.1 shows records of wind speeds recorded at three heights on a tall mast at Sale in southern Australia (as measured by sensitive cup anemometers, during a period of strong winds produced by a gale from a synoptic extra-tropical depression (Deacon, 1955)). The records show the main characteristics of a fully developed 'boundary-layer' flow in the atmosphere:

- The increase of the average wind speed as the height increases
- The gusty or turbulent nature of the wind speed at all heights
- The broad range of frequencies in the gusts in the air flow
- There is some similarity in the patterns of gusts at all heights, especially for the more slowly changing gusts, or lower frequencies

The term 'boundary-layer' means the region of wind flow affected by friction at the earth's surface, which can extend up to 1 km. The Coriolis forces (Section 1.2.2) become gradually less in magnitude as the wind speed falls near the earth's surface. This causes the geostrophic balance, as discussed in Section 1.2.3 to be disturbed, and the mean wind vector turns from being parallel to the isobars to having a component towards the low pressure, as the height above the ground reduces. Thus, the mean wind speed may change in direction slightly with height, as well as magnitude. This effect is known as the *Ekman spiral*. However, the direction change is small over the height range of normal structures, and is normally neglected in wind engineering.

The following sections will mainly be concerned with the characteristics of the mean wind and turbulence, near the ground, produced by severe gales in the higher latitudes. These winds have been studied in detail for more than 40 years and are generally well understood, at least over a flat homogeneous terrain. The wind and turbulence characteristics in tropical cyclones (Section 1.3.2) and thunderstorm downbursts (Section 1.3.5), which produce the extreme winds in the lower latitudes, are equally important, but are much less well understood. However, an existing knowledge of their characteristics is presented in Sections 3.2.5 and 3.2.6. Tornadoes are rare events, but can produce significant damage in some parts of the world. A simple horizontal profile of wind components in a tornado vortex is discussed in Section 3.2.7.

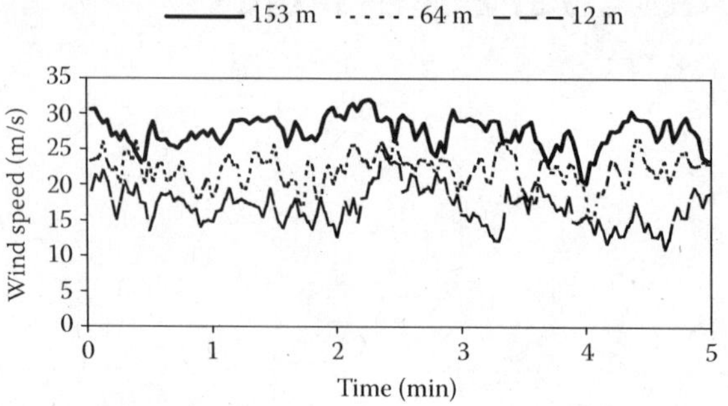

Figure 3.1 Wind speeds at three heights during a gale. (From Deacon, E.L. 1955. *Quarterly Journal of the Royal Meteorological Society*, 81: 562–73. With permission.)

3.2 MEAN WIND SPEED PROFILES

3.2.1 The 'Logarithmic Law'

In this section, we will consider the variation of the mean or time-averaged wind speed with height above the ground near the surface (say, in the first 100–200 m – the height range of most structures). In strong wind conditions, the most accurate mathematical expression is the 'logarithmic law'. The Logarithmic Law was originally derived for the turbulent boundary layer on a flat plate by Prandtl; however, it has been found to be valid in an unmodified form in strong wind conditions in the atmospheric boundary layer near the surface. It can be derived in a number of different ways. The following derivation is the simplest, and is a form of dimensional analysis.

We postulate that the wind shear, i.e. the rate of change of mean wind speed, $\bar{U}$, with height is a function of the following variables:

- The height above the ground, z
- The retarding force per unit area exerted by the ground surface on the flow – known as the surface shear stress, τ_o
- The density of air, ρ_a

Note that near the ground, the effect of the earth's rotation (Coriolis forces) is neglected. Also, because of the turbulent flow, the effect of molecular viscosity can be neglected.

Combining the wind shear with the above quantities, we can form a non-dimensional wind shear:

$$\frac{d\bar{U}}{dz} z \sqrt{\frac{\rho_a}{\tau_o}}$$

$\sqrt{(\tau_o/\rho_a)}$ has the dimensions of velocity, and is known as the *friction velocity*, u_* – (note that this is not a physical velocity). Then, since there are no other non-dimensional quantities involved,

$$\frac{d\bar{U}}{dz}\frac{z}{u_*} = \text{a constant}, \quad \text{say} \frac{1}{k} \tag{3.1}$$

Integrating

$$\bar{U}(z) = \frac{u_*}{k}(\log_e z - \log_e z_0) = \frac{u_*}{k}\log_e(z/z_0) \tag{3.2}$$

where z_o is a constant of integration, with the dimensions of length, known as the *roughness length*.

Equation 3.2 is the usual form of the Logarithmic Law. k is known as *von Karman's constant*, and has been found experimentally to have a value of about 0.4. z_o, the roughness length, is a measure of the roughness of the ground surface.

Another measure of the terrain roughness is the *surface drag coefficient*, κ, which is a non-dimensional surface shear stress, defined as

$$\kappa = \frac{\tau_0}{\rho \bar{U}_{10}^2} = \frac{u_*^2}{\bar{U}_{10}^2} \tag{3.3}$$

where $\bar{U}_{10}$ is the mean wind speed at 10-m height.

For urban areas and forests, where the terrain is very rough, the height, z, in Equation 3.2 is often replaced by an effective height, $(z - z_h)$, where z_h is a 'zero-plane displacement'. Thus, in this case,

$$\bar{U}(z) = \frac{u_*}{k}\log_e\left[\frac{z - z_h}{z_0}\right] \tag{3.4}$$

The zero-plane displacement can be taken as about three-quarters of the general rooftop height.

Usually, the most useful way of applying Equation 3.4 is to use it to relate the mean wind speeds at two different heights, as follows:

$$\frac{\bar{U}(z_1)}{\bar{U}(z_2)} = \frac{\log_e\left[(z_1 - z_h)/z_0\right]}{\log_e\left[(z_2 - z_h)/z_0\right]} \tag{3.5}$$

In the application of Equation 3.3, the 10-m reference height should be taken as 10 m above the zero-plane displacement, or $(10 + z_h)$ metres above the actual ground level.

By applying Equations 3.3 and 3.4 for z equal to 10 m, a relationship between the surface drag coefficient and the roughness length can be determined:

$$\kappa = \left[\frac{k}{\log_e\left(\dfrac{10}{z_0}\right)}\right]^2 \tag{3.6}$$

Table 3.1 gives the appropriate value of roughness length and surface drag coefficient, for various types of terrain.

Although the Logarithmic Law has a sound theoretical basis, at least for a fully developed wind flow over a homogeneous terrain, these ideal conditions are rarely met in practice.

Table 3.1 Terrain types, roughness length and surface drag coefficient

Terrain type	*Roughness length (m)*	*Surface drag coefficient*
Very flat terrain (snow, desert)	0.001–0.005	0.002–0.003
Open terrain (grassland, few trees)	0.01–0.05	0.003–0.006
Suburban terrain (buildings 3–5 m)	0.1–0.5	0.0075–0.02
Dense urban (buildings 10–30 m)	1–5	0.03–0.3

Also, the Logarithmic Law has some mathematical characteristics which may cause problems: firstly, since the logarithms of negative numbers do not exist, it cannot be evaluated for heights, z, below the zero-plane displacement z_h, and if $z - z_h$ is less than z_o, a negative wind speed is given. Secondly, it is less easy to integrate. To avoid some of these problems, wind engineers have often preferred to use the Power Law.

3.2.2 The 'Power Law'

The Power Law has no theoretical basis but is easily integrated over a height – a convenient property when wishing to determine bending moments at the base of a tall structure, for example.

To relate the mean wind speed at any height, z, with that at 10 m (adjusted if necessary for rougher terrains, as described in the previous section), the Power Law can be written as

$$\bar{U}(z) = \bar{U}_{10}\left(\frac{z}{10}\right)^{\alpha} \tag{3.7}$$

The exponent, α, in Equation 3.7 will change with the terrain roughness, and also with the height range, when matched to the Logarithmic Law. A relationship that can be used to relate the exponent to the roughness length, z_o, is as follows:

$$\alpha = \left(\frac{1}{\log_e(z_{ref}/z_0)}\right) \tag{3.8}$$

where z_{ref} is a reference height at which the two 'laws' are matched. z_{ref} may be taken as the average height in the range over which matching is required, or half the maximum height over which the matching is required.

Figure 3.2 shows a matching of the two laws for a height range of 100 m, using Equation 3.8, with z_{ref} taken as 50 m. It is clear the two relationships are extremely close, and that the Power Law is quite adequate for engineering purposes.

3.2.3 Mean wind profiles over the ocean

Over land, the surface drag coefficient, κ, is found to be nearly independent of mean wind speed. This is not the case over the ocean, where higher winds create higher waves, and hence higher surface drags coefficients. The relationship between κ and $\bar{U}_{10}$ has been the subject of much study, and a large number of empirical relationships have been derived.

Charnock (1955), using dimensional arguments, proposed a mean wind profile over the ocean, which implies that the roughness length, z_o, should be given by Equation 3.9.

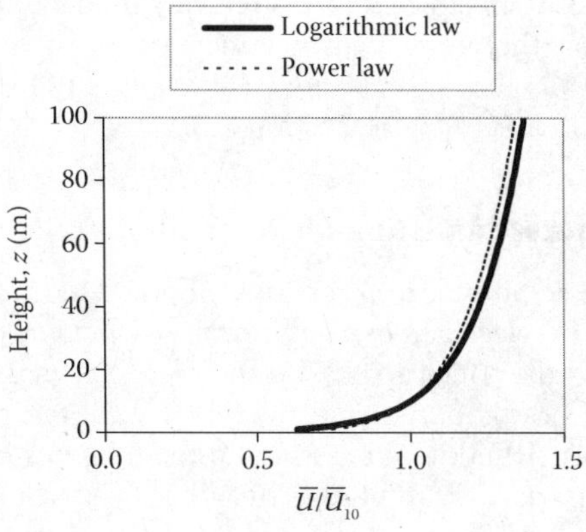

Figure 3.2 Comparison of the logarithmic ($z_o = 0.02$ m) and Power Laws ($\alpha = 0.128$) for mean velocity profile.

$$z_0 = \frac{au_*^2}{g} = \frac{a\kappa\bar{U}_{10}^2}{g} \tag{3.9}$$

where g is the gravitational constant, and a is an empirical constant.

Equation 3.9, with the constant a lying between 0.01 and 0.02, is valid over a wide range of wind speeds. It is not valid at very low wind speeds, under aerodynamically smooth conditions, and also may not be valid at very high wind speeds, during which the air–sea surface experiences intensive wave breaking and spray.

Substituting for the surface drag coefficient, κ, from Equation 3.6 into Equation 3.9, Equation 3.10 is obtained.

$$z_0 = \frac{a}{g}\left[\frac{k\bar{U}_{10}}{\log_e(10/z_0)}\right]^2 \tag{3.10}$$

(z_h is usually taken as zero over the ocean.)

The implicit nature of the relationship between z_o (or κ) and $\bar{U}_{10}$, in Equations 3.9 and 3.10 makes them difficult to apply, and several simpler forms have been suggested.

Garratt (1977) examined a large amount of experimental data and suggested a value for a of 0.0144. Using this value for a, taking g equal to 9.81 m/s² and k equal to 0.41, the relationship between z_o and $\bar{U}_{10}$ given in Table 3.2 is obtained.

Table 3.2 Roughness length over the ocean as a function of mean wind speed

$\bar{U}_{10}$ (m/s)	*Roughness length (mm)*
10	0.21
15	0.59
20	1.22
25	2.17
30	3.51

For values of $\bar{U}_{10}$ above 30 m/s, recent measurements in Atlantic hurricanes (Powell et al., 2003) have indicated that there is no further increase in the surface drag coefficient or roughness length. For mean wind speeds above 30 m/s, Equation 3.9 or Table 3.2 should be applied for $\bar{U}_{10}$ *equal to* 30 m/s (see also Section 3.2.5).

3.2.4 Relationship between upper level and surface winds

For large-scale atmospheric boundary layers in synoptic winds, dimensional analysis gives a functional relationship between a *geostrophic drag coefficient*, $C_g = u_*/U_g$, and the *Rossby number*, $Ro = U_g/fz_o$. u_* is the friction velocity and U_g is the geostrophic (Section 1.2.3), or gradient wind; f is the Coriolis parameter (Section 1.2.2) and z_o is the roughness length (Section 3.2.1). Davenport (1963), making use of measurements of Lettau (1959), proposed the following relationship based on a number of full-scale measurements:

$$C_g = 0.16\ Ro^{-0.09} \tag{3.11}$$

Applying the above relationship for a latitude of 40 degrees ($f = 0.935 \times 10^{-4} s^{-1}$), a value of U_g equal to 40 m/s, and a roughness length of 20 mm, gives a friction velocity of 1.40 m/s and from Equation 3.2, a value of $\bar{U}_{10}$ of 21.8 m/s. Thus, in this case, the wind speed near the surface is equal to 0.54 times the geostrophic wind – the upper-level wind away from the frictional effects of the earth's surface.

3.2.5 Mean wind profiles in tropical cyclones

A number of low-level flights into the Atlantic Ocean and Gulf of Mexico hurricanes have been made by the National Oceanic and Atmospheric Administration (NOAA) of the United States. However, the flight levels were not low enough to provide useful data on wind speed profiles below about 200 m. Measurements from fixed towers are also extremely limited. However, some measurements were made from a 390-m communications mast close to the coast near Exmouth, Western Australia, in the late 1970s (Wilson, 1979). SODAR (sonic radar) profiles have been obtained from typhoons on Okinawa, Japan (Amano et al., 1999). These show similar characteristics near the regions of maximum winds: a logarithmic-type profile up to a certain height (60–200 m), followed by a layer of strong convection, with nearly constant mean wind speed. More recently, probes known as 'dropwindsondes' have been dropped from aircraft flying through hurricanes, and their positions are continually tracked by Global Positioning System (GPS) satellites, enabling the estimation of horizontal wind speeds to be made (Hock and Franklin, 1999).

On the basis of averages of the dropwindsonde data, the following mean wind speed profile has been proposed for the eye wall region (Franklin et al., 2003):

$$\bar{U}_z = \bar{U}_{10}\frac{\log_e(z/0.0001)}{\log_e(10/0.0001)} \quad \text{for } z < 300 \text{ m}; \quad \bar{U}(z) = \bar{U}_{300} \quad \text{for } z \geq 300 \text{ m} \tag{3.12}$$

For regions over the ocean outside the eye wall, the relations given in Section 3.2.3 can be applied, noting that the roughness length does not increase with wind speed, for values of mean wind speed (i.e. $\bar{U}_{10}$). As the tropical cyclone crosses the coast, it weakens (see Section 1.3.2), and the mean wind profiles would be expected to adjust to the underlying ground roughness over the land.

3.2.6 Wind profiles in thunderstorm winds

The most common type of severe wind generated by a thunderstorm results from a severe downdraft, described by Fujita (1985) as a 'downburst', and discussed in Section 1.3.5. Downbursts produce severe winds for relatively short periods – typically a few minutes – and are transient in nature (see Figure 1.9). To define a 'mean' velocity requires a much shorter averaging time than 10 min to 1h typically used for synoptic-scale wind events. We can separate the slowly varying part (or 'running mean'), representing the downward airflow which becomes a horizontal 'outflow' near the ground, from any superimposed turbulence of higher frequency. A typical averaging time to produce quasi-stationary winds in a thunderstorm is in the range 30–60 s; for example, Holmes et al. (2008) found that 40 s was the optimum averaging period for a well-documented event in Texas in 2002.

Thanks to Doppler radar measurements in the United States, and in some tower anemometer measurements in Australia and the United States, there are some indications of the wind structure in the downburst type of thunderstorm wind, including the 'macroburst' and 'microburst' types identified by Fujita (1985). At the horizontal location where the maximum gust occurs, the wind speed increases from ground level up to a maximum value at a height of 50–100 m. Above this height, the wind speed reduces relatively slowly.

A useful model of the velocity profiles in the vertical and horizontal directions in a downburst was provided by Oseguera and Bowles (1988). This model satisfies the requirements of fluid mass continuity, but does not include any effect of storm movement. The horizontal velocity component is expressed as Equation 3.13.

$$U = \left(\frac{\lambda R^2}{2r}\right)\left[1 - e^{-(r/R)^2}\right]\left(e^{-z/z^*} - e^{-z/\varepsilon}\right) \tag{3.13}$$

where

r is the radial coordinate from the centre of the downburst
R is the characteristic radius of the downburst 'shaft'
z is the height above the ground
z^* is a characteristic height out of the boundary layer
ε is a characteristic height in the boundary layer
λ is a scaling factor, with dimensions of [time]$^{-1}$

The velocity profile at the radius of maximum winds ($r = 1.121R$) is shown in Figure 3.3. The profile clearly shows a maximum at the height of the boundary layer on the ground surface. Radar observations have shown that this height is 50–100 m in actual downbursts.

Recent measurements in thunderstorm outflows, using mobile Doppler radar of high resolution, have indicated similar wind profiles to that shown in Figure 3.3, at the peak wind speed in the event. However, as an outflow from a downdraft 'matures', it appears to adjust to the underlying terrain, and the profiles then appear more similar to those in synoptic winds, with maxima at a height of 500 m or greater (Gunter and Schroeder, 2013).

3.2.7 Wind profiles in tornadoes

There have been many studies of the wind structure in tornadoes based on full-scale studies using photogrammetry and portable Doppler radars (see also Section 1.3.4), laboratory studies of tornado-like vortices and theoretical analyses.

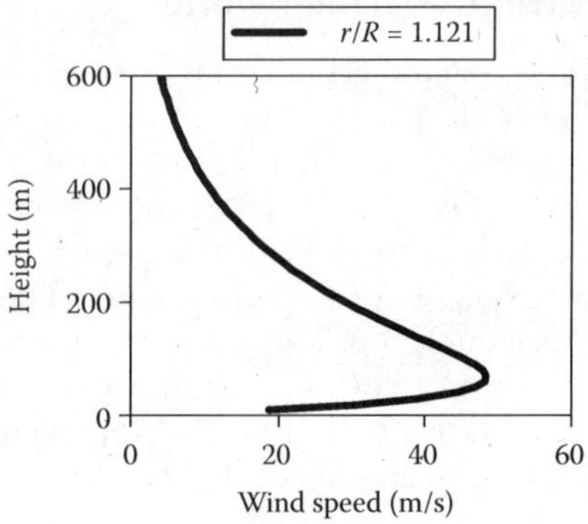

Figure 3.3 Profile of horizontal velocity near the ground during a stationary downburst.

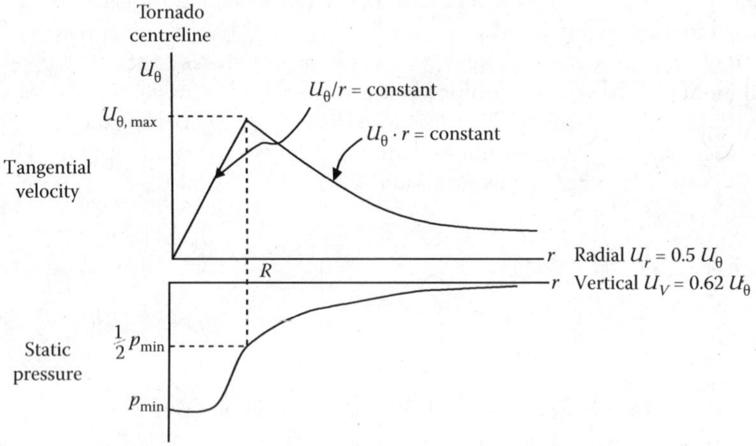

Figure 3.4 Velocity components in a tornado.

The simplest model of a horizontal wind profile in a tornado is based on the Rankine, or combined, vortex (Figure 3.4). This consists of an inner 'core' with solid body rotation, in which the tangential wind velocity component, U_θ, is proportional to the radius from the centreline of the tornado, r. In the outer region ($r > R$), the tangential velocity component is *inversely* proportional to the radius, r. This satisfies the equation of angular momentum (Lewellen, 1976), except the discontinuity at r equal to R.

This model does not define the radial, U_r, or vertical, U_v, velocity components, but empirical values of these are shown in Figure 3.4.

Several alternative theoretical models are discussed by Lewellen (1976).

3.3 TURBULENCE

The general level of turbulence or 'gustiness' in the wind speed, such as that shown in Figure 3.1, can be measured by its standard deviation, or root-mean square. First, we subtract out the steady or mean component (or the slowly varying component in the case of a transient storm, such as a thunderstorm), and then quantify the resulting deviations. Since both positive and negative deviations can occur, we first square the deviations before averaging them, and finally, the square root is taken to give a quantity with the units of wind speed. Mathematically, the formula for standard deviation can be written as

$$\sigma_u = \left\{ \frac{1}{T} \int_0^T \left[U(t) - \bar{U} \right]^2 dt \right\}^{\frac{1}{2}} \tag{3.14}$$

where $U(t)$ is the total velocity component in the direction of the mean wind, equal to $\bar{U} + u(t)$, where $u(t)$ is the 'longitudinal' turbulence component, that is the component of the fluctuating velocity in the mean wind direction.

Other components of turbulence in the lateral horizontal direction denoted by $v(t)$, and in the vertical direction denoted by $w(t)$, are quantified by their standard deviations, σ_v and σ_w, respectively.

3.3.1 Turbulence intensities

The ratio of the standard deviation of each fluctuating component to the mean value is known as the *turbulence intensity* of that component.

Thus,

$$I_u = \sigma_u / \bar{U} \text{ (longitudinal)} \tag{3.15}$$

$$I_v = \sigma_v / \bar{U} \text{ (lateral)} \tag{3.16}$$

$$I_w = \sigma_w / \bar{U} \text{ (vertical)} \tag{3.17}$$

Near the ground in gales produced by large-scale depression systems, measurements have found that the standard deviation of longitudinal wind speed, σ_u, is equal to 2.5 u_* to a good approximation, where u_* is the friction velocity (Section 3.2.1). Then the turbulence intensity, I_u, is given by Equation 3.18.

$$I_u = \frac{2.5 u_*}{(u_*/0.4)\log_e(z/z_0)} = \frac{1}{\log_e(z/z_0)} \tag{3.18}$$

Thus, the turbulence intensity is simply related to the surface roughness, as measured by the roughness length, z_0. For a rural terrain, with a roughness length of 0.04 m, the longitudinal turbulence intensities for various heights above the ground are given in Table 3.3.

Thus, the turbulence intensity decreases with height above the ground.

The lateral and vertical turbulence components are generally lower in magnitude than the corresponding longitudinal value. However, for well-developed boundary-layer winds, simple relationships between standard deviation and the friction velocity u_* have been suggested. Thus, approximately, the standard deviation of lateral (horizontal) velocity, σ_v, is equal to 2.20 u_*, and for the vertical component, σ_w is given approximately by 1.3 to 1.4 u_*.

Table 3.3 Longitudinal turbulence intensities for rural terrain ($z_o = 0.04$ m)

Height, z (m)	I_u
2	0.26
5	0.21
10	0.18
20	0.16
50	0.14
100	0.13

Then equivalent expressions to Equation 3.18 for the variation with height of I_v and I_w can be derived:

$$I_v \cong 0.88/\log_e(z/z_0) \tag{3.19}$$

$$I_w \cong 0.55/\log_e(z/z_0) \tag{3.20}$$

The turbulence intensities in tropical cyclones (typhoons and hurricanes) are believed by some to be higher than those in gales in temperate latitudes. Choi (1978) found that the longitudinal turbulence intensity was about 50% higher in tropical-cyclone winds compared to synoptic winds. From measurements on a tall mast in North-Western Australia during the passage of severe tropical cyclones, intermittent convective 'squall-like' turbulence was observed (Wilson, 1979). This was considerably more intense than the 'mechanical turbulence' seen closer to the ground, and was associated with the passage of bands of rain clouds.

Turbulence intensities in thunderstorm downburst winds are even less well defined than for tropical cyclones. However, the Andrews Air Force Base (AFB) event of 1983 (Figure 1.9) indicates a turbulence 'intensity' of the order of 0.1 (10%) superimposed on the underlying transient flow (see also Section 3.3.7).

3.3.2 Probability density

As shown in Figure 3.1, the variations of wind speed in the atmospheric boundary layer are generally random in nature, and do not repeat in time. The variations are caused by eddies or vortices within the air flow, moving along at the mean wind speed. These eddies are never identical, and we must use statistical methods to describe the gustiness.

The probability density, $f_u(u_o)$, is defined so that the proportion of time that the wind velocity, $U(t)$, spends in the range $u_o + du$ is $f_u(u_o) \cdot du$. Measurements have shown that the wind velocity components in the atmospheric boundary layer closely follow the Normal or Gaussian probability density function, given by Equation 3.21.

$$f_u(u) = \frac{1}{\sigma_u\sqrt{2\pi}}\exp\left[-\frac{1}{2}\left(\frac{u-\bar{U}}{\sigma_u}\right)^2\right] \tag{3.21}$$

This function has the characteristic bell shape. It is defined only by the mean value, $\bar{U}$, and standard deviation, σ_u (see also Section C.3.1 in Appendix C).

Thus, with the mean value and standard deviation, the probability of any wind velocity occurring can be estimated.

3.3.3 Gust wind speeds and gust factors

In many design codes and standards for wind loading (see Chapter 15), a peak gust wind speed is used for design purposes. The nature of wind as a random process means that the peak gust within a sample time, T, of, say, 10 min is itself also a random variable. However, we can define an *expected*, or average, value within the 10-min period.

Assuming that the longitudinal wind velocity has a Gaussian probability distribution, the expected peak gust, $\hat{U}$, is given by Equation 3.22.

$$\hat{U} = \bar{U} + g\sigma_u \tag{3.22}$$

The expected peak factor, g, depends on the effective averaging time of the gust, τ, as well as the sample time T. Table 3.4 shows the expected values of g for various values of τ, and for T equal to 600 and 3600 s. These values can be obtained theoretically (ESDU 83045; Holmes et al., 2014) by 'filtering' the wind spectrum (Equation 3.26) by the theoretical moving-average filter associated with the moving-average time, τ.

Meteorological instruments used for long-term wind measurements do not have a perfect response, and the peak gust wind speed they measure is dependent on their response characteristics. The response can be measured by an equivalent averaging time, τ. The equivalent averaging time for the output of a rotating-cup anemometer, the most common instrument used by meteorological agencies, depends on the distance constant of the instrument and on the mean wind speed, but typically lies in the range of 0.3–0.9 s at wind speeds of interest for the structural design (Holmes et al., 2014). However, it should be noted that after digitisation, most meteorological agencies apply a digital 'moving average' filter, with an averaging time, τ, of 3 s, as recommended by the World Meteorological Organization (Beljaars, 1987). This is the source of the commonly used '3-second' gust in codes and standards. However, it should be noted that the effective frontal area associated with this gust is quite large (equivalent to the area of a typical tall building in a city centre), and loads on small structures, calculated from gusts with that duration, need to be increased by a factor greater than 1.0 (Holmes et al., 2014).

The Dines pressure tube anemometer used in many countries up to the 1990s had an effective averaging time of around 0.2 s (depending on the type and on the mean wind speed (Miller et al., 2013)), and hence recorded generally higher wind gusts than modern automatic weather systems with cup anemometers and 3-s digital filtering (Holmes and Ginger, 2012).

Table 3.4 Expected peak factors for various τ and T

Averaging time, τ (s)	*Sample time, T (s)*	*g*
3	3600	3.0
3	600	2.5
1	3600	3.4
1	600	2.9
0.2	3600	3.8
0.2	600	3.4

For various terrains, a profile of peak gust with height can be obtained. Note, however, that gusts do not occur simultaneously at all heights, and such a profile would represent an envelope of the expected gust wind speed with height.

The *gust factor*, G, is the ratio of the expected maximum gust speed within a specified period to the mean wind speed. Thus,

$$G = \frac{\hat{U}}{\bar{U}} \tag{3.23}$$

The relationship between the expected gust factor, G, and the peak factor, g, is shown in Equation 3.24.

$$G = \frac{\hat{U}}{\bar{U}} = \frac{\bar{U} + \sigma_u}{\bar{U}} = 1 + gI_u \tag{3.24}$$

where I_u is the longitudinal turbulence intensity.

For gales (synoptic winds in temperate climates), the magnitude of gusts for various averaging times, τ, were studied by Durst (1960) and Deacon (1965). Deacon gave nominally 2-s gust factors at a height of 10 m, based on a 10-min mean wind speed, of about 1.45 for 'open country with few trees', and 1.96 for a suburban terrain.

Several authors have provided estimates of gust factors over land, for tropical cyclones or hurricanes, including, for example, Ishizaki (1983), Krayer and Marshall (1992), Black (1992) and Vickery and Skerjl (2005). The study by Krayer and Marshall (1992) of four U.S. hurricanes gave a value of 1.55 for the ratio of peak 2-s gust to a 10-min mean, for 10-m height in an open-country terrain. This value is based on tropical-cyclone winds with a wide range of wind speeds, to values as low as 10 m/s. The analysis by Black (1992), which appeared to be based on higher wind speeds in hurricanes, gave a higher value of 1.66 for the gust factor, $\hat{U}_{2\,s,10\,m}/\bar{U}_{10\,min,\,10\,m}$.

Gust factors in tropical cyclones were extensively reviewed by Harper et al. (2010), including the references mentioned previously. They found considerable scatter in the experimental data and relationships suggested. This is not surprising, as a theoretical approach to gust factors shows that they are dependent on the mean wind speed, the length scale of turbulence (Section 3.3.4), the turbulence intensity (Section 3.3.1) as well as the gust duration or averaging time, τ.

As an illustration of this, Figure 3.5 shows theoretical gust factors, G, for synoptic winds (including tropical cyclones) for a sample time, T, of 600 s, plotted against the gust duration, τ, for a value of the ratio of turbulence length scale, ℓ_u, to mean wind speed $\bar{U}$ of 10 (units

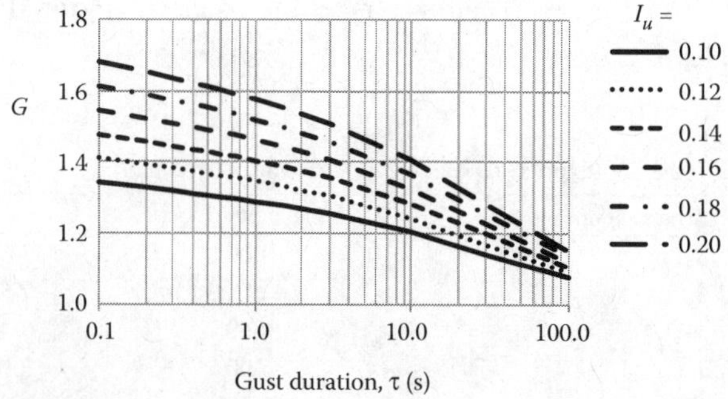

Figure 3.5 Gust factors ($\hat{U}_{\tau sec}/\bar{U}_{10\,min}$) as a function of turbulence intensity for $\ell_u/\bar{U}$ = 10 s.

in seconds). The values of G were derived using the approach described by Holmes et al. (2014). Figure 3.5 shows that gust factors are quite dependent on turbulence intensity (see Equation 3.24) which, in turn, depends on the height above the ground and the underlying terrain. Hence a single 'universal' gust factor curve does not exist.

3.3.4 Wind spectra

The probability density function (Section 3.3.2) tells us something about the magnitude of the wind velocity, but nothing about how slowly or quickly it varies with time. In order to describe the distribution of turbulence with frequency, a function called the *spectral density,* usually abbreviated to 'spectrum', is used. It is defined so that the contribution to the variance (σ_u^2, or square of the standard deviation), in the range of frequencies from n to $n + dn$, is given by $S_u(n) \cdot dn$, where $S_u(n)$ is the spectral density function for $u(t)$. Then integrating over all frequencies,

$$\sigma_u^2 = \int_0^\infty S_u(n)dn \tag{3.25}$$

There are many mathematical forms that have been used for $S_u(n)$ in meteorology and wind engineering. The most common and mathematically correct of these for the longitudinal velocity component (parallel to the mean wind direction) is the von Karman form (developed for laboratory turbulence by von Karman (1948), and adapted for wind engineering by Harris (1968)). This may be written in several forms; Equation 3.26 is a commonly used non-dimensional form.

$$\frac{n \cdot S_u(n)}{\sigma_u^{\,2}} = \frac{4\left(\dfrac{n\ell_u}{\bar{U}}\right)}{\left[1 + 70.8\left(\dfrac{n\ell_u}{\bar{U}}\right)^2\right]^{5/6}} \tag{3.26}$$

where ℓ_u is the turbulence integral-length scale.

In this form, the curve of $n \cdot S_u(n)/\sigma_u^{\,2}$ versus $n/\bar{U}$ has a peak; the value of ℓ_u determines the value of $(n/\bar{U})$ at which the peak occurs – the higher the value of ℓ_u, the higher the value of $(\bar{U}/n)$ at the peak, or λ, known as the 'peak wavelength'. For the von Karman spectrum, λ is equal to 6.85 ℓ_u. The length scale, ℓ_u, varies with both terrain roughness and height above the ground. The form of the von Karman spectrum is shown in Figure 3.6.

The other orthogonal components of atmospheric turbulence have spectral densities with somewhat different characteristics. The spectrum of vertical turbulence is the most important of these, especially for horizontal structures such as bridges. A common mathematical form for the spectrum of vertical turbulence (w') is the Busch and Panofsky (1968) form which can be written as Equation 3.27.

$$\frac{n \cdot S_w(n)}{\sigma_w^{\,2}} = \frac{2.15\left(\dfrac{nz}{\bar{U}}\right)}{\left[1 + 11.16\left(\dfrac{nz}{\bar{U}}\right)^{5/3}\right]} \tag{3.27}$$

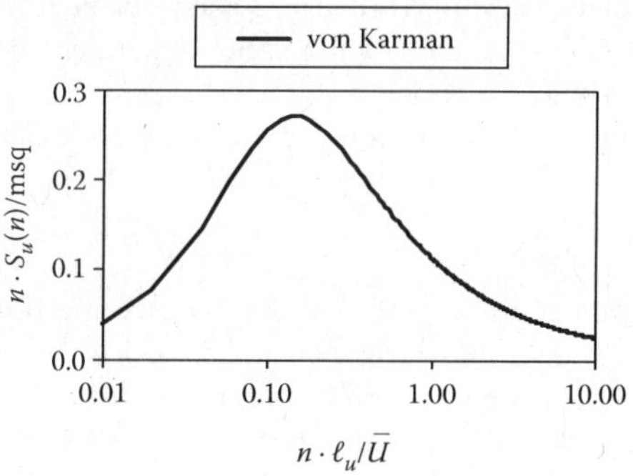

Figure 3.6 Normalised spectrum of a longitudinal velocity component. (From von Karman, T. 1948. *Proceedings of the National Academy of Sciences* (U.S.), 34: 530–9.)

In this case, the length scale is directly proportional to the height above the ground, z. The Busch and Panofsky spectrum for vertical turbulence (w') is shown in Figure 3.7.

3.3.5 Correlation

Covariance and *correlation* are two important properties of wind turbulence in relation to wind loading. The latter is the same quantity that is calculated in linear regression analysis. In this context, it relates the fluctuating wind velocities at two points in space, or wind pressures at two points on a building (such as a roof).

For example, consider the wind speed at two different heights on a tower (e.g. Figure 3.1). The covariance between the fluctuating (longitudinal) velocities at two different heights, z_1 and z_2, is defined according to Equation 3.28.

$$\overline{u'(z_1)u'(z_2)} = \frac{1}{T}\int_0^T \left[U(z_1,t) - \bar{U}(z_1)\right]\left[U(z_2,t) - \bar{U}(z_2)\right]dt \tag{3.28}$$

Thus, the covariance is the product of the fluctuating velocities at the two heights, averaged over time. Note that the mean values, $\bar{U}(z_1)$ and $\bar{U}(z_2)$, are subtracted from each velocity

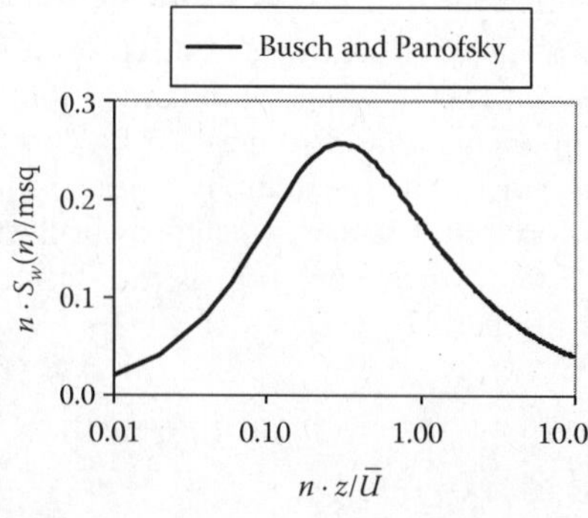

Figure 3.7 Normalised spectrum of a vertical velocity component. (From Busch, N. and Panofsky, H. 1968. *Quarterly Journal of the Royal Meteorological Society*, 94: 132–48. With permission.)

in the right-hand side of Equation 3.28. Note that in the special case when z_1 is equal to z_2, the right-hand side is then equal to the variance (σ_u^2) of the fluctuating velocity at the single height.

The correlation coefficient, ρ, is defined by Equation 3.29.

$$\rho = \frac{\overline{u'(z_1)u'(z_2)}}{\sigma_u(z_1) \cdot \sigma_u(z_2)} \tag{3.29}$$

When z_1 is equal to z_2, the value of ρ is +1 (i.e. we have full correlation). It can be shown that ρ must lie between –1 and +1. A value of 0 indicates no correlation (i.e. no statistical relationship between the wind velocities) – this usually occurs when the heights z_1 and z_2 are widely separated.

The covariance and correlation are very useful in calculating the fluctuating wind loads on tall towers, and for estimating span reduction factors for transmission lines. In the latter case, the points would be separated horizontally, rather than vertically.

A mathematical function which is useful for describing the correlation, ρ, is the exponential decay function:

$$\rho \approx \exp\left[-C|z_1 - z_2|\right] \tag{3.30}$$

This function is equal to +1 when z_1 is equal to z_2, and tends to zero when $|z_1 - z_2|$ becomes very large (very large separations).

Figure 3.8 shows Equation 3.30 with C equal to (1/40) m^{-1}. It is compared with some measurements of longitudinal velocity fluctuations in the atmospheric boundary, at a height of 13.5 m, with horizontal separations, over an urban terrain (Holmes, 1973).

3.3.6 Co-spectrum and coherence

When considering the resonant response of structures to wind (Chapter 5), the correlation of wind velocity fluctuations from separated points *at different frequencies* is important. For example, the correlations of vertical velocity fluctuations with span-wise separation, at the natural frequencies of vibration of a large-span bridge, are important in determining its response to buffeting.

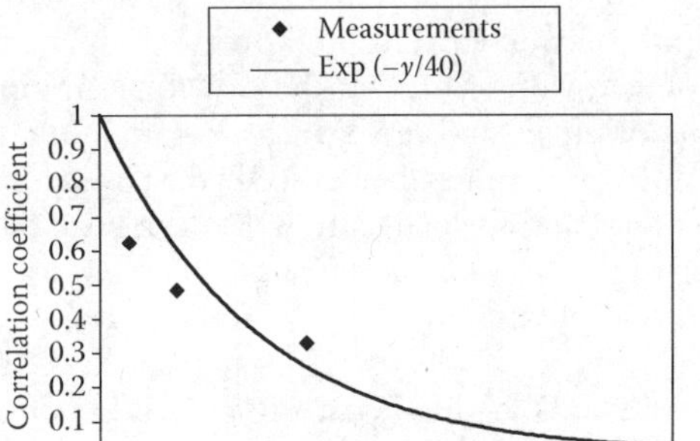

Figure 3.8 Cross-correlation of longitudinal velocity fluctuations in the atmospheric boundary layer at a height of 13.5 m. (From Holmes, J.D. 1973. Wind pressure fluctuations on a large building. PhD thesis, Monash University, Australia.)

The frequency-dependent correlation can be described by functions known as the *cross-spectral density, co-spectral density* and *coherence*. Mathematical definitions of these functions are given by Bendat and Piersol (1999), and others. The cross-spectral density as well as being a function of frequency is a complex variable, with real and imaginary components. The co-spectral density is the real part, and may be regarded as a frequency-dependent covariance (Section 3.3.5). The coherence is a normalised magnitude of the cross-spectrum, approximately equivalent to a frequency-dependent correlation coefficient. The normalised co-spectrum is very similar to coherence, but does not include the imaginary components; this is in fact the relevant quantity when considering the wind forces from turbulence on structures.

The normalised co-spectrum and coherence are often represented by an exponential function of separation distance and frequency:

$$\rho(\Delta z,n) = \exp\left[-\left(\frac{k \cdot n \cdot \Delta z}{\bar{U}}\right)\right] \tag{3.31}$$

where k is an empirical constant, used to fit measured data; a typical range of values for atmospheric turbulence is 10–20. Δz is the vertical separation distance. A similar function is used to represent the co-spectrum when lateral (horizontal) separations, Δy, are considered.

As for Equation 3.30, Equation 3.31 does not allow negative values – a theoretical problem – but of little practical significance. A more important disadvantage is that it implies full correlation at very low frequencies, no matter how large the separation distance, Δz. Since the equation only needs to be evaluated at the high frequencies corresponding to resonant frequencies, this is also not a great disadvantage.

More mathematically acceptable (but more complex) expressions for the normalised co-spectrum and coherence are available (e.g. Deaves and Harris, 1978).

3.3.7 Turbulence in a downdraft

The 'rear-flank' thunderstorm downdraft recorded by several towers near Lubbock, Texas on 4 June 2002, gave a unique opportunity to study the fluctuating wind characteristics, near the ground, in a severe event of this type (Orwig and Schroeder, 2007; Holmes et al., 2008).

An individual time history from one tower during this event is shown in Figure 3.9a. By applying a simple moving-average filter, a smoothed time history, that shows the main features of the event, can be extracted. This is shown in Figure 3.9b, in which a 40-s moving average has been applied; this record can be called a 'running mean'. Subtracting the 'filtered' history from the original 'unfiltered' history results in a residual time history that is more or less random in nature, and can be described as 'turbulence' (Figure 3.9c). This is a non-stationary time history, and the conventional 'turbulence intensity' (Section 3.3.1), as defined for stationary synoptic winds, cannot be used here in the same way. However, Figure 3.9a shows that the level of random fluctuation varies with the running mean (Figure 3.9b), with an approximate 'intensity' of 10%. This is somewhat lower than the level obtained in stationary boundary-layer winds at this height in an open country (e.g. Table 3.3 gives a value of 18%), but is similar to that obtained in the Andrews AFB downburst (Figure 1.9).

Analysis of the turbulence from the event shown in Figure 3.9 suggested that the turbulence intensity with respect to the running mean was lower than that for a similar terrain for synoptic boundary-layer winds. The high-frequency spectral density was similar to that in boundary-layer winds (i.e. Equation 3.26). However, correlations of the fluctuating components, with lateral separation, were considerably higher than those in boundary-layer winds,

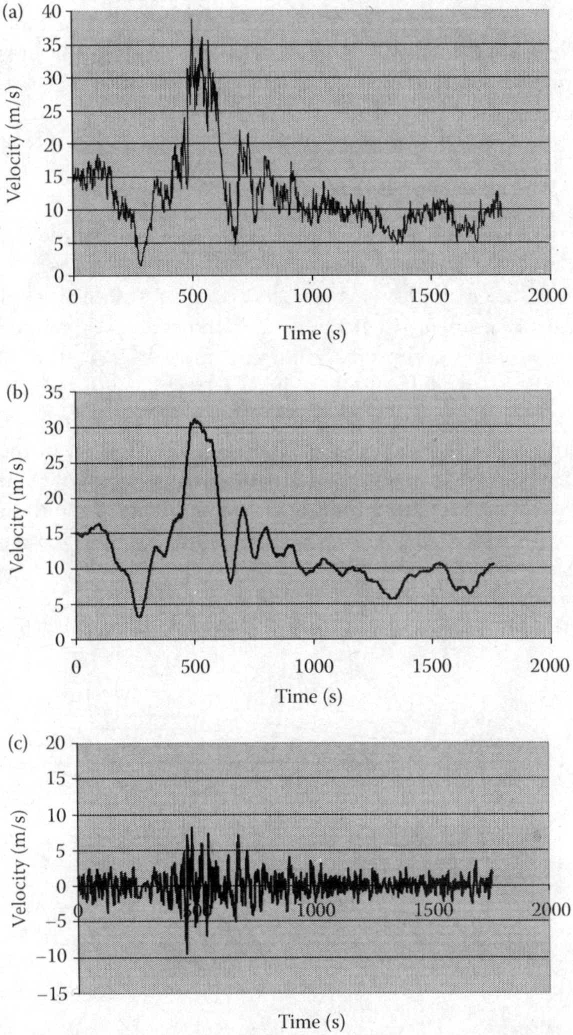

Figure 3.9 Time histories from a rear-flank downdraft, 4 June, 2002, Lubbock, Texas. (a) Velocities as recorded (unfiltered record); (b) time history filtered with a 40-s moving-average filter and (c) residual 'turbulence' obtained by subtraction.

as a result of the large underlying 'running mean' fluctuations. This leads to high values of 'span reduction factor' for transmission lines (see Section 13.2.3) and provides a convincing reason for many failures of transmission line structures in thunderstorm downdraft events (Holmes et al., 2008).

3.4 MODIFICATION OF WIND FLOW BY TOPOGRAPHY

Mean and gust wind speeds can be considerably increased by natural and man-made topography in the form of escarpments, embankments, ridges, cliffs and hills. These effects were the subject of considerable research in the 1970s and 1980s, with the incentive of the desire to exploit wind power, and to optimise the siting of wind turbines. This work greatly improved the prediction of mean wind speeds over shallow topography. Less well defined are the speed-up effects on turbulence and gust wind speeds, and the effects of steep topography – often of interest with respect to the structural design.

3.4.1 General effects of topography

Figure 3.10 shows the general features of boundary-layer wind flow over a shallow escarpment, a shallow ridge, a steep escarpment and a steep ridge.

As the wind approaches a shallow feature, its speed first reduces slightly as it encounters the beginning of the slope upwards. It then gradually increases in speed as it flows up the slope towards the crest. The maximum speedup occurs at the crest, or slightly upwind of it. Beyond the crest, the flow speed gradually reduces to a value close to that well upwind of the topographic feature; the adjustment is somewhat faster for a feature with a downwind slope, such as a ridge, than for an escarpment with a plateau downwind of the crest.

On steeper features, flow 'separation' (see also Section 4.1) may occur, as the flow is not able to overcome the increasing pressure gradients in the along-wind direction. Separations may occur at the beginning of the upwind slope, immediately downwind of the crest, and on the downwind slope for a ridge.

For steeper slopes (>about 0.3), the upwind separation 'bubble' presents an 'effective slope' of approximately constant value, independent of the actual slope underneath. This

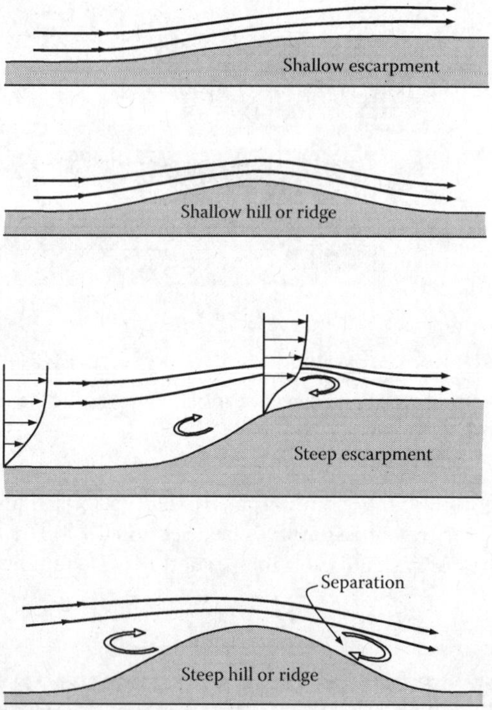

Figure 3.10 Flow over shallow and steep topography.

is often used in codes and standards to specify an upper limit to the speed-up effects of an escarpment or ridge.

The speed-up effects are greatest near the surface, and reduce with height above the ground. This can have the effect of producing mean velocity profiles, near the crest of a topographic feature, that are nearly constant, or have a peak (see Figure 3.10).

The above discussion relates to topographic features, which are two dimensional in nature, that is they extend for an infinite distance normal to the wind direction. This may be a sufficient approximation for many long ridges and escarpments. Three-dimensional effects occur when air flow can occur around the ends of a hill, or through gaps or passes. These alternative air paths reduce the air speeds over the top of the feature, and generally reduce the speed-up effects. For structural design purposes, it is often convenient, and usually conservative, to ignore the three-dimensional effects, and to calculate wind loads only for the speed-up effects of the upwind and downwind slopes parallel to the wind direction of interest.

3.4.2 Topographic multipliers

The definition of a topographic multiplier used in this chapter is as follows:

$$\text{Topographic multiplier} = \frac{\text{Wind speed at height, } z\text{, above the feature}}{\text{Wind speed at height, } z\text{, above the flat ground upwind}} \tag{3.32}$$

This definition applies to mean, peak gust and standard deviation wind speeds, and these will be denoted by $\bar{M}_t$, $\hat{M}_t$ and M_t', respectively.

Topographic multipliers measured in full-scale or in wind tunnels, or calculated by computer programmes, can be both greater or less than one. However, in the cases of most interest for the structural design, we are concerned with speed-up effects for which the topographic multiplier for mean or gust wind speeds will exceed unity.

3.4.3 Shallow hills

The analysis by Jackson and Hunt (1975) of the mean boundary-layer wind flow over a shallow hill produced the following form for the mean topographic multiplier:

$$\bar{M}_t = 1 + ks\phi \tag{3.33}$$

where

ϕ is the upwind slope of the topographic feature
k is a constant for a given shape of topography
s is a position factor

Equation 3.33 has been used in various forms for specifying topographic effects in several codes and standards. It indicates that the 'fractional speed-up', equal to $(\bar{M}_t - 1)$, is directly proportional to the upwind slope, ϕ. The latter is defined as $H/2L_u$, where H is the height of the crest above the ground level upwind, and L_u is the horizontal distance from the crest to where the ground elevation drops to $H/2$.

Taylor and Lee (1984) proposed the following values of the constant, k, for various types of topography:

- 4.0 for two-dimensional ridges
- 1.6 for two-dimensional escarpments
- 3.2 for three-dimensional (axisymmetric) hills

The position factor, s, is 1.0 close to the crest of the feature, and falls upwind and downwind and with height, z, above the local ground level. The reduction of s with height is more rapid near the ground, becoming more gradual as z increases.

To a first approximation, the longitudinal turbulence component, σ_u, does not change over the hill or escarpment. This results in the following equation for the gust topographic multiplier, $\hat{M}_t$:

$$\hat{M}_t = 1 + k' s \phi \tag{3.34}$$

where k' is a constant for the gust multiplier, related to k by Equation 3.35.

$$k' = \frac{k}{1 + g\left(\frac{\sigma_u}{\bar{U}}\right)} \tag{3.35}$$

$\left(\frac{\sigma_u}{\bar{U}}\right)$ is the longitudinal turbulence intensity (over flat level ground) defined in Section 3.3.1, and g is the peak factor (Section 3.3.3).

Equations 3.33 through 3.35 show that the gust topographic multiplier is lower than the mean topographic multiplier for the same type of topography and height above the ground.

There is a slight dependence of the topographic multipliers on the Jensen Number (Section 4.4.5) based on the hill height (H/z_o).

3.4.4 Steep hills, cliffs and escarpments

Once the upwind slope of a hill or escarpment reaches a value of about 0.3 (about 17 degrees), separations occur on the upwind face (Figure 3.10) and the simple formulae given in Section 3.4.3 cannot be applied directly.

For slopes between about 0.3 and 1 (17–45 degrees), the separation bubble on the upwind slope presents an effective slope to the wind which is relatively constant, as discussed in Section 3.4.1. The topographic multipliers, at or near the crest, are therefore also fairly constant with upwind slope in this range. Thus, for this range of slopes, Equations 3.33 and 3.34 can be applied with ϕ replaced by an effective slope ϕ', equal to about 0.3 (Figure 3.11).

For slopes >about 1, for example, steep cliffs, the flow stream lines near ground level at the crest, originate from the upwind flow at levels near the cliff height above the upwind ground level, rather than near the ground level upwind (Figure 3.12). The concept of the topographic multiplier as defined by Equation 3.32 is less appropriate in such cases. Some of the apparent speedup is caused by the upstream boundary-layer profile rather than a perturbation produced by the hill or cliff.

An additional complication for steep features is that separations can occur at or downwind of the crest (see Figure 3.10). Separated flow was found within the first 50-m height above the crest of a 480-m-high feature, with an upwind slope of only 0.48 (average angle

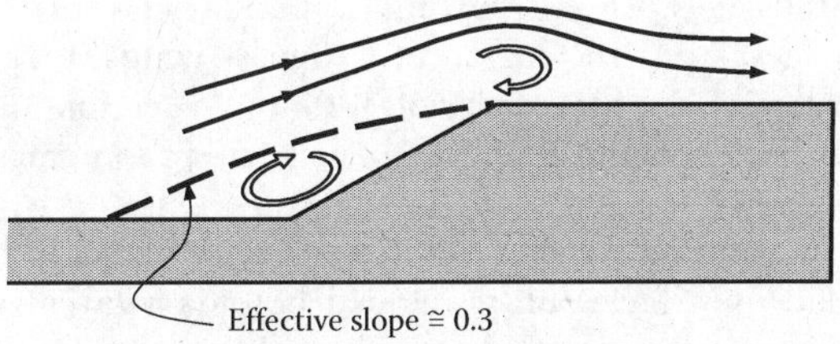

Figure 3.11 Effective upwind slope for steep escarpments.

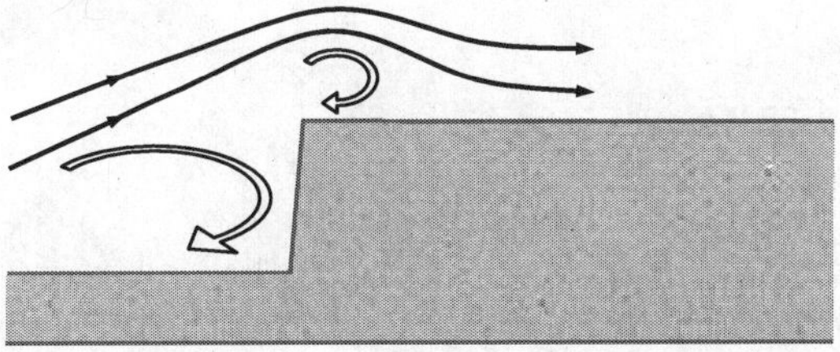

Figure 3.12 Wind flow over a steep cliff.

of 26 degrees), in both full-scale, and 1/1000-scale wind-tunnel measurements (Glanville and Kwok, 1997). This has the effect of decreasing the mean velocity, and increasing the turbulence intensity, as shown in Figure 3.13.

3.4.5 Effect of topography on tropical cyclones and thunderstorm winds

The effect of topographic features on wind near the ground in tropical cyclones and thunderstorm downbursts is much less clearly understood than those in the well-developed boundary layers of large-scale synoptic systems.

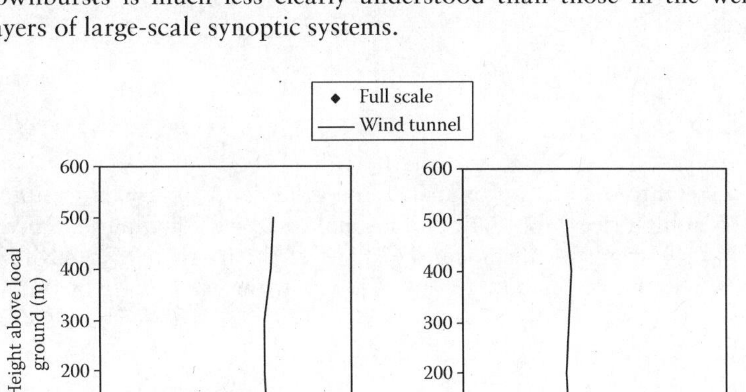

Figure 3.13 Mean velocity profile and rms longitudinal turbulence velocity near the crest of a steep escarpment ($H = 480$ m, upwind slope = 0.48).

Tropical cyclones are large storms with similar boundary layers to extra-tropical depressions on their outer edges. Near the region of strongest winds, they appear to have much lower boundary-layer heights – of the order of 300 m (see Section 3.2.5). Topographic features greater than this height would therefore be expected to interact with the structure of the storm itself.

Thunderstorm downdrafts have 'boundary-layers' with peaks in the velocity profiles at 50–100 m. They also do not have full-developed boundary-layer velocity profiles. There have been some basic studies using wind-tunnel jets impinging on a flat board (Letchford and Illidge, 1999; Wood et al., 1999), to indicate considerably lower topographic multipliers compared with developed thick boundary-layer flows. However, the effect of forward motion of the storm is uncertain.

3.5 CHANGE OF TERRAIN

When strong winds in a fully developed boundary layer encounter a change of surface roughness, for example, winds from an open country flowing over the suburbs of a town or city, a process of adjustment in the turbulent boundary-layer flow properties develops. The adjustment starts at ground level and gradually moves upwards. The result is the development of an internal boundary layer over the new terrain as shown in Figure 3.14.

Deaves (1981), from numerical studies, developed the following relationships for the horizontal position of the inner boundary layer as a function of its height, z.

For flow from a smooth terrain (roughness length z_{o1}) to a rougher terrain (z_{o2}) with $z_{o2} > z_{o1}$:

$$x_i(z) = z_{o2}\left(\frac{z}{0.36z_{o2}}\right)^{4/3} \tag{3.36}$$

For flow from a rough terrain (roughness length z_{o1}) to a smoother terrain (z_{o2}) with $z_{o1} > z_{o2}$:

$$x_i(z) = 14z\left(\frac{z_{o1}}{z_{o2}}\right)^{1/2} \tag{3.37}$$

Setting z_{o2} equal to 0.2 m, approximately the value for a suburban terrain with low-rise buildings 3–5 m high (see Table 3.1), and z equal to 10 m, Equation 3.36 gives a value for $x_i(10)$ of 144 m. Beyond this distance, the shape of the mean velocity profile below 10 m has the characteristics of the new terrain. However, the *magnitude* of the mean velocity

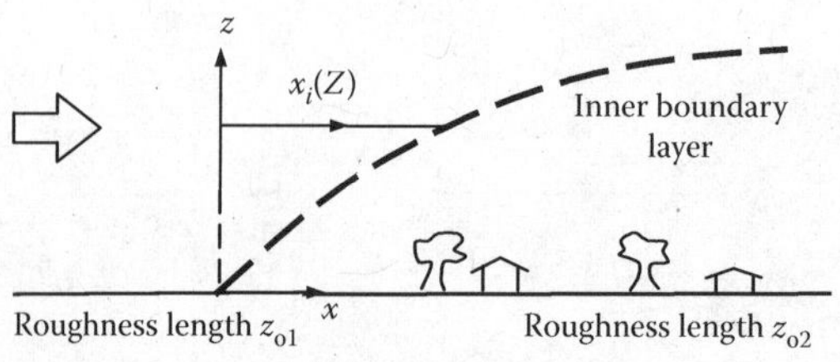

Figure 3.14 Internal boundary-layer development at a change of terrain roughness.

continues to reduce for many kilometres, until the complete atmospheric boundary layer has fully adjusted to the rougher terrain.

Melbourne (1992) found the *gust* wind speed, at a height of 10 m, adjusts to a new terrain approximately exponentially with a distance constant of about 2000 m. Thus, the peak gust at a distance x (in metres) into the new terrain (2) can be represented by Equation 3.38.

$$\hat{U}_{2,x} = \hat{U}_1 + (\hat{U}_2 - \hat{U}_1)\left[1 - \exp\left(\frac{-x}{2000}\right)\right] \tag{3.38}$$

where $\hat{U}_1$ and $\hat{U}_2$ are the asymptotic gust velocities over a fully developed terrain of type 1 (upstream) and 2 (downstream).

Equation 3.38 was found to fit data from a wind tunnel for flow from rough to smooth, as well as smooth to rough, and when there were several changes of roughness.

3.6 WEAKENING OF A TROPICAL CYCLONE AFTER A COAST CROSSING

As the high-convection regions of a tropical cyclone, including a hurricane or typhoon (see Section 1.3.2), cross a coastline from a warm ocean, the storm loses its source of energy and starts to weaken. There is also a reduction due to surface friction for winds blowing over land. These effects result in maximum wind speeds progressively falling with distance from the coastline.

Kaplan and DeMaria (1995, 2001) developed a simple empirical model for predicting the decay of tropical-cyclone winds after landfall. The model is based on a least-squares fit to the maximum sustained (1-min) surface wind estimates made by the National Hurricane Center to a land-falling tropical storm and hurricanes in the United States. The equation for the maximum sustained wind as a function of time takes the form

$$V(t) = V_b + (RV_0 - V_b)e^{-\alpha t} \tag{3.39}$$

where

V_b is a 'background' wind speed
V_0 is the maximum wind speed just before landfall
R is a reduction factor to account for the immediate effect of surface roughness immediately on landfall
α is a decay parameter

For storms making landfall closer to the Equator than 37°, Kaplan and DeMaria established values for V_b, α and R of 26.7 knots, 0.095 h^{-1} and 0.9, respectively. For storms land falling on the Atlantic coast north of 37°N, the values obtained were: $V_b = 29.6$ knots, $\alpha = 0.187$ h^{-1} and $R = 0.9$.

Equation 3.39 can be modified in terms of decay with *distance* instead of time:

$$V(x) = V_b + (RV_0 - V_b)e^{-\beta x}$$
$$V(w) = V_b + (RV_0 - V_b)e^{-(\beta/\cos\theta)w} \tag{3.40}$$

where

β is a decay constant based on the distance inland x

w is the shortest distance from the smoothed coastline equal to $x \cos \theta$

θ is the angle of the cyclone track from the normal to the smoothed coastline

β can be estimated from the decay constant of Kaplan and Maria as follows:

$$\beta = \frac{\alpha}{c_{av}} \tag{3.41}$$

where c_{av} is an average translation speed of the storm.

An average value of the angle to the normal can be calculated as follows:

$$(\cos\theta)_{av} = \left(\frac{1}{\pi}\right) \int_{\pi/2}^{+\pi/2} \cos\theta \, d\theta = \frac{2}{\pi} \tag{3.42}$$

Hence, substituting in Equation 3.40,

$$V(w) = V_b + (RV_0 - V_b)e^{-(\pi\alpha w/2c_{av})} \tag{3.43}$$

3.7 OTHER SOURCES

A well-documented and detailed description of the atmospheric boundary in temperate synoptic systems, for wind-loading purposes, is given in a series of data items published by the Engineering Sciences Data Unit (ESDU (1974–1999)). These include the effects of topographic and terrain changes. The mathematical model of atmospheric turbulence in temperate gale conditions of Deaves and Harris (1978), which used data only from measurements which satisfied rigorous conditions, such as very uniform upstream terrain, is also well known, and contains mathematically acceptable expressions for turbulence quantities in the atmospheric boundary layer. Cook (1985) has described, for the designer, a structure of the atmospheric boundary layer, which is consistent with the above models.

These references are strongly recommended for descriptions of a strong wind structure in temperate zones. However, as discussed in this chapter, the strong wind structure in tropical and semi-tropical locations, such as those produced by thunderstorms and tropical cyclones, is different, and such models should be used with caution in these regions.

3.8 SUMMARY

In this chapter, the structure of strong winds near the earth's surface, relevant to wind loads on structures, has been described. The main focus has been the atmospheric boundary layer in large synoptic winds over land. The mean wind speed profile and some aspects of the turbulence structure have been described. However, some aspects of wind over the oceans, and in tropical cyclones, thunderstorm downbursts and tornadoes, have also been discussed.

The modifying effects of topographic features and of changes in terrain have also been briefly covered.

REFERENCES

Amano, T., Fukushima, H., Ohkuma, T., Kawaguchi, A. and Goto, S. 1999. The observation of typhoon winds in Okinawa by Doppler SODAR. *Journal of Wind Engineering and Industrial Aerodynamics*, 83: 11–20.

Beljaars, A.C.M. 1987. The measurement of gustiness at routine wind stations – A review. Instruments and Observing Methods Report WMO No. 31, World Meteorological Organization, Geneva, Switzerland

Bendat, J.S. and Piersol, A.G. 1999. *Random Data: Analysis and Measurement Procedures*. 3rd edition. John Wiley, New York.

Black, P.G. 1992. Evolution of maximum wind estimates in typhoons. *ICSU/WMO Symposium on Tropical Cyclone Disasters*, Beijing, China, 12–18 October.

Busch, N. and Panofsky, H. 1968. Recent spectra of atmospheric turbulence. *Quarterly Journal of the Royal Meteorological Society*, 94: 132–48.

Charnock, H. 1955. Wind stress on a water surface. *Quarterly Journal of the Royal Meteorological Society*, 81: 639–40.

Choi, E.C.C. 1978. Characteristics of typhoons over the South China Sea. *Journal of Industrial Aerodynamics*, 3: 353–65.

Cook, N.J. 1985. The *Designer's Guide to Wind Loading of Building Structures. Part 1: Background, Damage Survey, Wind Data and Structural Classification*. Building Research Establishment and Butterworths, London.

Davenport, A.G. 1963. The relationship of wind structure to wind loading. *Proceedings, International Conference on Wind Effects on Buildings and Structures*, Teddington, UK, 26–28 June, 53–102.

Deacon, E.L. 1955. Gust variation with height up to 150 metres. *Quarterly Journal of the Royal Meteorological Society*, 81: 562–73.

Deacon, E.L. 1965. Wind gust speed: Averaging time relationship. *Australian Meteorological Magazine*, 51: 11–4.

Deaves, D.M. 1981. Computations of wind flow over changes in surface roughness. *Journal of Wind Engineering and Industrial Aerodynamics*, 7: 65–94.

Deaves, D.M. and Harris, R.I. 1978. A mathematical model of the structure of strong winds. Construction Industry Research and Information Association (UK), Report 76.

Durst, C.S. 1960. Wind speeds over short periods of time. *Meteorological Magazine*, 89: 181–6.

ESDU. 1974–1999. Wind speeds and turbulence. Engineering Sciences Data Unit (ESDU International), Wind Engineering Series Vols. 1a and 1b. ESDU Data Items 74030, 82026, 83045, 84011, 84031, 85020, 86010, 86035, 91043, 92032.

Franklin, J.L., Black, M.L. and Valde, K. 2003. GPS dropwindsonde wind profiles in hurricanes and their operational implications. *Weather and Forecasting*, 18: 32–44.

Fujita, T.T. 1985. *The Downburst*. Report on Projects NIMROD and JAWS. Published by the author at the University of Chicago.

Garratt, J.R. 1977. Review of drag coefficients over oceans and continents. *Monthly Weather Review*, 105: 915–29.

Glanville, M.J. and Kwok, K.C.S. 1997. Measurements of topographic multipliers and flow separation from a steep escarpment. Part II. Model-scale measurements. *Journal of Wind Engineering and Industrial Aerodynamics*, 69–71: 893–902.

Gunter, W.S. and Schroeder, J.L. 2013. High-resolution full-scale measurements of thunderstorm outflow winds. *12th Americas Conference on Wind Engineering*, Seattle, Washington, USA, 16–20 June.

Harper, B.A., Kepert, J.D. and Ginger, J.D. 2010. Guidelines for converting between various wind averaging periods in tropical cyclone conditions, World Meteorological Organization Report, WMO/TD No.1555, August 2010.

Harris, R.I. 1968. On the spectrum and auto-correlation function of gustiness in high winds. Electrical Research Association. Report 5273.

Hock, T.F. and Franklin, J.L. 1999. The NCAR GPS dropwindsonde. *Bulletin, American Meteorological Society*, 80: 407–20.

Holmes, J.D. 1973. Wind pressure fluctuations on a large building. PhD thesis, Monash University, Australia.

Holmes, J.D., Allsop, A. and Ginger, J.D. 2014. Gust durations, gust factors and gust response factors in wind codes and standards. *Wind and Structures*, 19: 339–52.

Holmes, J.D. and Ginger, J.D. 2012. The gust wind speed duration in AS/NZS 1170.2. *Australian Journal of Structural Engineering* (I.E. Aust.), 13: 207–17.

Holmes, J.D., Hangan, H.M., Schroeder, J.L., Letchford, C.W. and Orwig, K.D. 2008. A forensic study of the Lubbock–Reese downdraft of 2002. *Wind and Structures*, 11: 137–52.

Ishizaki, H. 1983. Wind profiles, turbulence intensities and gust factors for design in typhoon-prone regions. *Journal of Wind Engineering and Industrial Aerodynamics*, 13: 55–66.

Jackson, P.S. and Hunt, J.C.R. 1975. Turbulent flow over a low hill. *Quarterly Journal of the Royal Meteorological Society*, 101: 929–55.

Kaplan, J. and DeMaria, M. 1995. A simple empirical model for predicting the decay of tropical cyclone winds after landfall. *Journal of Applied Meteorology*, 34: 2499–512.

Kaplan, J. and DeMaria, M. 2001. On the decay of tropical cyclone winds after landfall in the New England area. *Journal of Applied Meteorology*, 40: 280–6.

Krayer, W.R. and Marshall, R.D. 1992. Gust factors applied to hurricane winds. *Bulletin, American Meteorological Society*, 73: 613–17.

Letchford, C.W. and Illidge, G. 1999. Turbulence and topographic effects in simulated thunderstorm downdrafts by wind tunnel jet. *Proceedings, 10th International Conference on Wind Engineering*, Copenhagen, Denmark, 21–24 June 1999, Balkema, Rotterdam.

Lettau, H.H. 1959. Wind profile, surface stress and geostrophic drag coefficients in the atmospheric boundary layer. *Proceedings, Symposium on Atmospheric Diffusion and Air Pollution*, Oxford, UK, Academic Press, New York.

Lewellen, W.S. 1976. Theoretical models of the tornado vortex. *Symposium on Tornadoes: Assessment of Knowledge and Implications for Man*, Texas Tech University, Lubbock, TX, 22–24 June 1976, pp.107–43.

Melbourne, W.H. 1992. Unpublished course notes. Monash University, Australia.

Miller, C.A., Holmes, J.D., Henderson, D.J., Ginger, J.D. and Morrison, M. 2013. The response of the Dines anemometer to gusts and comparisons with cup anemometers. *Journal of Atmospheric and Oceanic Technology*, 30: 1320–36.

Orwig, K.D. and Schroeder, J.L. 2007. Near-surface wind characteristics of extreme thunderstorm outflows. *Journal of Wind Engineering and Industrial Aerodynamics*, 95: 565–84.

Oseguera, R.M. and Bowles, R.L. 1988. A simple analytic 3-dimensional downburst model based on boundary layer stagnation flow. *NASA Technical Memorandum 100632*, National Aeronautics and Space Administration, Washington, DC.

Powell, M.D., Vickery, P.J. and Reinhold, T.A. 2003. Reduced drag coefficients for high wind speeds in tropical cyclones. *Nature*, 422: 279–83.

Taylor, P.A. and Lee, R.J. 1984. Simple guidelines for estimating windspeed variation due to small scale topographic features. *Climatological Bulletin* (Canada), 18: 3–32.

Vickery, P.J. and Skerjl, P.F. 2005. Hurricane gust factors revisited. *Journal of Structural Engineering*, 131: 825–32.

von Karman, T. 1948. Progress in the statistical theory of turbulence. *Proceedings of the National Academy of Sciences* (U.S.), 34: 530–9.

Wilson, K.J. 1979. Characteristics of the subcloud layer wind structure in tropical cyclones. *International Conference on Tropical Cyclones*, Perth, Western Australia, November 1979.

Wood, G.S., Kwok, K.C.S., Motteram, N. and Fletcher, D.F. 1999. Physical and numerical modelling of thunderstorm downbursts. *Proceedings, 10th International Conference on Wind Engineering*, Copenhagen, Denmark, 21–24 June 1999, Balkema, Rotterdam.

Chapter 4

Basic bluff-body aerodynamics

4.1 FLOW AROUND BLUFF BODIES

Structures of interest in this book can generally be classified as *bluff* bodies with respect to the air flow around them, in contrast to *streamlined* bodies, such as aircraft wings and yacht sails (when the boat is sailing across the wind). Figure 4.1 shows the flow patterns around an airfoil (at a low angle of attack), and around a two-dimensional body of a rectangular cross-section. The flow patterns are shown for steady free-stream flow; turbulence in the approaching flow, which occurs in the atmospheric boundary layer, as discussed in Chapter 3, can modify the flow around a bluff body, as will be discussed later.

It can be seen in Figure 4.1 that the flow streamlines around the airfoil closely follow the contours of the body. The free-stream flow is separated from the surface of the airfoil only by a thin boundary layer, in which the tangential flow is brought to rest at the surface. The flow around the rectangular section (a typical bluff body) in Figure 4.1 is characterised by a 'separation' of the flow at the leading-edge corners. The separated flow region is divided from the outer flow by a thin region of high shear and vorticity, a region known as a free shear layer, which is similar to the boundary layer on the airfoil, but not attached to a surface. These layers are unstable in a sheet form and will roll up towards the wake, to form concentrated vortices, which are subsequently shed downwind.

In the case of the bluff body with a long 'after-body' in Figure 4.1, the separated shear layer 're-attaches' onto the surface. However, the shear layer is not fully stabilised and vortices may be formed on the surface, and subsequently roll along the surface.

4.2 PRESSURE AND FORCE COEFFICIENTS

4.2.1 Bernoulli's equation

The region outside the boundary layers in the case of the airfoil and the outer region of the bluff-body flow, are regions of *inviscid* (zero viscosity) and *irrotational* (zero vorticity) flow, and the pressure, p, and velocity, U, in the fluid are related by *Bernoulli's equation*

$$p + \frac{1}{2}\rho_a U^2 = \text{a constant} \tag{4.1}$$

Denoting the pressure and velocity in the region outside the influence of the body by p_o and U_0, we have:

$$p + \frac{1}{2}\rho_a U^2 = p_o + \frac{1}{2}\rho_a U_0^2$$

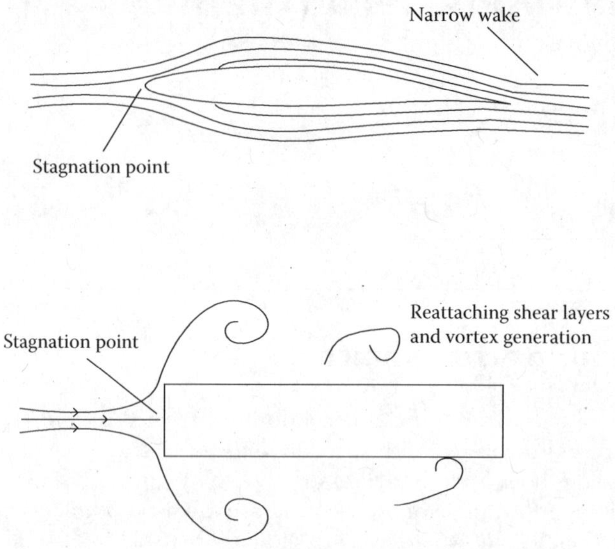

Figure 4.1 Flow around streamlined and bluff bodies.

Hence,

$$p - p_o = \frac{1}{2}\rho_a(U_0^2 - U^2)$$

The surface pressure on the body is usually expressed in the form a non-dimensional *pressure coefficient*:

$$C_p = \frac{p - p_o}{(1/2)\rho_a U_0^2} \tag{4.2}$$

In the region in which Bernoulli's equation holds:

$$C_p = \frac{(1/2)\rho_a(U_0^2 - U^2)}{(1/2)\rho_a U_0^2} = 1 - \left(\frac{U}{U_0}\right)^2 \tag{4.3}$$

At the stagnation point, where U is zero, Equation 4.3 gives a pressure coefficient of one. This is the value measured by a *total* pressure or *pitot* tube pointing into a flow. The pressure $(1/2)\rho_a U_0^2$ is known as the *dynamic* pressure. Values of pressure coefficient near 1.0 also occur at the stagnation point on a circular cylinder, but the largest (mean) pressure coefficients on the windward faces of buildings are usually less than this theoretical value.

In the regions where the flow velocity is greater than U_0, the pressure coefficients are negative. Strictly, Bernoulli's equation is not valid in the separated flow and wake regions, but reasonably good predictions of surface pressure coefficients can be obtained from Equation 4.3, by taking the velocity, U, as that just outside the shear layers and wake region.

4.2.2 Force coefficients

Force coefficients are defined in a similar non-dimensional way to pressure coefficients:

$$C_F = \frac{F}{(1/2)\rho_a U_0^2 A} \tag{4.4}$$

where F is the total aerodynamic force, and A is a reference area (not necessarily the area over which the force acts). Often, A is a projected frontal area.

In the case of long, or two-dimensional, bodies, a force coefficient per unit length is usually used:

$$C_f = \frac{f}{(1/2)\rho_a U_0^2 b} \tag{4.5}$$

where f is the aerodynamic force per unit length, and b is a reference length, usually the breadth of the structure normal to the wind.

Aerodynamic forces are conventionally resolved into two orthogonal directions. These may be parallel and perpendicular to the wind direction (or mean wind direction in the case of turbulent flow), in which case, the axes are referred to as *wind axes*, or are parallel and perpendicular to a direction related to the geometry of the body (*body axes*). These axes are shown in Figure 4.2.

Following the terminology of aeronautics, the terms 'lift' and 'drag' are commonly used in wind engineering for cross-wind and along-wind force components, respectively. Substituting 'L' and 'D' for 'F' in Equation 4.4 gives the definition of *lift and drag coefficients*.

The relationship between the forces and force coefficients resolved with respect to the two axes can be derived using trigonometry, in terms of the angle, α, between the sets of axes, as shown in Figure 4.3. α is called the *angle of attack* (or sometimes angle of incidence).

4.2.3 Dependence of pressure and force coefficients

Pressure and force coefficients are non-dimensional quantities, which are dependent on a number of variables related to the geometry of the body, and to the upwind flow characteristics. These variables can be grouped together into non-dimensional groups, using processes of dimensional analysis, or by inspection.

Assuming that we have a number of bluff bodies of a geometrically similar shape, which can be characterised by a single length dimension (e.g. buildings with the same ratio of

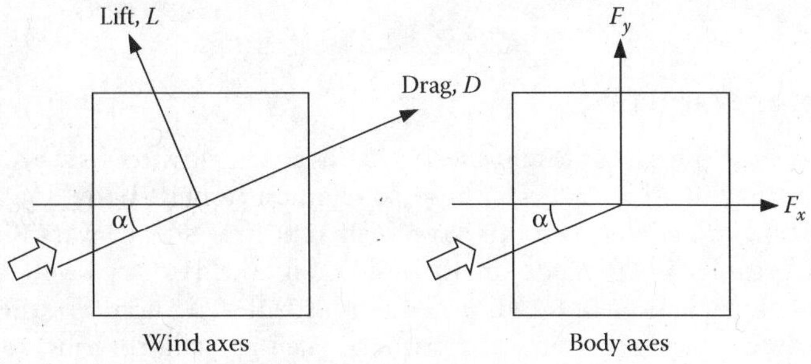

Figure 4.2 Wind axes and body axes.

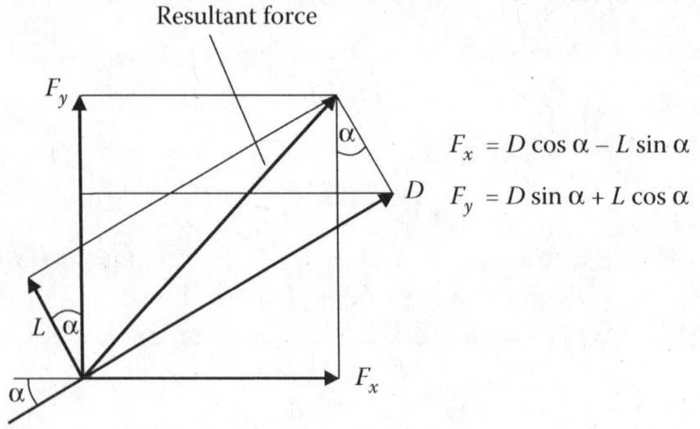

Figure 4.3 Relationship between resolved forces.

height, width and length, and with the same roof pitch, characterised by their height, h). Then the pressure coefficients for pressures at corresponding points on the surface of the body may be a function of a number of other non-dimensional groups: π_1, π_2, π_3 and so on.

$$\text{Thus, } C_p = f(\pi_1, \pi_2, \pi_3, \text{etc.}) \tag{4.6}$$

Examples of relevant non-dimensional groups are

- h/z_0 (Jensen Number)

(where z_0 is the roughness length as discussed in Section 3.2.1)

- I_u, I_v and I_w are the turbulence intensities in the approaching flow.
- (ℓ_u/h), (ℓ_v/h) and (ℓ_w/h) representing ratios of turbulence length scales in the approaching flow, to the characteristic body dimension.
- (Uh/ν), Reynolds Number, where ν is the kinematic viscosity of air.

Equation 4.6 is relevant to the practice of wind-tunnel model testing, in which geometrically scaled models are used to obtain pressure (or force) coefficients for application to full-scale prototype structures (see Section 7.4). The aim should be to ensure that all relevant non-dimensional numbers (π_1, π_2, π_3, etc.) should be equal in both model and full scale. This is difficult to achieve for all the relevant numbers, and methods have been devised for minimising the errors resulting from this. Wind-tunnel-testing techniques are discussed in Chapter 7.

4.2.4 Reynolds Number

Reynolds Number is the ratio of fluid inertia forces in the flow to viscous forces, and is an important parameter in all branches of fluid mechanics. In bluff-body flows, viscous forces are only important in the surface boundary layers and free shear layers (Section 4.1). The dependence of pressure coefficients on Reynolds Number is often overlooked for sharp-edged bluff bodies, such as most buildings and industrial structures. For these bodies, separation of flow occurs at sharp edges and corners, such as wall–roof junctions, over a very wide range of Reynolds Number. However, for bodies with curved surfaces such as circular cylinders or arched roofs, the separation points *are* dependent on Reynolds Number, and this parameter should be considered. Surface roughness has significant effects on flow around circular cylinders (see Section 4.5.1) and other bodies with curved surfaces. This may sometimes be used to an advantage to modify the flow around these shapes on wind-tunnel models to approximate the flow around the full-scale bode (see Section 7.4.4). The addition of turbulence to the flow also reduces the Reynolds Number dependence for bodies with curved surfaces.

In most references to Reynolds Number in this book, the breadth of the body, b, (i.e. the diameter in the case of a circular cylinder) is used to form the Reynolds Number, denoted by Re_b. However, the average roughness height, k, is also used as a length scale in Section 4.5.1 – forming the 'roughness Reynolds Number' Re_k, equal to Uk/ν.

4.3 FLAT PLATES AND WALLS

4.3.1 Flat plates and walls normal to the flow

The flat plate, with its plane normal to the air stream, represents a common situation for wind loads on structures. Examples are elevated hoardings and signboards, which are mounted so that their plane is vertical. Solar panels are another example but, in this case, they are generally inclined to the vertical to maximise the collection of solar radiation. Free-standing walls are another example, but the fact that they are attached to the ground has a considerable effect on the flow and the resulting wind loading. In this section, some fundamental aspects of flow and drag forces on flat plates and walls are discussed.

For a flat plate or wall with its plane normal to the flow, the only aerodynamic force will be one parallel to the flow, i.e. a drag force. Then if p_W and p_L are the average pressures on the front (windward) and rear (leeward) faces respectively, the drag force, D, will be given by

$$D = (p_W - p_L)\,A$$

where A is the frontal area of the plate or wall.

Then dividing both sides by $(1/2)\rho_a U^2 A$, we have

$$\begin{aligned} C_D &= C_{p,W} - C_{p,L} \\ &= C_{p,W} + (-C_{p,L}) \end{aligned} \qquad (4.7)$$

In practice, the windward wall pressure, p_W, and pressure coefficient, $C_{p,W}$, vary considerably with position on the front face. The leeward (or 'base') pressure, however, is nearly uniform over the whole rear face, as this region is totally exposed to the wake region, with relatively slow-moving air.

The mean drag coefficients for various plate and wall configurations are shown in Figure 4.4. The drag coefficient for a square plate in a smooth, uniform approach flow is about 1.1, slightly greater than the total pressure in the approach flow, averaged over the face of the plate. Approximately 60% of the drag is contributed by positive pressures (above static pressure) on the front face, and 40% is contributed by negative pressures (below static pressure) on the rear face (ESDU, 1970).

The effect of free-stream turbulence is to increase the drag on the normal plate slightly. The increase in drag is caused by a decrease in leeward, or base, pressure, rather than an increase in the front face pressure. The hypothesis is that the free-stream turbulence causes an increase in the rate of entrainment of air into the separated shear layers. This leads to a reduced radius of curvature of the shear layers, and to a reduced base pressure (Bearman, 1971).

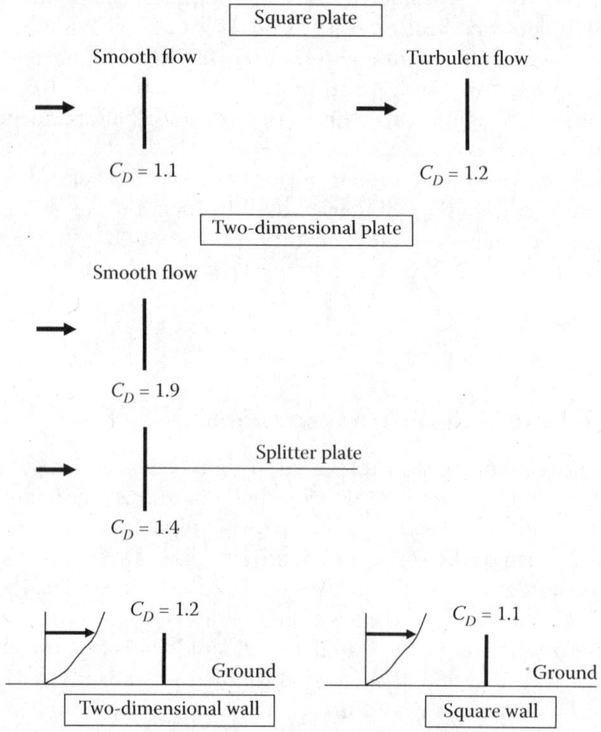

Figure 4.4 Drag coefficients for normal plates and walls.

Figure 4.4 also shows the drag coefficient on a long flat plate with a theoretically infinite width into the paper – the 'two-dimensional' flat plate. The drag coefficient of 1.9 is higher than that for the square plate. The reason for the increase on the wide plates can be explained as follows. For a square plate, the flow is deflected around the plate equally around the four sides. The extended width provides a high-resistance flow path into (or out of) the paper, thus forcing the flow to travel faster over the top edge, and under the bottom edge. This faster flow results in more entrainment from the wake into the shear layers, thus generating lower base, or leeward face, pressure and higher drag.

Rectangular plates with intermediate values of width to height have intermediate values of drag coefficient. A formula given by ESDU (1970) for the drag coefficient on plates of height/breadth ratio in the range $1/30 < h/b < 30$, in smooth uniform flow normal to the plate, is reproduced in Equation 4.8.

$$C_D = 1.10 + 0.02[(h/b) + (b/h)] \tag{4.8}$$

In the case of the two-dimensional plate, strong vortices are shed into the wake alternately from the top and bottom, in a similar way to the bluff-body flow shown in Figure 4.1. These greatly contribute to the increased entrainment into the wake of the two-dimensional plate. Suppression of these vortices by a splitter plate has the effect of reducing the drag coefficient to a lower value, as shown in Figure 4.4.

This suppression of vortex shedding is nearly complete when a flat plate is attached to a ground plane, and becomes a wall, as shown in the lower sketch in Figure 4.4. In this

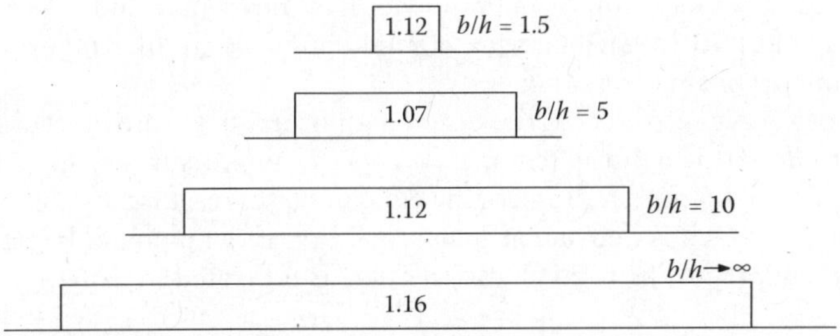

Figure 4.5 Mean drag coefficients on walls in boundary-layer flow.

case, the approach flow will be of a boundary-layer form with a wind speed increasing with height as shown. The value of the drag coefficient, with U taken as the mean wind speed at the top of the wall, U_h, is very similar for the two-dimensional wall, and finite wall of a square planform, that is a drag coefficient of about 1.2 for an infinitely long wall. The effect of finite length of the wall is shown in Figure 4.5. Little change in the mean drag coefficient occurs, although a slightly lower value occurs for an aspect ratio (length/height) of about 5 (Letchford and Holmes, 1994).

The case of two thin normal plates in series, normal to the flow, as shown in Figure 4.6, is an interesting one. At zero spacing, the two plates act like a single plate with a combined drag coefficient (based on the frontal area of one plate) of about 1.1, for a square plate. For spacings in the range of 0 to about 2 h, the combined drag coefficient is actually *lower* than that for a single plate, reaching a value of about 0.8 at a spacing of about 1.5 h, for two square plates. As the spacing is allowed to increase, the combined drag coefficient then increases, so that, for very high spacings, the plates act like individual plates with no interference with each other, and a combined drag coefficient of about 2.2. The mechanism that produces the reduced drag at the critical spacing of 1.5 h has not been studied in detail, but clearly, there is a large interference in the wake and in the vortex shedding, generated by the downstream plate.

The drag forces on two flat plates separated by small distances normal to the flow are also a relevant situation in wind engineering, with applications for clusters of lights or antennas

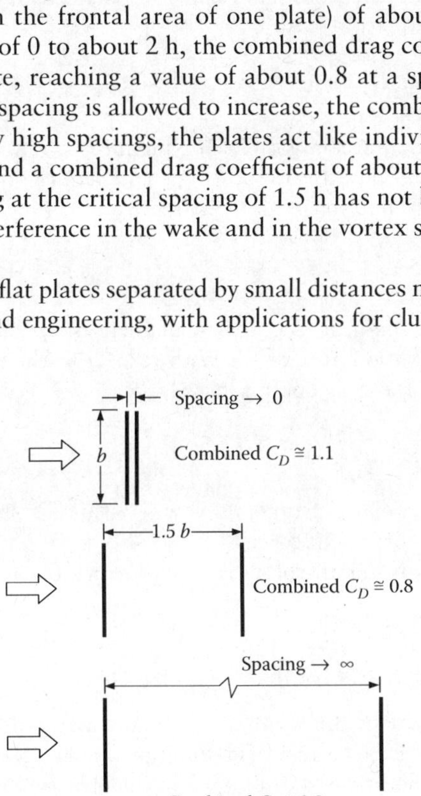

Figure 4.6 Drag coefficients for two square plates in series.

together on a frame, for example. Experiments by Marchman and Werme (1982) found increases in drag of up to 15% when square, rectangular or circular plates were within half a width (or diameter) from each other.

If uniform porosity is introduced, the drag on a normal flat plate or wall reduces as some air is allowed to flow through the plate, and reduces the pressure difference between front and rear faces. The reduction in drag coefficient can be represented by the introduction of a porosity factor, K_p, which is dependent on the solidity of the plate, δ, being the ratio of the 'solid' area of the plate, to the total elevation area, as indicated in Equation 4.9.

$$C_{D,\delta} = C_D \cdot K_p \tag{4.9}$$

K_p is not linearly related to the solidity. An approximate expression for K_p, which fits the data quite well for plates and walls with ratios of height to breadth between about 0.2 and 5, is given by Equation 4.10.

$$K_p \cong 1 - (1 - \delta)^2 \tag{4.10}$$

Equation 4.10 has the required properties of equalling one for a value of δ equal to 1, that is an impermeable plate or wall, and tending to zero as the solidity tends to zero. For very small values of δ (e.g. an open-truss plate made up of individual members), K_p tends to a value of 2δ, since, from Equation 4.10,

$$K_p = 1 - (1 - 2\delta + \delta^2) \cong 2\delta$$

since δ^2 is very small in comparison with 2δ for small δ.

Considering the application of this to the drag coefficient for an open-truss plate of square planform, we have from Equations 4.9 and 4.10,

$$C_{D,At} \cong 1.1\ (2\delta) = 2.2\delta$$

where $C_{D,At}$ denotes that the drag coefficient, defined as in Equation 4.4, is with respect to the total (enclosed) elevation area of A_t. With respect to the elevation area of the actual members in the truss A_m, the drag coefficient is larger, being given by

$$C_{D,Am} = C_{D,At}\,(A_t/A_m) = C_{D,At} \cdot (1/\delta) \cong 2.2$$

In this case of a very open plate, the members will act like isolated bluff bodies with individual values of the drag coefficient of 2.2.

Cook (1990) discusses in detail the effect of porosity on aerodynamic forces on bluff bodies.

4.3.2 Flat plates and walls inclined to the flow

Figure 4.7 shows the case with the wind at an oblique angle of attack, α, to a two-dimensional flat plate. In this case, the resultant force primarily remains at right angles to the plate surface, that is it is no longer a drag force in the direction of the wind. There is also a tangential component, or 'skin friction' force. However, this is not significant in comparison with the normal force, for angles of attack greater than about 10 degrees.

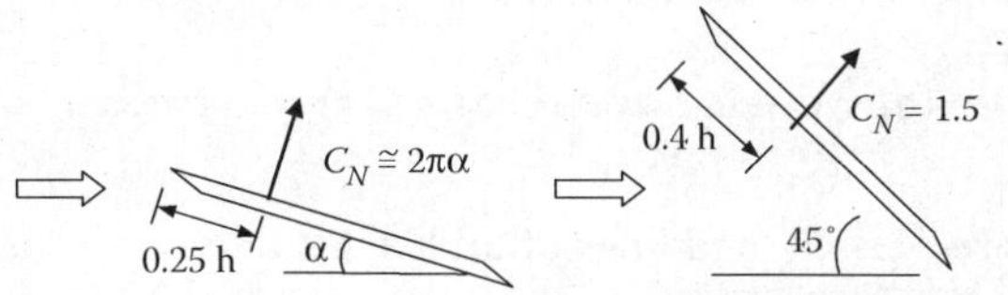

Figure 4.7 Normal force coefficients for an inclined two-dimensional plate.

For small angles of attack, α, (<10 degrees), the normal force coefficient, C_N, with respect to the total plan area of the plate viewed normal to its surface, is given approximately by

$$C_N \cong 2\pi\,\alpha \tag{4.11}$$

(where α is measured in radians, not in degrees).

Equation 4.11 comes from theory used in aeronautics. The 'centre of pressure', denoting the position of the line of action of the resultant normal force, is at, or near, one-quarter of the height h, from the leading edge, again a result from aeronautical theory.

As the angle of attack, α, increases, the normal force coefficient, C_N, progressively increases towards the normal plate case (α = 90°), discussed in Section 4.3.1, with the centre of pressure at a height of 0.5 h. For example, the normal force coefficient for an angle of attack of 45 degrees is about 1.5, with the centre of pressure at a distance of about 0.4 h from the leading edge, as shown in Figure 4.7. The corresponding values for α equal to 30 degrees are about 1.2, and 0.38 h (ESDU, 1970).

Now, we will consider finite-length walls and hoardings, at or near the ground level, and hence in a highly sheared and turbulent boundary-layer flow. The mean net pressure coefficients at the windward end of the wall, for an oblique wind blowing at 45 degrees to the normal, are quite high due to the presence of a strong vortex system behind the wall. Some values of area-averaged mean pressure coefficients are shown in Figure 4.8; these high values are usually the critical cases for the design of free-standing walls and hoardings for wind loads.

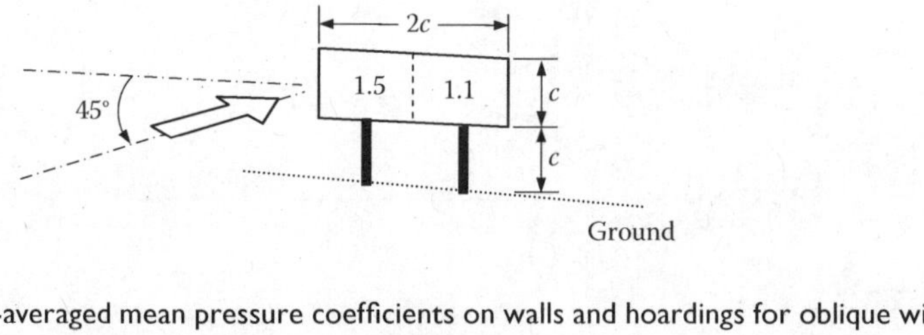

Figure 4.8 Area-averaged mean pressure coefficients on walls and hoardings for oblique wind directions.

4.4 RECTANGULAR PRISMATIC SHAPES

4.4.1 Drag on two-dimensional rectangular prismatic shapes

Understanding of the wind forces on rectangular prismatic shapes is clearly of importance for many structures, especially buildings of all heights and bridge decks. We will consider first the drag coefficients for two-dimensional rectangular prisms.

Figure 4.9 shows how the drag coefficient varies for two-dimensional rectangular prisms with sharp corners, as a function of the ratio, *d*/*b*, where *d* is the along-wind or after-body length, and *b* is the cross-wind dimension. The flow is normal to a face of width *b*, and is 'smooth', that is the turbulence level is low. As previously shown in Figure 4.4, the value of the drag coefficient is 1.9 for (*d*/*b*) close to zero, that is for a flat plate normal to a flow stream. As (*d*/*b*) increases to 0.65–0.70, the drag coefficient increases to about 2.9 (e.g. Bearman and Trueman, 1972). The drag coefficient then decreases with increasing (*d*/*b*), reaching 2.0 for a square cross-section. The drag coefficient continues to decrease with further increases in (*d*/*b*), reaching about 1.0 for values of (*d*/*b*) of 5 or greater.

These variations can be explained by the behaviour of the free shear layers separating from the upstream corners. These shear layers are unstable, as was shown in Figure 4.1, and eventually form discrete vortices. During the formation of these vortices, air is entrained from the wake region behind the prism; it is this continual entrainment process which sustains a base pressure lower than the static pressure. As (*d*/*b*) increases to the range 0.65–0.70, the size of the wake decreases simply because of the increased volume of the prism occupying a part of the wake volume. Thus, the same entrainment process acts on a smaller volume of wake air, causing the base pressure to decrease further, and the drag to increase. However, as (*d*/*b*) increases beyond 0.7, the rear, or downstream, corners interfere with the shear layers, and if the length *d* is long enough, the shear layers will stabilise, or 're-attach', onto the sides of the prisms. Although the attached shear layers will eventually separate again from the *rear* corners of the prism, the wake is smaller for prisms with long after-bodies (high *d*/*b*), and the entrainment is weaker. The result is a lower drag coefficient, as shown in Figure 4.9.

4.4.2 Effect of aspect ratio

The effect of a finite aspect ratio (height/breadth) is to introduce an additional flow path around the end of the body, and a means of increasing the pressure in the wake cavity. The reduced air flow normal to the axis results in a lower drag coefficient for finite-length bodies in comparison to two-dimensional bodies of an infinite aspect ratio. Figure 4.10 shows

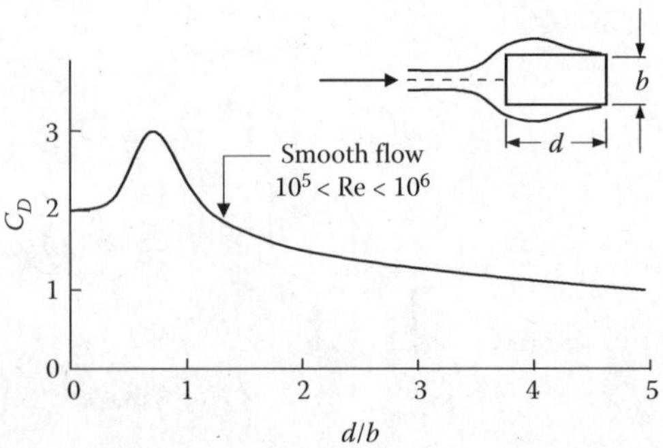

Figure 4.9 Drag coefficients for two-dimensional rectangular prisms in smooth flow.

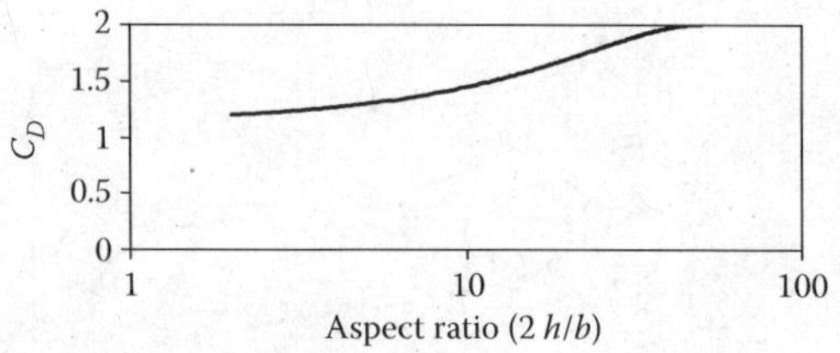

Figure 4.10 Effect of aspect ratio on the drag coefficient for a square cross-section.

the drag coefficient for a square cross-section with one free end exposed to the flow, which was smooth (Scruton and Rogers, 1972). The aspect ratio in this case is calculated as 2 h/b, where h is the height, since it is assumed that the flow is equivalent to that around a body with a 'mirror image' added to give an overall height of 2 h, with two free ends.

4.4.3 Effect of turbulence

Free-stream turbulence containing scales of the prism dimensions or smaller can have significant effects on the mean drag coefficients of rectangular prisms, as well as producing fluctuating forces. As shown in Figure 4.4, the effect of free-stream turbulence on a flat plate normal to an air stream is to increase the drag coefficient slightly (Bearman, 1971). This results from increased mixing and entrainment into the free shear layers induced by the turbulence. Observations have also shown a reduction in the radius of curvature of the mean shear layer position (Figure 4.11). As the after-body length increases, the drag first increases and then decreases, as it occurs in smooth flow. However, because of the decrease in the mean radius of curvature of the shear layers caused by the free-stream turbulence, the

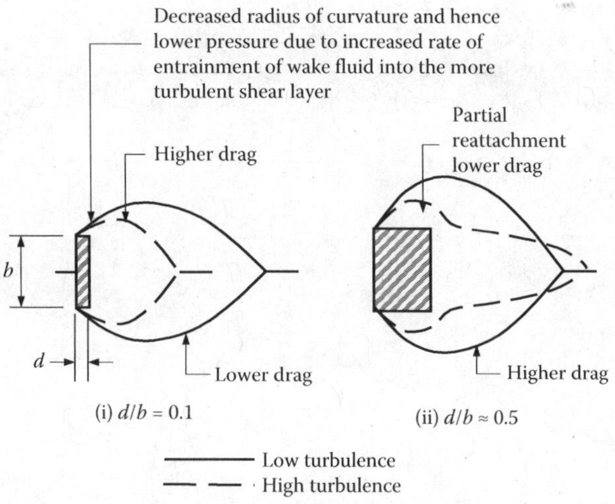

Figure 4.11 Effect of turbulence on shear layers from rectangular prisms. (From Gartshore, I.S. 1973. The effects of freestream turbulence on the drag of rectangular two-dimensional prisms. Boundary Layer Wind Tunnel Report BLWT-4-73, University of Western Ontario; Laneville, A., Gartshore, I.S. and Parkinson, G.V. 1975. An explanation of some effects of turbulence on bluff bodies. *Fourth International Conference on Wind Effects on Buildings and Structures*, London, UK, September. With permission.)

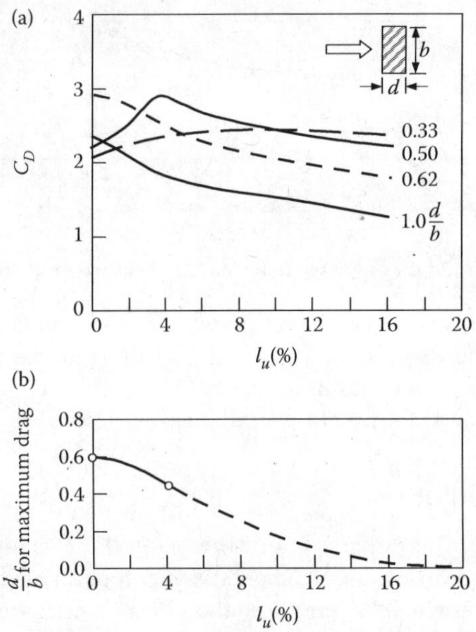

Figure 4.12 Effect of turbulence on drag coefficients for rectangular prisms. (a) Drag coefficient versus turbulence intensity, (b) side ratio (*d*/*b*) for maximum drag versus turbulence intensity. (From Laneville, A., Gartshore, I.S. and Parkinson, G.V. 1975. An explanation of some effects of turbulence on bluff bodies. *Fourth International Conference on Wind Effects on Buildings and structures*, London, UK, September. With permission.)

(*d*/*b*) ratio for maximum drag will decrease with increasing turbulence intensity, as shown in Figure 4.12 (Gartshore, 1973; Laneville et al., 1975).

The drag coefficients for two-dimensional rectangular prisms on the ground in turbulent boundary-layer flow are shown in Figure 4.13. In comparison with rectangular prisms in smooth uniform flow (Figure 4.9), the drag coefficients, based on the mean wind speed at

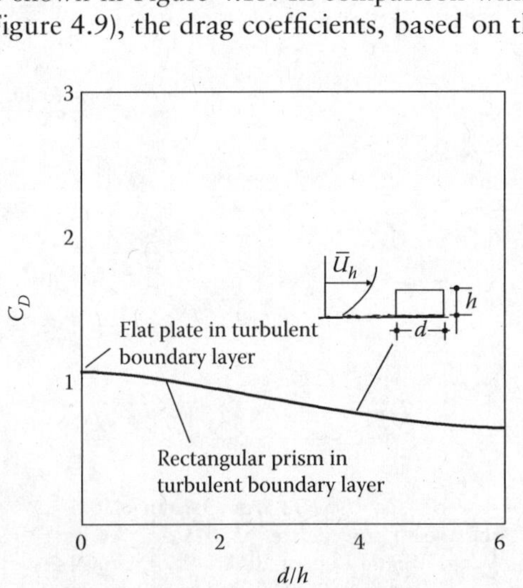

Figure 4.13 Mean drag coefficients for rectangular prisms in turbulent boundary-layer flow.

the height of the top of the prism, are much lower; because of the high turbulence in the boundary-layer flow, they do not show any maximum value.

Melbourne (1995) has discussed the important effects of turbulence on flow around bluff bodies in more detail.

4.4.4 Drag and pressures on a cube and finite-height prisms

The mean pressure distributions on a cube in a wind-tunnel flow with a mean velocity profile, similar to that in the atmospheric boundary layer, are shown in Figure 4.14 (Baines, 1963). However, no attempt was made to model the turbulence properties of the natural wind in these early tests. The pressure coefficients are based on the mean wind speed at the height of the top of the cube. The drag coefficient of 0.8 is lower than that of the two-dimensional square-section prism (*d*/*h* equal to 1.0 in Figure 4.13). This is due to the three-dimensional flows that occur around the side walls of the block which increase the base pressure (decrease the negative pressure).

The mean pressure distribution on a tall prism of square cross-section in a flow with a varying mean velocity profile, similar to that in the atmospheric boundary layer, is shown in Figure 4.15 (Baines, 1963). The mean pressure coefficients are again based on the dynamic pressure calculated from the mean wind speed at the top of the prism. The effect of the vertical velocity profile on the windward wall pressure is clearly seen. The maximum pressure occurs at about 85% of the height. On the windward face of unshielded tall buildings, the

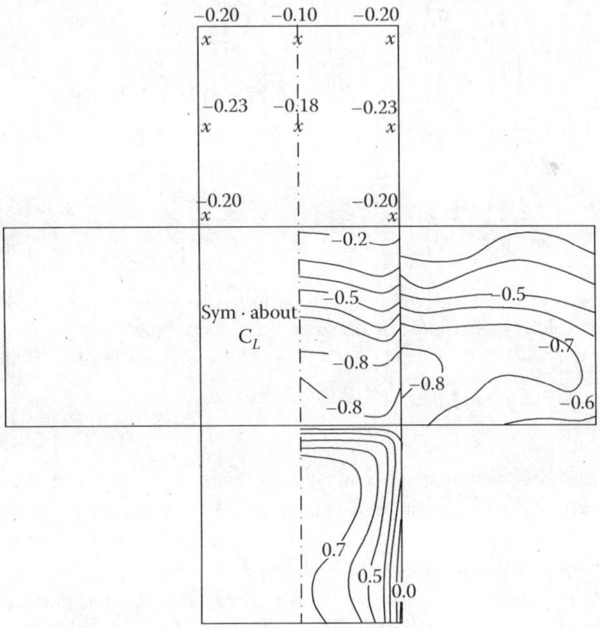

Figure 4.14 Mean pressure coefficients on a cube in boundary-layer flow. (From Baines, W.D. 1963. Effects of velocity distributions on wind loads and flow patterns on buildings. *Proceedings, International Conference on Wind Effects on Buildings and Structures*, Teddington, UK, 26–28 June, pp. 197–225. With permission.)

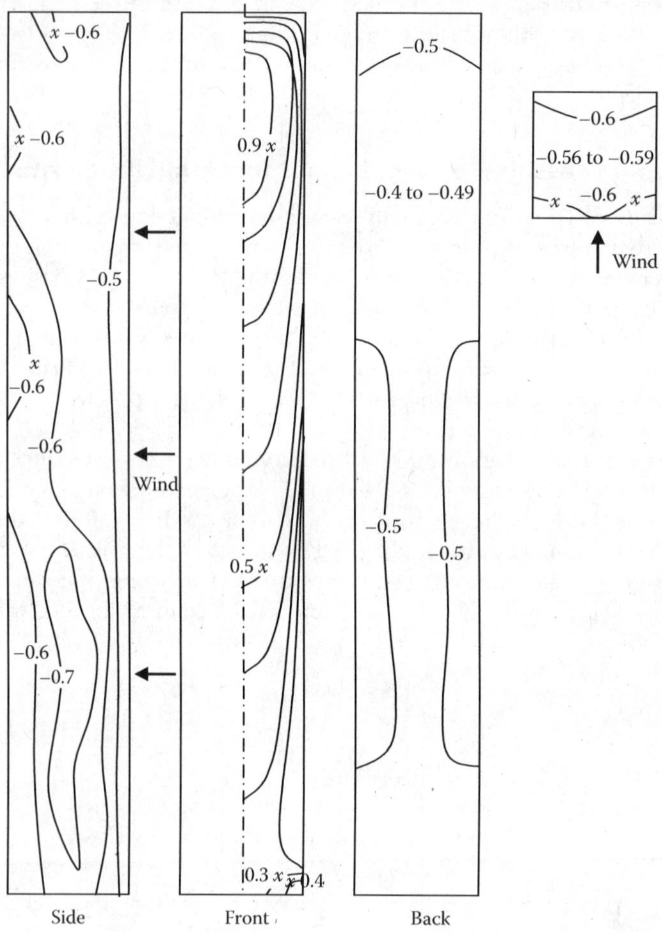

Figure 4.15 Mean pressure coefficients on a tall prism in boundary-layer flow. (From Baines, W.D. 1963. Effects of velocity distributions on wind loads and flow patterns on buildings. *Proceedings, International Conference on Wind Effects on Buildings and Structures*, Teddington UK, 26–28 June, pp. 197–225. With permission.)

strong pressure gradient can cause a strong downwards flow often causing high wind speeds which may cause problems for pedestrians at ground level.

Figure 4.16 shows effective sectional drag coefficients for finite-height prisms with rectangular cross-sections, representative of tall buildings, in a turbulent atmospheric boundary-layer simulation, representing wind flow over a suburban terrain. The effective drag coefficient is defined as the average sectional drag coefficient referenced to the upstream mean wind speed varying with height over the prism, which gives the measured or predicted base shear. It is of interest to compare these data with those in Figures 4.9 and 4.12, which show drag coefficients for two-dimensional rectangular cross-sections in smooth and turbulent flow, respectively. Clearly, the effect of the finite aspect ratio (defined as $h/\sqrt{(bd)}$ in Figure 4.16) is to greatly reduce the effective drag coefficient, as is also seen in Figure 4.10. In Figure 4.16, the maximum drag coefficients generally occur for the lowest value of d/b

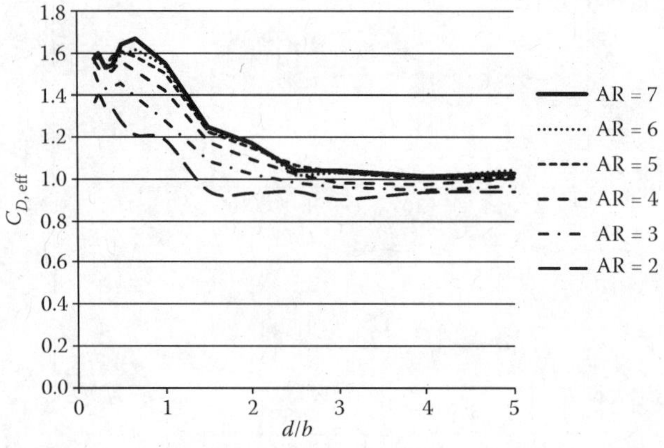

Figure 4.16 Effective drag coefficients for rectangular prisms in a boundary-layer flow. Aspect ratio, AR = $h/\sqrt{(bd)}$. (Data courtesy of Professor C.-M. Cheng of Tamkang University, Taiwan.)

of 0.2, although for the higher aspect ratios, the peak occurs at d/b equal to 0.67, a similar value as that for the peak drag on two-dimensional bodies in smooth flow (Figure 4.9).

4.4.5 Jensen Number

For bluff bodies such as buildings, immersed in a turbulent boundary-layer flow, the ratio of a characteristic body dimension, usually the height, h, in the case of a building, to the characteristic boundary-layer length, represented by the roughness length, z_o, is known as the Jensen Number. In a classic series of experiments, Jensen (1958) established the need for equality of (h/z_o) in order for wind-tunnel mean pressure measurements on a model of a small building to match those in full scale. The effect is greatest on the roof and side walls, where the increased turbulence in the flow over the rougher ground surfaces promotes shorter flow re-attachment lengths.

For a given height, h, greater values of roughness length, z_o, and lower values of Jensen Number imply a rougher ground surface and hence greater turbulence intensities at the height of the body. Thus, fluctuating pressure coefficients also depend on Jensen Number – decreasing Jensen Number generally giving increasing root-mean-square (rms) pressure coefficients.

4.5 CIRCULAR CYLINDERS

4.5.1 Effects of Reynolds Number and surface roughness

For bluff bodies with curved surfaces such as the circular cylinder, the positions of the separation of the local surface boundary layers are much more dependent on viscous forces than is the case with sharp-edged bodies. This results in a variation of drag forces with Reynolds Number, which is the ratio of inertial forces to viscous forces in the flow (see Section 4.2.4). Figure 4.17 shows the variation of the drag coefficient with Reynolds Number for square-section bodies with various corner radii (Scruton, 1981). The appearance of a 'critical' Reynolds Number, at which there is a sharp fall in drag coefficient, occurs at a relatively low corner radius.

The various flow regimes for a circular cylinder with a smooth surface finish in smooth (low turbulence) flow are shown in Figure 4.18. The sharp fall in the drag coefficient at a

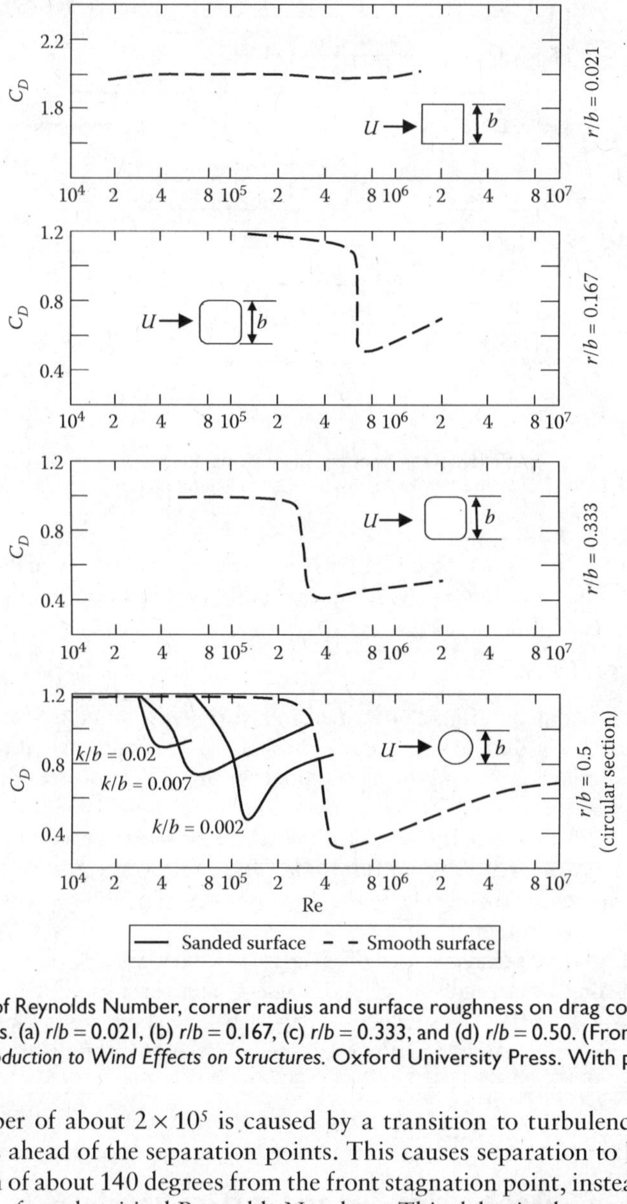

Figure 4.17 Effect of Reynolds Number, corner radius and surface roughness on drag coefficients of square sections. (a) $r/b = 0.021$, (b) $r/b = 0.167$, (c) $r/b = 0.333$, and (d) $r/b = 0.50$. (From Scruton, C. 1981. *An Introduction to Wind Effects on Structures*. Oxford University Press. With permission.)

Reynolds Number of about 2×10^5 is caused by a transition to turbulence in the surface boundary layers ahead of the separation points. This causes separation to be delayed to an angular position of about 140 degrees from the front stagnation point, instead of 90 degrees, which is the case for sub-critical Reynolds Numbers. This delay in the separation results in a narrowing in the wake, and an increased (less negative) base pressure, and a lower drag coefficient. The pressure distributions at sub-critical and super-critical Reynolds Numbers are shown in Figure 4.19.

Although information on flow and force coefficients for two-dimensional circular cylinders at sub-critical Reynolds Numbers (Re_b below about 1×10^5 for cylinders with nominally 'smooth' surfaces) is widely available, there is less data in the critical range up to the minimum drag coefficient, and very few studies in the super-critical range up to Re_b equal to 10^6 and beyond. However, the flow around most structures with circular cross-sections such as lighting poles, chimneys and observation towers at wind speeds for the structural design

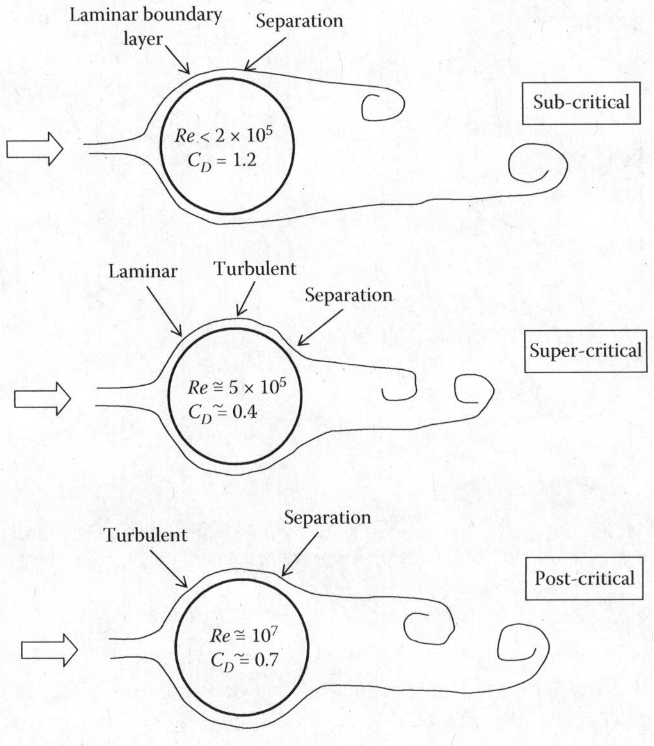

Figure 4.18 Flow regimes for a circular cylinder in smooth flow.

at ultimate limit states falls into the critical or super-critical ranges, where the information is most sparse.

As shown in Figure 4.17, the presence of a rough surface on a circular cylinder causes the critical Reynolds Number range to be lower than that for a smooth cylinder, as noted in the pioneering experiments of Fage and Warsap (1930). The minimum drag coefficient is also higher for the rougher surfaces.

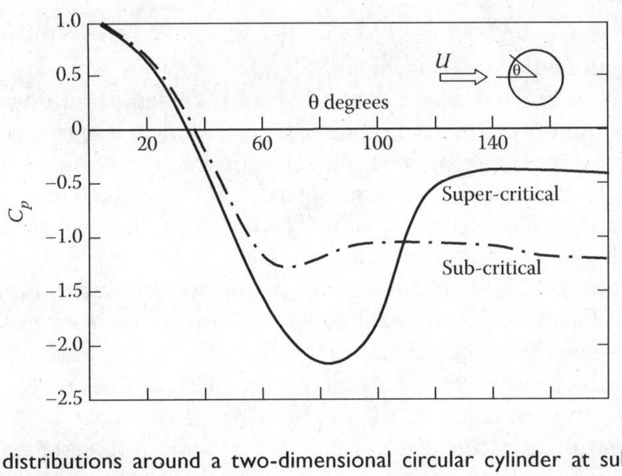

Figure 4.19 Pressure distributions around a two-dimensional circular cylinder at sub-critical and trans-critical Reynolds Numbers.

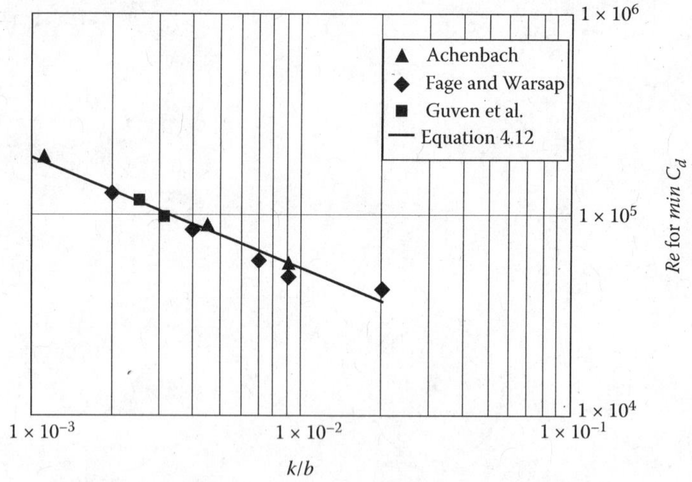

Figure 4.20 Reynolds Number (based on diameter) for the minimum drag coefficient of roughened circular cylinders, as a function of the ratio of average roughness height to diameter.

The Reynolds Number based on diameter (Re_b) defining the lower limit of the super-critical flow range can be taken to be that at which the drag coefficient reaches a minimum value, Re_{minC_d}. The latter value and the minimum drag coefficient C_{dmin} are both functions of the ratio of the average surface roughness height, k, to the diameter, b. As the ratio k/b increases, Re_{minC_d} decreases, and C_{dmin} increases.

Figure 4.20 shows Re_{minC_d} plotted against k/b for the results of Achenbach (1971), Fage and Warsap (1930) and Guven et al. (1980). The values are well fitted with a straight line on log–log axes, suggesting the relationship in Equation 4.12 (Holmes et al., 2012):

$$Re_{minCd} \cong 4210\ (k/b)^{-0.555} \tag{4.12}$$

For $(k/b) < 10^{-3}$, the value of Re_{minC_d} is nearly constant – that is it is independent of (k/b) for relatively smooth cylinders (Adachi, 1995).

Szechenyi (1975) re-plotted the results of Fage and Warsap, together with some new data, and showed that the drag coefficients, in the super-critical range (i.e. for $Re_b > Re_{minC_d}$), effectively collapsed when they were plotted against a *roughness Reynolds Number*, Re_k, defined as (Uk/ν), where U is the wind speed, k is the average roughness height and ν is the kinematic viscosity. Data from a nominally smooth cylinder were included by Szechenyi by assuming a relative roughness (k/b) of 3.5×10^{-5}.

Figure 4.21a shows some more recent data (Holmes et al., 2012) on drag coefficients of three roughened circular cylinders of high aspect ratio plotted against Re_b. Figure 4.21b shows the data from the super-critical range from the same tests re-plotted against the roughness Reynolds Number, Re_k. This method of plotting does indeed 'collapse' the data indicating the effectiveness of the Szechenyi (1975) approach.

The effectiveness of surface roughness in the simulation of super-critical flow in wind-tunnel tests on structures with a circular cross-section is further explored in Chapter 7.

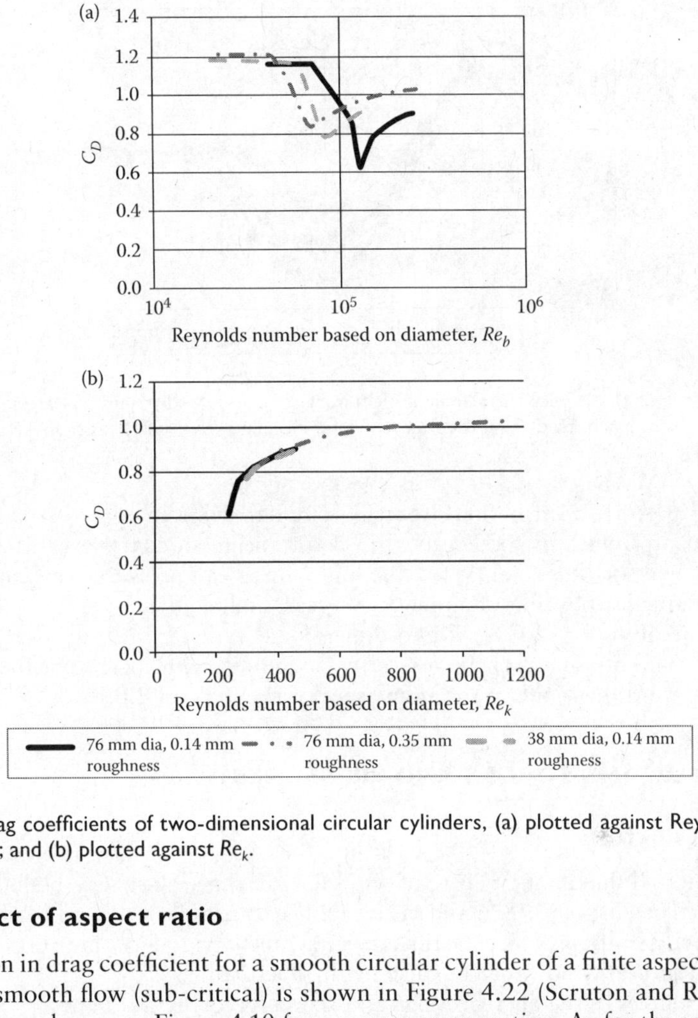

Figure 4.21 Drag coefficients of two-dimensional circular cylinders, (a) plotted against Reynolds Number Re_b; and (b) plotted against Re_k.

4.5.2 Effect of aspect ratio

The reduction in drag coefficient for a smooth circular cylinder of a finite aspect ratio (single free end) in smooth flow (sub-critical) is shown in Figure 4.22 (Scruton and Rogers, 1972). This figure is analogous to Figure 4.10 for a square cross-section. As for the square section, the reduction in drag for a circular cylinder results from the additional flow path provided by the free end on the body.

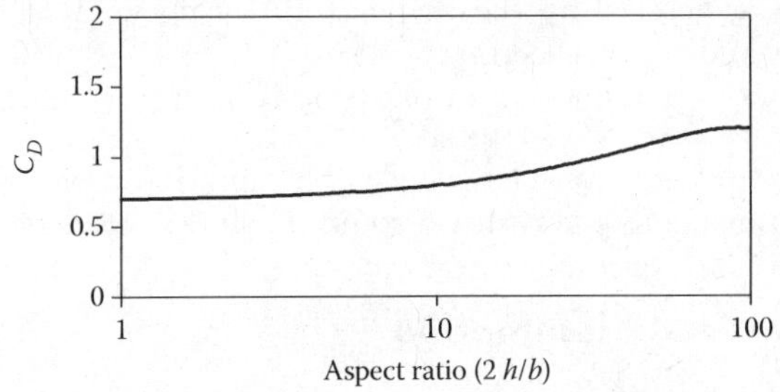

Figure 4.22 Effect of the aspect ratio on the drag coefficient of a circular cylinder (sub-critical Reynolds Number).

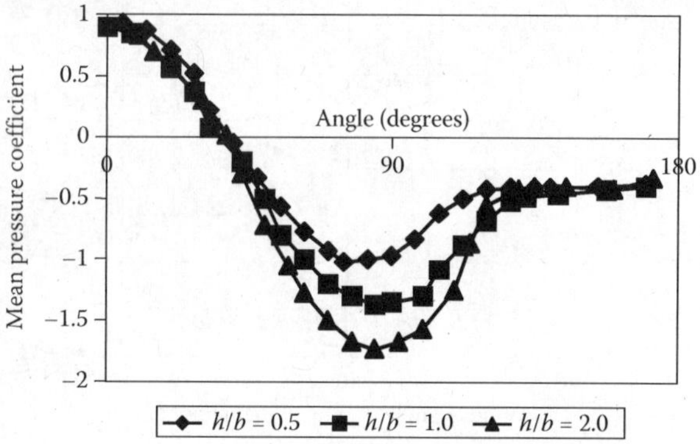

Figure 4.23 Effect of the aspect ratio (height/diameter) on pressure distributions around circular cylinders. (Adapted from Macdonald, P.A., Kwok, K.C.S. and Holmes, J.D. 1988. *Journal of Wind Engineering and Industrial Aerodynamics*, 31: 165–88.)

The mean pressure distribution around a circular cylinder with a height-to-diameter (aspect) ratio of 1, with its axis vertical in a turbulent boundary-layer flow is shown in Figure 4.23 (Macdonald et al., 1988). The minimum mean pressure coefficient on the side occurs at an angular position of about 80 degrees, and is about −1.2, lower in magnitude than the value of about −2.0 for a two-dimensional cylinder in a super-critical flow (see Figure 4.18). The minimum $\bar{C}_p$ increases in magnitude with increasing the aspect ratio, reaching the two-dimensional value at an aspect ratio of about 2.0.

4.6 FLUCTUATING FORCES AND PRESSURES

4.6.1 Introduction

The turbulent and fluctuating nature of wind flow in the atmospheric boundary layer has been described in Chapter 3. This and the unstable nature of flow around bluff bodies which results in flow separations, and sometimes re-attachments, produces pressures and forces on bodies in the natural wind which are also highly fluctuating.

The main sources of the fluctuating pressures and forces are as follows:

- Natural turbulence or gustiness in the free-stream flow. This is often called 'buffeting'. If the body dimensions are small relative to the length scales of the turbulence, the pressure and force variations will tend to follow the variations in velocity (see Section 4.6.2).
- Unsteady flow generated by the body itself, by phenomena such as separations, re-attachments and vortex shedding.
- Fluctuating forces due to movement of the body itself (e.g. aerodynamic damping).

The third source arises only for relatively flexible vibration-prone 'aeroelastic structures'. In the following sections, only the first two sources will be considered.

4.6.2 The Quasi-steady assumption

The 'Quasi-steady' assumption is the basis of many wind-loading codes and standards. The fluctuating pressure on a structure is assumed to follow the variations in longitudinal wind velocity upstream. Thus,

$$p(t) = C_{po}(1/2)\,\rho_a\left[U(t)\right]^2 \tag{4.13}$$

where C_{po} is a quasi-steady pressure coefficient

Expanding $U(t)$ into its mean and fluctuating components,

$$\begin{aligned} p(t) &= C_{po}\left(1/2\right)\rho_a[\bar{U} + u'(t)]^2 \\ &= C_{po}\left(1/2\right)\rho_a[\bar{U}^2 + 2\bar{U}\,u'(t) + u'(t)^2] \end{aligned} \tag{4.14}$$

Taking mean values,

$$\bar{p} = C_{po}\,(1/2)\,\rho_a\,[\bar{U}^2 + \sigma_u^2]$$

For small turbulence intensities, σ_u^2 is small in comparison with $\bar{U}^2$. Then the quasi-steady pressure coefficient, C_{po}, is approximately equal to the mean pressure coefficient, $\bar{C}_p$.

Then,

$$\bar{p} \cong C_{po}(1/2)\,\rho_a\bar{U}^2 = \bar{C}_p(1/2)\,\rho_a\bar{U}^2 \tag{4.15}$$

Subtracting the mean values from both sides of Equation 4.14, we have

$$\tilde{p}(t) = C_{po}(1/2)\,\rho_a[2\bar{U}u'(t) + u'(t)^2]$$

Neglecting the second term in the square brackets (valid for low turbulence intensities), squaring and taking mean values,

$$\overline{p'^2} \cong C_p^2\,(1/4)\,\rho_a^2[4\,\bar{U}^2\,\overline{u'^2}] = \bar{C}_p^2\rho_a^2\bar{U}^2\,\overline{u'^2} \tag{4.16}$$

Equation 4.16 is a quasi-steady relationship between mean square pressure fluctuations and mean square longitudinal velocity fluctuations.

To predict peak pressures by the quasi-steady assumption,

$$\hat{p},\check{p} = C_{po}(1/2)\,\rho_a[\hat{U}^2] \cong \bar{C}_p(1/2)\rho_a[\hat{U}^2] \tag{4.17}$$

Thus according to the quasi-steady assumption, we can predict peak pressures (maxima and minima) by using mean pressure coefficients with a peak gust wind speed. This is the basis of many codes and standards that use a peak gust as the basic wind speed (see Chapter 15). Its main disadvantage is that building-induced pressure fluctuations (the second source described in Section 4.6.1) are ignored. Also, when applied to wind pressures over large areas, it is conservative because full correlation of the pressure peaks is implied.

4.6.3 Body-induced pressure fluctuations and vortex-shedding forces

The phenomena of separating shear layers and vortex shedding have already been introduced in Sections 4.1, 4.3.1, 4.4.1 and 4.5 in descriptions of the flow around some basic bluff-body shapes. These phenomena occur whether the flow upstream is turbulent or not, and the resulting surface pressure fluctuations on a bluff body can be distinguished from those generated by the flow fluctuations in the approaching flow.

The regular vortex shedding into the wake of a long bluff body results from the rolling up of the separating shear layers alternately on one side, then the other, and occurs on bluff bodies of all cross-sections. A regular pattern of decaying vortices, known as the von Karman vortex 'street', appears in the wake. Turbulence in the approaching flow tends to make the shedding less regular, but the strengths of the vortices are maintained, or even enhanced. Vibration of the body may also enhance the vortex strength, and the vortex-shedding frequency may change to the frequency of vibration, in a phenomenon known as *lock-in*.

As each vortex is shed from a bluff body, a strong cross-wind force is induced towards the side of the shed vortex. In this way, the alternate shedding of vortices induces a nearly harmonic (sinusoidal) cross-wind force variation on the structure.

For a given cross-sectional shape, the frequency of vortex shedding, n_s, is proportional to the approaching flow speed, and inversely proportional to the width of the body. It may be expressed in a non-dimensional form, known as the *Strouhal Number, St*.

$$St = \frac{n_s b}{\bar{U}} \tag{4.18}$$

where

b is the cross-wind body width

$\bar{U}$ is the mean flow speed

The Strouhal Number varies with the shape of the cross-section, and for circular and other cross-sections with curved surfaces, it varies with the Reynolds Number. Some representative values of the Strouhal Number for a variety of cross-sections are shown in Figure 4.24.

An inclined, a two-dimensional flat plate with an angle of attack, α, (Figure 4.7), has a Strouhal Number of about 0.15 based on the breadth, b, normal to the flow, or (0.15/sin α) based on the chord, c, where $b = c \cdot \sin\alpha$ (Chen and Fang, 1996).

The variation with Reynolds Number for a smooth circular cylinder is shown in Figure 4.25 (Scruton, 1963; Schewe, 1983). In the sub-critical range, up to a Reynolds Number of 2×10^5, the Strouhal Number is quite constant at a value of 0.20. In the critical Reynolds Number range, coinciding with the sharp fall in the drag coefficient (see Figure 4.16), the Strouhal Number jumps to 0.3 and then 0.48, although in this range, the vortex shedding is random, and not clearly defined. A slightly decreasing Strouhal Number to about 0.4,

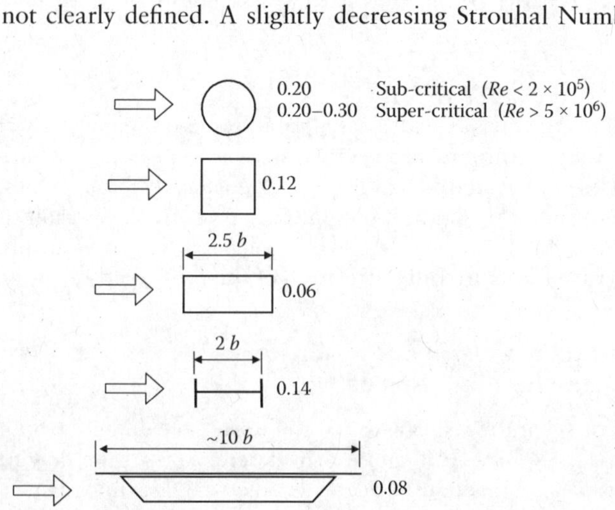

Figure 4.24 Strouhal Numbers for vortex shedding for various cross-sections.

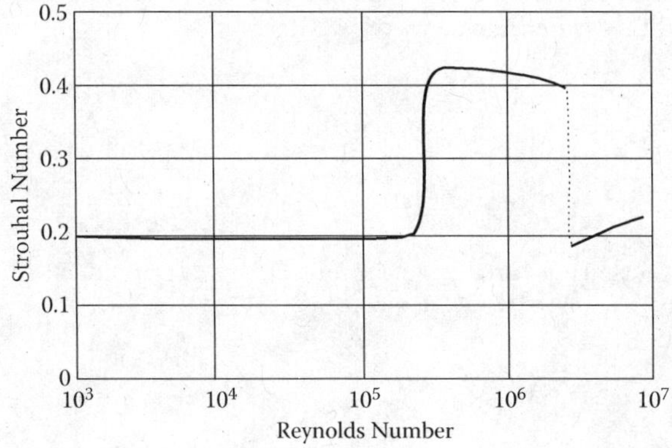

Figure 4.25 Strouhal Number versus Reynolds Number (based on diameter) for circular cylinders.

in the super-critical range, is followed by a fall to about 0.2 again, at a Reynolds Number of 2×10^6. Helical strakes (Figure 4.26) are often used to inhibit vortex shedding and the resulting cross-wind forces on structures with circular sections such as chimney stacks (Scruton and Walshe, 1957).

4.6.4 Fluctuating pressure and force coefficients

The rms fluctuating (standard deviation) pressure coefficient at a point on a bluff body is defined by

$$C_p' = \frac{\sqrt{\overline{p'^2}}}{\frac{1}{2}\rho_a \bar{U}^2} \tag{4.19}$$

$\sqrt{\overline{p'^2}}$ is the rms fluctuating, or standard deviation, pressure (also denoted by σ_p).

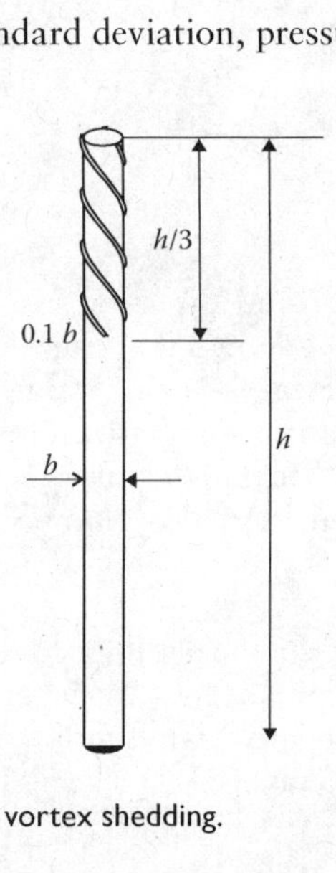

Figure 4.26 Helical strakes for inhibiting vortex shedding.

The rms fluctuating sectional force coefficient per unit length of a two-dimensional cylindrical or prismatic body is defined by

$$C'_f = \frac{\sqrt{\overline{f'^2}}}{(1/2)\rho_a \bar{U}^2 b} \tag{4.20}$$

$\sqrt{\overline{f'^2}}$ is the rms fluctuating force per unit length. b is a reference dimension – usually the cross-wind breadth.

For a whole body,

$$C'_F = \frac{\sqrt{\overline{F'^2}}}{\frac{1}{2}\rho_a \bar{U}^2 A} \tag{4.21}$$

$\sqrt{\overline{F'^2}}$ is the rms fluctuating force acting on the complete body. A is a reference area – usually the frontal area.

The total fluctuating force acting on a cylindrical body of finite length can be calculated from the fluctuating sectional force, knowing the correlation function, or correlation length.

With the quasi-steady assumption (Section 4.6.2), the rms fluctuating pressure coefficient can be estimated from Equations 4.16 and 4.19:

$$C'_p = \frac{\sqrt{\overline{p'^2}}}{\frac{1}{2}\rho_a \bar{U}^2} \approx \frac{\bar{C}_p \rho_a \bar{U} \sqrt{\overline{u'^2}}}{\frac{1}{2}\rho_a \bar{U}^2} = 2\bar{C}_p I_u \tag{4.22}$$

where I_u is the longitudinal turbulence intensity $\left(= \sqrt{\overline{u'^2}}/\bar{U} \text{ or } \sigma_u/\bar{U}\right)$ as defined in Section 3.3.1.

Similarly, the rms fluctuating drag coefficient can be estimated using the quasi-steady assumption:

$$C'_D \approx 2\bar{C}_D I_u \tag{4.23}$$

Fluctuating forces in the cross-wind direction are usually determined by an experiment, however. Measurements have shown that square cross-sections experience stronger cross-wind fluctuating forces due to vortex shedding, than do circular cross-sections. Figure 4.27 shows the variation of rms fluctuating cross-wind force per unit length, for a circular cylinder, as a function of Reynolds Number (Wootton and Scruton, 1970). The value is around 0.5 at sub-critical Reynolds Numbers, falling to much lower values in the critical and super-critical ranges, coinciding with a reduction in the drag coefficient (Section 4.5.1).

The fluctuating cross-wind force coefficient for a square cross-section with sharp corners is higher than that for a circular section, due to the greater strength of the shed vortices. In smooth flow, the rms fluctuating cross-wind force coefficient is about 1.3; this drops to about 0.7 in turbulent flow of 10% intensity (Vickery, 1966).

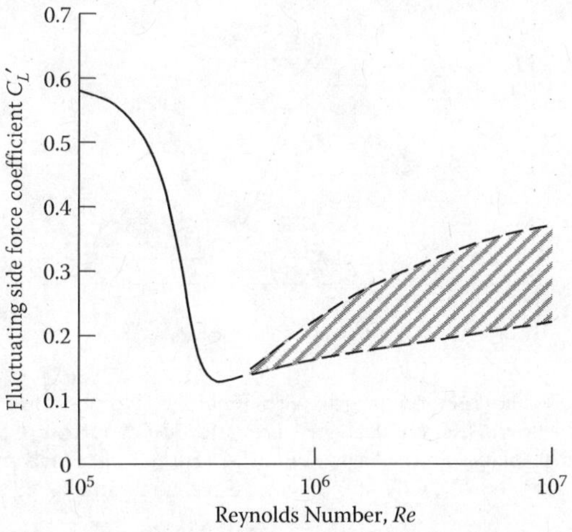

Figure 4.27 Variation of the fluctuating cross-wind force coefficient per unit length with Reynolds Number for a circular cylinder (smooth flow). (Reproduced by permission of Wootton, L.R. and Scruton, C.P. 1970. Aerodynamic stability. *Proceedings, CIRIA Seminar on the Modern Design of Wind-Sensitive Structures*, 18 June, 65 81, CIRIA, 6 Storey's Gate, London, UK, 65–81.)

4.6.5 Correlation length

The spatial correlation coefficient for fluctuating forces at two points along a cross-section is defined by

$$\rho = \frac{\overline{f_1'(t)f_2'(t)}}{\overline{f'^2}} = \frac{\overline{f_1'(t)f_2'(t)}}{\sigma_f^2} \tag{4.24}$$

where $f_1'(t)$, $f_2'(t)$ are the fluctuating forces per unit length at two sections along a cylindrical or prismatic body. (This was previously discussed in relation to atmospheric turbulence in Section 3.3.5.)

We have assumed that the mean square fluctuating force per unit length is constant along the body, so that:

$$\overline{f_1'^2} = \overline{f_2'^2} = \overline{f'^2}$$

As the separation distance, y, between the two sections 1 and 2 approaches zero, the correlation function, $\rho(y)$, approaches 1. As the separation distance becomes very large, $\rho(y)$ tends to zero; this means there is no statistical relationship between the fluctuating forces.

The *correlation length*, ℓ, is then defined as

$$\ell = \int_0^\infty \rho(y)\, dy \tag{4.25}$$

The correlation length is thus the area under the graph of $\rho(y)$ plotted against y.

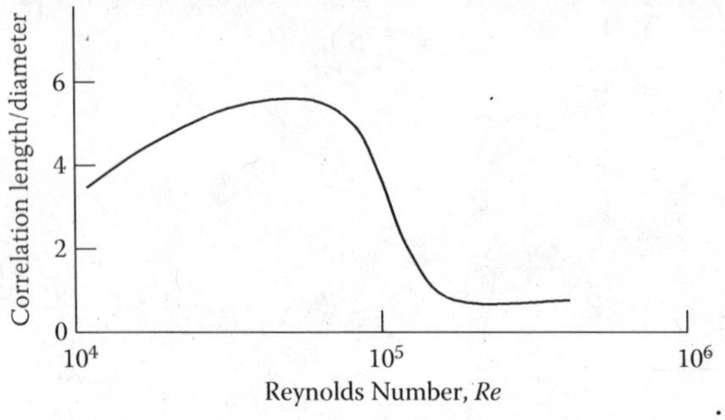

Figure 4.28 Variation of the correlation length with Reynolds Number for a stationary circular cylinder (smooth flow). (Reproduced by permission of Wootton, L.R. and Scruton, C.P. 1970. Aerodynamic stability. *Proceedings, CIRIA Seminar on the Modern Design of Wind-Sensitive Structures*, 18 June, 65–81, CIRIA, 6 Storey's Gate, London, UK, 65–81.)

Measurements of correlation length for a smooth circular cylinder in smooth flow are shown in Figure 4.28 (Wootton and Scruton, 1970). The correlation length falls from about five diameters to one diameter over the critical Reynolds Number range.

4.6.6 Total fluctuating forces on a slender body

Consider a long cylindrical, or prismatic, body of length, L, subjected to fluctuating wind forces along its length. Divide the body into a large number, N, of sections of width, δy_1, $\delta y_2, \ldots, \delta y_N$, as shown in Figure 4.29. Assume that the mean square fluctuating force is the same at all sections.

At any section, i, the total force per unit length can be separated into a mean, or time-averaged, component, and a fluctuating component with a zero mean:

$$f_i'(t) = \bar{f}_i + f_i' : (t) \tag{4.26}$$

The total mean force acting on the whole body is given by

$$\bar{F} = \Sigma \bar{F}_i \delta y_i$$

where the summation is taken from i equal to 1 to N.

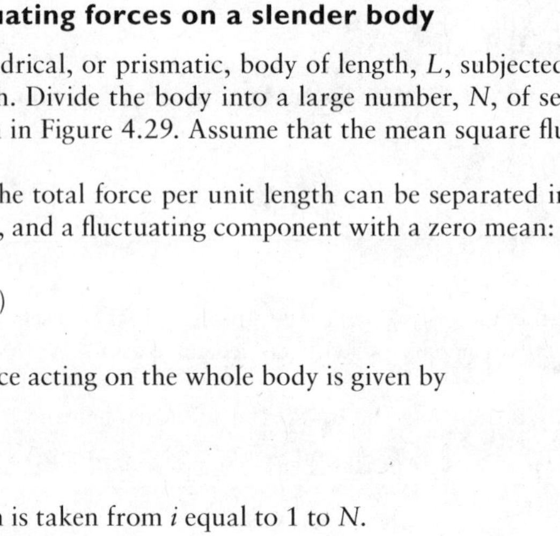

Figure 4.29 Sectional force fluctuations on a long cylinder.

As we let the number of sections tend to infinity, δy_i tends to zero, and the right-hand side becomes an integral:

$$\bar{F} = \int_0^L \bar{f}_i \, dy_i \tag{4.27}$$

The instantaneous fluctuating force on the body as a whole is

$$F'(t) = \Sigma fi'(t)\ \delta yi$$

$$= f_1'(t)\ \delta y_1 + f_2'(t)\ \delta y_2 + \cdots + f'_N\ (t)\ \delta y_N$$

Squaring both sides,

$$[F'(t)]^2 = [f_1'(t)\ \delta y_1 + f_2'(t)\ \delta y_2 + \cdots + f'_N(t)\ \delta y_N]^2$$

$$= [f_1'(t)\ \delta y_1]^2 + [f_2'(t)\ \delta y_2]^2 + \cdots + [f'_N(t)\ \delta y_N]^2 + f_1'(t)\ f_2'(t)\ \delta y_1 \delta y_2 + f_1'(t)\ f_3'(t)\ \delta y_1 \delta y_3 + \cdots$$

$$= \sum_i^N \sum_j^N f_i'(t) f_j'(t)\, \delta y_i \delta y_j$$

Now, taking the means (time averages) of both sides,

$$\overline{F'^2} = \sum_i^N \sum_j^N \overline{f_i'(t) f_j'(t)}\ \delta y_i \delta y_j \tag{4.28}$$

As δy_i, δy_j tend to zero,

$$\overline{F'^2} = \int_0^L \int_0^L \overline{f_i'(t) f_j'(t)}\, dy_i dy_j \tag{4.29}$$

Equations 4.28 and 4.29 are important equations, which illustrate how to obtain a total fluctuating force from the fluctuating force on small elements. The integrand in Equation 4.29 is the *covariance* of the sectional force fluctuations.

Now, assume that the integrand can be written in the form:

$$\overline{f_i'(t) f_j'(t)} = \overline{f'^2} \rho(y_i - y_j)$$

where $\rho(y_i - y_j)$ is the correlation coefficient for the fluctuating sectional forces, which is assumed to be a function of the separation distance, $(y_i - y_j)$, but not of the individual positions y_i and y_j – that is we assume that the wind forces are horizontally, or vertically, *homogeneous*.

$$\overline{F'^2} = \overline{f'^2(t)} \int_0^L \int_0^L \rho(y_i - y_j)\ dy_i dy_j \tag{4.30}$$

This is the fundamental equation for the total mean square fluctuating force on the body, in terms of the mean square fluctuating force per unit length.

By introducing a new variable equal to $(y_i - y_j)$, Equation 4.30 can be written as

$$\overline{F'^2} = \overline{f'^2} \int_0^L dy_j \int_{-y_j}^{L-y_j} \rho(y_i - y_j)\, d(y_i - y_j) \tag{4.31}$$

Equations 4.30 and 4.31 can be evaluated in two special cases:

i. *Full correlation*: This assumption implies that $\rho(y_i - y_j)$ equals 1 for all separations, $(y_i - y_j)$. Then Equation 4.30 becomes

$$\overline{F'^2} = \overline{f'^2(t)}\, L^2$$

In this case, the fluctuating forces are treated like static forces.

ii. *Rapidly decreasing correlation length*: In this case, ℓ is much less than L, and the second part of Equation 4.31 can be approximated by

$$\int_{-y_j}^{L-y_j} \rho(y_i - y_j)\, d(y_i - y_j) = \int_{-\infty}^{\infty} \rho(y_i - y_j)\, d(y_i - y_j) = 2\ell \text{ from Equation 4.25.}$$

Then from Equation 4.31,

$$\overline{F'^2} = \overline{f'^2(t)}\, L \cdot 2\ell \tag{4.32}$$

Thus, the mean square total fluctuating force is directly proportional to the correlation length, ℓ. This is an important result that is applicable to structures such as slender towers.

4.7 SUMMARY

This chapter has summarised the relevant aspects of bluff-body aerodynamics, which is itself a large subject with applications in many fields, to wind loads on structures. The basic fluid mechanics of stagnation, separation and wakes has been described, and pressure and force coefficients are defined. The characteristics of pressures and forces on the basic shapes of flat plates and walls, cubes and rectangular prisms and circular cylinders have been described. The effect of turbulence and the ground surface is covered.

Fluctuating pressures and forces, particularly those generated by upwind turbulence, and the regular shedding of vortices by a bluff body are discussed. The concept of correlation length and the averaging process by which fluctuating total forces on a body can be calculated are described.

REFERENCES

Achenbach, E. 1971. Influence of surface roughness on the cross-flow around a circular cylinder. *Journal of Fluid Mechanics*, 46: 321–35.

Adachi, T. 1995. The effect of surface roughness of a body in high Reynolds Number flow. *Journal of Rotating Machinery*, 1: 187–97.

Baines, W.D. 1963. Effects of velocity distributions on wind loads and flow patterns on buildings. *Proceedings, International Conference on Wind Effects on Buildings and Structures*, Teddington, UK, 26–28 June, pp. 197–225.

Bearman, P.W. 1971. An investigation of the forces on flat plates normal to a turbulent flow. *Journal of Fluid Mechanics*, 46: 177–98.

Bearman, P.W. and Trueman, D.H. 1972. An investigation of the flow around rectangular cylinders. *Aeronautical Quarterly*, 23: 229–37.

Chen, J.M. and Fang, Y.-C. 1996. Strouhal Numbers of inclined flat plates. *Journal of Wind Engineering and Industrial Aerodynamics*, 61: 99–102.

Cook, N.J. 1990. *The Designer's Guide to Wind Loading of Building Structures. Part 2: Static Structures*. Building Research Establishment and Butterworths, London.

ESDU. 1970. Fluid forces and moments on flat plates. Engineering Sciences Data Unit (ESDU International), ESDU Data Item 70015.

Fage, A. and Warsap, J.H. 1930. The effects of turbulence and surface roughness on the drag of a circular cylinder, Reports and Memoranda No. 1283, Aeronautical Research Council, UK.

Gartshore, I.S. 1973. The effects of freestream turbulence on the drag of rectangular two-dimensional prisms. Boundary Layer Wind Tunnel Report BLWT-4-73, University of Western Ontario.

Guven, O., Farell, C. and Patel, V.C. 1980. Surface-roughness effects on the mean flow past circular cylinders. *Journal of Fluid Mechanics*, 98: 673–701.

Holmes, J.D., Burton, D. and Fricke, H.W. 2012. Drag coefficients for roughened circular cylinders in super-critical flow. *15th Australasian Wind Engineering Workshop*, Sydney, Australia, 23–24 February.

Laneville, A., Gartshore, I.S. and Parkinson, G.V. 1975. An explanation of some effects of turbulence on bluff bodies. *Fourth International Conference on Wind Effects on Buildings and Structures*, London, UK, September.

Letchford, C.W. and Holmes, J.D. 1994. Wind loads on free-standing walls in turbulent boundary layers. *Journal of Wind Engineering and Industrial Aerodynamics*, 51: 1–27.

Jensen, M. 1958. The model law for phenomena in the natural wind. *Ingenioren*, 2: 121–8.

Macdonald, P.A., Kwok, K.C.S. and Holmes, J.D. 1988. Wind loads on circular storage bins, silos and tanks: I point pressure measurements on isolated structures. *Journal of Wind Engineering and Industrial Aerodynamics*, 31: 165–88.

Marchman, J.F. and Werme, T.D. 1982. Mutual interference drag on signs and luminaires. *ASCE Journal of the Structural Division*, 108: 2235–44.

Melbourne, W.H. 1995. Bluff body aerodynamics for wind engineering. In P. Krishna (ed.), *A State of the Art in Wind Engineering*. Wiley Eastern Limited, New Delhi.

Schewe, G. 1983. On the force fluctuations acting on a circular cylinder in crossflow from subcritical up to transcritical Reynolds Numbers. *Journal of Fluid Mechanics*, 133: 265–85.

Scruton, C. 1963. On the wind-excited oscillations of stacks, towers and masts. *Proceedings, International Conference on Wind Effects on Buildings and Structures*, Teddington, UK, 26–28 June, pp. 798–832.

Scruton, C. 1981. *An Introduction to Wind Effects on Structures*. Oxford University Press, Oxford, UK.

Scruton, C. and Rogers, E.W.E. 1972. Steady and unsteady wind loading of buildings and structures. *Philosophical Transactions, Royal Society, A*, 269: 353–83.

Scruton, C. and Walshe, D.E.J. 1957. A means for avoiding wind-excited oscillations of structures of circular or near-circular cross section. National Physical Laboratory (UK), NPL. Aero Report 335 (unpublished).

Szechenyi, E. 1975. Supercritical Reynolds Number simulation for two-dimensional flow over circular cylinders. *Journal of Fluid Mechanics*, 70: 529–42.
Vickery, B.J. 1966 Fluctuating lift and drag on a long cylinder of square cross-section in a smooth and turbulent flow. *Journal of Fluid Mechanics*, 25: 481–94.
Wootton, L.R. and Scruton, C.P. 1970. Aerodynamic stability. *Proceedings, CIRIA Seminar on the Modern Design of Wind-Sensitive Structures*, 18 June, 65–81, CIRIA, 6 Storey's Gate, London, UK, 65–81.

Chapter 5

Resonant dynamic response and effective static load distributions

5.1 INTRODUCTION

Owing to the turbulent nature of the wind velocities in storms of all types, the wind loads acting on structures are also highly fluctuating. There is a potential to excite the resonant dynamic response for structures, or parts of structures, with natural frequencies less than about 1 Hz. The resonant response of a structure introduces the complication of a time-history effect, in which the response at any time depends not just on the instantaneous wind gust velocities, or pressures, acting along the structure, but also on the previous time history of wind gusts or pressures.

This chapter will introduce the principles and analysis of the dynamic response to wind. Some discussion of aeroelastic and fatigue effects is included. Also, in this chapter, the method of equivalent or effective static wind-loading distributions is introduced.

Treatment of the dynamic response is continued in Chapters 9 through 12 on tall buildings, large roofs and sports stadiums, slender towers and masts and bridges, with emphasis on the particular characteristics of these structures. In Chapter 15, code approaches to the dynamic response are considered.

5.2 PRINCIPLES OF DYNAMIC RESPONSE

The fluctuating nature of wind velocities, pressures and forces, as discussed in Chapters 3 and 4, may cause the excitation of a significant resonant vibratory response in structures or parts of structures, provided their natural frequencies and damping are low enough. This resonant dynamic response should be distinguished from the background-fluctuating response to which all structures are subjected. Figure 5.1 shows the response spectral density of a dynamic structure under wind loading; the area under the entire curve represents the total mean-square-fluctuating response (note that the mean response is not included in this plot). The resonant responses in the first two modes of vibration are shown hatched in this diagram. The background response, largely made up of low-frequency contributions below the lowest natural frequency of vibration, is the largest contributor in Figure 5.1, and, in fact, is usually the dominant contribution in the case of along-wind loading. Resonant contributions become more and more significant, and will eventually dominate, as structures become taller or longer in relation to their width, and their natural frequencies become lower.

Figure 5.2a shows the characteristics of the time histories of an along-wind (drag) force; the structural response for a structure with a *high* fundamental natural frequency is shown in Figure 5.2b, and the response with a *low* natural frequency is shown in Figure 5.2c. In the former case, the resonant, or vibratory component, clearly plays a minor role in the

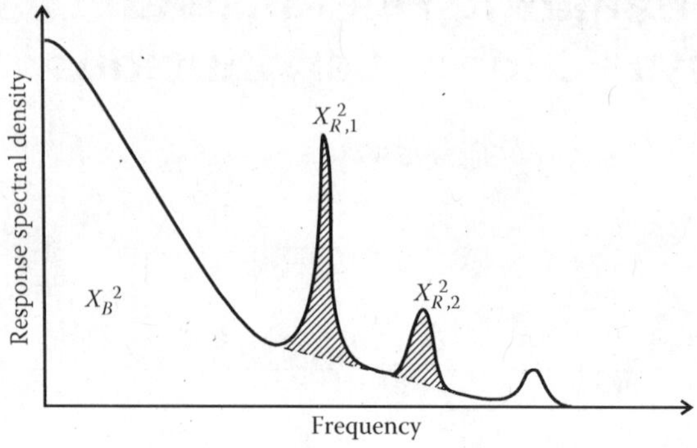

Figure 5.1 Response spectral density for a structure with significant resonant contributions.

response, which generally closely follows the time variation of the exciting forces. However, in the latter case, the resonant response, in the fundamental mode of vibration, is important, although the response in higher modes than the first case can usually be neglected.

In fact, the majority of structures fall into the category of Figure 5.2b, and will *not* experience a significant resonant dynamic response. A well-known rule of thumb is that the lowest natural frequency should be below 1 Hz for the resonant response to be significant. However, the amount of resonant response also depends on the damping, aerodynamic or structural, present. For example, high-voltage transmission lines usually have fundamental sway frequencies which are well below 1 Hz; however, the *aerodynamic* damping is very high – typically around 25% of the critical – so that the resonant response is largely damped out. Lattice towers, because of their low mass, also have high aerodynamic damping ratios. Slip-jointed steel-lighting poles have high *structural* damping due to friction at the joints – this energy-absorbing mechanism will limit the resonant response to wind.

Resonant response, when it does occur, may occasionally produce complex interactions, in which the movement of the structure itself results in additional aeroelastic forces being produced (Section 5.5). In some extreme cases, for example, the Tacoma Narrows Bridge failure of 1940 (see Chapter 1), catastrophic failure has resulted. These are exceptional cases, which of course must be avoided, but in the majority of structures with significant resonant dynamic response, the dynamic component is superimposed on a significant, or dominant, mean and background-fluctuating response.

The two major sources of fluctuating wind loads are discussed in Section 4.6. The first and obvious source, exciting resonant dynamic response, is the natural unsteady or turbulent flow in the wind, produced by shearing actions as the air flows over the rough surface of the earth, as discussed in Chapter 3. The other main source of fluctuating loads is the alternate vortex shedding which occurs behind bluff cross-sectional shapes, such as circular cylinders or square cross-sections. A further source is buffeting forces from the wakes of other structures upwind of the structure of interest.

When a structure experiences resonant dynamic response, counter-acting structural forces come into play to balance the wind forces:

- Inertial forces proportional to the mass of the structure.
- Damping or energy-absorbing forces; in their simplest form, these are proportional to the velocity, but this is not always the case.
- Elastic or stiffness forces proportional to the deflections or displacements.

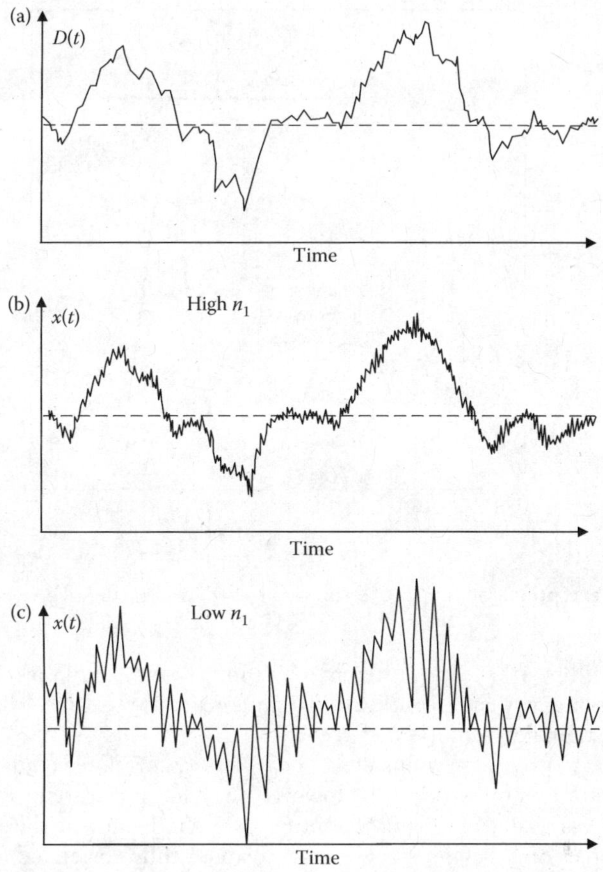

Figure 5.2 Time histories of (a) wind force, (b) response of a structure with a high natural frequency, and (c) response of a structure with a low natural frequency.

When a structure does respond dynamically, i.e. the resonant response is significant, an important principle to remember is that the condition of the structure, that is stresses, deflections, at any given time depends not only on the wind forces acting at the time, but also on the *past history* of wind forces. In the case of quasi-static loading, the structure directly responds to the forces acting instantaneously at any given time.

The effective load distribution due to the resonant part of the loading (Section 5.4.4) is given to a good approximation by the distribution of inertial forces along the structure. This is based on the assumption that the fluctuating wind forces at the resonant frequency approximately balance the damping forces once a stable amplitude of vibration is established.

At this point, it is worth noting the essential differences between the dynamic response of structures to wind and earthquake. The main differences between the excitation forces due to these two natural phenomena are:

1. Earthquakes are of a much shorter duration than wind storms (with the possible exception of the passage of a tornado), and are thus treated as transient loadings.
2. The predominant frequencies of the earthquake ground motions are typically 10–50 times those of the frequencies in fully developed wind storms. This means that

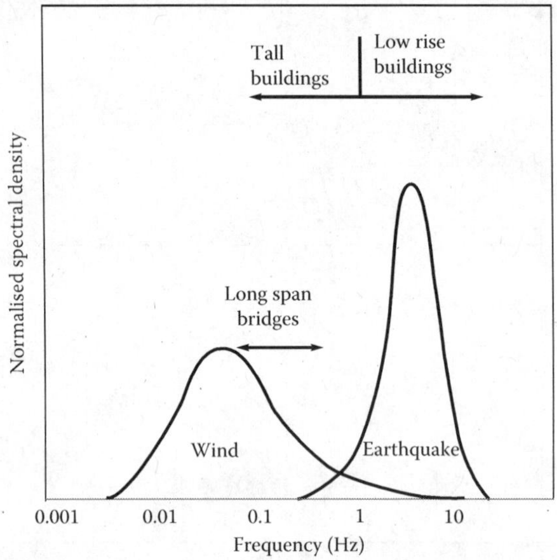

Figure 5.3 Dynamic excitation frequencies of structures by wind and earthquake.

structures will be affected in different ways, for example, buildings in a certain height range may not experience significant dynamic response to wind loadings, but may be prone to earthquake excitation.

3. The earthquake ground motions will appear as *fully* correlated equivalent forces acting over the height of a tall structure. However, the eddy structure in wind storms results in *partially* correlated wind forces acting over the height of the structure. Vortex-shedding forces on a slender structure are also not fully correlated over the height.

Figure 5.3 shows the various frequency ranges for excitation of structures by wind and earthquake actions.

5.3 THE RANDOM VIBRATION OR SPECTRAL APPROACH

In some important papers in the 1960s, A.G. Davenport outlined an approach to the wind-induced vibration of structures based on random vibration theory (Davenport, 1961, 1963, 1964). Other significant early contributions to the development of this approach were made by R.I. Harris (1963) and B.J. Vickery (1965, 1966).

The approach uses the concept of the stationary random process to describe wind velocities, pressures and forces. This assumes that the complexities of nature are such that we can never describe, or predict, perfectly (or 'deterministically') the forces generated by wind storms. However, we are able to use averaged quantities such as standard deviations, correlations and spectral densities (or 'spectra') to describe the main features of both the exciting forces and the structural response. The *spectral density*, which has already been introduced in Section 3.3.4 and Figure 5.1, is the most important quantity to be considered in this approach, which primarily uses the *frequency domain* to perform calculations, and is alternatively known as the *spectral approach*.

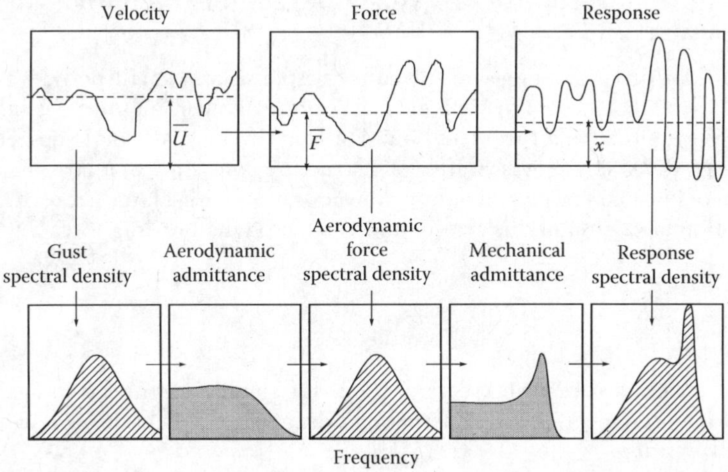

Figure 5.4 The random vibration (frequency domain) approach to the resonant dynamic response. (From Davenport, A.G. 1963. The buffeting of structures by gusts. *Proceedings, International Conference on Wind Effects on Buildings and Structures*, Teddington, UK, 26–28 June, 358–91. With permission.)

Wind speeds, pressures and the resulting structural response have generally been treated as stationary random processes in which the time-averaged or mean component is separated from the fluctuating component. Thus

$$X(t) = \overline{X} + x'(t) \tag{5.1}$$

where $X(t)$ denotes a wind velocity component, a pressure (measured with respect to a defined reference static pressure) or a structural response such as the bending moment, stress resultant, deflection and so on; $\overline{X}$ is the mean or time-averaged component; and $x'(t)$ is the fluctuating component such that $\overline{x'(t)} = 0$. If x is a response variable, $x'(t)$ should include any resonant dynamic response resulting from excitation of any natural modes of vibration of the structure.

Figure 5.4 (after Davenport, 1963) graphically illustrates the elements of the spectral approach. The main calculations are done in the bottom row, in which the total mean-square-fluctuating response is computed from the spectral density, or 'spectrum', of the response. The latter is calculated from the spectrum of the aerodynamic forces, which are, in turn, calculated from the wind turbulence, or gust spectrum. The frequency-dependent *aerodynamic* and *mechanical admittance* functions form links between these spectra. The amplification at the resonant frequency, for structures with a low fundamental frequency, will result in a higher mean-square-fluctuating and peak response, than is the case for structures with a higher natural frequency, as previously illustrated in Figure 5.2.

The use of stationary random processes and Equation 5.1 is appropriate for large-scale wind storms such as gales in temperate latitudes and tropical cyclones. It may not be appropriate for some short-duration, transient storms, such as downbursts or tornadoes associated with thunderstorms. Methods for these types of storms are still under development.

5.3.1 Along-wind response of a single-degree-of-freedom structure

We will consider first the along-wind dynamic response of a small body, whose dynamic characteristics are represented by a simple mass-spring damper (Figure 5.5), and which does not disturb the approaching turbulent flow significantly. This is a single-degree-of-freedom system, and is reasonably representative of a structure consisting of a large mass supported by a column of low mass, such as a lighting tower or mast with a large array of lamps on top.

The equation of motion of this system under an aerodynamic drag force, $D(t)$, is given by Equation 5.2

$$m\ddot{x} + c\dot{x} + kx = D(t) \tag{5.2}$$

The quasi-steady assumption (Section 4.6.2) for small structures allows the following relationship between mean-square-fluctuating drag force, and fluctuating longitudinal wind velocity to be written:

$$\overline{D'^2} = C_{Do}^2 \rho_a^2 \bar{U}^2 \overline{u'^2} A^2 \cong \overline{C_D}^2 \rho_a^2 \bar{U}^2 \overline{u'^2} A^2 = \frac{4\bar{D}^2}{\bar{U}^2} \overline{u'^2} \tag{5.3}$$

Equation 5.3 is analogous to Equation 4.16 for pressures.

Writing Equation 5.3 in terms of spectral density,

$$\int_0^\infty S_D(n) \cdot dn = \frac{4\bar{D}^2}{\bar{U}^2} \int_0^\infty S_u(n) \cdot dn$$

Hence,

$$S_D(n) = \frac{4\bar{D}^2}{\bar{U}^2} S_u(n) \tag{5.4}$$

To derive the relationship between the fluctuating force, and the response of the structure, represented by the simple dynamic system of Figure 5.5, the deflection is first separated into mean and fluctuating components, as in Equation 5.1:

$$X(t) = \bar{X} + x'(t) \tag{5.1}$$

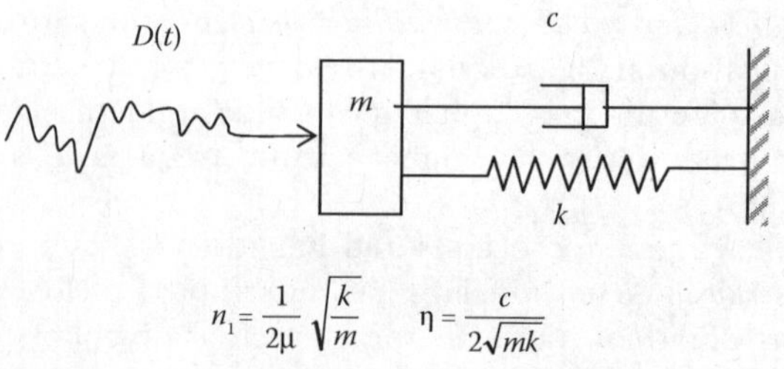

Figure 5.5 Simplified dynamic model of a structure.

The relationship between mean drag force, $\bar{D}$, and mean deflection, $\bar{X}$, is as follows:

$$\bar{D} = k\bar{X} \tag{5.5}$$

where k is the spring stiffness in Figure 5.5.

The spectral density of the deflection is related to the spectral density of the applied force as follows:

$$S_x(n) = \frac{1}{k^2}|H(n)|^2 S_D(n) \tag{5.6}$$

where $|H(n)|^2$ is known as the *mechanical admittance* for the single-degree-of-freedom dynamic system under consideration, given by Equation 5.7.

$$|H(n)|^2 = \frac{1}{\left[1-(n/n_1)^2\right]^2 + 4\eta^2(n/n_1)^2} \tag{5.7}$$

$|H(n)|$, that is the square root of the mechanical admittance, may be recognised as the *dynamic amplification factor*, or *dynamic magnification factor*, which arises when the response of a single-degree-of-freedom system to a harmonic, or sinusoidal, excitation force is considered. n_1 is the undamped natural frequency, and η is the ratio of the damping coefficient, c, to critical damping, as shown in Figure 5.5.

By combining Equations 5.4 and 5.6, the spectral density of the deflection response can be related to the spectral density of the wind velocity fluctuations.

$$S_x(n) = \frac{1}{k^2}|H(n)|^2 \frac{4\bar{D}^2}{\bar{U}^2} S_u(n) \tag{5.8}$$

Equation 5.8 applies to structures which have small frontal areas in relation to the length scales of atmospheric turbulence.

For larger structures, the velocity fluctuations do not occur simultaneously over the windward face and their correlation over the whole area, A, must be considered. To allow for this effect, an *aerodynamic admittance*, $X^2(n)$, is introduced.

$$S_x(n) = \frac{1}{k^2}|H(n)|^2 \frac{4\bar{D}^2}{\bar{U}^2} \cdot X^2(n) \cdot S_u(n)$$

Substituting for $\bar{D}$ from Equation 5.5,

$$S_x(n) = \frac{4\bar{X}^2}{\bar{U}^2}|H(n)|^2 \cdot X^2(n) \cdot S_u(n) \tag{5.9}$$

For open structures, such as lattice frame towers, which do not disturb the flow greatly, $X^2(n)$ can be determined from the correlation properties of the upwind velocity fluctuations (see Section 3.3.6). This assumption is also made for solid structures, but $X^2(n)$ has also been obtained experimentally.

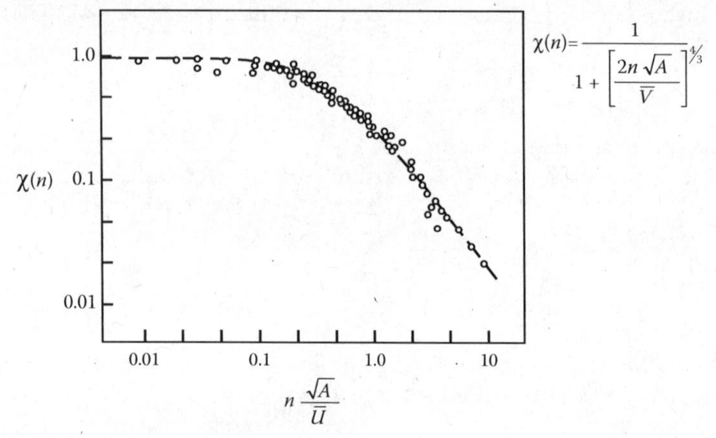

Figure 5.6 Aerodynamic admittance – experimental data and the fitted function. (From Vickery, B.J. 1965. On the flow behind a coarse grid and its use as a model of atmospheric turbulence in studies related to wind loads on buildings. Aero Report 1143, National Physical Laboratory, UK; Vickery, B.J. 1968. *ASCE Journal of the Engineering Mechanics Division*, 94: 31–46.)

Figure 5.6 shows some experimental data with an empirical function fitted. Note that $X(n)$ tends towards 1.0 at low frequencies and for small bodies. The low-frequency gusts are nearly fully correlated, and fully envelope the face of a structure. For high frequencies, or very large bodies, the gusts are ineffective in producing total forces on the structure, due to their lack of correlation, and the aerodynamic admittance tends towards zero.

To obtain the mean-square-fluctuating deflection, the spectral density of deflection, given by Equation 5.9, is integrated over all frequencies.

$$\sigma_x^2 = \int_0^\infty S_x(n)\cdot dn = \int_0^\infty \frac{4\bar{X}^2}{\bar{U}^2}\left|H(n)\right|^2 \cdot X^2(n)\cdot S_u(n)\cdot dn \tag{5.10}$$

The area underneath the integrand in Equation 5.10 can be approximated by two components, B and R, representing the 'background' and resonant parts, respectively (Figure 5.7).

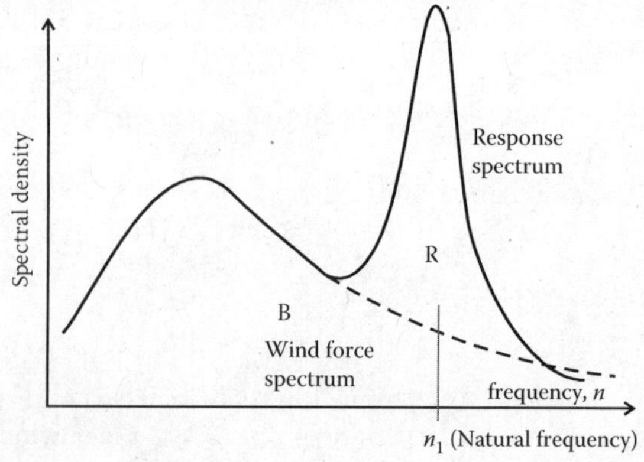

Figure 5.7 Background and resonant components of response.

Thus,

$$\sigma_x^2 = \frac{4\bar{X}^2\sigma_u^2}{\bar{U}^2}\int_0^\infty |H(n)|^2 \cdot X^2(n) \cdot \frac{S_u(n)}{\sigma_u{}^2} \cdot dn \cong \frac{4\bar{X}^2\sigma_u^2}{\bar{U}^2}[\mathrm{B}+\mathrm{R}] \tag{5.11}$$

where

$$\mathrm{B} = \int_0^\infty X^2(n) \cdot \frac{S_u(n)}{\sigma_u^2} \cdot dn \tag{5.12}$$

$$\mathrm{R} = X^2(n_1) \cdot \frac{S_u(n_1)}{\sigma_u^2}\int_0^\infty |H(n)|^2 \cdot dn \tag{5.13}$$

The approximation of Equation 5.11 is based on the assumption that over the width of the resonant peak in Figure 5.7, the functions $X^2(n)$, $S_u(n)$ are constant at the values $X^2(n_1)$, $S_u(n_1)$. This is a good approximation for the flat spectral densities characteristic of wind loading, and when the resonant peak is narrow, it occurs when the damping is low (Ashraf Ali and Gould, 1985). The integral $\int |H(n)|^2 \cdot dn$ integrated for n from 0 to ∞ can be evaluated by the method of poles (Crandall and Mark, 1963) and shown to be equal to $(\pi n_1/4\eta)$.

The approximation of Equation 5.11 is widely used in code methods of evaluating along-wind response, and is discussed further in Chapter 15.

The background factor, B, represents the quasi-static response caused by gusts below the natural frequency of the structure. Importantly, it is independent of frequency, as shown by Equation 5.12, in which the frequency appears only in the integrand, and is thus 'integrated out'. For many structures under wind loading, B is considerably greater than R, that is the background response is dominant in comparison with the resonant response. An example of such a structure is that whose response is shown in Figure 5.2b.

5.3.2 Gust response factor

A commonly used term in wind engineering is *gust response factor.* The term *gust-loading factor* was used by Davenport (1967), and *gust factor* was used by Vickery (1966). These essentially have the same meaning, although sometimes, the factor is applied to the effective applied loading, and sometimes to the response of the structure. The term 'gust factor' is better applied to the wind speed itself (Section 3.3.3).

The *gust response factor,* G, may be defined as the ratio of the expected maximum response (e.g. deflection or stress) of the structure in a defined time period (e.g. 10 min or 1 h), to the mean, or time-averaged response, in the same time period. It really only has meaning in stationary or near-stationary winds such as those generated by large-scale synoptic wind events such as gales from depressions in temperate latitudes, or tropical cyclones (see Chapter 2).

The expected maximum response of the simple system described in Section 5.3.1 can be written as

$$\hat{X} = \bar{X} + g\sigma_x$$

where g is a *peak factor*, which depends on the time interval for which the maximum value is calculated, and the frequency range of the response.

From Equation 5.11,

$$G = \frac{\hat{X}}{\overline{X}} = 1 + g\frac{\sigma_x}{\overline{X}} = 1 + 2g\frac{\sigma_u}{\overline{U}}\sqrt{B + R} \tag{5.14}$$

Equation 5.14 or variations of it are used in several codes and standards for wind loading, and for simple estimations of the along-wind dynamic loading of structures. The usual approach is to calculate G for the modal coordinate in the first mode of vibration, a_1, and then to apply it to a mean load distribution on the structure, from which all responses, such as bending moments, are calculated. This is an approximate approach which works reasonably well for some structures and load effects, such as the base-bending moment of tall buildings. However, in other cases, it gives significant errors and should be used with caution (e.g. Holmes, 1994; Vickery, 1995 – see also Chapter 11).

5.3.3 Peak factor

The along-wind response of structures to wind has a probability distribution which is closely Gaussian. For this case, Davenport (1964) derived the following expression for the expected peak factor, g:

$$g = \sqrt{2\log_e(\nu T)} + \frac{0.577}{\sqrt{2\log_e(\nu T)}} \tag{5.15}$$

where ν is the 'cycling rate' or effective frequency for the response; this is often conservatively taken as the natural frequency, n_1. T is the time interval over which the maximum value is required.

5.3.4 Dynamic response factor

In transient or non-stationary winds such as downbursts from thunderstorms, for example, the use of a gust factor, or gust response factor, is rather meaningless. The gust response factor is also meaningless in cases when the mean response is very small or zero (such as cross-wind response). In these cases, the use of a 'dynamic response factor' is more appropriate. This approach has been recently adopted in some codes and standards for wind loading. The dynamic response factor may be defined in the following way:

Dynamic response factor = (maximum response including resonant and correlation effects)/(maximum response calculated ignoring both resonant and correlation effects).

The denominator is in fact the response calculated using 'static' methods in codes and standards. The dynamic response factor defined as above, will usually have a value close to 1. A value greater than 1 can only be caused by a significant resonant response.

The use of the gust response factor and dynamic response factor in wind-loading codes and standards will be discussed further in Chapter 15.

5.3.5 Influence coefficient

When considering the action of a time-dependent and spatially varying load such as wind loading on a continuous structure, the *influence coefficient* or *influence line* is an important

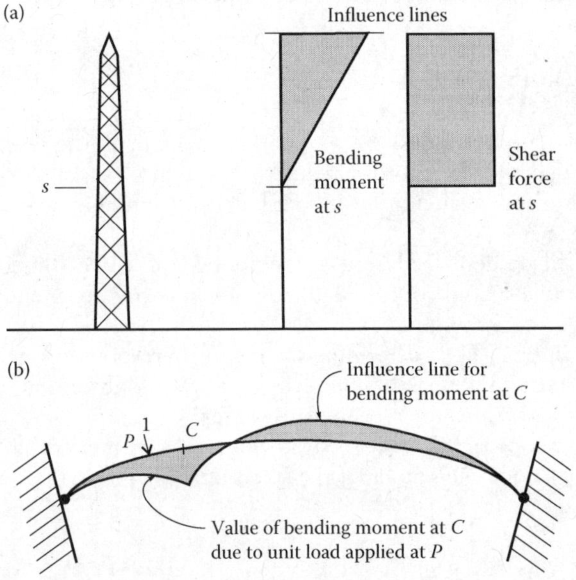

Figure 5.8 Examples of influence lines for (a) a tower, and (b) an arch roof.

parameter. To appreciate the need for this, we must understand the concept, familiar to structural designers, of 'load effect'. A load effect is not the load itself but a parameter resulting from the loading which is required for comparison with the design criteria. Examples are internal forces or moments such as bending moments or shear forces, stresses or deflections. The influence line represents the value of a single-load effect as a unit (static) load is moved around the structure.

Two examples of influence lines are given in Figure 5.8. Figure 5.8a shows the influence lines for the bending moment and shear force at a level, *s*, halfway up a lattice tower. These are relatively simple functions; in the case of the shear force, loads (or wind pressures) above the level *s* have a uniform effect on the shear force at that level. The influence line for the bending moment linearly varies from unity at the top to zero at the level *s*; thus, wind pressures at the top of the structure have a much larger effect than those lower down, on the bending moment, which, in turn, is closely related to the axial forces in the leg members of the tower. It should be noted that loads or wind pressures below the level *s* have *no* effect on the shear force or bending moment at that level.

Figure 5.8b shows the influence line for the bending moment at a point in an arch roof. In this case, the sign of the influence line changes along the arch. Thus, wind pressures applied in the same direction at different parts of the roof may have opposite effects on the bending moment at C, M_c.

It is important to take into account these non-uniform influences when considering the structural effects of wind loads, even for apparently simple structures, especially for the fluctuating part of the loading.

5.3.6 Along-wind response of a structure with distributed mass: modal analysis

The usual approach to the calculation of the dynamic response of multi-degree-of-freedom structures to dynamic forces, including resonance effects, is to expand the complete displacement response as a summation of components associated with each of the natural modes of vibration:

$$x(z,t) = \Sigma a_j(t)\phi_j(z) \tag{5.16}$$

where j denotes the natural modes; z is a spatial coordinate on the structure; $a_j(t)$ is a time-varying modal (or generalised) coordinate and $\phi_j(z)$ is a mode shape for the jth mode.

Modal analysis is discussed in most texts on structural dynamics (e.g. Clough and Penzien, 1975; Warburton, 1976).

The approach will be described here in the context of a two-dimensional or 'line-like' structure, with a single spatial coordinate, z, but it can easily be extended to more complex geometries.

Equation 5.16 can be used to determine the complete response of a structure to random forcing, that is including the mean component, $\bar{x}$, and the sub-resonant (background) fluctuating component, as well as the resonant responses.

The result of this approach is that separate equations of motion can be written for the modal coordinate $a_j(t)$, for each mode of the structure:

$$G_j\ddot{a}_j + C_j\dot{a}_j + K_j a_j = Q_j(t) \tag{5.17}$$

where

G_j is the generalised mass equal to $\int_0^L m(z)\phi_j^2(z)\,dz$
$m(z)$ is the mass per unit length along the structure
L is the length of the structure
C_j is the modal damping $(=2\eta_j G_j \omega_j)$
K_j is the modal stiffness
η_j is the damping as a fraction of critical for the jth mode
ω_j is the natural undamped circular frequency for the jth mode $(=2\pi n_j = \sqrt{(K_j/G_j)})$
$Q_j(t)$ is the generalised force, equal to $\int_0^L f(z,t)\phi_j(z)\,dz$
$f(z,t)$ is the force per unit length along the structure

$f(z,t)$ can be taken as along-wind or cross-wind forces. For along-wind forces, applying a 'strip' assumption, which relates the forces on a section of the structure with the flow conditions upstream of the section, it can be written as

$$f(z,t) = C_d(z)\cdot b(z)\frac{1}{2}\rho_a U^2(z,t) \tag{5.18}$$

where

$C_d(z)$ is a local drag coefficient
$b(z)$ is the local breadth

$u(z,t)$ is the longitudinal velocity upstream of the section. If the structure is moving, this should be a relative velocity, which then generates an aerodynamic damping force (Section 5.5.1 and Holmes, 1996a). However, at this point, we will assume the structure is stationary, in which case, $u(z,t)$ can be written as

$$U(z,t) = \bar{U}(z) + u'(z,t)$$

where $u'(z,t)$ is the fluctuating component of longitudinal velocity (zero mean).

Then from Equation 5.18,

$$f(z,t) = C_d(z) \cdot b(z)\rho_a\left[\frac{1}{2}\bar{U}^2(z) + \bar{U}(z)u'(z,t) + \frac{1}{2}u'^2(z,t)\right]$$

Neglecting the third term within the square brackets, the fluctuating sectional along-wind force is given by

$$f'(z,t) = C_d(z) \cdot b(z)\rho_a\bar{U}(z)u'(z,t)$$

and the instantaneous fluctuating generalised force is therefore

$$Q_j'(t) = \int_0^L f'(z,t)\varphi_j(z)\,dz = \int_0^L C_d(z) \cdot b(z)\rho_a\bar{U}(z)u'(z,t)\phi_j(z)\,dz$$

Applying the same procedure used in Section 4.6.6, the mean-square-generalised force is

$$\overline{Q_j'^2} = \int_0^L\int_0^L \overline{f'(z_1)f'(z_2)}\,\phi_j(z_1)\phi_j(z_2)\,dz_1\,dz_2$$

$$= \rho_a{}^2\int_0^L\int_0^L \overline{u'(z_1)u'(z_2)}\,C_d(z_1)\cdot C_d(z_2)b(z_1)b(z_2)\bar{U}(z_1)\bar{U}(z_2)\phi_j(z_1)\phi_j(z_2)\,dz_1\,dz_2$$

This can be simplified for a uniform cross-section, with $C_d(z)$ and $b(z)$ constant with z:

$$\overline{Q_j'^2} = (\rho_a C_d b)^2\int_0^L\int_0^L \overline{u'(z_1)u'(z_2)}\bar{U}(z_1)\bar{U}(z_2)\phi_j(z_1)\phi_j(z_2)\,dz_1\,dz_2 \tag{5.19}$$

where $\overline{u'(z_1)u'(z_2)}$ is the covariance for the fluctuating velocities at heights z_1 and z_2. If the standard deviation of velocity fluctuations is constant with z, then the covariance can be written as

$$\overline{u'(z_1)u'(z_2)} = \sigma_u{}^2\rho_{uu}(z_1,z_2)$$

where $\rho_{uu}(z_1,z_2)$ is the correlation coefficient for fluctuating velocities at heights z_1 and z_2, defined in Section 3.3.5.

The spectral density of $Q'_j(t)$ can be obtained in an analogous way to the mean square value:

$$S_{Qj}(n) = (\rho C_d b)^2\int_0^L\int_0^L Co(z_1,z_2,n)\,\bar{U}(z_1)\bar{U}(z_2)\phi_j(z_1)\phi_j(z_2)\,dz_1\,dz_2 \tag{5.20}$$

where $Co(z_1, z_2, n)$ is the co-spectral density of the longitudinal velocity fluctuations (Section 3.3.6) (defined in random process theory – e.g. Bendat and Piersol (2010)).

Analogously with Equation 5.6, the spectral density of the modal coordinate $a_j(t)$ is given by

$$S_{aj}(n) = \frac{1}{K_j^2}\left|H_j(n)\right|^2 S_{Qj}(n) \tag{5.21}$$

where the mechanical admittance for the jth mode is

$$\left|H_j(n)\right|^2 = \frac{1}{\left[1-(n/n_j)^2\right]^2 + 4\eta_j^2(n/n_j)^2} \tag{5.22}$$

The mean square value of $a_j(t)$ can then be obtained by integration of Equation 5.21 with respect to frequency:

$$\overline{a_j'^2} = \int_0^\infty S_{aj}(n)\cdot dn$$

Applying Equation 5.16, the mean square displacement is obtained from:

$$\overline{x'^2} = \sum_{j=1}^{N}\sum_{k=1}^{N}\overline{a_j' a_k'}\phi_j(z)\phi_k(z)$$

If cross-coupling between modes can be neglected (however, see Section 5.3.7), the above equation becomes:

$$\overline{x'^2} = \sum_{j=1}^{N}\overline{a_j'^2}\phi_j^2(z) \tag{5.23}$$

The mean square value of any other response, r, (e.g. bending moment, stress) can similarly be obtained if the response, R_j for a unit value of the modal coordinate, a_j, is known. That is

$$\overline{r'^2} = \sum_{j=1}^{N}\overline{a_j'^2}R_j^2 \tag{5.24}$$

5.3.7 Along-wind response of a structure with distributed mass: separation of background and resonant components

In the case of wind loading, the method described in the previous section is not an efficient one. For the vast majority of structures, the natural frequencies are at the high end of the range of forcing frequencies from wind loading. Thus, the resonant components as j

increases in Equation 5.16 become very small. However, the contributions to the mean and background-fluctuating components for $j > 1$ in Equation 5.16 *may not be small*. Thus, it is necessary to include higher modes ($j > 1$) in Equation 5.16 not for their resonant contributions, but to accurately determine the mean and background contributions. For example, Vickery (1995) found that more than 20 modes were required to determine the mean value of a response, and more than 10 values were needed to compute the variance. Also, for the background response, cross-coupling of modes cannot be neglected, that is Equation 5.23 is not valid.

A much more efficient approach is to separately compute the mean and background components, as for a quasi-static structure. Thus, the total peak response, $\hat{r}$, can be taken to be

$$\hat{r} = \bar{r} + \sqrt{\hat{r}_B^2 + \sum_j (\hat{r}_{R,j}^2)} \tag{5.25}$$

where $\hat{r}_B$ is the peak background response equal to $g_B\sigma_B$; and $\hat{r}_{R,j}$ is the peak resonant response computed for the jth mode, equal to $g_j\sigma_{R,j}$. This approach is illustrated in Figure 5.1.

g_B and g_j are peak factors which can be determined from Equation 5.15; in the case of the resonant response, the cycling rate, ν, in Equation 5.15, can be taken as the natural frequency, n_j.

The mean square value of the quasi-static fluctuating (background) value of any response, r, is

$$\begin{aligned} \overline{r_B'^2} = \sigma_B^2 &= \int_0^L \int_0^L \overline{f'(z_1)f'(z_2)}\, I_r(z_1)I_r(z_2)\,dz_1\,dz_2 \\ &= \rho_a^2 \int_0^L \int_0^L \overline{u'(z_1)u'(z_2)}\, C_d(z_1)\cdot C_d(z_2)b(z_1)b(z_2)\bar{U}(z_1)\bar{U}(z_2)I_r(z_1)I_r(z_2)\,dz_1\,dz_2 \end{aligned} \tag{5.26}$$

where $I_r(z)$ is the influence line for r, that is the value of r when a unit load is applied at z.

The resonant component of the response in mode j can be written, to a good approximation, as

$$\overline{r_{R,j}'^2} = \frac{S_{Qj}(n_j)R_j^2}{K_j^2}\int_0^\infty \left|H_j(n)\right|^2 \cdot dn = \frac{\pi n_1 \cdot S_{Qj}(n_j)R_j^2}{4\eta_j K_j^2} \tag{5.27}$$

because the integral $\int_0^\infty \left|H_j(n)\right|^2 \cdot dn$, evaluated by the method of poles (Crandall and Mark, 1963), is equal to $(\pi n_j/4\eta_j)$.

5.3.8 Along-wind response to non-stationary (transient) winds

It is clear that downburst winds as generated by severe thunderstorms produce time histories which are non-stationary, as shown in Figure 5.9. Calculation of the dynamic response to such winds requires a different approach to those described earlier in this chapter, for turbulent winds generated within the boundary layers of synoptic winds – which can be considered statistically stationary. One such approach has been ‘borrowed’ from earthquake engineering.

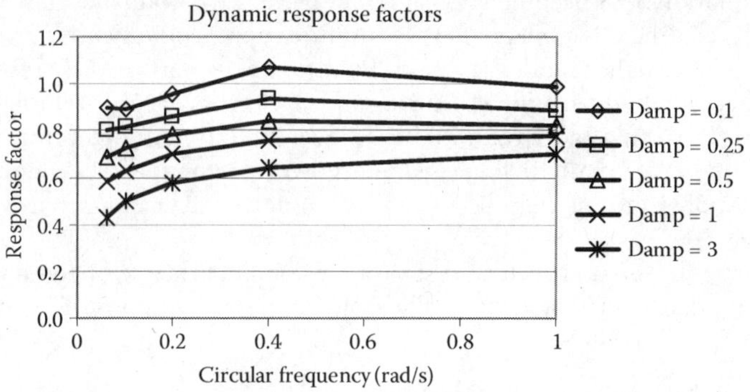

Figure 5.9 Dynamic response factors for the response of structures of various frequencies and damping to the Lubbock downburst of 2002. (From Holmes, J.D., Forristall, G. and McConochie, J. 2005. Dynamic response of structures to thunderstorm winds. *10th Americas Conference on Wind Engineering*, Baton Rouge, Louisiana, 1–4 June, 2005.)

The use of Duhamel's Integral is a standard technique for calculation of the dynamic response of structures, to transient loadings such as blast loadings, or earthquakes. Since it represents the response to an arbitrary loading as the superposition of the response to many discrete impulses, this technique is limited to linear structures. However, structures with non-linear characteristics (e.g. stiffness and damping) can usually be linearised with sufficient accuracy to make use of this very convenient technique.

The displacement response of any linear system to an arbitrary force input $D(t)$ can be written as

$$x(t) = \int_0^t h(t-\tau) \cdot D(\tau)\, d\tau \tag{5.28}$$

where $h(t)$ is the unit impulse response function. Equation 5.28 is a 'convolution integral'.

The unit impulse response function for a simple mass-spring-damper system (Figure 5.5), with an equation of motion given by Equation 5.2, depends on the value of the damping ratio, η, and the natural circular frequency, ω_1, where $\eta = c/2\sqrt{(mk)}$, and $\omega_1 = \sqrt{(k/m)}$.

For $\eta < 1$,

$$h(t) = \left(\frac{1}{m \cdot \omega_1\sqrt{1-\eta^2}}\right)\exp\left[-\eta\omega_1 t\right] \cdot \sin\left[\omega_1\sqrt{1-\eta^2} \cdot (t)\right] \tag{5.29}$$

Hence, from Equation 5.28,

$$x(t) = \left(\frac{1}{m \cdot \omega_1\sqrt{1-\eta^2}}\right)\int_0^t \exp\left[-\eta\omega_1(t-\tau)\right] \cdot \sin\left[\omega_1\sqrt{1-\eta^2} \cdot (t-\tau)\right] \cdot D(\tau)\, d\tau \tag{5.30}$$

The right-hand side of Equation 5.30 is known as 'Duhamel's Integral', for example, Clough and Penzien (1975).

The effective quasi-static along-wind force, $D_{eff}(t)$, is then given by the product of the displacement response, $x(t)$, and the stiffness, k:

$$D_{eff}(t) = k \cdot x(t) = \left(\frac{\omega_1}{\sqrt{1-\eta^2}} \right) \int_0^t \exp[-\eta\omega_1(t-\tau)] \cdot \sin\left[\omega_1\sqrt{1-\eta^2} \cdot (t-\tau)\right] \cdot D(\tau) d\tau \qquad (5.31)$$

The dynamic response factor (Section 5.3.4) can then be obtained as the ratio of the maximum value of the effective static force, $D_{eff}(t)$, in the time history, to the maximum value of the applied force $D(t)$ in the same time history. Note that these maxima will generally not occur at the same time, t.

Figure 5.9 shows calculated dynamic response factors to the downburst measured in 2002 at Lubbock, Texas (Figure 3.9), for structures with periods of 6–100 s (circular frequencies of 0.06–1 rad/s), and damping ratios from 0.1 to 3 (Holmes et al., 2005). For structures with shorter periods, more closely spaced time intervals are required in the recorded wind time histories. The information in Figure 5.9 closely resembles that provided in the 'response spectra' for an earthquake design.

5.4 EFFECTIVE STATIC LOADING DISTRIBUTIONS

5.4.1 Introduction

Effective static wind load distributions are those loadings that produce the correct expected values of peak load effects, such as bending moments, axial forces in members or deflections, generated by the fluctuating wind loading. The effective peak-loading distributions associated with the mean wind loading, the fluctuating quasi-static or background response and the resonant response are identified, and combined to give a total effective peak wind-loading distribution.

Following the procedure described in the previous sections, effective static peak-loading distributions can be separately derived for the following three components:

1. Mean component
2. Background or sub-resonant component
3. Resonant components

The background component is derived by making use of a formula derived by Kasperski and Niemann (1992), and depends on the load effect in question. The resonant component comprises an inertial loading, similar to that used in earthquake engineering.

The approach will be illustrated by examples of buildings with long-span roofs and free-standing lattice towers and chimneys. Simplifications will be suggested to make the method more palatable to structural engineers used to analysing and designing with static loadings.

The main advantage of the effective static load distribution approach is that the distributions can be applied to static structural analysis computer programmes for use in a detailed structural design. The approach can be applied to any type of structure (Holmes and Kasperski, 1996).

5.4.2 Mean load distributions

The mean wind loading on a structure which does not distort the air flow significantly can be obtained simply by relating the mean local pressure or force per unit length to the mean wind speed. Thus, for the mean along-wind force per unit height acting on a tower:

$$\bar{f}(z) = [(\tfrac{1}{2})\,\rho_a \bar{U}(z)^2]\; C_d b(z) \tag{5.32}$$

where ρ_a is the density of air; $\bar{U}(z)$ is the mean wind speed at height z; C_d is a drag coefficient and $b(z)$ is the reference breadth at the height z.

The mean value of any load effect (e.g. shear force, bending moment and deflection) can be obtained by integrating the local load with the influence line over the appropriate height. However, if the purpose is to derive an equivalent static loading, then Equation 5.32 is already in this form.

In the case of 'solid' structures (such as cooling towers and most buildings) with at least two dimensions comparable to the size of turbulent eddies in the atmosphere, Equation 5.32 cannot be used, but wind-tunnel tests can be employed to determine mean pressure coefficients, $\bar{C}_p$, which can then be used with a reference wind speed, $\bar{U}_h$, to determine local mean pressures on the structure:

$$\bar{p}(z) = [(\tfrac{1}{2})\,\rho_a \bar{U}_h^2]\; \bar{C}_p \tag{5.33}$$

5.4.3 Background-loading distributions

As discussed previously, the background wind loading is the quasi-static loading produced by fluctuations due to turbulence, but with frequencies too low to excite any resonant response. Over the duration of a wind storm, because of the incomplete correlations of pressures at various points on a structure, loadings varying both in space and time will be experienced. It is necessary to identify those instantaneous loadings which produce the critical load effects in a structure. The formula which enables this to be done is the 'Load-Response Correlation' formula derived by Kasperski and Niemann (1992).

This formula gives the expected 'instantaneous' pressure distribution associated with the maximum or minimum load effect. Thus, for the maximum value, $\hat{r}$, of a load effect, r:

$$(p_i)_{\hat{r}} = \bar{p}_i + g_B \cdot \rho_{r,pi} \cdot \sigma_{pi} \tag{5.34}$$

where $\bar{p}_i$ and σ_{pi} are the mean and rms pressures at point or panel, i. $\rho_{r,pi}$ is the correlation coefficient between the fluctuating load effect, and the fluctuating pressure at point i (this can be determined from the correlation coefficients for the fluctuating pressures at all points on the tributary area, and from the influence coefficients); and g_B is the peak factor for the background response which normally lies in the range 2.5–5.

A simple example of the application of this formula is given in Appendix F.

The second term on the right-hand side of Equation 5.34 represents the background-fluctuating load distribution. This term can also be written in the form of a continuous distribution:

$$f_B(z) = g_B\, \rho(z)\, \sigma_p(z) \tag{5.35}$$

where $\rho(z)$ denotes the correlation coefficient between the fluctuating load at position z on the structure, and the load effect of interest; and $\sigma_p(z)$ is the rms fluctuating load at position z.

In Equation 5.34, the correlation coefficient, $\rho_{r,pi}$, can be shown to be given by

$$\rho_{r,pi} \sum_k \left[\overline{p_i(t) \quad p_k(t)}\; I_k \right] \Big/ (\sigma_{pi}\ \sigma_r) \tag{5.36}$$

where I_k is the influence coefficient for a pressure applied at position, k.

The standard deviation of the structural load effect, σ_r, is given by (Holmes and Best, 1981)

$$\sigma_r^2 = \sum_i \sum_k \overline{p_i(t) \quad p_k(t)}\; I_i I_k \tag{5.37}$$

When the continuous form is used, Equations 5.36 and 5.37 are replaced by an integral form (Holmes, 1996b):

$$\rho(z) = \frac{\int_s^h \overline{f'(z)\, f'(z_1)}\; I_r(z_1) b(z_1)\, dz_1}{\left\{ \int_s^h \int_s^h \overline{f'(z_1) f'(z_2)}\; I_r(z_1) I_r(z_2)\, b(z_1)\, b(z_2)\, dz_1\, dz_2 \right\}^{(1/2)} \sqrt{\overline{f'^2(z)}}} \tag{5.38}$$

where $I_r(z)$ now denotes the influence function for the load effect, r, as a function of position z, and $b(z)$ is the breadth of the structure at position z. For a vertical structure, the integrations in Equation 5.38 are carried out for the height range from s, the height at which the load effect (e.g. bending moment, shearing force, and member force) is being evaluated, and the top of the structure, h.

Clearly, since the correlation coefficient, $\rho_{r,pi}$, calculated by Equation 5.36, or $\rho(z)$ calculated by Equation 5.38, is dependent on the particular load effect, then the background load distribution will also depend on the nature of the load effect.

Figures 5.10 and 5.11 give examples of background-loading distributions calculated using these methods. Figure 5.10 shows examples of peak load (mean+background) distributions for a support reaction (dashed) and a bending moment (dotted) in an arch roof. These distributions fall within an envelope formed by the maximum and minimum pressure distributions along the arch. It should also be noted that the distribution for the bending moment at C includes a region of positive pressure.

Figure 5.11 shows the background pressure distribution for the base shear force and base-bending moment on a lattice tower 160 m high, determined by calculation using Equation 5.35 (Holmes, 1996b). The maxima for these distributions occur at around 70 m height for the base shear and about 120 m for the base-bending moment. An approximation (Holmes, 1996b) to these distributions, which is independent of the load effect but dependent on the height at which the load effect is evaluated, is also shown in Figure 5.11.

5.4.4 Load distributions for resonant response (single resonant mode)

The equivalent load distribution for the resonant response in the first mode can be represented as a distribution of inertial forces over the length of the structure. Thus, an equivalent load distribution for the resonant response, $f_R(z)$, is given by

$$f_R(z) = g_R m(z)\, (2\pi n_1)^2 \sqrt{\overline{a'^2}}\, \phi_1(z) \tag{5.39}$$

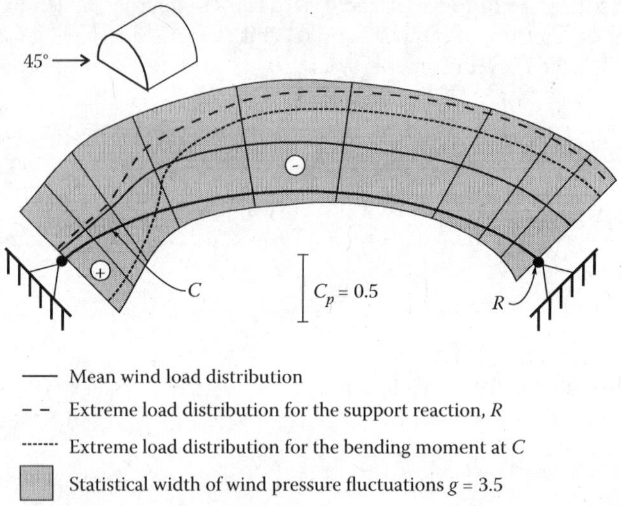

Figure 5.10 Mean and effective background-loading distributions for an arch roof. (After Kasperski, M. and Niemann, H.-J. 1992. *Journal of Wind Engineering and Industrial Aerodynamics*, 43: 1753 – 63.)

where g_R is the peak factor for resonant response; $m(z)$ is a mass per unit length; n_1 is the first mode natural frequency; $\sqrt{\overline{a'^2}}$ is the rms modal coordinate (only resonant contribution) and $\phi_1(z)$ is the mode shape for the first mode of vibration.

Determination of the rms modal coordinate requires knowledge of the spectral density of the exciting forces, the correlation of these forces at the natural frequency (or aerodynamic admittance) and the modal damping and stiffness, as discussed in Sections 5.3.1 and 5.3.6.

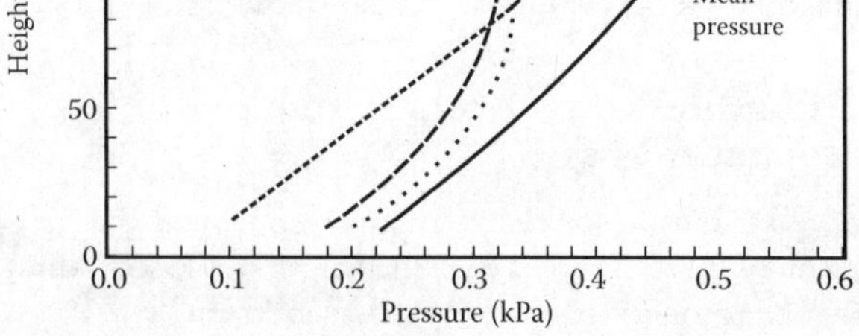

Figure 5.11 Mean and effective background load distributions for a 160-m tower. (After Holmes, J.D. 1996b. *Engineering Structures*, 18: 489–94.)

5.4.5 Combined load distribution

The combined effective static load distribution for mean, background and resonant components (one mode) is obtained as follows:

$$f_c(z) = \bar{f}(z) + W_{back}\, f_B(z) + W_{res}(z)\, f_R(z) \tag{5.40}$$

where the absolute values of the weighting factors W_{back} and W_{res} are given by

$$\left|W_{back}\right| = \frac{g_B \sigma_{r,B}}{(g_B^2 \sigma_{r,B}^2 + g_R^2 \sigma_{r,R}^2)^{1/2}} \qquad \left|W_{res}\right| = \frac{g_R \sigma_{r,R}}{(g_B^2 \sigma_{r,B}^2 + g_R^2 \sigma_{r,R}^2)^{1/2}} \tag{5.41}$$

Equation 5.40 assumes that the fluctuating background and resonant components are uncorrelated with each other, so that Equation 5.25 applies. W_{back} and W_{res} will be positive if the both influence line of the load effect, r, and the mode shape are all positive, but either of them could be negative in many cases.

By multiplying the influence coefficient and summing over the whole structure, Equation 5.40 will give Equation 5.25 for the total peak load effect.

An alternative to Equation 5.40 is to combine the background and resonant distributions in the same way that the load effects themselves were combined (Equation 5.25), that is

$$f_c(z) = \bar{f}(z) + \sqrt{[f_B(z)]^2 + [f_R(z)]^2} \tag{5.42}$$

The second term on the right-hand side is an approximation to the correct combination formula (Equation 5.40), and is independent of the load effect or its influence line. Equation 5.42 with positive and negative signs taken in front of the square root is, in fact, an 'envelope' of the combined distributions for all load effects. However, it is a good approximation for cases where the influence line $I_r(z)$ and the mode shape have the same sign for all z (Holmes, 1996b).

Examples of the combined distribution, calculated using Equation 5.42, are given in Figure 5.12 for a 160-m lattice tower (Holmes, 1996b). When the resonant component is included, the combined loading can exceed the 'peak gust pressure envelope', that is the expected limit of non-simultaneous peak pressures, as is the case in Figure 5.12 for the bending moment at 120 m.

Equations 5.40 and 5.41 can be extended to cover more than one resonant mode by introducing an additional term for each participating mode of vibration. An example of combined equivalent static load distributions, when more than one resonant mode contributes significantly, is discussed in Section 12.3.4.

5.5 AEROELASTIC FORCES

For very flexible, dynamically wind-sensitive structures, the motion of the structure may itself generate aerodynamic forces. In extreme cases, the forces may be of such a magnitude and act in a direction to sustain, or increase, the motion; in these cases, an unstable situation may arise such that a small disturbance may initiate a growing amplitude of vibration. This is known as 'aerodynamic instability' – examples of which are the 'galloping' of iced-up transmission lines and the flutter of long suspension bridges (such as the Tacoma Narrows Bridge failure of 1940).

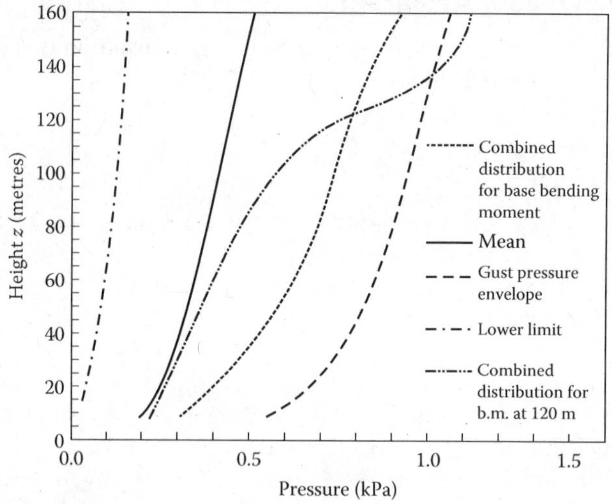

Figure 5.12 Combined mean, background and resonant load distributions for a 160- m tower. (From Holmes, J.D. 1996b. *Engineering Structures*, 18: 489–94.)

On the other hand, 'aerodynamic damping' forces may act to reduce the amplitude of vibration induced by wind. This is the case with the along-wind vibration of tall structures such as lattice towers of relatively low mass.

The subject of aeroelasticity and aerodynamic stability is a complex one, and one which most engineers will not need to be involved with. However, some discussion of the principles will be given in this section. A number of general reviews are available for this aspect of wind loads (e.g. Scanlan, 1982).

5.5.1 Aerodynamic damping

Consider the along-wind motion of a structure with a square cross-section, as shown in Figure 5.13. Initially ignoring the effects of turbulence, we will consider only the mean wind speed, $\bar{U}$. If the body itself is moving in the along-wind direction with a velocity, $\dot{x}$, the relative velocity of the air with respect to the moving body is $(\bar{U} - \dot{x})$. We then have a drag force per unit length of the structure equal to

$$D = C_D \frac{1}{2}\rho_a b(\bar{U} - \dot{x})^2 \cong C_D \frac{1}{2}\rho_a b\bar{U}^2\left(1 - \frac{2\dot{x}}{\bar{U}}\right)$$

$$= C_D \frac{1}{2}\rho_a b\bar{U}^2 - C_D\rho_a b\bar{U}\dot{x}$$

for small values of $\dot{x}/\bar{U}$. The second term on the right-hand side is a quantity proportional to the structure velocity, $\dot{x}$, and this represents a form of damping. When transferred to the left-hand side of the equation of motion (Equation 5.2), it combines with the structural damping term, $c\dot{x}$, to reduce the aerodynamic response.

For a continuous structure, the along-wind aerodynamic damping coefficient in mode j can be shown to be (Holmes, 1996a):

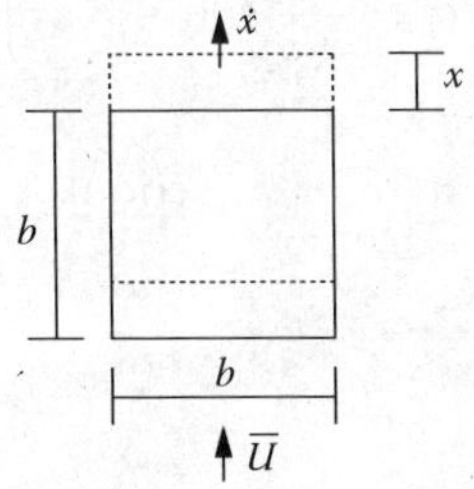

Figure 5.13 Along-wind relative motion and aerodynamic damping.

$$C_{\text{aero},j} = \rho_a \int_0^L C_d(z)b(z)\bar{U}(z)\phi_j^2(z)\,dz$$

giving a critical aerodynamic damping ratio, $\eta_{\text{aero},j}$, equal to

$$\eta_{\text{aero},j} = \frac{\rho_a \int_0^L C_d(z)b(z)\bar{U}(z)\phi_j^2(z)\,dz}{4\pi n_j G_j} \tag{5.43}$$

where G_j is the generalized mass (see Equation 5.17).

5.5.2 Galloping

Galloping is a form of single-degree-of-freedom aerodynamic instability, which can occur for long bodies with certain cross-sections. It is a pure translational, cross-wind vibration. Consider a section of a body with a square cross-section as shown in Figure 5.14.

The aerodynamic force per unit length, in the z-direction, is obtained from the lift and drag by a change of axes (Figure 4.3).

$$F_z = D\sin\alpha + L\cos\alpha = \frac{1}{2}\rho_a\bar{U}^2 b(C_D \sin\alpha + C_L \cos\alpha)$$

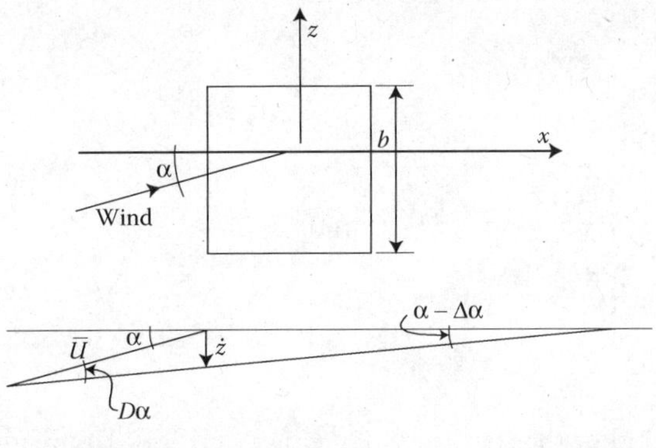

Figure 5.14 Cross-wind relative motion and galloping.

$$\text{Hence, } \frac{dF_z}{d\alpha} = \frac{1}{2}\rho_a\bar{U}^2 b\left(C_D\cos\alpha + \frac{dC_D}{d\alpha}\sin\alpha - C_L\sin\alpha + \frac{dC_L}{d\alpha}\cos\alpha\right)$$

Setting α equal to zero (for flow in the x-direction),

$$\frac{dF_z}{d\alpha} = \frac{1}{2}\rho_a\bar{U}^2 b\left(C_D + \frac{dC_L}{d\alpha}\right) \tag{5.44}$$

If the body is moving in the z-direction with velocity, $\dot{z}$, there will be a reduction in the apparent angle of attack of the flow by $\dot{z}/\bar{U}$, or an increase in the angle of attack by $-\dot{z}/\bar{U}$.

From Equation 5.44,

$$\Delta F_z \cong \frac{1}{2}\rho_a\bar{U}^2 b\left(C_D + \frac{dC_L}{d\alpha}\right)\Delta\alpha$$

Substituting $\Delta\alpha = -\dot{z}/\bar{U}$,

$$\begin{aligned}\Delta F_z &\cong \frac{1}{2}\rho_a\bar{U}^2 b\left(C_D + \frac{dC_L}{d\alpha}\right)\left(-\frac{\dot{z}}{\bar{U}}\right)\\ &= -\frac{1}{2}\rho_a\bar{U}b\left(C_D + \frac{dC_L}{d\alpha}\right)\dot{z}\end{aligned} \tag{5.45}$$

If $(C_D + dC_L/d\alpha) < 0$, there will be an aerodynamic force in the z-direction, proportional to the velocity of the motion, $\dot{z}$, or a *negative* aerodynamic damping term when it is transposed to the left-hand side of the equation of motion. This is known as 'den Hartog's criterion'.

This situation can arise for a square section, which has a negative slope $(dC_L/d\alpha)$, with a magnitude greater than C_D, for α equal to zero (Figure 5.14).

Galloping instability will be initiated when the negative aerodynamic force overcomes the positive damping force due to structural damping. Hence, a critical wind speed for galloping instability, $\bar{U}_{crit}$, can be determined by equating the negative damping force from Equation 5.45, to the positive velocity-dependent structural damping:

$$\frac{1}{2}\rho_a\bar{U}_{crit}b\left(C_D + \frac{dC_L}{d\alpha}\right)\dot{z} + c\dot{z} = 0$$

Hence,

$$\begin{aligned}\bar{U}_{crit} &= \frac{-2c}{\rho_a b\left(C_D + \dfrac{dC_L}{d\alpha}\right)} = \frac{-2\eta_{struct}c_c}{\rho_a b\left(C_D + \dfrac{dC_L}{d\alpha}\right)}\\ \bar{U}_{crit} &= \frac{-8\pi\eta_{struct}n_1 m}{\rho_a b\left(C_D + \dfrac{dC_L}{d\alpha}\right)}\end{aligned} \tag{5.46}$$

where η_{struct} is the ratio of the structural damping to critical damping ($c_c = 4\pi m n_1$), m is the mass per unit length of the body and n_1 is the natural frequency in the cross-wind (z) direction.

Equation 5.46 can be re-written as

$$\frac{\bar{U}_{\text{crit}}}{n_1 b} = Sc \cdot \frac{2}{-\left(C_d + \dfrac{dC_l}{d\alpha}\right)} \tag{5.47}$$

where Sc is known as the 'Scruton Number', or 'mass-damping' parameter:

$$Sc = \frac{4\eta_{\text{struct}} m}{\rho_a b^2} \tag{5.48}$$

Low values of Scruton Number are a general indication of a propensity to wind-induced vibrations (see also Sections 11.5.1 and 11.5.2 which are concerned with the vortex-induced vibrations of towers and chimneys).

Equation 5.46 or 5.47 can be used to estimate the wind speed for the initiation of galloping instability, knowing the various aerodynamic and structural parameters on the right-hand side.

5.5.3 Flutter

Consider now a two-dimensional bluff body able to move, with elastic restraint, in both vertical translation and rotation (i.e. bending and torsion deflections).

The body shown in Figure 5.15 is being twisted, and the section shown is rotating with an angular velocity, $\dot{\theta}$, radians per second. This gives the relative wind, with respect to the rotating body, a vertical component of velocity at the leading edge of $\dot{\theta}d/2$ and hence a relative angle of attack between the apparent wind direction and the rotating body of $-\dot{\theta}d/2\bar{U}$. This effective angle of attack can generate both a vertical force, and a moment if the centre of pressure is not collinear with the centre of rotation of the body. These aeroelastic forces can generate instabilities, if they are not completely opposed by the structural damping in the rotational mode. Aerodynamic instabilities involving rotation are known as 'flutter', using aeronautical parlance, and are a potential problem with the suspended decks of long-span bridges.

The equations of motion (per unit mass or moment of inertia) for the two degrees of freedom of a bluff body can be written as (Scanlan and Tomko, 1971; Scanlan and Gade, 1977; Matsumoto, 1996)

$$\ddot{z} + 2\eta_z\omega_z\dot{z} + \omega_z^2 z = \frac{F_z(t)}{m} + H_1\dot{z} + H_2\dot{\theta} + H_3\theta \tag{5.49}$$

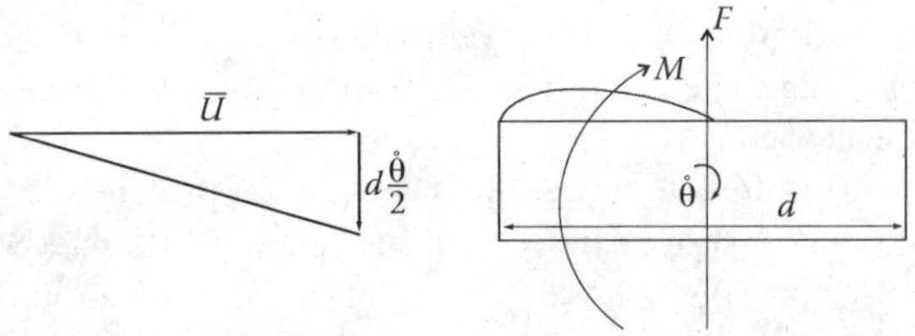

Figure 5.15 Aeroelastic forces generated by rotation of a cross-section.

Table 5.1 Types of aerodynamic instabilities

Name	*Conditions*	*Type of motion*	*Type of section*
Galloping	$H_1 > 0$	Translational	Square section
'Stall' flutter	$A_2 > 0$	Rotational	Rectangle, *H*-section
'Classical' flutter	$H_2 > 0, A_1 > 0$	Coupled	Flat plate, airfoil

Coupled aeroelastic instabilities in relation to long-span bridge decks, and flutter derivatives, are further discussed in Chapter 12.

$$\ddot{\theta} + 2\eta_\theta\omega_\theta\dot{\theta} + \omega_\theta^2\theta = \frac{M(t)}{I} + A_1\dot{z} + A_2\dot{\theta} + A_3\theta \tag{5.50}$$

The terms A_i and H_i are linear aeroelastic coefficients, or *flutter derivatives*, which are usually determined experimentally for particular cross-sections. They are functions of non-dimensional or *reduced* frequency. $F_z(t)$ and $M(t)$ are forces and moments due to other mechanisms which act on a static body (e.g. turbulent buffeting or vortex shedding). ω_z ($=2\pi n_z$) and ω_θ ($=2\pi n_\theta$) are the undamped circular frequencies in still air for vertical motion and rotation, respectively.

Note that Equations 5.49 and 5.50 have been 'linearised', that is they only contain terms in $\dot{z}$, θ, $\dot{\theta}$ and so on. There could be smaller terms in $\dot{z}^2, \theta^2, \dot{\theta}^3$ and so on. The two equations are 'coupled' second-order linear differential equations. The coupling arises from the occurrence of terms in z and θ, or their derivatives in both equations. This can result in coupled aeroelastic instabilities, which are a combination of vertical (bending) and rotational (torsion) motions, depending on the signs and magnitudes of the A_i and H_I derivatives. All bridge decks will reach this state at a high-enough wind speed.

Several particular types of instability for bluff bodies have been defined. Three of these are summarised in Table 5.1.

5.5.4 Lock-in

Motion-induced forces can occur during vibration produced by vortex shedding (Section 4.6.3). Through a feedback mechanism, the frequency of the shedding of vortices can 'lock-in' to the frequency of motion of the body. The strength of the vortices shed, and the fluctuating forces resulting are also enhanced. *Lock-in* has been observed many times during the vibration of lightly damped cylindrical structures such as steel chimneys, and occasionally during the vortex-induced vibration of long-span bridges.

5.6 FATIGUE UNDER WIND LOADING

5.6.1 Metallic fatigue

The 'fatigue' of metallic materials under cyclic loading has been well researched, although the treatment of fatigue damage under the random dynamic loading characteristic of wind loading is less well developed.

In the usual engineering failure model for the fatigue of metals, it is assumed that each cycle of a sinusoidal stress response inflicts an increment of damage which depends on the amplitude of the stress. Each successive cycle then generates additional damage which accumulates in proportion to the number of cycles until failure occurs. The results of constant amplitude fatigue tests are usually expressed in the form of an S–N curve, where S is the

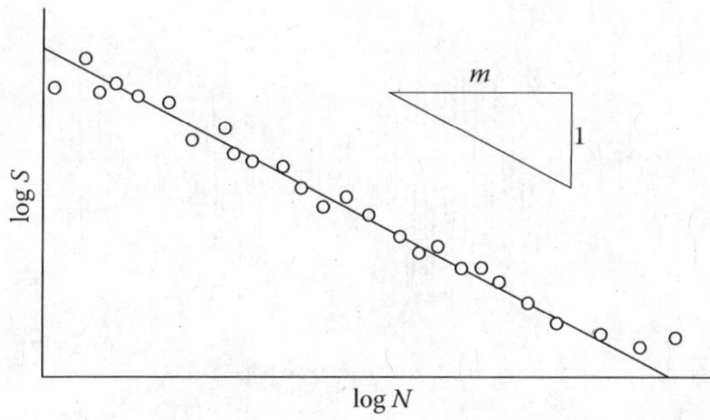

Figure 5.16 Form of a typical S–N curve relating stress range to cycles to failure.

stress range (equal to twice the amplitude, s), and N is the number of cycles until failure. For many materials, the S–N curve is well approximated by a straight line when log S is plotted against log N (Figure 5.16). This implies an equation of the form:

$$NS^m = N(2s)^m = K \tag{5.51}$$

where K is a constant which depends on the material, and the exponent m varies between about 5 and 20.

A criterion for failure under repeated loading for various stress ranges is Miner's Rule:

$$\sum\left(\frac{n_i}{N_i}\right) = 1 \tag{5.52}$$

where n_i is the number of stress cycles with a range for which N_i cycles are required to cause failure. Thus, failure is expected when the sum of the fractional damage for all stress levels is unity.

Note that there is no restriction on the *order* in which the various stress ranges, or amplitudes, are applied in Miner's Rule. Thus, we may apply it to a random loading process which can be considered as a series of cycles with randomly varying amplitudes.

5.6.2 Narrow-band fatigue loading

Some wind-loading situations produce resonant 'narrow-band' vibrations. For example, the along-wind response of structures with low natural frequencies (Section 5.3.1), and cross-wind vortex-induced response of circular cylindrical structures with low damping. In these cases, the resulting stress variations can be regarded as quasi-sinusoidal with randomly varying amplitudes, as shown in Figure 5.17.

For a narrow-band random stress $s(t)$, the proportion of cycles with amplitudes in the range from s to $s + \delta s$ is $f_p(s) \cdot \delta s$, where $f_p(s)$ is the probability density of the peaks. The total number of cycles in a time period, T, is $\nu_0^+ T$, where ν_0^+ is the rate of crossing of the mean stress. For narrow-band resonant vibration, ν_0^+, may be taken to be equal to the natural frequency of vibration.

Then the total number of cycles with amplitudes from s to δs is

$$n(s) = \nu_0^+ T\, f_p(s) \cdot \delta s \tag{5.53}$$

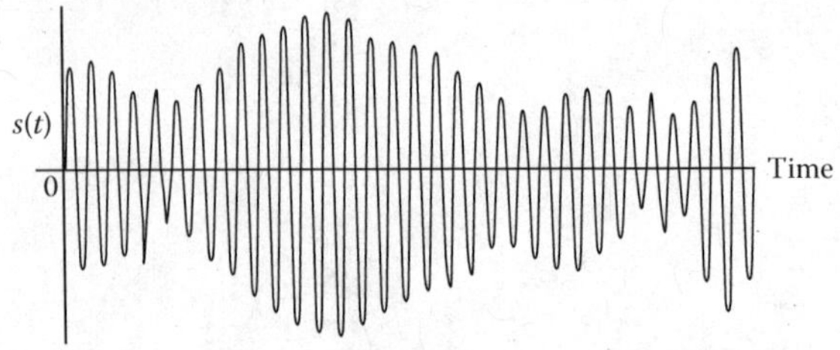

Figure 5.17 Stress-time history under narrow-band random vibrations.

If $N(s)$ is the number of cycles at amplitude, s, to cause failure, then the fractional damage is at this stress level

$$= \frac{n(s)}{N(s)} = \frac{\nu_0^+ T f_p(s)(2s)^m s}{K}$$

where Equation 5.53 has been used for $n(s)$, and Equation 5.51 has been used for $N(s)$.

The total expected fractional damage over all stress amplitudes is then by Miner's Rule

$$D = \sum_0^{\infty} \frac{n(s)}{N(s)} = \frac{\nu_0^+ T \int_0^{\infty} f_p(s)(2s)^m \, ds}{K} \qquad (5.54)$$

Wind-induced narrow-band vibrations can be taken to have a Normal or Gaussian probability distribution (Appendix C, Section C.3.1). If this is the case, then the peaks or amplitudes, s, have a Rayleigh distribution (e.g. Crandall and Mark, 1963):

$$f_p(s) = \frac{s}{\sigma^2} \exp\left(-\frac{s^2}{2\sigma^2}\right) \qquad (5.55)$$

where σ is the standard deviation of the entire stress history. Derivation of Equation 5.55 is based on the level-crossing formula of Rice (1944–1945) (see Appendix C, Section 5.2).

Substituting into Equation 5.54,

$$D = \frac{\nu_0^+ T}{K^2} \int_0^{\infty} 2^m s^{m+1} exp\left(-\frac{s^2}{2^2}\right) ds = \frac{\nu_0^+ T}{K} \left(2\sqrt{2}\right)^m \left(\frac{m}{2} + 1\right) \qquad (5.56)$$

Here, the following mathematical result has been used (Dwight, 1961):

$$\int_0^{\infty} x^q \, exp\left[-(rx)^p\right] dx = \frac{1}{pr^{q+1}} \left(\frac{q+1}{p}\right) \qquad (5.57)$$

where $\Gamma(x)$ is the Gamma Function.

Equation 5.56 is a very useful 'closed-form' result, but it is restricted by two important assumptions:

1. 'High-cycle' fatigue behaviour in which steel is in the elastic range, and for which an *S–N* curve of the form of Equation 5.51 is valid, has been assumed.
2. Narrow-band vibration in a single resonant mode of the form shown in Figure 5.17 has been assumed. In wind loading, this is a good model of the behaviour for vortex-shedding-induced vibrations in low turbulence conditions. For along-wind loading, the background (sub-resonant) components are always important and result in a random wide-band response of the structure.

5.6.3 Wide-band fatigue loading

Wide-band random vibration consists of contributions over a broad range of frequencies, with a large resonant peak – this type of response is typical for wind loading (Figure 5.7). A number of cycle-counting methods for wide-band stress variations have been proposed (Dowling, 1972). One of the most realistic of these is the 'rainflow'-counting method, proposed by Matsuishi and Endo (1968). In this method, which uses the analogy of rain flowing over the undulations of a roof, cycles associated with complete hysteresis cycles of the metal are identified. The use of this method rather than a simple level-crossing approach which is the basis of the narrow-band approach described in Section 5.6.2 invariably results in fewer cycle counts.

A useful empirical approach has been proposed by Wirsching and Light (1980). They proposed that the fractional fatigue damage under a wide-band random stress variation can be written as

$$D = \lambda D_{nb} \tag{5.58}$$

where D_{nb} is the damage calculated for narrow-band vibration with the same standard deviation, σ, (Equation 5.56). λ is a parameter determined empirically. The approach used to determine λ was to use simulations of wide-band processes with spectral densities of various shapes and bandwidths, and rainflow counting for fatigue cycles.

The formula proposed by Wirsching and Light to estimate λ was

$$\lambda = a + (1 - a)(1 - \varepsilon)^b \tag{5.59}$$

where a and b are functions of the exponent m (Equation 5.51), obtained by least-squares fitting, as follows:

$$a \cong 0.926 - 0.033m \tag{5.60}$$

$$b \cong 1.587m - 2.323 \tag{5.61}$$

ε is a spectral bandwidth parameter equal to

$$\varepsilon = 1 - \frac{\mu_2^2}{\mu_0 \mu_4} \tag{5.62}$$

where μ_k is the kth moment of the spectral density defined by

$$\mu_k = \int_0^\infty n^k S(n)\,dn \tag{5.63}$$

For narrow-band vibration, ε tends to zero, and, from Equation 5.59, λ approaches 1. As ε tends to its maximum possible value of 1, λ approaches a, given by Equation 5.60. These values enable upper and lower limits on the damage to be determined. Alternatively, the spectral bandwidth parameter, ε, can be calculated from the spectral density of the response, and Equation 5.59 can be used to calculate λ.

5.6.4 Effect of varying wind speed

Equation 5.56 applies to a particular standard deviation of stress, σ, which in turn is a function of mean wind speed, $\bar{U}$. This relationship can be written in the form:

$$\sigma = A\bar{U}^n \tag{5.64}$$

The mean wind speed, $\bar{U}$, itself, is a random variable. Its probability distribution can be represented by a Weibull Distribution (see Sections 2.5 and C.3.4):

$$f_U(\bar{U}) = \frac{w\bar{U}^{w-1}}{c^w} exp\left[-\left(\frac{\bar{U}}{c}\right)^w\right] \tag{5.65}$$

The total damage from narrow-band vibration for all possible mean wind speeds is obtained from Equations 5.56, 5.64 and 5.65 and integrating.

The fraction of the time T during which the mean wind speed falls between U and $U + \delta U$ is $f_U(U) \cdot \delta U$.

Hence, the amount of damage generated while this range of wind speed occurs is, from Equations 5.56 and 5.64:

$$D_U = \frac{\nu_0^+ T f_U(U) U}{K} (2\sqrt{2}AU^n)^m \Gamma\left(\frac{m}{2} + 1\right)$$

For wide-band loading, the average cycling rate of the stress cycles can be assumed to vary with mean wind speed according to Equation 5.66:

$$\nu_o^+ = \nu_1 \bar{U}^p = \nu_c \left(\frac{\bar{U}}{c}\right)^p \tag{5.66}$$

where ν_c is the cycling rate when the mean wind speed is equal to c, and ν_1 is notionally the cycling rate when the mean wind speed is equal to 1 m/s. p is an exponent with a value of 0.1–0.5.

Then the expected amount of damage in a time, T, at mean wind speed, $\bar{U}$, is

$$D_U = \frac{\nu_c T f_{\bar{U}}\left(\bar{U}\right)\bar{U}}{Kc^p}(2\sqrt{2}A)^m \bar{U}^{mn+p}\,\Gamma\left(\frac{m}{2}+1\right)$$

5.6.5 Total accumulated fatigue damage, and fatigue life estimation

The total damage in time T during all mean wind speeds between 0 and ∞ is

$$\begin{aligned} D &= \frac{\nu_c T(2\sqrt{2}A)^m}{Kc^p}\Gamma\left(\frac{m}{2}+1\right)\int_0^\infty \bar{U}^{mn+p} f_{\bar{U}}(\bar{U})d\bar{U} \\ &= \frac{\nu_c T(2\sqrt{2}A)^m}{Kc^p}\Gamma\left(\frac{m}{2}+1\right)\int_0^\infty \bar{U}^{mn+p+w-1}\frac{w}{c^w}\exp\left[-\left(\frac{\bar{U}}{c}\right)^w\right]d\bar{U} \end{aligned} \tag{5.67}$$

Hence,

$$D = \frac{w\nu_c T(2\sqrt{2}A)^m}{Kc^{p+w}}\Gamma\left(\frac{m}{2}+1\right)\int_0^\infty \bar{U}^{mn+p+w-1}\exp\left[-\left(\frac{\bar{U}}{c}\right)^w\right]d\bar{U}$$

This is now of the form of Equation 5.57, so that:

$$\begin{aligned} D &= \frac{w\nu_c T(2\sqrt{2}A)^m}{Kc^{p+w}}\Gamma\left(\frac{m}{2}+1\right)\frac{c^{mn+p+w}}{w}\Gamma\left(\frac{mn+p+w}{w}\right) \\ &= \frac{\nu_c T(2\sqrt{2}A)^m c^{mn}}{K}\Gamma\left(\frac{m}{2}+1\right)\Gamma\left(\frac{mn+p+w}{k}\right) \end{aligned} \tag{5.68}$$

This is a useful closed-form expression for the fatigue damage over a lifetime of wind speeds, assuming narrow-band vibration.

For wide-band vibration, Equation 5.68 can be modified, following Equation 5.58, to

$$D = \frac{\nu_c T(2\sqrt{2}A)^m c^{mn}}{K}\Gamma\left(\frac{m}{2}+1\right)\Gamma\left(\frac{mn+p+w}{k}\right) \tag{5.69}$$

By setting D equal to 1 in Equations 5.68 and 5.69, we can obtain lower and upper limits to the fatigue life as follows:

$$T_{\text{lower}} = \frac{K}{\nu_c(2\sqrt{2}A)^m c^{mn}\Gamma(m/2+1)\Gamma\left((mn+p+w)/k\right)} \tag{5.70}$$

$$T_{\text{upper}} = \frac{K}{\nu_c(2\sqrt{2}A)^m c^{mn}(m/2+1)\Gamma\left((mn+p+w)/k\right)} \tag{5.71}$$

EXAMPLE

To enable the calculation of fatigue life of a welded connection at the base of a steel pole, using Equations 5.70 and 5.71, the following values are assumed:

$$m = 5; \quad n = 2; \quad p = 0.2; \quad \nu_c = 0.42\ \text{Hz}$$

$$K = 6.4 \times 10^{16} \left[\text{MPa}\right]^5; \quad c = 8\,\text{m/s}; \quad w = 2; \; A = 0.1 \frac{\text{MPa}}{(\text{m/s})^2}$$

$$\Gamma\left(\frac{m}{2} + 1\right) = \Gamma(3.5) = e^{1.201} = 3.323$$

$$\Gamma\left(\frac{mn + p + w}{k}\right) = \Gamma(6.1) = e^{4.959} = 142.5$$

Then from Equation 5.70,

$$T_{\text{lower}} = \frac{6.4 \times 10^{16}}{0.42\left(2\sqrt{2} \times 0.1\right)^5 \times 8^{10} \times 3.323 \times 142.5} = 1.656 \times 10^8 s = 5.25 \text{ years}$$

From Equation 5.60, $a = 0.926 - 0.033\,m = 0.761$
From Equation 5.59, this is a lower limit for λ.

$$T_{\text{upper}} = \frac{T_{\text{lower}}}{\lambda} = 6.90 \text{ years}$$

This example illustrates the sensitivity of the estimates of fatigue life to the values of both A and c. For example, increasing A to 0.15 (MPa/(m/s)^2) would decrease the fatigue life by 7.6 times, that is (1.5^5). Decreasing c from 8 to 7 m/s will increase the fatigue life by 3.8 times, that is $(8/7)^{10}$.

5.6.6 Number of cycles above a defined stress level

The number of cycles of stress with amplitude, s, or range, S, is often required when the fatigue damage relationship is undefined, or cannot be described by Equation 5.52. For a narrow-band process, application of Equations 5.53, 5.55 and 5.64 gives the expected number of cycles exceeding a stress range, S, at a mean wind speed, $\bar{U}$, in a time period, T:

$$N(S) = \nu_0^+ T \exp\left(-\frac{s^2}{2^2}\right) = \nu_0^+ T \exp\left(-\frac{S^2}{8A^2\bar{U}^{2n}}\right) \tag{5.72}$$

Then, for a wind speed represented by the Weibull Distribution (Equation 5.63), the total number of cycles with a range exceeding, S, in a time period, T, for all wind speeds, is obtained by integration:

$$N(S) = \frac{w\nu_c T}{c^{p+w}} \int_0^{\infty} \bar{U}^{p+w-1} exp\left[-\left(\frac{\bar{U}}{c}\right)^w - \left(\frac{S^2}{8A^2\bar{U}^{2n}}\right) \right] d\bar{U} \tag{5.73}$$

Changing the variable of integration from $\bar{U}$ to $(\bar{U}/c)$ gives

$$\frac{N(S)}{w\nu_c T} = \int_0^{\infty} \left(\frac{\bar{U}}{c}\right)^{p+w-1} exp\left[-\left(\frac{\bar{U}}{c}\right)^w - \left(\frac{S^2/(8A^2c^{2n})}{(\bar{U}/c)^{2n}}\right) \right] d\left(\frac{\bar{U}}{c}\right) \tag{5.74}$$

From Equation 5.74, it can be shown that the number of stress cycles in the time period, T, is a function of the following non-dimensional parameters:

$$N(s) = F\{b, w, n, \bar{N}, (S/S_{max})\} \tag{5.75}$$

where

$\bar{N}$ is equal to $(\nu_c T)$ $[=(\nu_1 c^p T)]$ and is a characteristic number of cycles (or mean crossings) in the time period T.

S_{max} is the expected largest value of stress in the time T, and can be closely approximated by setting $N(S)$ equal to 1 in Equation 5.74, and solving it numerically.

Thus, for a fixed set of the four parameters, p, w, n and $\bar{N}$, a single relationship between $N(S)$ and (S/S_{max}) can be obtained. Note that Equation 5.75 is independent of the parameter A in Equation 5.64.

It can also be shown that (S/S_{max}) is relatively insensitive to T, c, n and p (Holmes, 2012). There is more sensitivity to the Weibull shape factor, w (Kemper and Feldmann, 2013).

However, assuming a conservative value for w of about 2.5 provides a basis for a 'universal' relationship between the normalised stress range (S/S_{max}), and the expected number of cycles under wind loading (Holmes, 2012). Such a relationship has been provided in several design codes and standards, such as Eurocode 1 (British Standards Institution, 2005). This is shown in Figure 5.18, which shows the variation of the stress-range cycle count with w with some typical values of the other parameters, based on numerical calculations of Equations 5.74 and 5.75, and compared with that given in Eurocode 1.

5.7 SUMMARY

This chapter has covered a wide range of topics relating to the dynamic response of structures to wind forces. For wind loading, the sub-resonant or background response should be distinguished from the contributions at the resonant frequencies, and calculated separately.

The along-wind response of structures that can be represented as single- and multi-degree-of-freedom systems has been considered. The effective static load approach in which the distributions of the mean, background and resonant contributions to the loading are considered separately, and assembled as a combined effective static wind load, has been presented.

Aeroelastic effects such as aerodynamic damping, and the instabilities of galloping and flutter have been introduced. Finally, wind-induced fatigue has been treated resulting in usable formulae for the calculation of fatigue life of a structure under turbulent wind loading.

Cross-wind dynamic response from vortex shedding has not been treated in this chapter, but is discussed in Chapters 9 and 11.

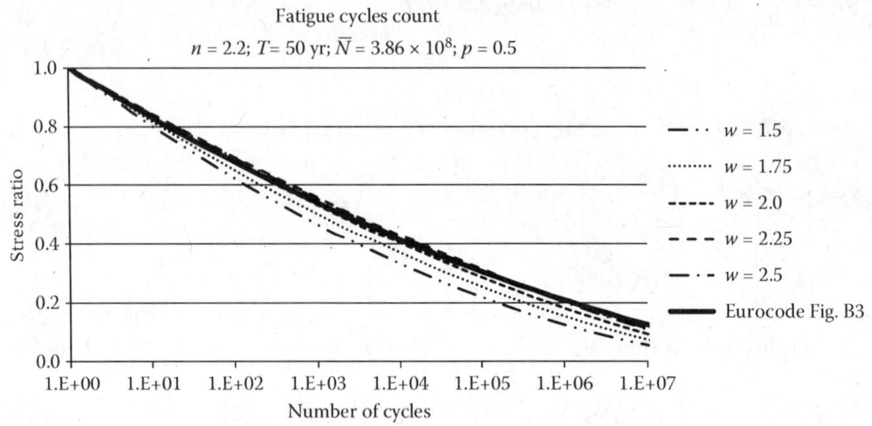

Figure 5.18 Wind-induced fatigue—normalised stress-range ratio versus the expected number of cycles, and variation with Weibull shape factor, *w*. (From Holmes, J.D. 2012. Wind-induced fatigue cycle counts—Sensitivity to wind climate and dynamic response. *7th International Congress of the Croatian Society of Mechanics (7th ICCSM)*, 22–25 May, Zadar, Croatia.)

REFERENCES

Ashraf Ali, M. and Gould, P.L. 1985. On the resonant component of the response of single degree-of-freedom systems under wind loading. *Engineering Structures*, 7: 280–2.

Bendat, J.S. and Piersol, A.G. 2010. *Random Data: Analysis and Measurement Procedures*. 4th edition. Wiley, New York.

British Standards Institution. 2005. *Eurocode 1: Actions on Structures—Parts 1–4: General Actions—Wind Actions*. BS EN 1991-1-4.6, BSI, London, UK.

Clough, R.W. and Penzien, J. 1975. *Dynamics of Structures*. McGraw-Hill, New York.

Crandall, S.H. and Mark, W.D. 1963. *Random Vibration in Mechanical Systems*. Academic Press, New York.

Davenport, A.G. 1961. The application of statistical concepts to the wind loading of structures. *Proceedings, Institution of Civil Engineers*, 19: 449–71.

Davenport, A.G. 1963. The buffeting of structures by gusts. *Proceedings, International Conference on Wind Effects on Buildings and Structures*, Teddington, UK, 26–28 June, 358–91.

Davenport, A.G. 1964. Note on the distribution of the largest value of a random function with application to gust loading. *Proceedings, Institution of Civil Engineers*, 28: 187–96.

Davenport, A.G. 1967, Gust loading factors. *ASCE Journal of the Structural Division*, 93: 11–34.

Dowling, N.E. 1972. Fatigue failure predictions for complicated stress–strain histories. *Journal of Materials*, 7: 71–87.

Dwight, H.B. 1961. *Tables of Integrals and Other Mathematical Data*. Macmillan, Toronto, Canada.

Harris, R.I. 1963. The response of structures to gusts. *Proceedings, International Conference on Wind Effects on Buildings and Structures*, Teddington, UK, 26–28 June, 394–421.

Holmes, J.D. 1994. Along-wind response of lattice towers: Part I— Derivation of expressions for gust response factors. *Engineering Structures*, 16: 287 – 92.

Holmes, J.D. 1996a. Along-wind response of lattice towers: Part II— Aerodynamic damping and deflections. *Engineering Structures*, 18: 483–8.

Holmes, J.D. 1996b. Along-wind response of lattice towers: Part III—Effective load distributions. *Engineering Structures*, 18: 489–94.

Holmes, J.D. 2012. Wind-induced fatigue cycle counts—Sensitivity to wind climate and dynamic response. *7th International Congress of the Croatian Society of Mechanics (7th ICCSM)*, 22–25 May, Zadar, Croatia.

Holmes, J.D. and Best, R.J. 1981. An approach to the determination of wind load effects for low-rise buildings. *Journal of Wind Engineering and Industrial Aerodynamics*, 7: 273–87.

Holmes, J.D., Forristall, G. and McConochie, J. 2005. Dynamic response of structures to thunderstorm winds. *10th Americas Conference on Wind Engineering*, Baton Rouge, Louisiana, 1–4 June, 2005.

Holmes, J.D. and Kasperski, M. 1996. Effective distributions of fluctuating and dynamic wind loads. *Civil Engineering Transactions, Institution of Engineers, Australia*, CE38: 83–8.

Kasperski, M. and Niemann, H.-J. 1992. The L.R.C. (load-response-correlation) method: A general method of estimating unfavourable wind load distributions for linear and non-linear structural behaviour. *Journal of Wind Engineering and Industrial Aerodynamics*, 43: 1753 – 63.

Kemper, F.H. and Feldmann, M. 2013. Stress range spectra and damage equivalence factors for along-wind responses of arbitrary structures. *12th Americas Conference on Wind Engineering*, Seattle, Washington, USA, 16–20 June, 2013.

Matsuishi, M. and Endo, T. 1968. Fatigue of metals subjected to varying stress. *Japan Society of Mechanical Engineers Meeting*, Fukuoka, March.

Matsumoto, M. 1996. Aerodynamic damping of prisms. *Journal of Wind Engineering and Industrial Aerodynamics*, 59: 159–75.

Rice, S.O. 1944–1945. Mathematical analysis of random noise. *Bell System Technical Journal*, 23: 282–332 and 24: 46–156. Reprinted in Wax, N. *Selected Papers on Noise and Stochastic Processes.* Dover, New York, 1954.

Scanlan, R.H. 1982. Developments in low-speed aeroelasticity in the civil engineering field. *American Institute of Aeronautics and Astronautics Journal*, 20: 839–44.

Scanlan, R.H. and Gade, R.H. 1977. Motion of suspended bridge spans under gusty winds. *ASCE Journal of the Structural Division*, 103: 1867–83.

Scanlan, R.H. and Tomko, J.J. 1971. Airfoil and bridge deck flutter derivatives. *ASCE Journal of the Engineering Mechanics Division*, 97: 1717–37.

Vickery, B.J. 1965. On the flow behind a coarse grid and its use as a model of atmospheric turbulence in studies related to wind loads on buildings. Aero Report 1143, National Physical Laboratory, UK.

Vickery, B.J. 1966. On the assessment of wind effects on elastic structures. *Australian Civil Engineering Transactions*, CE8: 183 – 92.

Vickery, B.J. 1968. Load fluctuations in turbulent flow. *ASCE Journal of the Engineering Mechanics Division*, 94: 31–46.

Vickery, B.J. 1995. The response of chimneys and tower-like structures to wind loading. In P. Krishna (ed.), *A State of the Art in Wind Engineering*, Wiley Eastern Limited, New Delhi.

Warburton, G.B. 1976. *The Dynamical Behaviour of Structures*. 2nd edition. Pergamon Press Ltd, Oxford, UK.

Wirsching, P.H. and Light, M.C. 1980. Fatigue under wide band random stresses. *Journal of the Structural Division, ASCE*, 106: 1593–607.

Chapter 6

Internal pressures

6.1 INTRODUCTION

Internal pressures produced by wind can form a high proportion of the total design wind load in some circumstances – for example, for low-rise buildings when there are dominant openings in the walls. On high-rise buildings, a critical design case for a window at a corner may be an opening in the adjacent wall at the same corner – perhaps caused by glass failure due to flying debris.

In this chapter, the fundamentals of the prediction of wind-induced internal pressures within enclosed buildings are discussed. A number of cases are considered: a single dominant opening in one wall, multiple wall openings and the effect of background wall porosity. The possibility of Helmholtz resonance occurring is also discussed.

6.2 SINGLE WINDWARD OPENING

We will first consider the case of a dominant windward wall opening, a situation which often arises in severe wind storms – for example, after failure of a glass window due to flying debris. In a steady flow situation, the internal pressure will quickly build up to an equal external pressure on the windward wall in the vicinity of the opening – there may be some oscillations in internal pressure (Section 6.2.4), but these will die out after a short time. However, when a building is immersed in a turbulent boundary-layer wind, the external pressure will be highly fluctuating, and the internal pressure will respond in some way to these fluctuations. Since there is only a single opening, flow into the building resulting from an increase in external pressure will cause an increase in the density of the air within the internal volume; this, in turn, will produce an increase in internal pressure. The pressure changes produced by wind are only about 1% of the atmospheric pressure (1000 Pa compared to the atmospheric pressure of about 100,000 Pa), and the relative density changes are of the same order. These small density changes can be maintained by small mass flows in and out of the building envelope, and consequently, the internal pressure can be expected to respond quite quickly to external pressure changes, except for very small opening areas.

6.2.1 Dimensional analysis

It is useful to first carry out a dimensional analysis for the fluctuating internal pressures, resulting from a single windward opening to establish the non-dimensional groups involved.

The fluctuating internal pressure coefficient, $C_{pi}(t)$, can be written as

$$C_{pi} = \frac{p_i - p_o}{\frac{1}{2}\rho_a \bar{U}^2} = F(\Phi_1, \Phi_2, \Phi_3, \Phi_4, \Phi_5) \tag{6.1}$$

where p_o is the reference atmospheric, or static, pressure, ρ_a is the density of air and $\bar{U}$ is the mean wind velocity in the approach flow.

$\Phi_1 = A^{3/2}/V_o$ – where A is the area of the opening, and V_o is the internal volume

$$\Phi_2 = a_s/\bar{U} = \sqrt{\gamma p_o/\rho_a}\Big/\bar{U}$$

where

a_s is the speed of sound, and γ is the ratio of specific heats of air

$\Phi_3 = \rho_a \bar{U} A^{1/2}/\mu$ – where μ is the dynamic viscosity of air (Reynolds Number)

$\Phi_4 = \sigma_u/\bar{U}$ – where σ_u is the standard deviation of the longitudinal turbulence velocity upstream – i.e. Φ_4 is the turbulence intensity (Section 3.3.1)

$\Phi_5 = l_u/\sqrt{A}$ – where l_u is the length scale of turbulence (Section 3.3.4)

Φ_1 is a non-dimensional parameter related to the geometry of the opening and the internal volume, Φ_3 is a Reynolds Number (Section 4.2.4) based on a characteristic length of the opening and Φ_5 is a ratio between characteristic length scales in the turbulence in the approaching flow and of the opening. Φ_2 is the Mach number.

Among these parameters, Φ_1, Φ_2 and Φ_5 are important (see Section 6.2.7).

6.2.2 Response time

If the inertial (i.e. mass times acceleration) effects are initially neglected, an expression for the time taken for the internal pressure to become equal to a sudden increase in pressure outside the opening such as that caused by a sudden window failure can be derived (Euteneur, 1970).

For conservation of mass, the rate of mass flow through the opening must equal the rate of mass increase inside the volume:

$$\rho_i Q = \left(\frac{d\rho_i}{dt}\right) V_o \tag{6.2}$$

where ρ_i denotes the air density within the internal volume, Q is the volume flow rate and V_o is the internal volume.

For turbulent flow through an orifice, the following relationship between flow rate, Q, and the pressure difference across the orifice, $p_e - p_i$, applies:

$$Q = kA\sqrt{\frac{2(p_e - p_i)}{p_e - p_i}} \tag{6.3}$$

where k is an orifice constant, about 0.6 for steady uni-directional flow.

Assuming an adiabatic law relating the internal pressure and density,

$$\frac{p_i}{\rho_i^{\gamma}} = \text{constant} \tag{6.4}$$

where γ is the ratio of specific heats of air.

Substituting from Equations 6.2 and 6.4 into Equation 6.3, and integrating the differential equation, the following expression for the response, or equilibrium, time, τ, when the internal pressure becomes equal to the external pressure, can be obtained:

$$\tau = \frac{\rho_a V_o \bar{U}}{\gamma k A p_o} \sqrt{C_{pe} - C_{pio}} \tag{6.5}$$

where the pressures have been written in terms of pressure coefficients:

$C_{pe} = \dfrac{p_e - p_o}{(1/2)\rho_a \bar{U}^2}$ and $C_{pi} = \dfrac{p_i - p_o}{(1/2)\rho_a \bar{U}^2}$ and C_{pio} is the initial value of C_{pi} (i.e. at $t = 0$)

EXAMPLE

It is instructive to apply Equation 6.5 to a practical example. The following numerical values will be substituted:

$\rho_a = 1.20\,\text{kg/m}^3$; $V_o = 1000\,\text{m}^3$; $\bar{U} = 40\,\text{m/s}$

$\gamma = 1.4$; $k = 0.6$; $A = 1.0\,\text{m}^2$; $p_o = 10^5\,\text{Pa}$

$C_{pe} = +0.7$; $C_{pio} = -0.2$

Then the response time,

$$\tau = \frac{1.2 \times 1000 \times 40}{1.4 \times 0.6 \times 1.0 \times 10^5} \sqrt{0.7 - (-0.2)} = 0.54\,\text{s}$$

Thus, for an internal volume of 1000 cubic metres, Equation 6.5 predicts a response time of just over half a second for the internal pressure to adjust to the external pressure, following the creation of an opening on the windward face of 1 square metre.

6.2.3 Helmholtz resonator model

In the previous section, inertial effects on the development of internal pressure following a sudden opening were neglected. These will now be included in a general model of internal pressure, which can be used for the prediction of the response to turbulent external pressures (Holmes, 1979).

The Helmholtz resonator is a well-established concept in acoustics (Rayleigh, 1896; Malecki, 1969), which describes the response of small volumes to the fluctuating external pressures. Although originally applied to the situation where the external pressures are caused by acoustic sources, it can be applied to the case of external wind pressures 'driving' the internal pressures within a building. It also describes the low-frequency fluctuations felt by occupants of a travelling motor vehicle, with an open window. Acoustic resonators made from brass or earthenware, based on this principle, were used to improve the acoustic quality in the amphitheatres of ancient Greece and Rome (Malecki, 1969).

Figure 6.1 illustrates the concept as applied to internal pressures in a building. It is assumed that a defined 'slug' of air moves in and out of the opening in response to the external pressure changes. Thus, mixing of the moving air, either with the internal air or the external air, is disregarded in this model of the situation.

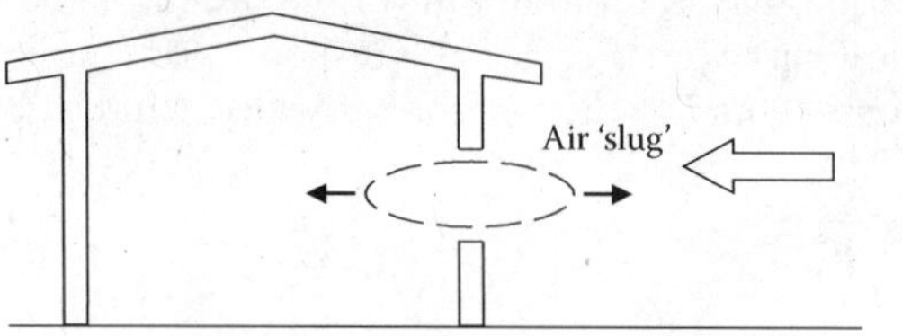

Figure 6.1 The Helmholtz resonator model of fluctuating internal pressures with a single dominant opening.

A differential equation for the motion of the slug of air can be written as follows:

$$\rho_a A \ell_e \ddot{x} + \frac{\rho_a A}{2k^2} \dot{x}|\dot{x}| + \frac{\gamma p_o A^2}{V_o} x = A \Delta p_e(t) \tag{6.6}$$

The dependent variable, x, in this differential equation is the displacement of the air 'slug' from its initial or equilibrium position. The first term on the left-hand side of Equation 6.6 is an inertial term proportional to the acceleration, $\ddot{x}$, of the air slug, whose mass is $\rho_a A \ell_e$, in which ℓ_e is an effective length for the slug. The second term is a loss term associated with energy losses for flow through the orifice, and the third term is a 'stiffness' associated with the resistance of the air pressure already in the internal volume to the movement of the 'slug'.

A movement x in the air slug can be related to the change in density $\Delta\rho_i$, and hence pressure, Δp_i, within the internal volume:

$$\rho_a A x = V_o \Delta\rho_i = \frac{\rho_i V_o}{\gamma p_o} \Delta p_i \tag{6.7}$$

Making use of Equation 6.4 and converting the internal and external pressures into pressure coefficients, Equation 6.6 can be re-written in the form of a differential equation for the fluctuating internal pressure coefficient, $C_{pi}(t)$:

$$\frac{\rho_a \ell_e V_o}{\gamma p_o A} \ddot{C}_{pi} + \left(\frac{\rho_a V_o \bar{U}}{2k\gamma\, A p_0} \right)^2 \dot{C}_{pi} \left| \dot{C}_{pi} \right| + C_{pi} = C_{pe} \tag{6.8}$$

Equation 6.8 can also be derived (Vickery, 1986) by writing the discharge equation for unsteady flow through the orifice in the form:

$$p_e - p_i = \left(\frac{1}{k^2} \right) \frac{1}{2} \rho_a u_o^2 + \rho_a \ell_e \frac{du_o}{dt} \tag{6.9}$$

where ρ_a is taken as the air density within the volume (ρ_i), and u_o is taken as the (unsteady) spatially averaged velocity through the opening.

Equations 6.6 and 6.8 give the following equation for the (undamped) natural frequency for the resonance of the movement of the air slug, and of the internal pressure fluctuations. This frequency is known as the Helmholtz frequency, n_H.

$$n_H = \frac{1}{2\pi} \sqrt{\frac{\gamma A p_o}{\rho_a \ell_e V_o}} \tag{6.10}$$

Internal pressure resonances at, or near, the Helmholtz frequency, have been measured both in wind-tunnel studies (Holmes, 1979; Liu and Rhee, 1986), and in full scale.

The effective length, ℓ_e, varies with the shape and depth of the opening, and is theoretically equal to $\sqrt{(\pi A/4)}$ for a thin circular orifice. For practical purposes (openings in thin walls), it is sufficiently accurate to take ℓ_e as equal to $1.0\sqrt{A}$ (Vickery, 1986).

6.2.4 Sudden windward opening with inertial effects

Equation 6.8 can be solved numerically for the case of a step change in the external pressure coefficient, C_{pe} (representative of the situation after a sudden window failure). Figures 6.2a and b show the response of a 600 m³ volume (rigid walls and roof) with opening areas of 1 and 9 m², respectively (Holmes, 1979). For these simulations, the effective length, ℓ_e, was equivalent to $0.96\sqrt{A}$, and the discharge coefficient, k, was taken as 0.6.

It is apparent from Figure 6.2b that the inertial effects are significant for the larger opening when the damping term in Equation 6.8 is much smaller (note that the area, A, is in the denominator in this term). Many oscillatory cycles in internal pressure occur before equilibrium conditions are reached in this case. However, the flexibility of the walls and roof of real buildings, discussed in the following section, also increases the damping term (Vickery, 1986), and hence causes more rapid attenuation of the oscillations.

6.2.5 Effect of roof flexibility

Equations 6.8 and 6.10 assume that the building or enclosure has rigid walls and a roof. Real buildings may have considerable flexibility.

Sharma (2008) studied the effect of the building flexibility. The case of roof flexibility results in a modified equation for internal pressure fluctuations:

$$\frac{\rho_a l_e V_0}{\gamma p_0 A}\left(\ddot{C}_{pi} + \dot{v}\dot{C}_{pi} + \frac{\gamma p_0}{q}\ddot{v}\right) + \frac{\rho_a q V_0^2}{2(k\gamma A p_0)^2}\left(\dot{C}_{pi} + \frac{\gamma p_0}{q}\dot{v}\right)\left|\dot{C}_{pi} + \frac{\gamma p_0}{q}\dot{v}\right| + C_{pi} = C_{pe} \quad (6.11)$$

where $q = \frac{1}{2}\rho_a \bar{U}^2$

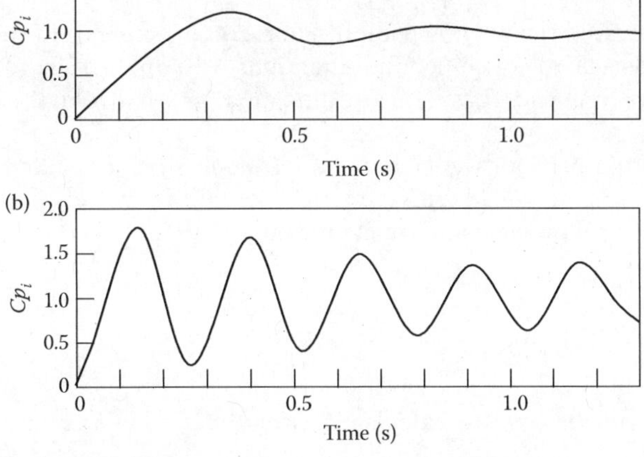

Figure 6.2 Response to a step change in external pressure, $V_o = 600$ m³, $\bar{U} = 30$ m/s. (a) $A = 1$ m²; (b) $A = 9$ m².

The fractional change in volume

$$v = \frac{\Delta V}{V_0} = \frac{q}{K_B}(C_{pi} - C_{pr})$$

where C_{pr} is an area-averaged-fluctuating external roof pressure coefficient.

K_B is the bulk modulus for the building – that is the internal pressure for a unit change in the relative internal volume. Theoretically, it is the internal pressure required to double the internal volume.

Simplifying and neglecting the second-order terms,

$$\frac{\rho_a l_e V_0}{\gamma p_0 A}\left(\ddot{C}_{pi} - \frac{b}{(1+b)}\ddot{C}_{pr}\right) + \frac{\rho_a q V_0^2 (1+b)^2}{2(k\gamma A p_0)^2}\left(\dot{C}_{pi} - \frac{b}{(1+b)}\dot{C}_{pr}\right)\left|\dot{C}_{pi} - \frac{b}{(1+b)}\dot{C}_{pr}\right| + C_{pi} = C_{pe} \quad (6.12)$$

where b ($=K_A/K_B$) is the ratio of the bulk modulus of air to the bulk modulus for the building. K_A, the bulk modulus of air, $(\rho_a \Delta p)/\Delta\rho$, is equal to $\gamma\, p_o$.

The typical range of K_A/K_B for low-rise buildings is 0.2–5.

Equation 6.12 in general requires numerical solutions.

From Equation 6.12, the equation for the Helmholtz frequency becomes (Vickery, 1986):

$$n_H = \frac{1}{2\pi}\sqrt{\frac{\gamma A p_o}{\rho_a \ell_e V_o [1+b]}} \quad (6.13)$$

It can be seen from Equation 6.13 that the effect of the building flexibility is equivalent to an increase in the effective internal volume.

6.2.6 Helmholtz resonance frequencies

Section 6.2.3 discussed the phenomenon of Helmholtz resonance in the interior of buildings, when there is a single opening, and Equations 6.10 and 6.13 gave formulas to calculate the Helmholtz frequency, given the opening area, internal volume and flexibility of the roof.

Applying Equation 6.13 for the Helmholtz resonance frequency, and setting $p_o = 10^5$ Pa (atmospheric pressure), $\rho_a = 1.2$ kg/m^3 (air density), $\gamma = 1.4$ (ratio of specific heats) and ℓ_e to $1.0\sqrt{A}$, we have the following approximate formula for n_H:

$$n_H \approx 55\frac{A^{1/4}}{V_o^{1/2}[1 + (K_A/K_B)]^{1/2}} \quad (6.14)$$

Equation 6.14 can be used to calculate n_H for typical low-rise buildings in Table 6.1 (Vickery, 1986).

Table 6.1 Helmholtz resonance frequencies for some typical buildings

Type	*Internal volume* *(m^3)*	*Opening area* *(m^2)*	*Stiffness ratio* *K_A/K_B*	*Helmholtz frequency* *(Hz)*
House	600	4	0.2	2.9
Warehouse	5000	10	0.2	1.3
Concert hall	15,000	15	0.2	0.8
Arena (flexible roof)	50,000	20	4	0.23

Table 6.1 indicates that for the two smallest buildings, the Helmholtz frequencies are >1 Hz, and hence a significant resonant excitation of internal pressure fluctuations by natural wind turbulence is unlikely. However, for the large arena, this would certainly be possible. In this case, the structural frequency of the roof is likely to be considerably greater than the Helmholtz resonance frequency of the internal pressures, and the latter will therefore not excite any structural vibration of the roof (Liu and Saathoff, 1982). It is clear, however, that there could be an intermediate combination of area and volume (such as the 'concert hall' in Table 6.1), for which the Helmholtz frequency is similar to the natural structural frequency of the roof, and in a range which could be excited by the natural turbulence in the wind. However, such a situation has not yet been recorded.

6.2.7 Non-dimensional formulation

In order to compare full-scale and wind-tunnel data on internal pressures produced by large openings (see Section 7.4.2), and to assist in the development of reduction factors for large internal volumes for design purposes (Section 6.2.8), a non-dimensional formulation of the governing equation is useful, making use of the non-dimensional groups defined in Section 6.2.1.

Holmes (1979), and Holmes and Ginger (2012), showed that Equation 6.8 can be re-written in the following form, in which all the terms are non-dimensional:

$$C_I \frac{1}{\Phi_1 {\Phi_2}^2 {\Phi_5}^2} \frac{d^2 C_{pi}}{dt^{*2}} + \left(\frac{1}{4k^2}\right)\left[\frac{1}{\Phi_1 {\Phi_2}^2 \Phi_5}\right]^2 \frac{dC_{pi}}{dt^*}\left|\frac{dC_{pi}}{dt^*}\right| + C_{pi} = C_{pe} \tag{6.15}$$

where $C_I = l_e/\sqrt{A}$ that is an 'inertial coefficient' representing the effective length of the 'slug' discussed in Section 6.2.3, non-dimensionalised by the area of the opening. t^* is a non-dimensional time given by $(t\bar{U}/l_u)$.

Φ_1, Φ_2 and Φ_5 were defined in Section 6.2.1, and k is the loss coefficient defined in Section 6.2.2.The product $\Phi_1\Phi_2^2$ can be replaced by a single non-dimensional variable and defined as the non-dimensional opening size to the volume parameter, $S^* = (a_s/\bar{U})^2\left(A^{3/2}/V_0\right)$. S^* is related to the parameter, S, adopted by Yu et al. (2006), by $S^* = S^{3/2}$. Vickery and Bloxham (1992) defined an alternative parameter, also denoted by S, equal to the square root of S^*.

S^* is the single most important non-dimensional parameter affecting internal pressure fluctuations. For example, for correct scaling, it is desirable to have an equality of this parameter between model and full scale, when carrying out wind-tunnel studies of fluctuating internal pressures (see Sections 6.5 and 7.4.2).

6.2.8 Reduction factors for large volumes and small opening areas

After fitting the available experimental data, Holmes and Ginger (2012) proposed the following expressions for the ratio of the standard deviation of the internal pressure to that of the external pressure at the opening:

$$\frac{\sigma_{pi}}{\sigma_{pe}} = 1.1 \quad \text{for } S^* \geq 1.0 \tag{6.16}$$

$$\frac{\sigma_{pi}}{\sigma_{pe}} = 1.1 + 0.2\log_{10}(S^*) \quad \text{for } S^* < 1.0 \tag{6.17}$$

where the parameter S^* was defined in Section 6.2.7.

The slight amplification for values of S^* greater than 1.0 (i.e. for small volumes and relatively large opening areas) occurs because of Helmholtz resonance within the internal volume (see Section 6.2.6).

The ratios of the expected peak internal pressure to the peak external pressure at the opening on a windward wall can be written as (Holmes and Ginger, 2012)

$$\frac{\hat{p}_i}{\hat{p}_e} = \frac{1 + 2gI_u\left(\sigma_{pi}/\sigma_{pe}\right)}{1 + 2gI_u} \tag{6.18}$$

where I_u is the longitudinal turbulence intensity (Section 3.3.1), and g is a peak factor for the pressures (from 3.5 to 4).

Combining Equation 6.17 with 6.18, it can be shown that for $(A^{3/2}/V_0)$ less than about 0.006:

$$\frac{\hat{p}_i}{\hat{p}_e} = a_1 + a_2 \log_{10}\left(\frac{A^{3/2}}{V_0}\right) \tag{6.19}$$

where

$$a_1 = \frac{1 + 2gI_u\left[1.1 + 0.4\log_{10}\left(\frac{a_s}{\overline{U}}\right)\right]}{1 + 2gI_u} \quad \text{and} \quad a_2 = \frac{0.4gI_u}{1 + 2gI_u}$$

It can be shown that the values of a_1 and a_2 are relatively insensitive to a range of typical design values of $\overline{U}$ (20–35 m/s), and of I_u (0.15–0.30), and to the peak factor, g (3.5–4.0). Then, taking average values of a_1 and a_2 of 1.33 and 0.12, respectively:

$$\frac{\hat{p}_i}{\hat{p}_e} \approx 1.33 + 0.12\log_{10}\left(\frac{A^{3/2}}{V_0}\right) \quad \text{for } (A^{3/2}/V_0) < 0.006 \tag{6.20}$$

$\hat{p}_i/\hat{p}_e \approx 1.06$ for $(A^{3/2}/V_0) \geq 0.006$. Although a number of approximations have been made to derive it, Equation 6.20 can be used to indicate the approximate reduction in peak internal pressure expected within a building of large internal volume, as is illustrated by the following example.

It should be noted that Equation 6.20 has been derived for full-scale design wind speeds, and hence generally is not applicable to wind-tunnel data obtained at lower wind speeds, for which Equation 6.19 should be used directly.

EXAMPLE

Consider a rigid building of internal volume of 120,000 m², with a single dominant opening of 10 m² on the windward wall.

$A^{3/2}/V_0 = 10^{1.5}/120{,}000 = 2.6 \times 10^{-4}$, that is less than 0.006

By Equation 6.20, $\dfrac{\hat{p}_i}{\hat{p}_e} \approx 1.33 + 0.12 \log_{10}\left(2.6 \times 10^{-4}\right) = 1.33 - 0.43 = 0.90$

Thus, a reduction factor of about 0.9 should be applicable to obtain the peak internal pressure from the external pressure in the vicinity of the opening.

6.3 MULTIPLE WINDWARD AND LEEWARD OPENINGS

6.3.1 Mean internal pressures

The mean internal pressure coefficient inside a building with total areas (or effective areas if permeability is included) of openings on the windward and leeward walls of A_W and A_L, respectively, can be derived by using, Equation 6.3, and applying mass conservation. The latter relation can be written for a total of N openings in the envelope:

$$\sum_{1}^{N} \rho_a Q_j = 0 \tag{6.21}$$

If quasi-steady and incompressible flow is assumed initially, we can assume the density, ρ_a, to be constant. Then, applying Equation 6.3 for the flow through each of the N openings, Equation 6.21 becomes

$$\sum_{1}^{N} A_j \sqrt{\left|p_{e,j} - p_i\right|} = 0 \tag{6.22}$$

where the modulus, $\left|p_{e,j} - p_i\right|$, allows for the fact that for some openings, the flow is from the interior to the exterior.

Figure 6.3 shows a building (or a floor of a high-rise building) with five openings in the envelope. Applying Equation 6.22 to this case:

$$A_1\sqrt{\left|p_{e,1} - p_i\right|} + A_2\sqrt{\left|p_{e,2} - p_i\right|} + A_3\sqrt{\left|p_{e,3} - p_i\right|} = A_4\sqrt{\left|p_{e,4} - p_i\right|} + A_5\sqrt{\left|p_{e,5} - p_i\right|} \tag{6.23}$$

In Equation 6.23, the inflows through the windward openings on the left-hand side balance the outflow through openings on the leeward and side walls, on the right-hand side. Equation 6.23, or similar equations for a large number of openings, can be solved by iterative numerical methods.

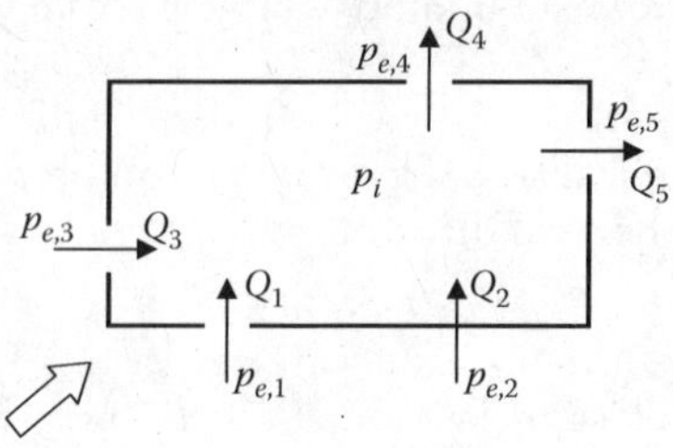

Figure 6.3 Inflows and outflows for multiple openings.

For the simpler case of a single windward opening with a single leeward opening, Equation 6.22 can be applied, with a conversion to pressure coefficients, to give:

$$A_W\sqrt{C_{pW} - C_{pi}} = A_L\sqrt{C_{pi} - C_{pL}}$$

This can be re-arranged to give Equation 6.24 for the coefficient of internal pressure:

$$C_{pi} = \frac{C_{pW}}{1 + (A_L/A_W)^2} + \frac{C_{pL}}{1 + (A_W/A_L)^2} \tag{6.24}$$

Equation 6.24 can be applied with A_W taken as the combined open area for several openings on a windward wall, and C_{pW} taken as an average mean pressure coefficient, with similar treatment for the leeward/side walls. It has been applied to give specified values of internal pressures in design codes and standards (see Chapter 15), in which case the coefficients are used with mean pressure coefficients to predict peak internal pressures, making use of the quasi-steady assumption (see Section 4.6.2).

Measurements of mean internal pressure coefficients for a building model with various ratios of windward/leeward opening area are shown in Figure 6.4. The solid line in this figure is Equation 6.24 with C_{pW} taken as +0.7, and C_{pL} taken as −0.2. These values were the values of mean external pressure coefficients on the walls at or near the windward and leeward openings, respectively. It may be seen that the agreement between the measurements and Equation 6.24 is good.

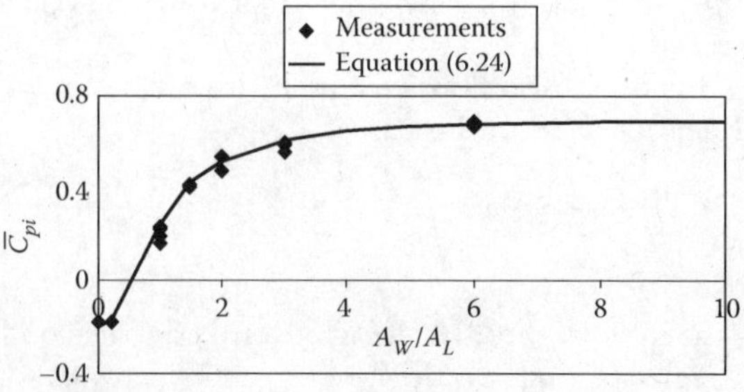

Figure 6.4 Mean internal pressure coefficient as a function of windward/leeward open area.

6.3.2 Fluctuating internal pressures

The analysis of fluctuating internal pressures when there are openings on more than one wall of a building is more difficult than for a single opening. In general, numerical solutions are required (Saathoff and Liu, 1983). However, some useful results can be obtained if the inertial terms are neglected, and the damping term is linearised (Vickery, 1986, 1991; Harris, 1990). The neglect of the inertial term in comparison to the damping term is justified when there is background porosity on the walls of a building, but may not be so when there are one or more large openings.

It can be shown (Harris, 1990) that, when there is a combined open area on a windward wall of A_W and external pressure coefficient C_{pW}, and on a leeward wall with the total open area A_L and external pressure coefficient C_{pL}, then there is a characteristic response time given by

$$\tau = \frac{\rho_a V_o \bar{U} A_W A_L}{\gamma k p_o (A_W^2 + A_L^2)^{3/2}} \sqrt{C_{pW} - C_{pL}} \qquad (6.25)$$

There is some similarity between Equation 6.25 and Equation 6.5 for a single opening, but they are not exactly equivalent. External pressure fluctuations which have periods much greater than τ are transmitted as internal pressures in a quasi-steady manner – that is they will follow Equation 6.24. Fluctuations with periods of the same order as τ will be significantly attenuated; those with periods less than τ will have a negligible effect on the fluctuating internal pressures.

The effect of the building wall and roof flexibility is such as to increase the response time according to Equation 6.26 (Vickery, 1986):

$$\tau = \frac{\rho_a V_o \bar{U} A_W A_L [1 + (K_A/K_B)]}{\gamma k p_o (A_W^2 + A_L^2)^{3/2}} \sqrt{C_{pW} - C_{pL}} \qquad (6.26)$$

For 'normal' low-rise building construction, K_A/K_B is about 0.2 (Vickery, 1986, and, Table 6.1), and the response time therefore increases by about 20%.

6.4 NOMINALLY SEALED BUILDINGS

The situation of buildings that are nominally sealed, but have some leakage distributed over all surfaces, can be treated by neglecting the inertial terms, and lumping together windward and leeward leakage areas (Vickery, 1986, 1994; Harris, 1990).

A characteristic frequency, n_c, is obtained. Pressure fluctuations below this frequency are effectively communicated to the interior of the building. n_c is given by Equation 6.27 (Vickery, 1994).

$$\frac{n_c V_o}{\bar{U} A_{W,\text{total}}} = \frac{\phi}{2\pi} \frac{k}{1 + (K_A/K_B)} \left(\frac{a_s}{\bar{U}}\right)^2 \frac{r^{1/2}\left(r + (1/r)\right)^{3/2}}{\sqrt{C_{pw} - C_{pL}}} \qquad (6.27)$$

where r is the ratio of the total leeward wall surface area to the windward wall surface area a_s is the speed of sound and the other parameters were defined previously. $A_{W,\text{total}}$ is the total

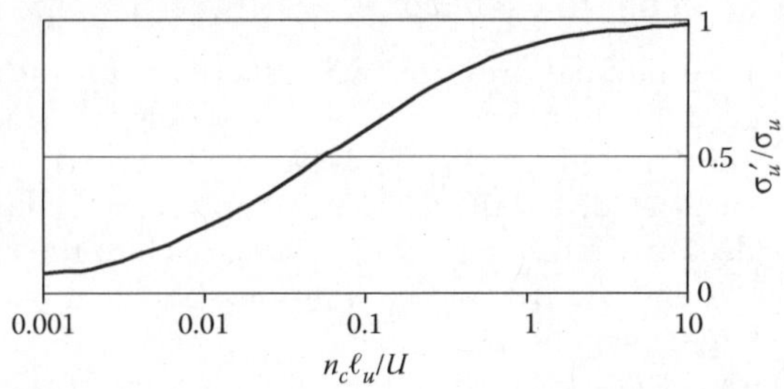

Figure 6.5 Reduction factor for fluctuating internal pressures for a building with distributed porosity. (From Vickery, B.J. 1994. *Journal of Wind Engineering and Industrial Aerodynamics*, 53: 125–44. With Permission.)

surface area of the windward wall, and φ is the wall porosity. Equation 6.27 is essentially the same as Equation 6.26, with τ equal to $(1/2\pi n_c)$.

The peak internal pressure coefficient can be estimated by

$$\hat{C}_{pi} \cong \bar{C}_{pi}\left[1 + 2g\frac{\sigma'_u}{\bar{U}}\right] \tag{6.28}$$

where σ'_u is an effective, filtered standard deviation of velocity fluctuations that are capable of generating internal pressure fluctuation, given by

$$\sigma'^2_u = \int_0^\infty S_u(n)\Big/\left[1 + (n/n_c)\right]^2 dn \tag{6.29}$$

Equation 6.29 has been evaluated using Equation 3.26 for the longitudinal turbulence spectrum, and σ'_u/σ_u is shown plotted against $(n_c\ell_u/\bar{U})$ in Figure 6.5 (Vickery, 1994). g is a peak factor which lies between 3.0 and 3.5. The mean internal pressure coefficient in Equation 6.28 can be evaluated using Equation 6.24.

Evaluation of Equation 6.29 for a large warehouse building with a wall porosity of 0.0005 gave a value of σ'_u/σ_u equal to 0.7, that is there is a 30% reduction in the effective velocity fluctuations resulting from the filtering effect of the porosity of the building (Vickery, 1994).

6.5 MODELLING OF INTERNAL PRESSURES

To correctly model internal pressures in wind-tunnel tests, it is necessary to ensure that the frequencies associated with the internal pressure fluctuations are scaled correctly with respect to the frequencies in the external flow. The relevant internal pressure frequencies are the Helmholtz resonance frequency (Sections 6.2.3 and 6.2.6), and the 'characteristic frequency' (Section 6.4).

For correct scaling of internal pressure fluctuations at full-scale design wind speeds, it is usually necessary to increase the internal volume above that obtained from normal geometric scaling. The details of the scaling rules for internal pressures are discussed in Chapter 7 (Section 7.4.2).

6.6 SUMMARY

The topic of internal pressures produced by wind has been covered in this chapter. The relevant non-dimensional parameters are introduced, and the response time of the interior of a building or a single room to a sudden increase in external pressure at an opening has been evaluated.

The dynamic response of an internal volume to excitation by a sudden generation of a windward wall opening, or by turbulence, using the Helmholtz resonator model, which includes inertial effects, has been considered. The effect of roof flexibility on the governing equation, and on the Helmholtz resonance frequency has been discussed. An approximate formula for the reduction in peak internal pressure resulting from a large internal volume, and/or a small opening area has been derived.

The effect of multiple windward and leeward openings on mean and fluctuating internal pressures has been introduced. The case of a nominally sealed building with distributed porosity was also considered.

The requirements for modelling of internal pressures in wind-tunnel studies have also been mentioned; however, the full details of this are given in Chapter 7.

Most of the results in this chapter have been validated by wind-tunnel studies, and, more importantly, by full-scale measurements (e.g. Ginger et al., 1997).

REFERENCES

Euteneur, G.A. 1970. Druckansteig im Inneren von Gebauden bei Windeinfall. *Der Bauingenieur*, 45: 214–6.

Ginger, J.D., Mehta, K.C. and Yeatts, B.B. 1997. Internal pressures in a low-rise full-scale building. *Journal of Wind Engineering and Industrial Aerodynamics*, 72: 163–74.

Harris, R.I. 1990. The propagation of internal pressures in buildings. *Journal of Wind Engineering and Industrial Aerodynamics*, 34: 169–84.

Holmes, J.D. 1979. Mean and fluctuating internal pressures induced by wind. *Proceedings, 5th Internal Conference on Wind Engineering*, Pergamon Press, Fort Collins, Colorado, 435–50.

Holmes, J.D. and Ginger, J.D. 2012. Internal pressures—The dominant windward opening case—A review. *Journal of Wind Engineering and Industrial Aerodynamics*, 100: 70–6.

Liu, H. and Rhee, K.H. 1986. Helmholtz oscillation in building models. *Journal of Wind Engineering and Industrial Aerodynamics*, 24: 95–115.

Liu, H. and Saathoff, P.J. 1982. Internal pressure and building safety. *ASCE Journal of the Structural Division*, 108: 2223–34.

Malecki, I. 1969. *Physical Foundations of Technical Acoustics*. Pergamon Press, Oxford.

Rayleigh, L. 1896. *Theory of Sound*, Vol. 2. Macmillan, London UK. (Reprinted by Dover Publications, New York, USA. 1945).

Saathoff, P.J. and Liu, H. 1983. Internal pressure of multi-room buildings. *Journal of the Engineering Mechanics Division, American Society of Civil Engineers*, 109: 908–19.

Sharma, R.N. 2008. Internal and net envelope pressures in a building having quasi-static flexibility and a dominant opening. *Journal of Wind Engineering and Industrial Aerodynamics*, 96: 1074–83.

Vickery, B.J. 1986. Gust factors for internal pressures in low-rise buildings. *Journal of Wind Engineering and Industrial Aerodynamics*, 23: 259–71.

Vickery, B.J. 1991. Discussion of the propagation of internal pressures in buildings, by R.I. Harris. *Journal of Wind Engineering and Industrial Aerodynamics*, 37: 209–12.

Vickery, B.J. 1994. Internal pressures and interaction with the building envelope. *Journal of Wind Engineering and Industrial Aerodynamics*, 53: 125–44.

Vickery, B.J. and Bloxham, C. 1992. Internal pressure dynamics with a dominant opening. *Journal of Wind Engineering and Industrial Aerodynamics*, 41: 193–204.

Yu, S., Lou, W. and Sun, B. 2006. Wind-induced internal pressure fluctuations of structure with single windward opening. *Journal of Zhejiang University Science B*, 7: 415–23.

Chapter 7

Laboratory simulation of strong winds and wind loads

7.1 INTRODUCTION

Practising structural engineers will not generally themselves operate wind-tunnels, or other laboratory equipment, for simulation of strong wind effects on structures, but they may be clients of specialist groups who will provide wind-loading information for new or existing structures, usually by means of model tests. For this reason, this chapter will not attempt to describe in detail wind-tunnel, or other simulation, techniques. There are detailed references, guide books and manuals of practice available which perform this function (e.g. Cermak, 1977; Reinhold, 1982; Australasian Wind Engineering Society, 2001; American Society of Civil Engineers, 2012). However, sufficient detail is given here to enable the educated client to be able to 'ask the right questions' of their wind-tunnel contractors.

In the following section, a brief description of wind-tunnel layouts is given, and methods of simulation of natural wind flow and experimental measurement techniques are discussed.

7.2 WIND-TUNNEL LAYOUTS

7.2.1 Historical

The first use of a wind tunnel to measure wind forces on buildings is believed to have been made by Kernot in Melbourne, Australia (1893). A sketch of the apparatus, which he called a 'blowing machine', is given in Figure 7.1 (Aynsley et al., 1977). This would now be described as an 'open-circuit, open test-section' arrangement. With this equipment, Kernot studied wind forces on a variety of bluff bodies – cubes, pyramids, cylinders and so on, and on roofs of various pitches.

At about the same time, Irminger (1894) in Copenhagen, Denmark used the flow in a flue of a chimney to study wind pressures on some basic shapes (Larose and Franck, 1997).

Wind tunnels for aeronautical applications developed rapidly during the first half of the twentieth century, especially during and between the two World Wars. The two basic wind-tunnel layouts: the *open circuit*, or 'N.P.L. (National Physical Laboratory) – type', and the *closed circuit*, or 'Göttingen-type' were developed during this period, named after the research establishments in the United Kingdom and Germany where they originated. These two types are outlined in the following sections.

7.2.2 Open-circuit type

The simplest type of wind-tunnel layout is the open-circuit or NPL type. The main components are shown in Figure 7.2. The contraction, usually with a flow straightener, and fine

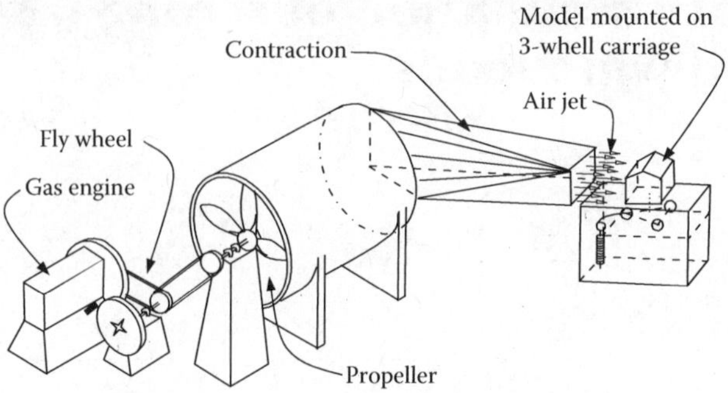

Figure 7.1 Sketch of W.C. Kernot's 'blowing machine' of 1893. (From Aynsley, R.D., Melbourne, W.H. and Vickery, B.J. 1977. *Architectural Aerodynamics*. Applied Science Publishers, London. With permission.)

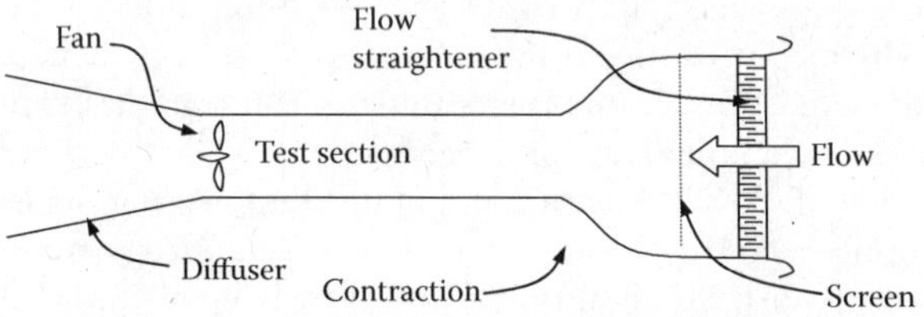

Figure 7.2 Layout of an open-circuit wind tunnel.

mesh screens, has the function of smoothing out mean flow variations, and reducing turbulence in the test section. For modelling atmospheric boundary-layer flows, which are themselves very turbulent, as described in Chapter 3, it is not essential to include a contraction, although it is better to start with a reasonably uniform and smooth flow before commencing to simulate atmospheric profiles and turbulence.

The function of the diffuser, shown in Figure 7.2, is to conserve power by reducing the amount of kinetic energy that is lost with the discharging air. Again, this is not an essential item, but omission will be at the cost of higher power consumption.

Figure 7.2 shows an arrangement with an axial-flow fan downstream of the test section. This arrangement is conducive to better flow, but, since the function of the fan is to produce a pressure rise to overcome the losses in the wind tunnel, there will be a pressure drop across the walls and floor of the test section that can be a problem if leaks exist. An alternative is a 'blowing' arrangement in which the test section is downstream of the fan (see Figure 7.5). Usually, a centrifugal blower is used, and a contraction with screens is essential to eliminate the swirl downstream of the fan. However, in this arrangement, the test section is at or near atmospheric pressure.

Both the arrangements described above have been used successfully in wind-engineering applications.

7.2.3 Closed-circuit type

In the closed-circuit, or Göttingen-type, wind tunnel, the air is continually recirculated, instead of being expelled. The advantages of this arrangement are as follows:

- It is generally less noisy than the open-circuit type.
- It is usually more efficient. Although the longer circuit gives higher frictional losses, there is no discharge of kinetic energy at the exit.
- More than one test section with different characteristics can be incorporated.

However, this type of wind tunnel has a higher capital cost, and the air heats up over a long period of operation before reaching a steady-state temperature. This can be a problem when operating temperature-sensitive instruments, such as hot-wire or other types of thermal anemometers, which use a cooling effect of the moving air for their operation.

7.3 SIMULATION OF THE NATURAL WIND FLOW

In this section, methods of simulation of strong wind characteristics in a wind tunnel are reviewed. Primarily, the simulation of the atmospheric boundary layer in gale, or large-scale synoptic conditions, is discussed. This type of large-scale storm is dominant in the temperate climates, for latitudes greater than about 40 degrees, as discussed in Chapter 1.

Even in large-scale synoptic wind storms, flows over sufficiently long homogeneous fetch lengths, so that the boundary layer is fully developed, are relatively uncommon. They will occur over an open sea with consistent wave heights, and follow large fetches of a flat open country or desert terrain. Buildings or other structures, which are exposed to these conditions, are few in number, however. Urban sites, with flat homogeneous upwind roughness of sufficient length to produce full development of the boundary layer, are also relatively uncommon. However, there have been sufficient measurements in conditions that are close to ideal to produce generally accepted semi-theoretical models of the strong wind atmospheric boundary layer for engineering purposes. These models have been validly used as the basis for wind-tunnel modelling of phenomena in the atmosphere, and the salient points have been discussed in Chapter 3.

In the case of the wind loading and response of structures, such as buildings, towers, bridges and so on, gales produced by large, mature, extra-tropical, depressions are adequately described by these models, and they form a benchmark by which wind-tunnel flows are usually assessed. However, there are significant differences of opinion regarding some turbulence properties, such as length scales and spectra, which are important in determining wind forces and dynamic response. These uncertainties should be considered when assessing the reliability of wind-tunnel tests as a predictor of wind effects on real structures.

As outlined in Chapter 3, these models are also not good ones for storm winds produced by localised thermal mechanisms, namely tropical cyclones (hurricanes, typhoons), thunderstorms (including tornadoes) and monsoons. Winds produced by these storms are the dominant ones for the design of structures in latitudes within about 40 degrees from the Equator.

The following sections consider natural growth methods requiring long test sections, methods used for wind tunnels with short test sections and methods developed for simulating only the inner or surface layer of the atmospheric boundary layer. Finally, some possibilities for simulations of strong winds in tropical cyclone and thunderstorm conditions are discussed. Laboratory modelling of these phenomena is still in an early stage of development, but some ideas on the subject are presented in Section 7.3.4.

7.3.1 Similarity criteria and natural growth methods

The 'ideal' neutral atmospheric boundary layer has two characteristic length scales – one for the outer part of the flow which depends on the rate of rotation of the earth and the latitude and on a velocity scale, and one for the flow near the surface itself which depends on the size and density of the roughness on the surface. The region near the surface, which is regarded as being independent of the effects of the earth's rotation, has a depth of about 100 m, and is known as the *inner* or *surface layer.*

The first deliberate use of a boundary-layer flow to study wind pressure on buildings was apparently by Flachsbart (1932). However, the work of Martin Jensen in Denmark provided the foundation for modern boundary-layer wind-tunnel-testing techniques. Jensen (1958) suggested the use of the inner layer length scale, or roughness length z_o (see Section 3.2.1), as the important length scale in the atmospheric boundary-layer flow, so that for modelling phenomena in the natural wind, ratios such as building height to roughness length (h/z_o) – later known as the Jensen Number, are important. Jensen (1965) later described model experiments carried out in a small wind tunnel in Copenhagen, in which natural boundary layers were allowed to grow over a fetch of uniform roughness on the floor of the wind tunnel. In the 1960s, larger 'boundary-layer' wind tunnels were constructed, and began to be used for wind-engineering studies of tall buildings, bridges and other large structures (Davenport and Isyumov, 1967; Cermak, 1971). These tunnels are either of a closed-circuit design (Section 7.2.3), or an open circuit of the 'sucking' type, with the axial-flow fan mounted downstream of the test section (Section 7.2.2). In more recent years, several open-circuit wind tunnels of the 'blowing' type have been constructed with a centrifugal fan upstream of the test section, supplying it through a rapid diffuser, a settling chamber containing screens and a contraction. As discussed in Section 7.2.2, the latter system has the advantage of producing nearly zero static pressure difference across the wind-tunnel walls at the end of the boundary-layer test section.

A naturally grown rough-wall boundary layer will continue to grow until it meets the boundary layer on the opposite wall or roof. In practical cases, this equilibrium situation is not usually reached, and tests of tall structures are carried out in boundary layers that are still developing, but are sufficient to envelop the model completely. In most cases of structural tests, more rapid boundary-layer growth must be promoted by a 'tripping' fence or grid at the beginning of the test section. Dimensional analysis indicates that the full height of the atmospheric boundary layer depends on the wind speed and the latitude. However, the typical height is about 1000 m. Assuming a geometric scaling ratio of 1/500, this means that a *minimum* wind-tunnel height of 2 m is required to model the full atmospheric boundary layer. Usually, a lower boundary-layer height is accepted, but the turbulent boundary-layer flow should completely envelop any structure under a test.

In the early days of boundary-layer wind tunnels, it was common to install a roof of adjustable height for the purpose of maintaining a constant pressure gradient in the along-wind direction. This allows for the increasing velocity deficit in the flow direction, and maintains the 'free-stream' velocity outside of the boundary layer approximately constant. This should also reduce the errors due to blockage for large models. For smaller models with lower blockage ratios, the errors in the measurements when the roof is maintained at a constant height, or with a fixed slope are quite small, and it has been found to be unnecessary to continually adjust the roof, in most situations. Blockage errors and corrections are discussed in Section 7.7.

As noted previously, the real atmospheric boundary layer is affected by the earth's rotation, and apparent forces of the Coriolis type must be included when considering the equations of motion of air flow in the atmosphere. One effect of this is to produce a mean velocity vector which is not constant in direction with height; it is parallel to the pressure gradient at the top of the boundary layer (or 'gradient' height), and rotates towards the lower static pressure side as the ground level is approached. This effect is known as the 'Ekman Spiral', (although the original solution by Ekman was obtained by assuming a shear stress in the flow proportional to the vertical velocity gradient – an assumption later shown to be unrealistic), and it has been shown to occur in full scale, with mean flow direction changes up to 30 degrees having been measured. This effect cannot be achieved in conventional wind tunnels, and the direction change is usually regarded as unimportant over the heights of most structures.

7.3.2 Methods for short test sections

In the 1960s and 1970s, to avoid the costs of constructing new boundary-layer wind tunnels, several methods of simulating the atmospheric boundary layer in existing (aeronautical) wind tunnels with test sections of low aspect ratio, i.e. short with respect to their height and width, were investigated. These usually make use of tapered fins or spires, which produce an immediate velocity gradient downstream, and which develops into a mean velocity profile representative of that in the atmosphere within a short downstream distance. Other bluff devices, such as grids or barriers, are required upstream, together with roughness on the floor of the wind tunnel, to increase the turbulence intensities to full-scale values.

Flows produced by these methods are likely to be still in a process of rapid development at the end of the short test section, and the interaction of the vortex structures produced in the wakes of the various devices may well result in unwanted characteristics in the turbulence at the measurement position. Unless detailed fluctuating velocity measurements, including spatial correlations, are made, such characteristics may never be detected. Fortunately, wind pressures and forces on structures appear to be dependent mainly on single-point statistics, such as turbulence intensities, and integral length scales in the along-wind direction, and not on the detailed eddy structures within the turbulence, in the approach flow.

Among the several methods developed in the late 1960s and early 1970s, that of Counihan (1969) is perhaps the best documented. The upstream devices consisted of a castellated fence, or barrier, several elliptical 'sharks-fins' and a short fetch of surface roughness (Figure 7.3). Detailed measurements of mean velocity and turbulence intensity profiles at various span-wise stations, and of cross-correlations and spectra were made.

7.3.3 Simulation of the surface layer

For simulation of wind forces and other wind effects on low-rise buildings, say <10 m in height, geometric scaling ratios of 1/400 result in extremely small models and do not allow any details on the building to be reproduced. The large differences in Reynolds Numbers between the model and full scale may mean that the wind-tunnel test data are quite unreliable. For this type of structure, no attempt should be made to model the complete atmospheric boundary layer. Simulation of the inner or surface layer, which is approximately

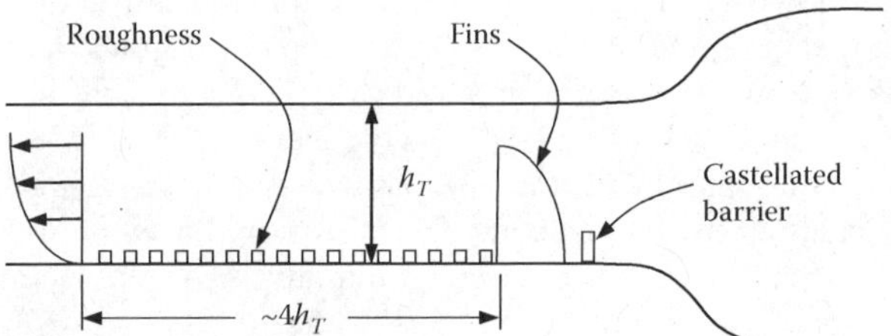

Figure 7.3 The Counihan method for short test sections.

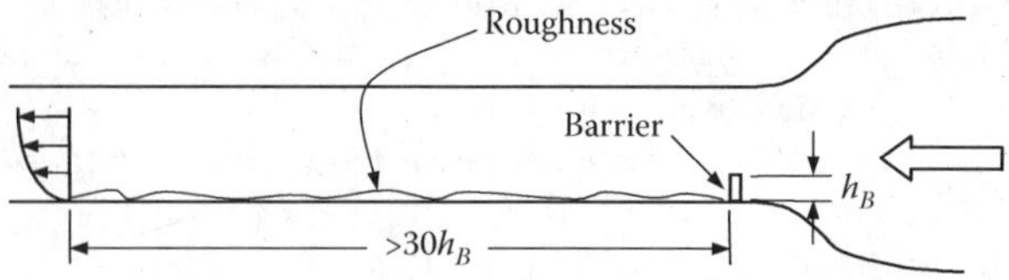

Figure 7.4 The barrier-roughness technique.

100 m thick in full scale, is sufficient for such tests. If this is done, larger and more practical, scaling ratios in the range of 1/50–1/200 can be used for the models.

Cook (1973) developed a method for simulation of the lower third of the atmospheric boundary layer. This system consists of a castellated barrier, a mixing grid and surface roughness. A simpler system consisting of a plain barrier, or wall, at the beginning of the test section, followed by several metres of uniform surface roughness has also been used (Figure 7.4) (Holmes and Osonphasop, 1983). This system has the advantage that simultaneous control of the longitudinal turbulence intensity and the longitudinal length scale of turbulence, to match the model-scaling ratio, is obtained by adjustment of the height of the barrier. Larger scales of turbulence can be produced by this method than by other approaches – large horizontal vortices with their axes normal to the flow are generated in the wake of the barrier. Studies of the development of the flow in the wake of the barrier (Holmes and Osonphasop, 1983) showed that a fetch length of at least 30 times the barrier height is required to obtain a stable and monotonically increasing mean velocity profile. However, there is still a residual peak in the shear stress profile at the height of the barrier at this downstream position; this shows that the flow is still developing at the measurement position, but the effect of this on pressures on and flow around single buildings should not be significant.

7.3.4 Simulation of tropical cyclone and thunderstorm winds

As discussed in Chapter 1, strong winds produced by tropical cyclones and thunderstorms dominate the populations of extreme winds in most locations with latitudes <40 degrees, including many sites in the United States, Australia, India and South Africa. Unfortunately, full-scale measurements of such events are few in number, and there are no reliable analytical models for the surface wind structures in these storms. However, a few full-scale measurements, and some meso-scale numerical models, have enabled qualitative characteristics of the winds to be determined.

Tropical cyclones, also known as 'hurricanes' and 'typhoons' in some parts of the world, are circulating systems with a complex three-dimensional wind structure near their centre (Section 1.3.2). At the outer radii, where the wind speeds are lower, a boundary-layer structure should exist and conventional boundary-layer wind tunnels should be quite adequate for flow modelling. However, the region of maximum horizontal winds occurs just outside the eye wall. In that region, measurements with 'dropwindsondes' have indicated a maximum in the velocity profile at a height of about 300 m. Above that height, the mean wind velocity is approximately constant or falls slowly (Section 3.2.5). Measurements of turbulence intensities in typhoons have sometimes shown intermittent higher values than they occur at the same site in non-cyclonic conditions (Section 3.3.1). However, generally, tropical cyclone wind flow can be represented by a boundary-layer flow similar to that used for simulating extra-tropical conditions.

The laboratory modelling of thunderstorm winds is a more difficult problem for a number of reasons. Firstly, there are a number of different types of local wind storms associated with

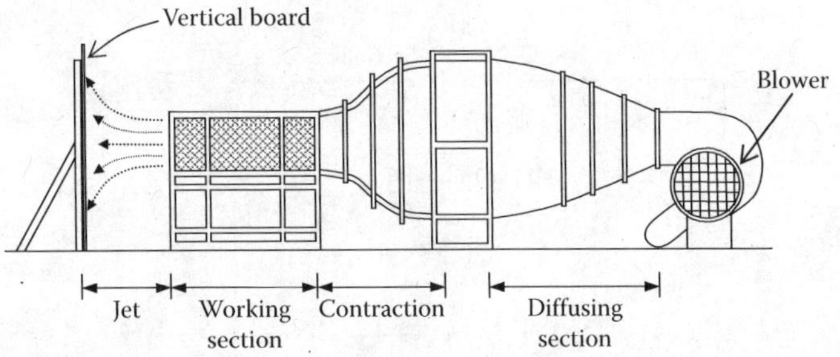

Figure 7.5 Simulation of thunderstorm downburst by an impinging jet.

thunderstorms, although some of these have similar characteristics. Secondly, these storms are individually transient, although a number of them may occur sequentially in the same day. The length of an individual storm rarely exceeds 30 min. Thirdly, thunderstorm winds are driven by thermodynamic processes which probably cannot be reproduced in a laboratory simulation.

The velocity profile in a thunderstorm downdraft is quite similar to a wall jet. The latter has been proposed as a laboratory model of the flow in a downdraft, and some studies have been conducted using the outlet jet from a wind tunnel impinging on a vertical board, as shown in Figure 7.5. Measurements can be carried out at various radial positions from the centre of the board. This system gives velocity profiles which are quite similar to those measured by radar in microbursts, but the transient characteristics of the real downdraft flow are not reproduced, and the turbulence characteristics in the two flows could be quite different.

7.3.5 Laboratory simulation of tornadoes

Some characteristics of tornadoes and their effects on structures were discussed in Chapter 1 (Section 1.3.4) and Chapter 3 (Section 3.2.7). Davies-Jones (1976) gave a detailed review of the simulation of tornadoes, or 'tornado-like vortices' in laboratories. These have produced reasonable kinematic and dynamic similarity with full-scale tornadoes.

Chang (1971) and Ward (1972) used a ducted fan above a flat board, with rotary motion imparted to the air flowing into a convective chamber above the board by means of a rotating screen. In the Ward type, the rising air exits the apparatus to an upper plenum, through a fine mesh honeycomb, which prevents fan-induced vorticity from entering the apparatus (Figure 7.6). In these systems, the rotational velocity is controlled by the rotational speed of the screen, and the core radius is controlled by the size of the opening to the upper plenum.

However, although methods of simulating tornadoes in laboratories are quite well developed, relatively few studies of wind pressures or forces on structures in laboratory simulations of tornadoes have been carried out, and virtually, none of them have been carried out since the 1970s.

7.4 MODELLING OF STRUCTURES FOR WIND EFFECTS

7.4.1 General approach for structural response

The modelling of structures for wind effects, in boundary-layer winds, requires knowledge of dimensional analysis and the theory of modelling (e.g. Whitbread, 1963).

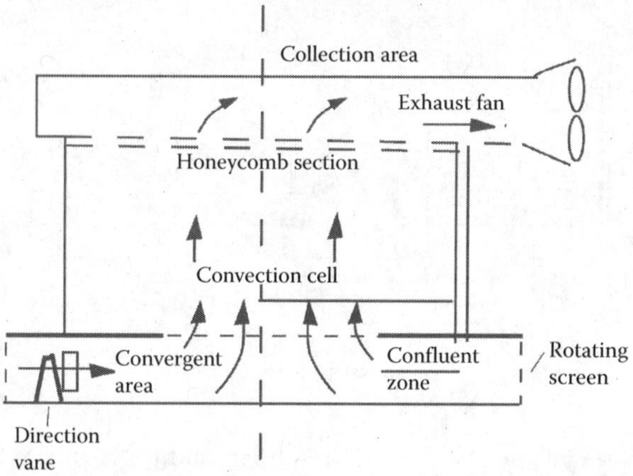

Figure 7.6 Laboratory simulation of a tornado-like vortex. (From Ward, N.B. 1972. *Journal of Atmospheric Sciences*, 29: 1194–204. With permission.)

The general approach is as follows. It may be postulated that the response of a structure to wind loading, including resonant dynamic response, is dependent on a number of basic variables, such as the following (not necessarily exclusive):

$\bar{U}$ – The mean wind speed at some reference position
z_o – Roughness length defining the approaching terrain and velocity profile (Section 3.2.1)
σ_u – Standard deviation of longitudinal turbulence
σ_v – Standard deviation of lateral turbulence
σ_w – Standard deviation of vertical turbulence
ℓ_u – Length scale of longitudinal turbulence (Section 3.3.4)
ℓ_v – Length scale of lateral turbulence
ℓ_w – Length scale of vertical turbulence
ρ_a – Density of air
υ – Viscosity of air
g – Acceleration due to gravity
ρ_s – Density of the structure
E – Young's Modulus for the structural material
G – Shear modulus for the structural material
η – Structural damping ratio
L – Characteristic length of the structure

The above list has been simplified considerably. For example, for a bridge, there will usually be different structural properties for the deck, the towers, the cables and so on. However, the above list will suffice to illustrate the principles of structural modelling.

The above sixteen dimensioned variables can be reduced to thirteen (16–3) independent dimensionless groups, according to the Buckingham-Pi Theorem. A possible list of these is as follows:

L/z_o – Jensen Number
$\sigma_u/\bar{U}$ – Longitudinal turbulence intensity

$\sigma_v/\bar{U}$ – Lateral turbulence intensity
$\sigma_w/\bar{U}$ – Vertical turbulence intensity
ℓ_u/L – Length ratio
ℓ_v/L – Length ratio
ℓ_w/L – Length ratio
$\bar{U}L/\upsilon$ – Reynolds Number (Section 4.2.4)
ρ_s/ρ_a – Density ratio
$\bar{U}/\sqrt{Lg}$ – Froude Number (inertial forces (air)/gravity forces (structure))
$E/\rho_a\bar{U}^2$ – Cauchy Number (normal internal forces in the structure/inertial forces (air))
$G/\rho_a\bar{U}^2$ – Cauchy Number (internal shear forces in the structure/inertial forces (air))
η – Critical damping ratio

For correct scaling, or similarity in behaviour between the model and full-scale structure, these non-dimensional groups should be numerically equal for the model (wind tunnel) and prototype situation.

The 13 groups are not a unique set. Other non-dimensional groups can be formed from the 16 basic variables, but there are only 13 *independent* groups, and it will be found that the additional groups can be formed by taking products of the specified groups or their powers.

For example, it is often convenient to replace a Cauchy Number by a reduced frequency ($n_sL/\bar{U}$), where n_s is a structural frequency. For structures or structural members in bending, n_s is proportional to $\sqrt{(E/\rho_sL^2)}$.

Then the reduced frequency,

$$\frac{n_sL}{\bar{U}} = K\sqrt{\frac{E}{\rho_sL^2}}\cdot\frac{L}{\bar{U}} = K\sqrt{\frac{E}{\rho_a\bar{U}^2}}\cdot\sqrt{\frac{\rho_a}{\rho_s}} \tag{7.1}$$

where K is a constant.

Thus, the reduced frequency is proportional to the square root of the Cauchy Number divided by the square root of density ratio.

7.4.2 Modelling of internal pressures

The phenomenon of Helmholtz resonance of internal pressures when the interior of a building is vented at a single opening was described in Chapter 6 (Sections 6.2.3 and 6.2.6). The 'characteristic' frequency of a building with distributed openings on windward and leeward walls was also discussed (Section 6.4). It is clearly important when simulating internal pressures in a wind-tunnel model of a building that these frequencies be scaled correctly with respect to the frequencies in the external flow. The scaling requirements to ensure this are derived in the following equations.

For a single dominant opening (area A), the Helmholtz resonance frequency is given by Equation 6.10:

$$n_H = \frac{1}{2\pi}\sqrt{\frac{\gamma Ap_o}{\rho_a\ell_eV_o}} \tag{6.10}$$

i.e. $n_H \propto \sqrt{\sqrt{A}p_o/\rho_aV_o}$

$$n_H^2 \propto \frac{\sqrt{A}\cdot p_o}{\rho_aV_o}$$

Denoting the ratio of the model to full-scale quantities by $[\,]_r$, the ratio of the model to a full-scale frequency is given by

$$[n_H]_r^2 = \frac{[L]_r[p_o]_r}{[\rho_a]_r[V_o]_r} = \frac{[L]_r}{[V_o]_r}$$

since $[p_o]_r = [\rho_a]_r = 1.0$, for testing in air at normal atmospheric pressures.

However, for scaling with frequencies in the external flow:

$$[n_H]_r = \frac{[U]_r}{[L]_r}$$

Hence, for correct scaling,

$$[n_H]_r^2 = \frac{[L]_r}{[V_o]_r} = \frac{[U]_r^2}{[L]_r^2}$$

that is

$$[V_o]_r = \frac{[L]_r^3}{[U]_r^2} \tag{7.2}$$

Thus, if the velocity ratio, $[U]_r$, is equal to 1.0, that is when the wind-tunnel speed is the same as full-scale design speeds, then the internal volume should be scaled according to the geometrical scaling ratio

$$[V_o]_r = [L]_r^3 \times 1.0$$

However, usually, in wind-tunnel testing, the wind speed is considerably less than full-scale design wind speeds. Thus, $[U]_r$ is usually <1.0, and the internal volume should then be increased by a factor of $1/[U]_r^2$. For example, if the velocity ratio is 0.5, then the internal volume, V_o, should be increased by a factor of 4.

The characteristic response time for internal pressures in a building with distributed openings on the windward side A_W, and on the leeward side, A_L, is given by Equation 6.25 (neglecting inertial effects):

$$\tau = \frac{\rho_a V_o \bar{U} A_W A_L}{\gamma k p_o (A_W^2 + A_L^2)^{3/2}} \sqrt{C_{pW} - C_{pL}} \tag{6.25}$$

For $A_W = A_L = A$ and fixed C_{Pw} and C_{PL},

$$\tau \propto \frac{\rho_a V_o \bar{U}}{p_o A}$$

and the characteristic frequency is

$$n_C \propto \frac{p_o A}{\rho_a V_o \bar{U}}$$

Then, the ratio of the model to the full-scale frequency is given by

$$[n_C]_r = \frac{[p_o]_r[L]_r^2}{[\rho_a]_r[V_o]_r[U]_r} = \frac{[L]_r^2}{[V_o]_r[U]_r}$$

For correct scaling with frequencies in the external flow,

$$[n_C]_r = \frac{[U]_r}{[L]_r} = \frac{[L]_r^2}{[V_o]_r[U]_r}$$

Hence,

$$[V_o]_r = \frac{[L]_r^{\,3}}{[U]_r^{\,2}} \tag{7.2}$$

Thus, the same scaling criterion applies, as for Helmholtz resonance frequency – that is the internal volume needs to be distorted if the velocity ratio is not equal to 1.0.

It may be noted that satisfaction of Equation 7.2 amounts to ensuring equality of the non-dimensional parameter S^*, defined in Section 6.2.7, for the case when the speed of sound, a_s, is the same in full and model scale, that is when the atmospheric pressure and air density are the same in full- and model scale.

The additional internal volume required by Equation 7.2, when the velocity ratio is <1.0, can usually be provided beneath a wind-tunnel floor and connected to the interior of the model.

Failing to provide a sufficiently large volume will generally result in over-prediction of the fluctuating internal pressures, but it is difficult to quantify the errors involved. Thus, it is advisable to correctly scale the internal volume, unless it is particularly difficult or inconvenient to do this.

7.4.3 Simulation requirements for structures in tornadoes

The similarity requirements in laboratory models of tornadoes, for simulating wind pressures on model structures, were discussed by Chang (1971) and Jischke and Light (1979). The latter proposed that the following non-dimensional parameters should be made the same in full- and model scale for correct similarity:

$$\frac{h_i}{r_c}, \quad \frac{\Gamma r_u}{Q}, \quad \frac{z_0}{h_i}, \quad \frac{L}{r_c}$$

where the dependent variables are as follows:

h_i – Depth of the layer of horizontal inflow into the tornado
r_c – Radius of the core
Γ – Imposed circulation far from the axis of the tornado
r_u – Radius of the updraft region
Q – Volume flow rate
z_0 – Surface roughness length on the ground surface
L – Characteristic length of the structure

7.4.4 Reynolds Numbers and roughening techniques

Reproducing full-scale flow conditions around circular cylinders on wind-tunnel models has been attempted on many occasions since the 1960s (e.g. Armitt, 1968). A rational approach to this was introduced by Szechenyi (1975); this involves similarity of the *roughness* Reynolds Number, Re_k, in which the average roughness height, k, is used as the length parameter.

In order for Re_k similarity to be achieved, it must also be checked that super-critical flow is obtained on the model. This can be checked by the use of Equation 4.12. The following example illustrates the use of this approach.

7.4.4.1 Example

Consider a reinforced concrete observation tower with a diameter of 13 m. The full-scale design mean wind speed at the top of the tower is 30 m/s, and the average roughness height of the concrete surface is 250 μm (250 × 10⁻⁶ m).

The equivalent wind speed for a 1/100-scale wind-tunnel model, which has been roughened with coarse sand with an average roughness height of 350 μm, is required.

The roughness Reynolds Number for full scale is

$$Re_k = \frac{30 \times 250 \times 10^{-6}}{14.5 \times 10^{-6}} \cong 517$$

To give the same value of Re_k in model scale,

$$\bar{U} = \frac{517 \times 14.5 \times 10^{-6}}{350 \times 10^{-6}} = 21.3 \text{ m/s}$$

A check should now be made that super-critical flow is achieved on the model at this wind speed.

By Equation 4.12, the value of Re_b for the minimum drag coefficient is

$$44{,}210\left(\frac{k}{b}\right)^{-0.555} = 210\left(\frac{350 \times 10^{-6}}{0.13}\right)^{-0.555} = 1.12 \times 10^5$$

For the model test at 21.3 m/s,

$$Re_b = \frac{21.3 \times 0.13}{14.5 \times 10^{-6}} = 1.91 \times 10^5$$

Since this exceeds $Re_{b,min\,C_d}$, super-critical flow will be obtained.

Note that if the above calculations are repeated for a model scale of 1/500, sub-critical flow occurs on the model, and correct similarity cannot be obtained by roughening.

7.5 MEASUREMENT OF LOCAL PRESSURES

Modern cheap sensitive solid-state pressure sensors, either as individual transducers or as part of a multi-channel electronic-scanning system, enable near-simultaneous measurements of fluctuating wind pressures on wind-tunnel models of buildings and structures for up to several hundred measurement positions (Holmes, 1995).

For reasons of cost or geometric constraint, it is usually necessary to mount the pressure sensor or scanning unit remotely from the point where the pressure measurement is required. Then the fluctuating pressure must be transmitted by tubing between the measurement and sensing points. The dynamic frequency response of the complete pressure measurement system, including the sensor itself, the volume exposed to the diaphragm and the tubing, is an important consideration.

Inadequate response can lead to significant errors especially when measuring peak pressures or suctions on building models (e.g. Durgin, 1982; Holmes, 1984; Irwin, 1988). As a rule of thumb, the equivalent full-scale upper-frequency response limit should not be less than about 2 Hz. To convert this into model frequency, the frequency ratio is obtained by dividing the velocity ratio by the geometric length-scaling ratio, for example, for a typical velocity ratio of 1/3, and a geometric ratio of 1/300, the frequency ratio is 100, and the desirable upper limit is 200 Hz.

The transmission of pressure fluctuations is affected by the mass inertia, compressibility and energy dissipation in the transmitting fluid (e.g. Bergh and Tijdeman, 1965). Standing waves can produce unwanted resonant peaks in the amplitude frequency response characteristics of the system, and a non-linear variation of phase lag with frequency.

An ideal system would have an amplitude response which is constant over the frequencies of interest, and a linear-phase variation with frequency. The latter characteristic guarantees that there is no distortion of transient pressure 'signatures' by the system.

As well as pressure measurements at a single point, systems in which pressures from a number of points are connected to a common manifold or *pneumatic averager* have been widely used. In wind engineering, this arrangement has been used to obtain fluctuating and peak pressures appropriate to a finite area, or panel, on a building model in a turbulent wind-tunnel flow (e.g. Surry and Stathopoulos, 1977; Holmes and Rains, 1981; Gumley, 1984; Holmes, 1987; Kareem et al., 1989).

7.5.1 Single-point measurements

Three systems that have been used in the past are:

1. *'Short' tube systems.* This system uses a relatively short length of tubing to connect the measurement point to the sensor. Typically, for wind-tunnel testing, this may consist of tubing that is 20–100 mm long, and 1–2 mm internal diameter. The short tube lengths will result in resonant frequencies that are high, hopefully well above the range of interest for the measurements. However, the short tube also results in low dissipation of energy, and the amplitude response rises to a high value at the peak.
2. *'Restricted' tube systems.* Restricted-tube systems may be defined as those involving one or more changes in internal diameter along the tube length. Such systems often allow location of pressure sensors at distances of 150–500 mm from the measurement point, with good amplitude and phase characteristics up to 200 Hz, or more. The simplest system of this type is the two-stage type, in which a section of a narrower tube is inserted between the main tube section and the transducer. Restricted-tube systems are very effective in removing resonant peaks and giving linear-phase response characteristics (e.g. Surry and Isyumov, 1975; Irwin et al., 1979; Holmes and Lewis, 1987a). An effective frequency range can be obtained which is better than that for a constant diameter tubing with a fraction of the length.

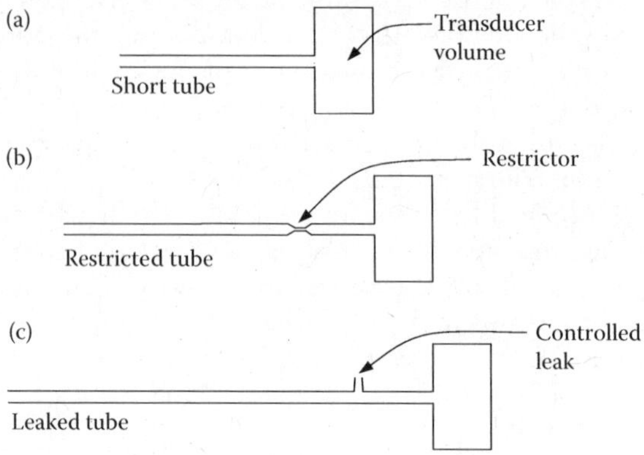

Figure 7.7 Tubing arrangements for measurement of point pressures. (a) Short tube, (b) restricted tube, and (c) leaked tube.

3. *'Leaked' tube systems.* The leaked-tube system was proposed by Gerstoft and Hansen (1987). A theoretical model was developed by Holmes and Lewis (1989). A relatively flat amplitude frequency response to frequencies of 500 Hz, with 1 m of connecting tubing, is possible with a system of this type. This is achieved by inserting a controlled side leak part way along the main connecting tube, usually close to the transducer. It has the effect of attenuating the amplitude response to low-frequency fluctuations, and to steady pressures, to the level of a conventional closed system at higher frequencies. Thus, the leak effectively introduces a high-pass filter into the system. The amplitude ratio at frequencies approaching zero is simply a function of the resistance to steady laminar flow of the main tube and leak tube. For multiple pressure tap measurements with this system, it is normally necessary to connect all the leaks to a common reference pressure, usually that inside a closed chamber, or plenum, to which the reference static pressure is also connected.

The general arrangement of the three types of single-point measurements is shown in Figure 7.7.

The systems described above were used in a period when computer-based data-acquisition systems were somewhat slower than they are now. The most common system in use today is digital correction, in which filtering, in the frequency domain, is used to correct the amplitude response to the one that is nearly flat over the frequency range of interest. Usually, the Fast Fourier Transform (FFT) is used for this purpose. The inverse FFT is then used to recreate the time series (e.g. Irwin et al., 1979). This method requires accurate knowledge of the frequency response characteristics of the tubing–transducer system before filtering. If the tubing dimensions and transducer volume are known accurately, the Bergh and Tijdeman (1965) theory can be used for this purpose; otherwise, experimental calibrations should be carried out.

7.5.2 Measurement of area-averaged pressures

Systems which average the pressure fluctuations from a number of measurement points, so that area-averaged wind loads on finite areas of a structure can be obtained, are now in common use. Averaging manifolds were first used in wind tunnels by Surry and Stathopoulos (1977). Gumley (1981, 1983) developed a theoretical model for their response.

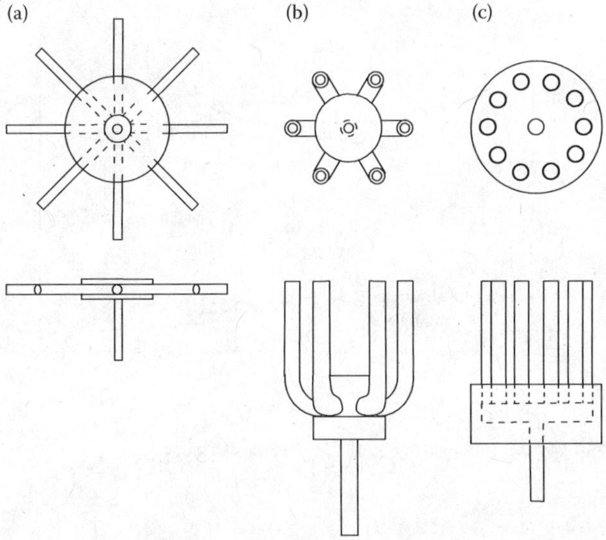

Figure 7.8 Manifolds for pressure averaging. (a) Radial, (b) axial–radial, and (c) axial.

Figure 7.8 shows the types of parallel tube and manifold arrangement that have been commonly used in wind-engineering work. Provided that the inlet tubes are identical in length and diameter, such a system should provide a true average in the manifold, of the fluctuating pressures at the entry to the input tubes, assuming that laminar flow exists in them. Usually, flatter amplitude response curves to higher frequencies can be obtained with the multi-tube–manifold systems, compared with single-point measurements using the same tube lengths, due to the reinforcement of the higher frequencies in the input tubes. However, once the number of input tubes exceeds to about five, there is little change to the response characteristics. The response is also not greatly sensitive to the volume of the averaging manifold.

The assumption that the average of discrete fluctuating point pressures, sampled within a finite area of a surface, adequately approximates the continuous average aerodynamic load on the surface and requires consideration (Surry and Stathopoulos, 1977; Holmes and Lewis, 1987b; Letchford, 1989).

Figure 7.9 shows the ratio of the variance of the averaged panel force to the variance of the point pressure, using firstly, the correct continuous averaging over the panel denoted by R_c, and secondly, the discrete averaging approximation performed using the pneumatic averaging system with the 10 pressure tappings within a panel, denoted by R_d. Calculations of these ratios were made, assuming a correlation coefficient for the fluctuating pressures of the form, $\exp(-Cr)$, where r is a separation distance, and C is a constant. The variance of the local pressure fluctuations across the panel of dimensions B by $B/2$ was assumed to be constant.

It can be seen that R_d exceeds R_c for all values of CB. This is due to the implied assumption, in the discrete averaging, that the pressure fluctuations are fully correlated in the tributary area around each pressure tap. Clearly, the errors increase with increasing C due to the lower correlation of the pressure fluctuations, and with increasing panel size, B. The errors can be decreased by increasing the number of pressure tappings within a panel of a certain size. However, it should be noted that the errors are larger at higher frequencies than at lower frequencies; a more detailed analysis of the errors requires knowledge of the coherence of the pressure fluctuations.

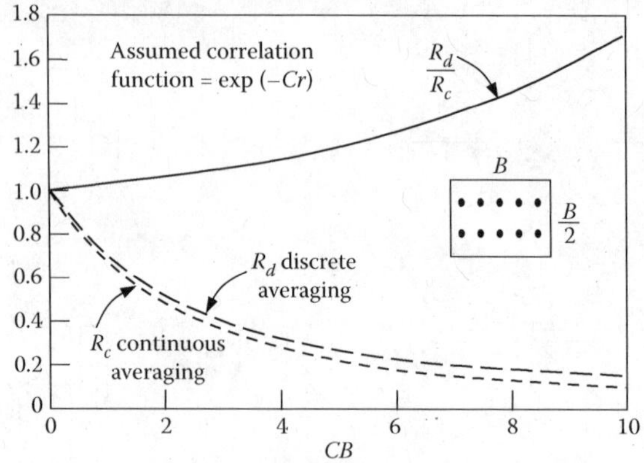

Figure 7.9 Discrete and continuous averaging of fluctuating pressures.

7.5.3 Equivalent time averaging

An alternative procedure for determining wind loads acting over finite surface areas from point pressures is known as 'equivalent time averaging'. In this approach, the time histories of fluctuating *point* pressures are filtered by means of a moving-average filter. As originally proposed by Lawson (1976), the averaging time, τ, was estimated to be given by the following formula:

$$\tau \cong 4.5\frac{L}{\bar{U}} \tag{7.3}$$

where L is usually taken as the length of the diagonal for the panel of interest.

However, a later analysis (Holmes, 1997) showed Equation 7.3 to be unconservative, and that a more correct relationship is

$$\tau \cong 1.0\frac{L}{\bar{U}} \tag{7.4}$$

However, the 'constants' in the above equations are likely to vary considerably depending on the location of the pressure measurement position on a building model – that is windward wall, roof and so on. This method is less accurate than the area-averaging technique by manifolding described in Section 7.5.2.

7.6 MODELLING OF OVERALL LOADS AND RESPONSE OF STRUCTURES

7.6.1 Base-pivotted model testing of tall buildings

This section describes the procedure for conducting the aeroelastic wind-tunnel testing of high-rise buildings, using rigid models.

The use of rigid body aeroelastic modelling of tall buildings is based on three basic assumptions:

1. The resonant response of the building to wind loads in torsional (twisting modes) can be neglected.
2. The response in sway modes higher than the first mode in each orthogonal direction can be neglected.
3. The mode shapes of the fundamental sway modes can be assumed to be linear.

With these assumptions, the motion of a rigid model of a building, pivotted at, or near, ground level, and located in a wind tunnel in which an acceptable model of the atmospheric boundary layer in strong winds has been set up, can be taken to represent the sway motion of the prototype building. The fact that a scaled reproduction of the building motion has been obtained means that fluctuating aerodynamic forces that depend on that motion have been reproduced in the wind tunnel. This is not the case when fixed models are used to measure the fluctuating wind pressures, or the 'base balance' technique is used.

Even buildings that have a non-linear mode shape can often be modelled by means of rigid body rotation, but in these cases, it may be appropriate to position the pivot point at a different level to the ground level. For example, a building supported on stiff columns near the ground level might be modelled by a rigid model pivotted at a height above the ground level (e.g. Isyumov et al., 1975). The disadvantage of this approach is that the bending moment at ground level cannot be measured.

There is a direct analogy between the generalised mass of the prototype building, G_1, and the moment of inertia of the model building, including the contributions from the support shaft and any other moving parts.

Assuming that the mode shape of the building is given by

$$\phi_1(z) = (z/h) \tag{7.5}$$

the generalised mass is given by

$$G_1 = \int_0^h m(z)\phi_1^2(z)\,dz \tag{7.6}$$

The equivalent prototype moment of inertia for rigid body rotation about the ground level is then

$$I_p = \int_0^h m(z)z^2 dz = = (1/h^2)\, G_1 \tag{7.7}$$

The equivalent model moment of inertia is then given by

$$I_m = M_r L_r^2 I_p = L_r^5 (1/h^2)\, G_1 \tag{7.8}$$

where M_r and L_r are the mass ratio and length ratio, respectively. In order to maintain a density ratio of unity in both model and full scale, assuming that air is the working fluid in both cases,

$$M_r = L_r^3 \tag{7.9}$$

Equation 7.8 can be used to establish the required model moment of inertia.

In order to obtain the correct moment of inertia, and at the same time to achieve a relatively rigid model, it is normally necessary to manufacture the model from a light material

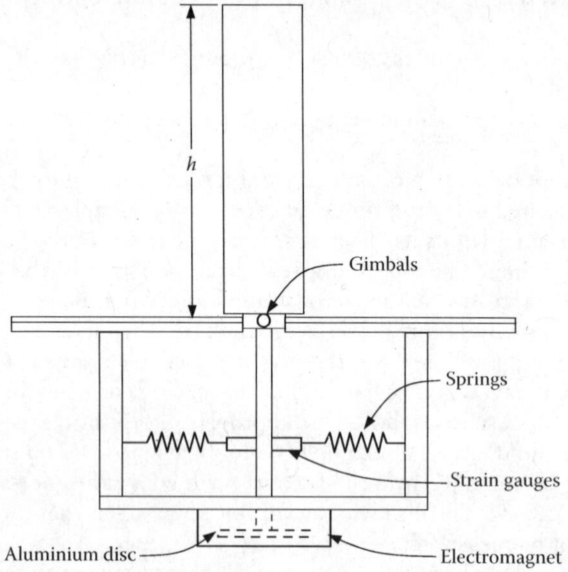

Figure 7.10 A base-pivotted tall building model.

such as expanded foam, or balsa wood. A typical mounting is shown in Figure 7.10. The model is supported by gimbals of low friction, and rotation about any horizontal axis is permitted. Elastic support can be provided by springs whose position can be adjusted vertically. In the case of the system shown in Figure 7.10, damping is provided by an eddy current device, but vanes moving in a container of viscous liquid can also be used.

The moment of inertia of the model and the supporting rod and damper plates can be determined in one or more of the following three ways:

1. By swinging the model, supporting rod and attachments, as a compound pendulum and measuring the period of oscillation
2. By measuring the frequency of vibration in the mounted position, and knowing the spring constants
3. By measuring the angular deflection of the supporting rod for known overturning moments applied to the model in position and using the measured frequencies

The support system shown in Figure 7.10 is the most common arrangement, but a method of support based on a cantilever support has also been used. The vertical position of the model on the cantilever is adjusted to minimise the rotation at ground level. The advantage of this method is that base shear, as well as base-bending moment, can be measured.

Testing of the model to determine either the base-bending moment or the tip deflection over a range of reduced velocities should be carried out. The assumptions made to justify the rigid model aeroelastic testing result in a relationship between the base-bending moment, M_b, and the tip deflection, x, as follows:

$$M_b = (\omega_1^2 I_p / h)\, x = (\omega_1^2 G_1 / h^3)\, x \cong (\omega_1^2 m\, h^2/3)\, x \tag{7.10}$$

where ω_1 is the natural circular frequency, and m is an average mass/unit height.

The relationship in Equation 7.10 implies that the mean and background wind loads are distributed over the height of the building in the same way as the resonant response, that is according to the distribution of inertial forces for the first mode response. This is a consequence of the neglect of the higher modes of vibration.

The upper limit of reduced velocity should correspond to a mean wind speed which is larger than any design value for any wind direction. As it will be required to fit a relationship between response (either peak or rms) and mean wind speed, testing should be carried out by at least three reduced velocities.

It is wise to conduct aeroelastic tests for at least two different damping ratios – a value representative of that expected at perceptible accelerations for the height and construction type, and a higher value that may be achieved at ultimate conditions, or at serviceability design conditions when an auxiliary system is added. If the resonant response is dominant, values outside these conditions can be estimated by assuming that the rms response varies as the inverse of the square root of the damping ratio.

The final stage of an aeroelastic investigation should be to provide the structural engineer with vertical distributions of loads which are compatible with the base-bending moments obtained from the experiments and subsequent processing. As discussed in Chapter 5, there are different distributions for the mean component, background or sub-resonant fluctuating component, and the resonant component of the peak response, for any wind direction. If wind-tunnel pressure measurements are available, these can be used to determine the mean load distribution. Pressure measurements could, in principle, also be used to determine the background-fluctuating loads, although this requires extensive correlation measurements; also, the loading distribution should also be 'tailored' to the particular load effect, such as a column load.

For tall buildings, a linear-loading distribution with a maximum at the top, reducing to zero at the pivot point, is often assumed. Then the load per unit height at the top of the building, w_0, is given by

$$w_0 = 3M_b/h^2 \tag{7.11}$$

For a linear mode of vibration, this is a realistic distribution for the inertial loading of the resonant part of the response (Section 5.4.4). However, this is not a realistic distribution for the mean (Section 5.4.2) or the background response (Section 5.4.3), when the loading is primarily along wind.

7.6.2 The high-frequency base-balance technique

For most tall buildings, the 'high-frequency base-balance' (HFBB) technique (Tschanz and Davenport, 1983) has now replaced aeroelastic model testing. This is alternatively known as the 'high-frequency *force* balance' (HFFB) method, or simply as the 'high-frequency balance' technique. In this method, there is no attempt to model the dynamic properties of the building – in fact, the support system is made deliberately stiff to put the building model above the range of the exciting forces of the wind. A rigid model, which reproduces the building shape, is used. The model is supported at the base by a measurement system, which is capable of measuring the mean and fluctuating wind forces and moments to a high frequency, without significant amplification or attenuation. The spectral densities of the base forces and moments are measured, and the resonant response of the building, with appropriate dynamic properties incorporated, is computed using a spectral or random vibration approach, similar to that described in Section 5.3 of Chapter 5. A range of damping ratios and mean wind speeds can be simulated using this approach.

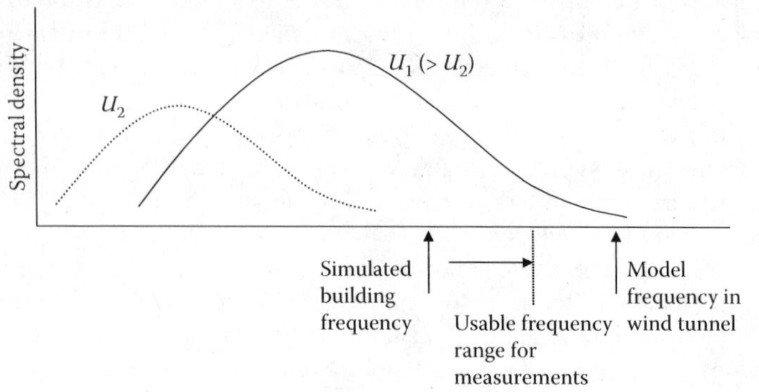

Figure 7.11 Frequency relationships for a high-frequency base balance.

Note that the HFBB measures the mean and background-fluctuating (quasi-static) base moments directly. Calculation is only required for the *resonant* components.

Figure 7.11 shows how the spectrum of wind force varies with different speeds in a wind tunnel. For a given design of balance, there will be an upper limit to the wind force (proportional to wind speed squared) that is capable of being measured by the balance; this will be proportional to the stiffness of the balance for a particular force component. Thus, the maximum wind-tunnel speed for which a balance can be used is proportional to the square root of the stiffness. Since the natural frequency of a model of a given mass is also proportional to the square root of the stiffness, the ratio of maximum wind speed to the maximum usable frequency will be a constant for a given design of balance.

When the prototype building does not have a linear sway mode shape, corrections are required to the computed resonant response, as they are for the base-pivotted aeroelastic model technique. Base torque can also be measured, and used to determine the response in the torsional mode of vibration, although quite large mode shape corrections are required as discussed in the following paragraph.

A variety of mode shape correction factors have been developed for the HFBB (e.g. Holmes, 1987; Boggs and Peterka, 1989). These depend on the assumptions made for the variation of the fluctuating wind forces (or torques) with height, and the correlation between the fluctuating sectional forces at different heights (Holmes et al., 2003). There appear to be considerable differences between various commercial laboratories with regard to the corrections made, especially for the torsional, or twist, modes. Some laboratories make use of the measured base shears, as well as the base-bending moments, available from a high-frequency base balance, and assume a linear variation of the instantaneous wind force with height (Xie and Irwin, 1998). However, such methods do not eliminate the need for mode shape corrections (Chen and Kareem, 2005a).

The base moments $M_x(t)$, $M_y(t)$ and $M_z(t)$ measured by the HFBB must be converted into generalised forces for the two fundamental sway modes and twist mode. For example, using mode shape corrections proposed by Holmes (1987) and Holmes et al. (2003):

$$Q_x(t) = \left(\frac{1}{h}\right)\sqrt{\frac{4}{1+3\beta_x}} \cdot M_y(t) \tag{7.12}$$

$$Q_y(t) = \left(\frac{1}{h}\right)\sqrt{\frac{4}{1+3\beta_y}} \cdot M_x(t) \tag{7.13}$$

$$Q_z(t) = \sqrt{\frac{1}{1+2\beta_z}} \cdot M_z(t) \tag{7.14}$$

where $Q_x(t)$, $Q_y(t)$ and $Q_z(t)$ are the generalised forces in the x- and y-sway modes, and twist modes, respectively. h is the building height.

It has been assumed that the mode shapes can be represented by power functions, that is

$$\phi_x = \left(\frac{z}{h}\right)^{\beta_x} \tag{7.15}$$

$$\phi_y = \left(\frac{z}{h}\right)^{\beta_y} \tag{7.16}$$

$$\phi_z = \left(\frac{z}{h}\right)^{\beta_z} \tag{7.17}$$

If the mode shape exponent for the twist mode, β_z, is 1.0 (i.e. the dynamic twist linearly varies with height from the ground to the top of the building), then from Equation 7.14, the mode shape correction term is √(1/3) or 0.58. This is significantly different from 1.0 because the HFBB measures the base torsional moment *uniformly* weighted with height, whereas the generalised force for the twist mode requires a *linear* weighting with height. On the other hand, the generalised forces for the sway modes usually have mode shape correction factors close to 1.0 (equal to 1.0 for linear mode shapes).

Many modern tall buildings have dynamic modes that involve *coupled* sway and twist motions. This often results from differences between the average positions of the centre of mass and centre of stiffness (shear centre) of the cross-sections of the building. It is extremely difficult (and expensive) to manufacture accurate aeroelastic wind-tunnel models of buildings with coupled modes. However, methods are available to make reasonable predictions of the resonant contributions from the coupled modes of tall buildings using the HFBB technique (Holmes et al., 2003; Chen and Kareem, 2005b).

The HFBB technique requires relatively simple models, and clearly reduces the amount of wind-tunnel-testing time by a large factor, at the expense of computing resources, which have rapidly become cheaper. There are methodologies to account for complex coupled sway and twist dynamic modes. Most tall buildings can adequately be studied using the HFBB technique – a very cost-effective method.

Recent developments in the HFBB technique, as applied to model studies of tall buildings, have been described in detail in several papers in a special issue of the international journal, '*Wind and Structures*' (Boggs, 2014; Chen et al., 2014; Ho et al., 2014; Tse et al., 2014; Xie and Garber, 2014).

Results (i.e. base moments and roof-top accelerations) from eight different laboratories for two 'benchmark' tall buildings, based on the HFBB method, have been compared in an international comparison reported by Holmes and Tse (2014). The dimensions and dynamic properties (i.e. frequencies, mode shapes and damping) of the two buildings were specified tightly, together with the flow properties in the approaching simulated atmospheric boundary layer. The agreement between the peak base moments (including base torque) was good, with coefficients of variation of <14% for the base-bending moments. There was greater variation in the predictions of the peak accelerations at the corner of the roofs; these are more sensitive to assumptions and correction methods for mode shape, coupled and closely spaced modes and combination of acceleration components. More detail of these comparisons is given, in a graphical form, on the web site of the International Association of Wind Engineering (http://iawe.org/).

7.6.3 Sectional and taut strip models of bridges

A common, and long-standing, technique to confirm the aerodynamic stability of the decks of long-span suspension or cable-stayed bridges is the section model test. This is another form of rigid body aeroelastic modelling. The technique dates back to the investigations following the failure of the first Tacoma Narrows Bridge (Farquarson et al., 1949–1954). A short section of the bridge deck is supported on springs and allowed to move in translation and rotation. By suitable adjustment of the springs, the model frequencies in rotation and vertical translation can be arranged to have the same ratio as those for the primary bending and torsional modes of the prototype bridge. Then in order to achieve similarity between model, m, and prototype, p, the reduced frequencies (Section 7.4.1) should be kept equal:

$$\left(\frac{n_s L}{U}\right)_m = \left(\frac{n_s L}{U}\right)_p \tag{7.18}$$

where n_s should be taken as the lowest frequencies both in vertical translation (bending), and in rotation (torsion).

The models are made as rigid as possible, but they are also required to satisfy the density-scaling requirement that the ratio ρ_s/ρ_a should be the same in model and full scale, where ρ_s is the average density of the structure, and ρ_a is the air density. The details of the deck at the leading edge – such as edge beams and guard railings, are usually modelled in some detail, as these have been found to affect the aeroelastic behaviour.

Section models are primarily used to determine the critical flutter speeds of the section in both smooth and turbulent flow. The static aerodynamic coefficients can also be determined for use in calculations of turbulent buffeting of the section. A more advanced use is for determination of the aeroelastic coefficients, or flutter derivatives (Sections 5.5.3 and 12.3.2), for subsequent use in more complete computational modelling of bridge behaviour; both free- (Scanlan and Tomko, 1971) and forced vibration (e.g. Matsumoto et al., 1992) methods have been developed.

Sectional models are primarily a two-dimensional simulation, and cannot easily be used in turbulent flow, which of course is more representative of atmospheric and three-dimensional flow in nature, without significant corrections. A more advanced test method for bridges, known as 'taut strip', involves the central span of the model bridge deck supported on two parallel wires, pulled into an appropriate tension and separated by an appropriate distance, so that the bending and torsional modes are approximately matched. The deck is made in elements or short sections, so that no stiffness is provided. Such a model can be tested in full-simulated boundary-layer flows, but is more economical than a full aeroelastic model test.

Scanlan (1983) and Tanaka (1990) have given useful reviews of the section model and taut-strip techniques for bridge decks, together with a discussion of full aeroelastic model testing of bridges.

7.6.4 Multi-mode aeroelastic modelling

For the modelling of structures with non-linear mode shapes, or for structures which dynamically respond to wind in several of their natural resonant modes of vibration, such as tall towers and long-span bridges, the rigid body-modelling technique is not sufficient. In the case of long-span bridges, the aerodynamic influences of the cables and the supporting towers, which are not included in the section model or taut-strip testing (Section 7.6.3), may often be significant. More complete aeroelastic and structural modelling techniques are then required.

There are three different types of these multi-mode models:

1. 'Replica' models, in which the construction of the model replicates that of the prototype structure.
2. 'Spine' models which reproduce the stiffness properties of the prototype structure by means of smaller central members or 'spines'. Added sections reproduce the mass and aerodynamic shape of the prototype.
3. 'Lumped mass' models, in which the mass of the model is divided into discrete 'lumps', connected together by flexible elements. The number of vibration modes that can be reproduced by this type of model is limited by the number of lumped masses.

The design of these models generally follows the scaling laws based on dimensional analysis, as outlined in Section 7.4. Full model testing of suspension bridges and cable-suspended roofs, where stiffness is, at least partially, provided by gravitational forces, requires equality of Froude Number, $U/\sqrt{(Lg)}$, (introduced in Section 7.4), between model and full scale. Thus

$$\left(\frac{U}{\sqrt{Lg}}\right)_m = \left(\frac{U}{\sqrt{Lg}}\right)_p$$

since the gravitational constant, g, is the same in model and full scale, this results in a velocity scaling given by

$$\frac{U_m}{U_p} = \sqrt{\frac{L_m}{L_p}} \qquad (7.19)$$

Thus, the velocity ratio is fixed at the square root of the length ratio (or model scale). Thus, for a 1/100-scale suspension bridge model, the velocity in the wind tunnel is one-tenth of the equivalent velocity in full scale.

For the majority of structures, in which the stiffness is provided by internal stresses (e.g. axial, bending and shear), Froude Number scaling is not required for aeroelastic models, and a free choice can be made of the velocity scaling when designing a model. Usually, a fine adjustment of the velocity scaling is made after the model is built, to ensure equality of reduced frequency (see Equation 7.18).

Examples of aeroelastic models are shown in Figures 11.6 (observation tower), and 12.7 (a bridge under construction). Both of these are 'spine' models.

7.6.5 Simulation requirements for tensioned and pneumatic structures

Large roofs comprising tensioned membranes made from fabric present particular challenges if it is required to completely reproduce the dynamic and aeroelastic responses in a wind-tunnel model. However, in many cases, it is not necessary to correctly scale the elastic forces in a membrane structure, since the geometric stiffness (i.e. resistance to deflection provided by changes in angle of the membrane material as it deforms) usually dominates over the elastic stiffness of the membrane material (Tryggvason, 1979).

In pneumatic, or air-supported, structures, the tension in the membrane surface is determined by the geometry, and the internal pressurisation, Δp_i (i.e. the difference between the internal and external pressures in no-wind conditions). In air-supported roofs, the pressurisation is comparable to the weight per unit area of the membrane material. For this situation, Tryggvason (1979) proposed a scaling of the wind velocity that ensures the correct scaling of wind pressures, the internal pressure and the weight per unit area of the membrane:

$$[U]_r^2 = [\Delta p_i - \mu g]_r \tag{7.20}$$

To correctly scale the fluctuating component of the internal pressure for pneumatic structures, it is necessary to scale the internal volume according to Equation 7.2. This will ensure that the 'acoustic stiffness' is correctly scaled. The additional volume can be provided beneath the floor of a wind tunnel (Tryggvason, 1979).

However, for large air-supported roofs, the 'pneumatic' damping, associated with the ventilation air flow within the building volume, is large, and sufficient to effectively damp any resonant response of a roof of this type to wind (see Kind, 1984, and Section 10.5). Hence, aeroelastic models are usually not necessary for roofs of this type.

7.6.6 Aeroelastic modelling of chimneys

Chimneys and other slender structures of circular cross-sections are vulnerable to cross-wind excitation by fluctuating pressures due to vortex shedding (Sections 4.6.3 and 11.5). In the 1950s and 1960s, it was quite common to investigate this behaviour with small-scale wind-tunnel models. However, the forces from vortex shedding are quite dependent on Reynolds Number (Section 4.2.4), and wind-tunnel tests will severely over-estimate the cross-wind response of prototype-large chimneys (Vickery and Daly, 1984). The prediction of full-scale response of such structures is better undertaken by the use of mathematical models of the response (Section 11.5) with input parameters derived from full-scale measurements at high Reynolds Numbers.

7.6.7 Distorted 'dynamic' models

In many cases, the resonant response of a structure may be significant, but the prototype structure may be stiff enough such that aeroelastic forces (i.e. the motion-dependent forces) are not significant. Furthermore, the scaling requirements (Section 7.4.1) and properties of the available modelling materials may make it difficult, or even impossible, to simultaneously scale the mass, stiffness and aerodynamic shape of a structure.

In such cases, both the mass and stiffness properties of the structure can be distorted by the same factor (usually >1.0). Then the correct frequency relationship for the applied fluctuating wind forces and the structural frequencies is obtained. That is, it retains the correct value of reduced frequency (Section 7.4.1) and preserves the correct relationship between the frequencies associated with the flow (e.g. turbulence and vortex shedding), and those related to the structure. Internal forces and moments are correctly modelled (including resonant effects), but the deflections, accelerations and motion-induced forces, such as aerodynamic damping (Section 5.5.1), are not scaled correctly.

Such 'distorted' dynamic models have been used on certain open-frame structures, where aeroelastic 'spine' models were not possible.

7.6.8 Structural loads through pressure measurements

For structures such as large roofs of sports stadiums, or large low-rise buildings, with structural systems that are well defined and for which resonant dynamic action is not dominant, or can be neglected, wind-tunnel pressure measurements on rigid models can be effectively used to determine load effects such as member forces and bending moments, or deflections. This method is often used in conjunction with the area-averaging pressure technique described in Section 7.5.2. Also required are *influence coefficients,* representing the values of a load effect under the action of a single uniformly distributed static 'patch load' acting on the area corresponding to a panel on the wind-tunnel model. Two methods are possible.

1. Direct on-line weighting of the fluctuating panel pressures recorded in the wind-tunnel test with the structural influence coefficients, to directly determine fluctuating and peak values of the load effects (Surry and Stathopoulos, 1977; Holmes, 1988).
2. Measurement of correlation coefficients between the fluctuating pressures on pairs of panels, and calculation of rms and peak load effects by integration (Holmes and Best, 1981; Holmes et al., 1997).

The latter method has advantages that the influence coefficients are not required at the time of the wind-tunnel testing, and also that the information can be used to determine equivalent static load distributions, as discussed in Chapter 5. When resonant response is of significance, as may be the case for the largest stadium roofs, time histories of the fluctuating pressures can be used to generate a time history of generalised force for each mode of significance. From the spectral density of the generalised force, the mean square-generalised displacement (modal coordinate), and effective inertial forces acting can be determined (Section 5.4.4). The application of pressure model studies to large roofs is discussed in Chapter 10.

Pressure-based methods can also be used for structural loads and response of tall buildings (ASCE, 2012). Although these methods require a large number of simultaneous pressure measurements and extensive post-processing of the wind-tunnel data, an accurate account of non-linear resonant mode shapes can be made, and in many cases, this method has replaced the HFBB technique. A significant advantage is that the same building model used to determine local cladding pressures can be used to determine the overall wind loads and response. However, a practical difficulty with this technique is the installation of a sufficient number of tubes for pressure measurement, within the available cross-section of a model.

7.7 BLOCKAGE EFFECTS AND CORRECTIONS

In a wind tunnel with a closed test section, the walls and roof of the wind tunnel provide a constraint on the flow around a model building or group of buildings, which depends on the blockage ratio. The blockage ratio is the maximum cross-sectional area of the model at any cross-section, divided by the area of the wind-tunnel cross-section. If this ratio is high enough, there may be significant increases in the flow velocities around, and pressures on, the model. In the case of an open test section, the errors are in the opposite direction; that is the velocities around the model are reduced. To deal with the blockage problem, several approaches are possible:

- Ensure that the blockage ratio is small enough that the errors introduced are small, and no corrections are required. The usual rule for this approach is that the blockage ratio should not exceed 5%.

- Accept a higher blockage ratio, and attempt to make corrections. The difficulty with this approach is that the appropriate correction factors may themselves be uncertain. Although there are well-documented correction methods for drag and base pressure on stalled airfoils, and other bluff bodies in the centre of a wind tunnel with uniform, or homogeneous turbulent flow, there is very little information for buildings or other structures, mounted on the floor of a wind tunnel in turbulent boundary-layer flow. McKeon and Melbourne (1971) provided corrections for *mean* windward and leeward pressures, and the total drag force, on simple plates and blocks. However, no corrections are readily available for pressures, mean or fluctuating, in separated flow regions, as they occur on roofs or side walls of building models.
- Design the walls and/or roof of the working section in such a way as to minimise the blockage errors. The most promising method for doing this appears to be the *slotted wall* concept (Parkinson, 1984; Parkinson and Cook, 1992). In this system, the walls and roof of the tests section are composed of symmetrical aerofoil slats, backed with a plenum chamber. The optimum open-area ratio is about 0.55, and it is claimed that blockage area ratios of up to 30% can be used without correction.

7.8 COMPUTATIONAL WIND ENGINEERING

Computational fluid dynamics (CFD) techniques as applied to wind engineering have been under development for a number of years. There have been several conferences on the subject. It is clear that wind flow around buildings is a very complex fluid mechanics problem, involving a large range of turbulence scales – varying from the very large eddy structures of atmospheric turbulence (see Chapter 3) to the small scales generated by the flow around the bluff-body shapes of buildings and other structures (Chapter 4). The result of this is that, at the time of writing, the most common CFD techniques are capable of predicting the *mean* pressures on buildings with reasonable accuracy, but are not sufficiently accurate for the fluctuating and peak pressures. As an example, mean pressures on arched-roof buildings generated by CFD are discussed in Section 10.3.

The poor representation of the pressure fluctuations is primarily because it is necessary to incorporate over-simplified representations of the turbulence in the fluid flow equations. At the current rate of progress, this situation is unlikely to change in the short term. A realistic and wide-ranging assessment of the penetration of CFD into wind engineering was provided by Cochran and Derickson (2011). This chapter describes the successes and failures of CFD in simulating atmospheric wind-flow problems. The former includes large-scale problems such as flow over mountain ranges for which turbulence plays a relatively small role. The technique has been much less successful in predicting wind pressures and forces on buildings and other structures, especially local pressures, for which small-scale and locally generated turbulence play a significant role, as discussed in the previous paragraph.

CFD techniques are, however, currently capable of providing useful insights into wind flow around buildings for environmental considerations. Useful reviews of such techniques are given by Baskaran and Kashev (1996) and Stathopoulos and Baskaran (1996).

7.9 SUMMARY

In this chapter, a review of methods of laboratory simulation of natural strong wind characteristics for the investigation of wind pressures, forces and structural response has been given. Early methods used natural growth of boundary layers on the floor of wind tunnels to simulate the mean flow and turbulence structure in the fully developed boundary layer in gale wind conditions. To make use of shorter test sections in aeronautical wind tunnels, rapid growth methods were developed and described. For investigations on smaller structures, such as low-rise buildings, methods of simulating only the lower part, or surface layer, of the atmospheric boundary layer were devised.

Laboratory methods of simulating tornadoes, which were quite advanced as early as the 1970s, are discussed. Methods of simulating strong winds in tropical cyclones and thunderstorms, which are the dominant types for structural design at locations in the tropics and sub-tropics at latitudes from 0 to 40 degrees, are still at an early stage of development. A major problem is the lack of good full-scale data of the wind structure, on which the simulations can be based.

Experimental methods of measuring local pressures, and the overall structural loads in wind-tunnel tests are described in Sections 7.5 and 7.6, and the problem of wind-tunnel blockage, and its correction is discussed in Section 7.7.

REFERENCES

American Society of Civil Engineers. 2012. *Wind Tunnel Testing for Buildings and Other Structures*. ASCE Standard ASCE/SEI 49-12. ASCE, Reston, Virginia.

Armitt, J. 1968. The effect of surface roughness and free-stream turbulence on the flow around a model cooling tower at critical Reynolds Numbers. *Proceedings, Symposium on Wind Effects on Buildings and Structures*, Loughborough, UK, 2–4 April.

Australasian Wind Engineering Society. 2001. *Wind-Engineering Studies of Buildings. Quality Assurance Manual*. AWES-QAM-1-2001, AWES, Australia.

Aynsley, R.D., Melbourne, W.H. and Vickery, B.J. 1977. *Architectural Aerodynamics*. Applied Science Publishers, London.

Baskaran, A. and Kashev, A. 1996. Investigation of air flow around buildings using computational fluid dynamics techniques. *Engineering Structures*, 18: 861–75.

Bergh, H. and Tijdeman, H. 1965. Theoretical and experimental results for the dynamic response of pressure measurement systems. National Aero- and Astronautical Research Institute (Netherlands), Report NLR-TR-F.238, January 1965.

Boggs, D.W. 2014. The past, present and future of high-frequency balance testing. *Wind and Structures*, 18: 323–45.

Boggs, D.W. and Peterka, J.A. 1989. Aerodynamic model tests of tall buildings. *Journal of Engineering Mechanics*, 115: 618–35.

Cermak, J.E. 1971. Laboratory simulation of the atmospheric boundary layer. *AIAA Journal*, 9: 1746–54.

Cermak, J.E. 1977. Wind-tunnel testing of structures. *ASCE Journal of the Engineering Mechanics Division*, 103: 1125–40.

Chang, C.C. 1971. Tornado effects on buildings and structures by laboratory simulation. *Proceedings, 3rd International Conference on Wind Effects on Buildings and Structures*, Tokyo, Japan, 6–9 September, pp. 231–40.

Chen, X. and Kareem, A. 2005a. Validity of wind load distribution based on high frequency force balance measurements. *Journal of Structural Engineering*, 131: 984–7.

Chen, X. and Kareem, A. 2005b. Dynamic wind effects on buildings with 3D coupled modes: Application of high frequency force balance measurements. *Journal of Engineering Mechanics*, 131: 1115–25.

Chen, X., Kwon, D.-K. and Kareem, A. 2014. High-frequency force balance technique for tall buildings: A critical review and some new insights. *Wind and Structures*, 18: 391–422.

Cochran, L.S. and Derickson, R.G. 2011. A physical modeler's view of computational wind engineering. *Journal of Wind Engineering and Industrial Aerodynamics*, 99: 139–53.

Cook, N.J. 1973. On simulating the lower third of the urban adiabatic boundary layer in a wind tunnel. *Atmospheric Environment*, 7: 691–705.

Counihan, J. 1969. An improved method of simulation of an atmospheric boundary layer in a wind tunnel. *Atmospheric Environment*, 3: 197–214.

Davenport, A.G. and Isyumov, N. 1967. The application of the boundary layer wind tunnel to the prediction of wind loading. *Proceedings, International Research Seminar on Wind Effects on Buildings and Structures*, Ottawa, Canada, 11–15 September, pp. 201–30.

Davies-Jones, R.F. 1976. Laboratory simulation of tornadoes. *Symposium on Tornadoes: Assessment of Knowledge and Implications for Man*, Texas Tech University, Lubbock, TX, 22–24 June 1976, pp. 151–74.

Durgin, F. 1982. Instrumentation requirements for measuring aerodynamic pressures and forces on buildings and structures. In T. Reinhold (ed.), *Wind Tunnel Modeling for Civil Engineering Applications*. Cambridge University Press, Cambridge, UK, and New York, USA.

Farquarson, F.B., Smith, F.C. and Vincent, G.S. 1949–1954. Aerodynamic stability of suspension bridges with special reference to the Tacoma Narrows Bridge. University of Washington Engineering Experiment Station. Bulletin No. 116, Parts I to V.

Flachsbart, O. 1932. Winddruck auf geschlossene und offene Gebäude. In L.L. Prandl and A. Betz (eds.), *Ergebnisse der aerodynamischen Versuchanstalt zu Göttingen, IV*. Verlag von R. Oldenbourg, Munich and Berlin.

Gerstoft, P. and Hansen, S.O. 1987. A new tubing system for the measurement of fluctuating pressures. *Journal of Wind Engineering and Industrial Aerodynamics*, 25: 335–54.

Gumley, S.J. 1981. Tubing systems for the measurement of fluctuating pressures in wind engineering. DPhil thesis, University of Oxford.

Gumley, S.J. 1983. Tubing systems for pneumatic averaging of fluctuating pressures. *Journal of Wind Engineering and Industrial Aerodynamics*, 12: 189–228.

Gumley, S.J. 1984. A parametric study of extreme pressures for the static design of canopy structures. *Journal of Wind Engineering and Industrial Aerodynamics*, 16: 43–56.

Ho, T.C.E., Jeong, U.Y. and Case, P.C. 2014. Components of wind tunnel analysis using force balance test data. *Wind and Structures*, 18: 347–73.

Holmes, J.D. 1984. Effect of frequency response on peak pressure measurements. *Journal of Wind Engineering and Industrial Aerodynamics*, 17: 1–9.

Holmes, J.D. 1987. Mode shape corrections for dynamic response to wind. *Engineering Structures*, 9: 210–2.

Holmes, J.D. 1988. Distribution of peak wind loads on a low-rise building. *Journal of Wind Engineering and Industrial Aerodynamics*, 29: 59–67.

Holmes, J.D. 1995. Methods of fluctuating pressure measurement in wind engineering. In P. Krishna (ed.), *A State of the Art in Wind Engineering*. Wiley Eastern Limited, New Delhi.

Holmes, J.D. 1997. Equivalent time averaging in wind engineering. *Journal of Wind Engineering and Industrial Aerodynamics*, 72: 411–9.

Holmes, J.D. and Best, R.J. 1981. An approach to the determination of wind load effects for low-rise buildings. *Journal of Wind Engineering and Industrial Aerodynamics*, 7: 273–87.

Holmes, J.D., Denoon, R.O., Kwok, K.C.S. and Glanville, M.J. 1997. Wind loading and response of large stadium roofs. *International Symposium on Shell and Spatial Structures*, Singapore, 10–14 November.

Holmes, J.D. and Lewis, R.E. 1987a. Optimization of dynamic-pressure-measurement systems. I. Single point measurements. *Journal of Wind Engineering and Industrial Aerodynamics*, 25: 249–73.

Holmes, J.D. and Lewis, R.E. 1987b. Optimization of dynamic-pressure-measurement systems. II. Parallel tube–manifold systems. *Journal of Wind Engineering and Industrial Aerodynamics*, 25: 275–90.

Holmes, J.D. and Lewis, R.E. 1989. A re-examination of the leaked-tube dynamic pressure measurement system. *10th Australasian Fluid Mechanics Conference*, Melbourne, Australia. December, pp. 5.39–5.42.

Holmes, J.D. and Osonphasop, C. 1983. Flow behind two-dimensional barriers on a roughened ground plane, and applications for atmospheric boundary-layer modelling. *Proceedings, 8th Australasian Fluid Mechanics Conference*, Newcastle, Australia.

Holmes, J.D. and Rains, G.J. 1981. *Wind Loads on Flat and Curved Roof Low Rise Buildings*. Colloque Construire avec le Vent, Nantes, France, July 1981.

Holmes, J.D., Rofail, A. and Aurelius, L. 2003. High frequency base balance methodologies for tall buildings with torsional and coupled resonant modes. *Proceedings, 11th International Conference on Wind Engineering*, Lubbock, TX, USA, 1–5 June.

Holmes, J.D. and Tse, T.K.T. 2014. International high-frequency base balance benchmark study. *Wind and Structures*, 18: 457–71.

Irminger, J.O.V. 1894. Nogle forsog over trykforholdene paa planer og legemer paavirkede af luftstrominger. *Ingenioren*, 17.

Irwin, H.P.A.H., Cooper, K.R. and Girard, R. 1979. Correction of distortion effects caused by tubing systems in measurements of fluctuating pressures. *Journal of Industrial Aerodynamics*, 5: 93–107.

Irwin, P.A. 1988. Pressure model techniques for cladding wind loads. *Journal of Wind Engineering and Industrial Aerodynamics*, 29: 69–78.

Isyumov, N., Holmes, J.D., Surry, D. and Davenport, A.G. 1975. A study of wind effects for the First National City Corporation Project, New York. University of Western Ontario, Boundary Layer Wind Tunnel Special Study Report, BLWT-SS1-75.

Jensen, M. 1958. The model law for phenomena in the natural wind. *Ingenioren* (International edition), 2: 121–8.

Jensen, M. 1965. *Model Scale Tests in the Natural Wind. (Parts I and II)*. Danish Technical Press, Copenhagen, Denmark.

Jischke, M.C. and Light, B.D. 1979. Laboratory simulation of tornadic wind loads on a cylindrical structure. *Proceedings, 5th International Conference on Wind Engineering*, Fort Collins, Colorado, USA, July, pp. 1049–59, Pergamon Press.

Kareem, A. Cheng, C.-M. and Lu, P.C. 1989. Pressure and force fluctuations on isolated circular cylinders of finite height in boundary layer flows. *Journal of Fluids and Structures*, 3: 481–508.

Kernot, W.C. 1893. Wind pressure. *Proceedings, Australasian Society for the Advancement of Science*, V: pp. 573–81, Adelaide, Australia, September 25–October 3.

Kind, R.J. 1984. Pneumatic stiffness and damping in air-supported structures. *Journal of Wind Engineering and Industrial Aerodynamics*, 17: 295–304.

Larose, G.L. and Franck, N. 1997. Early wind engineering experiments in Denmark. *Journal of Wind Engineering and Industrial Aerodynamics*, 72: 493–9.

Lawson, T.V. 1976. The design of cladding. *Building and Environment*, 11: 37–8.

Letchford, C.W. 1989.. On the discrete approximation in pneumatic averaging. *Proceedings, 2nd Asia-Pacific Conference on Wind Engineering*, Beijing, June, pp. 1159–67, Pergamon Press.

Matsumoto, M., Shirato, H. and Hirai, S. 1992. Torsional flutter mechanism of 2-D H-shaped cylinders and effect of flow turbulence. *Journal of Wind Engineering and Industrial Aerodynamics*, 41: 687–98.

McKeon, R. and Melbourne, W.H. 1971. Wind-tunnel blockage effects and drag on bluff bodies in rough wall turbulent boundary layers. *Third International Conference on Wind Effects on Buildings and Structures*, Tokyo, Saikon Shuppan Publishers.

Parkinson, G.V. 1984. A tolerant wind tunnel for industrial aerodynamics. *Journal of Wind Engineering and Industrial Aerodynamics*, 16: 293–300.

Parkinson, G.V. and Cook, N.J. 1992. Blockage tolerance of a boundary-layer wind tunnel. *Journal of Wind Engineering and Industrial Aerodynamics*, 42: 873–84.

Reinhold, T. (ed.) 1982. Wind tunnel modeling for civil engineering applications. *International Workshop on Wind Tunnel Modeling Criteria and Techniques in Civil Engineering Applications*, Cambridge University Press, Gaithersburg, MD, USA.

Scanlan, R.H. 1983. Aeroelastic simulation of bridges. *Journal of Structural Engineering, ASCE*, 109: 2829–37.

Scanlan, R.H. and Tomko, J.J. 1971. Airfoil and bridge deck flutter derivatives. *Journal of the Engineering Mechanics Division, ASCE*, 97: 1717–37.

Stathopoulos, T. and Baskaran, B.A 1996. Computer simulation of wind environmental conditions around buildings. *Engineering Structures*, 18: 876–85.

Surry, D. and Isyumov, N. 1975. Model studies of wind effects – A perspective on the problems of experimental technique and instrumentation. *6th International Congress on Aerospace Instrumentation*, Ottawa.

Surry, D. and Stathopoulos, T. 1977. An experimental approach to the economical measurement of spatially-averaged wind loads. *Journal of Industrial Aerodynamics*, 2: 385–97.

Szechenyi, E. 1975. Supercritical Reynolds Number simulation for two-dimensional flow over circular cylinders. *Journal of Fluid Mechanics*, 70: 529–42.

Tanaka, H. 1990. Similitude and modelling in wind tunnel testing of bridges. *Journal of Wind Engineering and Industrial Aerodynamics*, 33: 283–300.

Tryggvason, B.V. 1979. Aeroelastic modelling of pneumatic and tensioned fabric structures. *Proceedings, 5th Internal Conference on Wind Engineering*, Fort Collins, Colorado, pp. 1061–72, Pergamon Press.

Tschanz, T. and Davenport, A.G. 1983. The base balance technique for the determination of dynamic wind loads. *Journal of Wind Engineering and Industrial Aerodynamics*, 13: 429–39.

Tse, K.T., Yu, X.J. and Hitchcock, P.A. 2014. Evaluation of mode shape linearization for HFBB analysis of real tall buildings. *Wind and Structures*, 18: 423–41.

Vickery, B.J. and Daly, A. 1984. Wind tunnel modelling as a means of predicting the response to vortex shedding. *Engineering Structures*, 6: 363–8.

Ward, N.B. 1972. The exploration of certain features of tornado dynamics using a laboratory model. *Journal of Atmospheric Sciences*, 29: 1194–204.

Whitbread, R.E. 1963. Model simulation of wind effects on structures. *Proceedings, International Conference on Wind Effects on Buildings and Structures*, Teddington, UK, 26–28 June, pp. 284–302.

Xie, J. and Garber, J. 2014. HFFB technique and its validation studies. *Wind and Structures*, 18: 375–89.

Xie, J. and Irwin, P.A. 1998. Application of the force balance technique to a building complex. *Journal of Wind Engineering and Industrial Aerodynamics*, 77/78: 579–90.

Chapter 8

Low-rise buildings

8.1 INTRODUCTION

For the purposes of this chapter, *low-rise buildings* are defined as roofed low-rise structures less than fifteen metre in height. Large roofs on major structures such as sports stadia, including arched roofs, are discussed in Chapter 10; free-standing roofs and canopies are covered in Chapter 14.

The following factors make the assessment of wind loads for low-rise buildings as difficult as for taller buildings and other larger structures:

- They are usually immersed within the layer of aerodynamic roughness on the earth's surface, where the turbulence intensities are high, and interference and shelter effects are important, but difficult to quantify.
- Roof loadings, with all the variations due to changes in geometry, are of critical importance for low-rise buildings. The highest wind loadings on the surface of a low-rise structure are generally the suctions on the roof, and many structural failures are initiated there.
- Low-rise buildings often have a single internal space, and internal pressures can be very significant, especially when a dominant opening occurs in a windward wall. The magnitude of internal pressure peaks, and their correlation with peaks in external pressure, must be assessed.

However, resonant dynamic effects can normally be neglected for smaller buildings. The majority of structural damage in wind storms is incurred by low-rise buildings, especially family dwellings, which are often non-engineered and lacking in maintenance.

The following sections will discuss the history of research on wind loads on low-rise buildings, the general characteristics of wind pressures and model-scaling criteria and a summary of the results of many studies that were carried out in the 1970s, 1980s and 1990s.

Several comprehensive reviews of wind loads on low-rise buildings have been made by Holmes (1983), Stathopoulos (1984, 1995), Krishna (1995) and Surry (1999).

8.2 HISTORICAL

8.2.1 Early wind-tunnel studies

Some of the earliest applications of wind tunnels were in the study of wind pressures on low-rise buildings. The two earliest investigations were by Irminger (1894) in Copenhagen, Denmark, and Kernot (1893) in Melbourne, Australia. Irminger used a small tunnel driven by the suction of a factory chimney, and measured pressures on a variety of models, including one of a house. He demonstrated the importance of roof suction, a poorly understood concept at the time. Kernot used what would now be called an open-jet wind tunnel (see Section 7.2.1), as well as a whirling arm apparatus, and measured forces on a variety of building shapes. The effects of roof pitch, parapets and adjacent buildings were all examined.

Over the following 30 years, isolated studies were carried out in aeronautical wind tunnels at the National Physical Laboratory (NPL) in the United Kingdom, the Deutsches Zentrum für Luft- und Raumfahrt (DLR) laboratories at Göttingen, Germany, the National Bureau of Standards in the United States and the Central Aero-Hydrodynamical Institute of the Union of Soviet Socialist Republics. These early measurements showed some disagreement with each other, although they were all measurements of steady wind pressures in nominally steady flow conditions. This was probably due to small but different levels of turbulence in the various wind tunnels (Chapter 4 discusses the effect of turbulence on the mean flow and pressures on bluff bodies), and other effects, such as blockage.

In Denmark, Irminger, with Nokkentved (1930), carried out further wind-tunnel studies on low-rise buildings. These tests were again carried out in steady, uniform flow conditions, but included some innovative work on models with porous walls, and the measurement of internal as well as external pressures. Similar but less extensive, measurements were carried out by Richardson and Miller (1932) in Australia.

In 1936, the American Society of Civil Engineers (1936) surveyed the data available at that time on wind loads on steel buildings. This survey included consideration of 'rounded and sloping roofs'. These data consisted of a variety of early wind-tunnel measurements presumably carried out in smooth flow.

Flachsbart, at the Göttingen Laboratories in Germany, is well known for his extensive wind-tunnel measurements on lattice frames and bridge trusses, in the 1930s. Less well known, however, is the work he did in comparing wind pressures on a low-rise building in smooth and boundary-layer flow (Flachsbart, 1932). This work, probably the first boundary-layer wind-tunnel study, has been rediscovered, and reported, by Simiu and Scanlan (1996).

Recognition of the importance of boundary-layer flow was also made by Bailey and Vincent (1943) at the National Physical Laboratories in the United Kingdom. In doing so, they were able to make some progress in explaining differences, between wind-tunnel and full-scale measurements of pressures, on a low-rise shed.

However, it was not until the 1950s that Jensen (1958), at the Technical University of Denmark, satisfactorily explained the differences between full-scale and wind-tunnel model measurements of wind pressures. Figure 8.1 reproduces some of his measurements, which fully established the importance of using a turbulent boundary-layer flow to obtain pressure coefficients in agreement with full-scale values. The non-dimensional ratio of building height to roughness length, h/z_0, was later named the *Jensen Number* (see Section 4.4.5), in recognition of this work. Jensen and Franck (1965) later carried out extensive wind-tunnel measurements on a range of building shapes in a small boundary-layer wind tunnel.

The work of Jensen and Franck was the precursor to a series of generic, wind-tunnel studies of wind loads on low-rise buildings in the 1970s and 1980s, including those on industrial buildings by Davenport et al. (1977), and on houses by Holmes (1983, 1994). Results from these studies are discussed in the later sections.

Important contributions to the understanding of the effect of large groupings of bluff bodies in turbulent boundary layers, representative of large groups of low-rise buildings, were made by Lee and Soliman (1977), and Hussain and Lee (1980). Three types of flow were established, depending on the building spacing: *skimming* flow (close spacing), *wake-interference* flow (medium spacing) and *isolated-roughness* flow (far spacing).

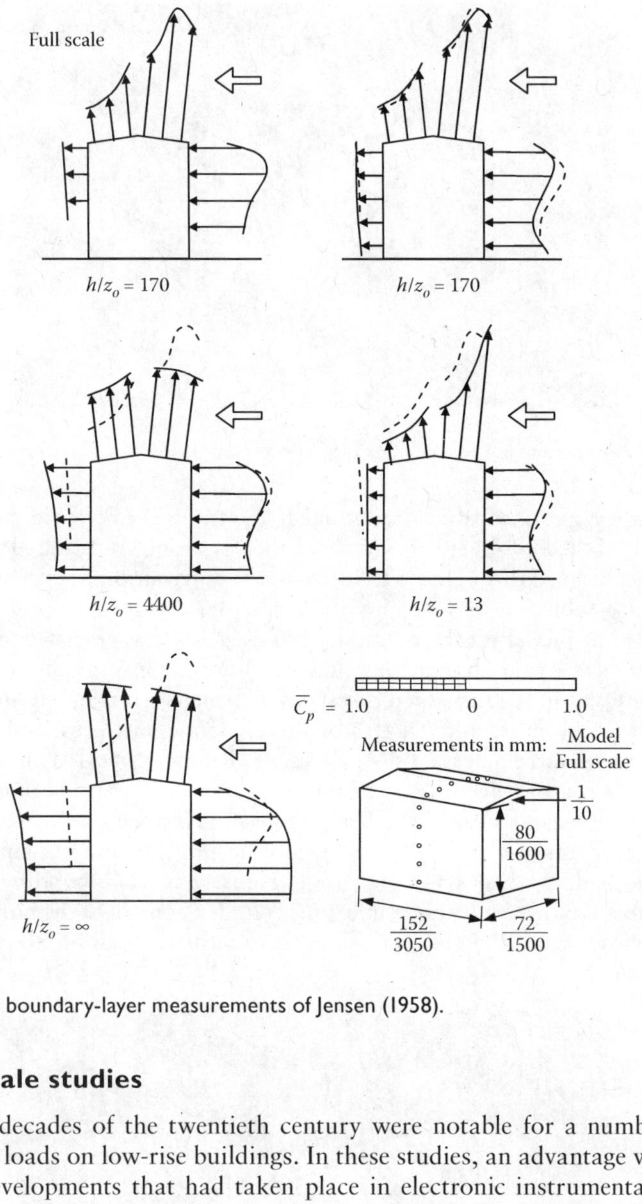

Figure 8.1 Pioneer boundary-layer measurements of Jensen (1958).

8.2.2 Full-scale studies

The last three decades of the twentieth century were notable for a number of full-scale studies of wind loads on low-rise buildings. In these studies, an advantage was taken of the considerable developments that had taken place in electronic instrumentation, and computer-based statistical analysis techniques, and provided a vast body of data which challenged wind-tunnel-modelling techniques.

In the early 1970s, the Building Research Establishment in the United Kingdom commenced a programme of full-scale measurements on a specially constructed experimental building, representative of a two-storey low-rise building at Aylesbury, England. The building had the unique feature of a roof pitch which was adjustable between 5 and 45 degrees (Figure 8.2).

The results obtained in the Aylesbury Experiment emphasised the highly fluctuating nature of the wind pressures, and the high-pressure peaks in separated flow regions near the roof eaves and ridge, and near the wall corners (Eaton and Mayne, 1975; Eaton et al., 1975).

Figure 8.2 Aylesbury Experimental Building (United Kingdom 1970–1975).

Unfortunately, the experiment was discontinued, and the experimental building dismantled after only 2 years at the Aylesbury site. However, interest from wind-tunnel researchers in the Aylesbury data continued through the 1980s, when an International Aylesbury Comparative Experiment was established. Seventeen wind-tunnel laboratories around the world tested identical 1/100-scale models of the Aylesbury building, using various techniques for modelling the upwind terrain and approaching flow conditions. This unique experiment showed significant differences in the measured pressure coefficients – attributed mainly to different techniques used to obtain the reference static and dynamic pressures, and in modelling the hedges in the upwind terrain at the full-scale site (Sill et al., 1989, 1992).

In the late 1980s, two new full-scale experiments on low-rise buildings were set up in Lubbock, Texas, USA, and Silsoe, UK. The Lubbock experiment, known as the Texas Tech Field Experiment, comprised a small steel shed of height, 4.0 m, and plan dimensions, 9.1 and 13.7 m; the building had a near-flat roof (Figure 8.3). The building had the unique capability of being mounted on a turntable, thus enabling control of the building orientation relative to the mean wind direction. Pressures were measured with high response pressure

Figure 8.3 Texas Tech Field Experiment (United States).

transducers mounted close to the pressure tappings on the roof and walls; the transducers were moved around to different positions at different times during the course of the experiments. A 50-m high mast upwind of the building, in the prevailing wind direction, had several levels of anemometers, enabling the approaching wind properties to be well defined. The upwind terrain was quite flat and open. The reference static pressure was obtained from an underground box, 23 m away from the centre of the test building (Levitan and Mehta, 1992a,b).

The Texas Tech Experiment has produced a large amount of wind pressure data for a variety of wind directions. External and internal pressures, with and without dominant openings in the walls, were recorded. Very high extreme pressures at the windward corner of the roof for 'quartering' winds blowing directly on to the corner, at about 45 degrees to the walls, were measured; these were considerably greater than those measured at equivalent positions on small (1/100)-scale wind-tunnel models. The internal pressures, however, showed similar characteristics to those measured on wind-tunnel models, and predicted by theoretical models.

The Silsoe Structures Building was a larger steel portal-framed structure, 24 m long, 12.9 m span and 4 m to the eaves, with a 10 degrees roof pitch, located in an open country. As well as 70 pressure-tapping points on the building roof and walls, the building was equipped with 12 strain gauge positions on the central portal frame to enable measurements of structural response to be made (Robertson, 1992).

The building could be fitted with both curved and sharp eaves. The curved eaves were found to give lower mean negative pressures immediately downwind of the windward wall, than those produced by the sharp eaves. Measurements of strain in the portal frame were found to be predicted quite well by a structural analysis computer programme when the correct column fixity was applied. Spectral densities of the strains were also measured – these showed the effects of Helmholtz resonance (Section 6.2.3) on the internal pressures, when there was an opening in the end wall of the building. Generally, these measurements justified a *quasi-steady* approach to wind loads on low-rise buildings (Section 4.6.2).

8.3 GENERAL CHARACTERISTICS OF WIND LOADS ON LOW-RISE BUILDINGS

Full-scale measurements of wind pressures on low-rise buildings, such as those described in Section 8.2.2, showed the highly fluctuating nature of wind pressures, area-averaged wind loads and load effects, or responses, on these structures. The fluctuations with time can be attributed to two sources (see also Section 4.6.1):

1. Pressure fluctuations induced by upwind turbulent velocity fluctuations (see Chapter 3). In an urban situation, the turbulence may arise from the wakes of upwind buildings.
2. Unsteady pressures produced by local vortex shedding, and other unsteady flow phenomena, in the separated flow regions near sharp corners, roof eaves and ridges (see Chapter 4).

These two phenomena may interact with each other to further complicate the situation.

It should be noted that, as well as a variation with time, as shown for a single point on a building in Figure 8.4, there is a variation with space, that is the same pressure or response variation with time, may not occur simultaneously at different points separated from each other, on a building.

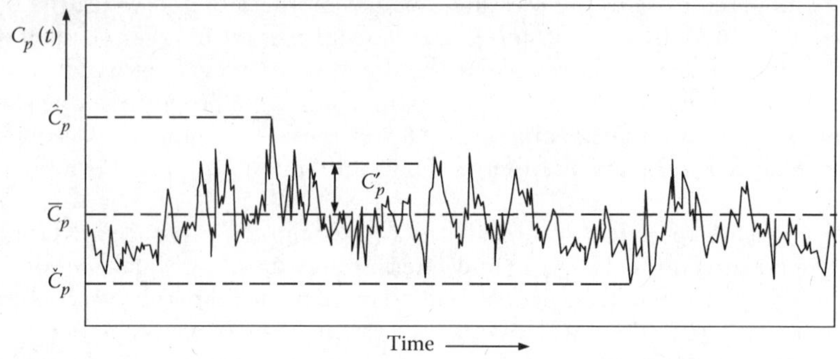

Figure 8.4 Typical variation of wind pressure and definition of pressure coefficients.

8.3.1 Pressure coefficients

The basic definition of a pressure coefficient for a bluff body was given in Section 4.2.1, and the rms fluctuating (standard deviation) pressure coefficient was defined in Section 4.6.4. A general time-varying pressure coefficient, $C_p(t)$, for buildings in stationary, or synoptic, wind storms is as follows:

$$C_p(t) = \frac{p(t) - p_o}{1/2\, \rho_a \bar{U}^2} \tag{8.1}$$

where p_o is a static reference pressure (normally atmospheric pressure measured at a convenient location near the building, but not affected by the flow around the building), ρ_a is the density of air, and $\bar{U}$ is the mean (time-averaged) velocity measured at an appropriate defined reference height; as in the atmospheric boundary layer, there is a variation of mean wind speed with height (Section 3.2). In the case of a low-rise building, this is usually taken to be at roof height, at eaves level, mid-height of the roof or at the highest level of the roof; as for the static pressure, this must be away from the direct influence of the building.

Figure 8.4 shows a typical variation of $C_p(t)$ on a low-rise building, and four significant values of the pressure coefficient:

$\bar{C}_p$ – the mean or time-averaged pressure coefficient
$C'_p (= \sigma_{Cp})$ – the rms fluctuating value, or standard deviation, representing the average departure from the mean
$\hat{C}_p$ (or $C_{\hat{p}}$) – the expected maximum value of the pressure coefficient in a given time period
$\check{C}_p$ (or $C_{\check{p}}$) – the expected minimum value of the pressure coefficient in a given time period

8.3.2 Dependence of pressure coefficients

The dependence of pressure coefficients on other non-dimensional quantities, such as Reynolds Number and Jensen Number, in the general context of bluff-body aerodynamics, was discussed in Section 4.2.3. This dependence is applicable to wind loads on low-rise buildings.

For bodies which are sharp edged, and on which points of flow separation are generally fixed, the flow patterns and pressure coefficients are *relatively* insensitive to viscous effects and hence the Reynolds Number. This means that provided an adequate reproduction of the turbulent flow characteristics in atmospheric boundary-layer flow is achieved, and the model is geometrically correct, wind-tunnel tests can be used to predict pressure and force coefficients on full-scale buildings. However, the full-scale studies from the Texas Tech Field Experiment have indicated that for certain wind directions, pressure peaks in some separated flow regions are not reproduced in wind-tunnel tests with small-scale models, and some Reynolds Number dependency *is* indicated.

As discussed in Section 8.2.1, Martin Jensen identified the Jensen Number, h/z_0, the ratio of building height to the aerodynamic roughness length in the logarithmic law (Sections 3.2.1 and 4.4.5), as the most critical parameter in determining mean pressure coefficients on low-rise buildings. The Jensen Number clearly directly influences the mean pressure distributions on a building through the effect of the mean velocity profile with height. However, in a fully developed boundary layer over a rough ground surface, the turbulence quantities such as intensities (Section 3.3.1) and spectra (Section 3.3.4) should also scale with the ratio z/z_0 near the ground. There is an indirect influence of the turbulence properties on the mean pressure coefficients (Section 4.4.3), which would have been responsible for some of the differences observed by Jensen (1958), and seen in Figure 8.1. In wind-tunnel tests, the turbulence intensity similarity will only be achieved with h/z_0 equality, if the turbulent inner surface layer in the atmospheric boundary layer has been correctly simulated in the boundary layer in the wind tunnel. Many researchers prefer to treat parameters such as turbulence intensities and ratios of turbulence length scale to building dimension as independent non-dimensional quantities (see Section 4.2.3), but unfortunately, it is difficult to independently vary these parameters in wind-tunnel tests.

Fluctuating and peak external pressures on low-rise buildings which are most relevant to the structural design, are highly dependent on the turbulence properties in the approach flow, especially turbulence intensities. Consequently, peak load effects, such as bending moments in framing members, are also dependent on the upwind turbulence. For 'correctly' simulated boundary layers, in which turbulence quantities near the ground scale as z/z_0, as discussed in the previous paragraph, peak load effects can be reduced to a variation with Jensen Number (e.g. Holmes and Carpenter, 1990).

Finally, the question of the dependency of pressures and load effects on low-rise buildings in wind storms of the downdraft type (Section 1.3.5) arises. As discussed in Section 3.2.6, these winds have boundary layers which are not strongly dependent on the surface roughness on the ground – hence, the Jensen Number may not be such an important parameter. Further research is required to identify non-dimensional parameters in the downdraft flow which are relevant to wind pressures on buildings in these types of storms.

8.3.3 Flow patterns and mean pressure distributions

Figure 8.5 shows the main features of flow over a building with a low-pitched roof, which has many of the features of flow around a two-dimensional bluff body described in Section 4.1. The flow separates at the top of the windward wall and *re-attaches* at a region further downwind on the roof, forming a separation zone or 'bubble'. However, this bubble exists only as a time average. The separation zone is bounded by a free shear layer, a region of high-velocity gradients and high turbulence. This layer rolls up intermittently to form vortices; as these are shed downwind, they may produce high negative pressure peaks on the roof

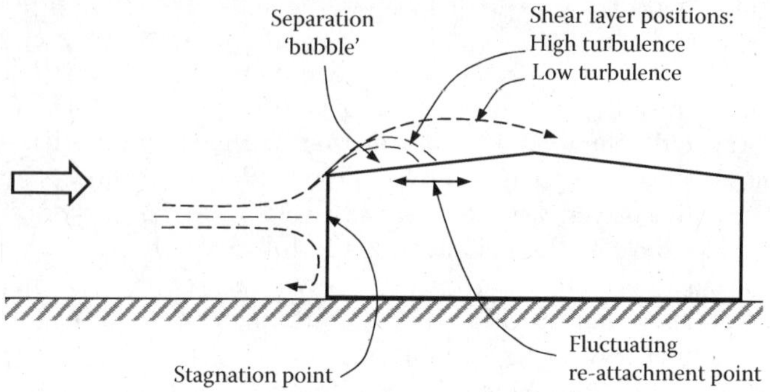

Figure 8.5 Wind flow around a low-rise building.

surface. The effect of turbulence in the approaching flow is to cause the vortices to roll up closer to the leading edge, and a shorter distance to the re-attachment zone results.

The longitudinal intensities of turbulence at typical roof heights of low-rise buildings, are 20% or greater, and separation zone lengths are shorter, compared to those in smooth, or low turbulence, flow. Small separation zones with high shear layer curvatures are associated with low pressures, i.e. high initial negative pressures, but rapid pressure recovery downwind.

Roof pitches up to about 10 degrees, for wind normal to a ridge or gable end, are *aerodynamically flat.* When the mean wind direction is parallel to a ridge line, the roof is also seen as aerodynamically flat, for any roof pitch. For winds normal to the ridge line, and roof pitches between 10 and 20 degrees, a second flow separation occurs at the ridge, producing regions of high negative pressures on both sides of the ridge. Downwind of the ridge, a second re-attachment of the flow occurs with an accompanying recovery in pressure. At roof pitches greater than about 20 degrees, positive mean pressures occur on the upwind roof face, and fully separated flows without re-attachment occur downwind of the ridge giving relatively uniform negative mean pressures on the downwind roof slope.

It should be noted that the above comments are applicable only to low-rise buildings with height/downwind depth (*h*/*d*) ratios less than about 0.5. As this ratio increases, roof pressures generally become more negative. This influence can be seen in Figure 8.6 which shows the mean pressure distribution along the centre line of low-rise buildings for various roof pitches and *h*/*d* ratios; the horizontal dimension across the wind (into the paper in Figure 8.6) is about twice the along-wind dimension. For higher buildings with *h*/*d* ratios of 3 or greater, the roof pressure will be negative on both faces, even for roof slopes greater than 20 degrees.

Similar flow separation and re-attachment, as described for roofs, occurs on the side walls of low-rise buildings, although the magnitude of the mean pressure coefficients is generally lower. The mean pressures on windward walls are positive with respect to the freestream static pressure. Leeward walls are influenced by the re-circulating wake, and generally experience negative pressures of lower magnitude; however, the values depend on the building dimensions, including the roof pitch angle.

When the wind blows obliquely on to the corner of a roof, a more complex flow pattern emerges as shown in Figure 8.7. *Conical* vortices similar to those found on delta wings of an aircraft occur. Figure 8.8 shows these vortices visualised by smoke – their axes are slightly inclined to the adjacent walls forming the corner. The pressures underneath these are the

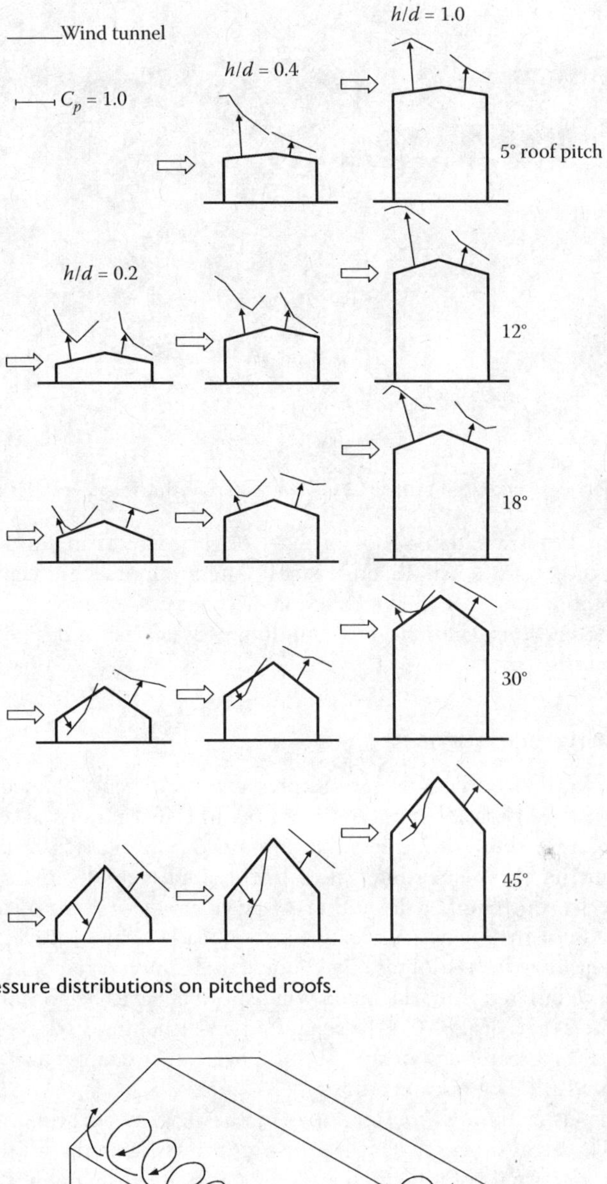

Figure 8.6 Mean pressure distributions on pitched roofs.

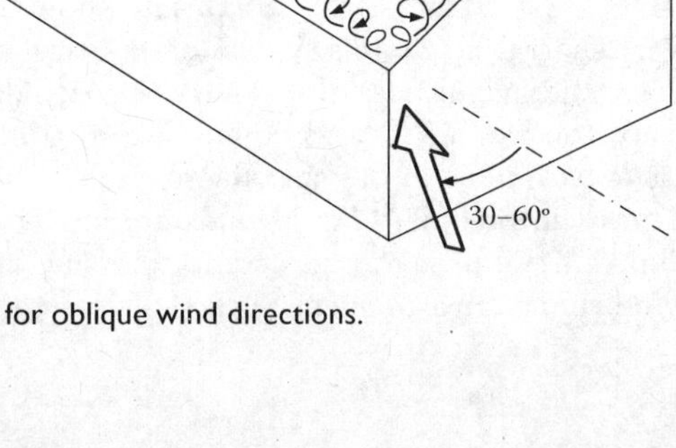

Figure 8.7 Conical vortices for oblique wind directions.

Figure 8.8 Corner vortices generated by quartering winds. (From the Texas Tech Field Experiment.)

largest to occur on the low-pitched roofs, square or rectangular in planform, although the areas over which they act are usually quite small, and are more significant for pressures on small areas of cladding than for the loads in major structural members.

In the following sections, the effects of building geometries on design loads will be discussed in more detail.

8.3.4 Fluctuating pressures

The rms fluctuating, or standard deviation, pressure coefficient, defined in Sections 4.6.4 and 8.3.1, is a measure of the general level of pressure fluctuations at a point on a building. As discussed in Section 8.3.2, the values obtained on a particular building are generally dependent on the turbulence intensities in the approaching flow, which in turn are dependent on the Jensen Number. In boundary-layer winds over an open-country terrain, for which longitudinal turbulence intensities are typically around 20%, at heights typical of eaves heights on low-rise buildings, the values of rms pressure coefficients (based on a dynamic pressure calculated from the mean wind speed at eaves height) on windward walls, are typically in the range of 0.3–0.4. In separated–re-attaching flow regions on side walls values of C_p' of 0.6 or greater can occur. Even higher values can occur at critical points on roofs, with values >1.0 being not uncommon.

High instantaneous peak pressures tend to occur at the same locations as high rms fluctuating pressures. The highest negative peak pressures are associated with the conical vortices generated at the roof corners of low-pitch buildings, for quartering winds blowing on to the corner in question (Figures 8.7 and 8.8). Figure 8.9 shows a short sample of pressure – time history, from a pressure measurement position near the formation point of one of these vortices, on the Texas Tech building (Mehta et al., 1992). This shows that high-pressure peaks occur as 'spikes' over very short time periods. Values of negative peak pressure coefficients as high as –10 often occur, and magnitudes of –20 have occasionally been measured.

The probability density function (pdf) and cumulative distribution function (cdf) are measures of the amplitude variations in pressure fluctuations at a point. Even though the upwind velocity fluctuations in boundary-layer winds are nearly Gaussian (Section 3.3.2 and Appendix C, Section C.3.1), this is not the case for pressure fluctuations on buildings. Figure 8.10 shows a wind-tunnel measurement of the cdf for pressure fluctuations on the

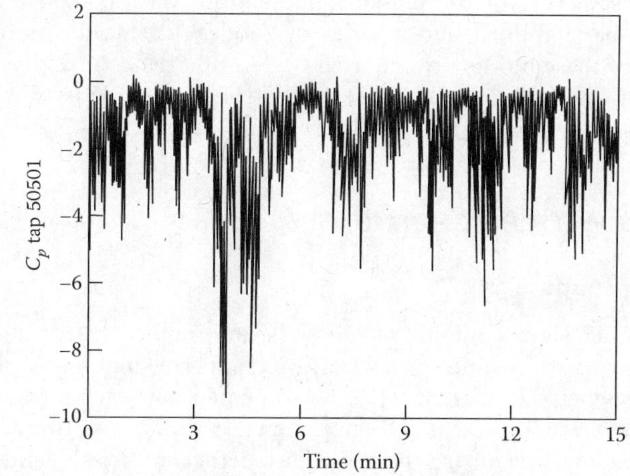

Figure 8.9 Pressure coefficient versus time from a corner pressure tap (Texas Tech Field Experiment). (From Mehta, K.C. et al. 1992. *Journal of Wind Engineering and Industrial Aerodynamics*, 41: 181–92. With permission.)

windward wall of a low-rise building model (Holmes, 1981, 1983). On this graph, a straight line indicates a Gaussian distribution. Clearly, the measurements showed upward curvature, or positive skewness (Appendix C, Figure C.3). This can, in part, be explained by the square-law relationship between pressure and velocity (see Equation 4.12) (Holmes, 1981 and Appendix C.3.3). Negative skewness occurs for pressure fluctuations in separated flow regions of a building.

The spatial structure of fluctuating pressures on low-rise buildings has been investigated in detail by a number of researchers, using a technique known as *Proper Orthogonal Decomposition* (e.g. Best and Holmes, 1983; Holmes, 1990a; Letchford and Mehta, 1993; Bienkiewicz et al., 1993; Ho et al., 1995; Holmes et al., 1997; Baker, 1999). The mathematics of this technique is beyond the scope of this book, but the method allows the complexity

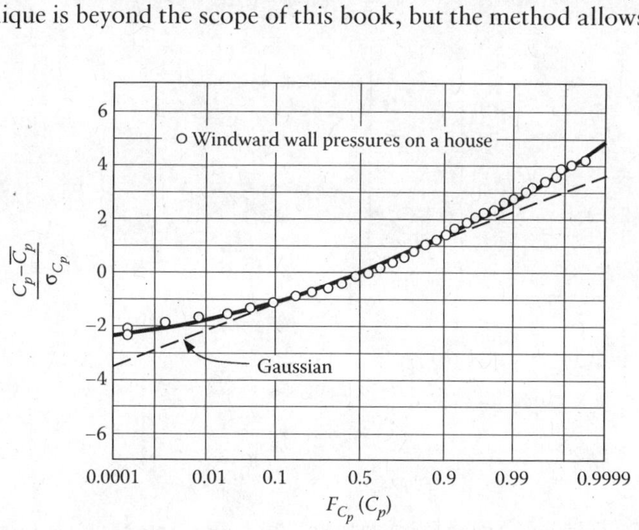

Figure 8.10 Cumulative probability distribution for pressure fluctuations on the windward wall of a house.

of the space-time structure of the pressure fluctuations on a complete roof, building or tributary area, to be simplified into a series of 'modes', each with its own spatial form. Surprisingly, few of these modes are required to describe the complexity of the variations. Invariably, for low-rise buildings, the first, and strongest, mode is 'driven' by the quasi-steady mechanism associated with upwind turbulence fluctuations.

8.4 BUILDINGS WITH PITCHED ROOFS

8.4.1 Cladding loads

Figures 8.11 and 8.12 show contours of the worst minimum pressure coefficients, for any wind direction, measured in wind-tunnel tests on models of single-storey houses with gable roofs of various pitches (Holmes, 1994). The simulated approach terrain in the approach boundary-layer flow was representative of an open country, and the wind direction was varied at 10-degree intervals during the tests. The coefficients are all defined with respect to the *eaves* height mean wind speed.

The highest magnitude coefficients occur on the roof. At the lowest pitch (10 degrees), the contours of highest negative pressures converge towards the corner of the roof; the effect of increasing the roof pitch is to emphasise the gable end as the worst-loaded region. The worst local negative peak pressures occur on the 20-degree pitch roof in this area. The highest magnitude minima on the walls occur near a corner.

Similar plots for shapes representative of industrial buildings with roof pitches of 5-, 18- and 45-degrees pitch are shown in Figures 8.13 through 8.15 (Davenport et al., 1977). In these figures, contours of maximum pressure coefficients, as well as minimum pressure coefficients, are plotted. Plots are given for three different eaves heights, for each roof pitch. Results from building models located in a simulated urban terrain are shown.

For any given roof pitch, there is not a large variation in the magnitudes of the minimum and maximum pressure coefficients with eaves height – however, the pressure coefficients are defined with respect to the mean dynamic pressure at eaves height in each case. Since the

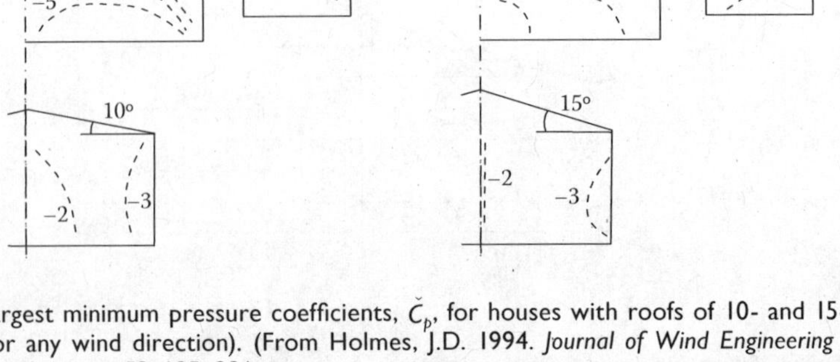

Figure 8.11 Largest minimum pressure coefficients, $\breve{C}_p$, for houses with roofs of 10- and 15-degree pitch (for any wind direction). (From Holmes, J.D. 1994. *Journal of Wind Engineering and Industrial Aerodynamics*, 53: 105–23.)

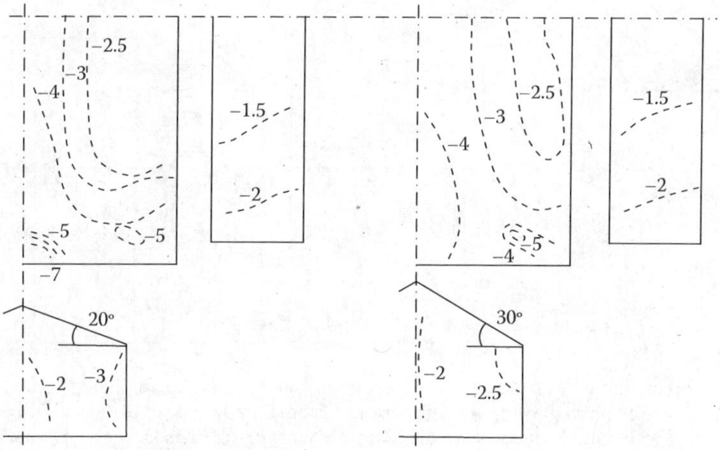

Figure 8.12 Largest minimum pressure coefficients, $\check{C}_p$, for houses with roofs of 20- and 30-degree pitch (for any wind direction). (From Holmes, J.D. 1994. *Journal of Wind Engineering and Industrial Aerodynamics*, 53: 105–23.)

mean velocity, and hence dynamic pressure, in a boundary layer increases with increasing height, the pressures themselves will generally increase with height of the building. Since the fluctuating pressure coefficients are closely related to the turbulence intensities in the approach flow, lower magnitudes might be expected at greater eaves heights, where the turbulence intensities are lower, and this can be seen in Figures 8.13 through 8.15. However, the local pressure peaks are also influenced by local flow separations, and hence by the relative building dimensions.

The worst minimum pressure coefficients for the 18-degree pitch roofs (Figure 8.14) occur near the ridge at the gable end (compare also the house with the 20-degree pitch roof in Figure 8.12). For the 5-degree pitch case (Figure 8.13), there is a more even distribution of the largest minimum (negative) pressure coefficients around the edge of the roof. For the 45-degree pitch, the corner regions of the roof generally experience the largest minima; the maximum pressure coefficients are also significant in magnitude on the 45-degree pitch roof.

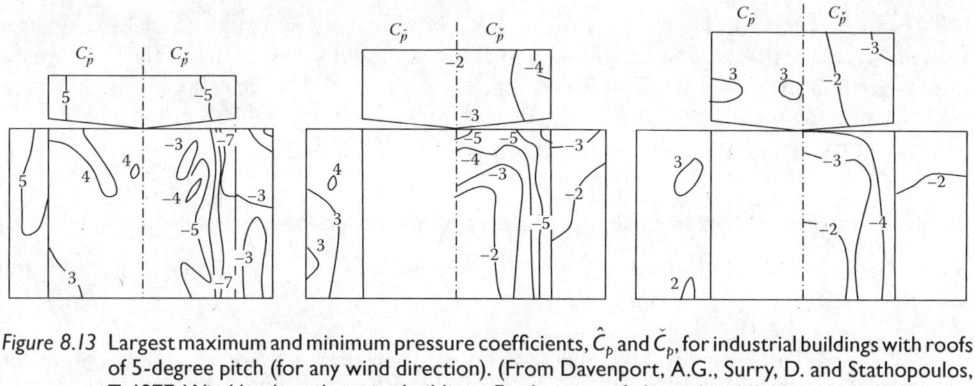

Figure 8.13 Largest maximum and minimum pressure coefficients, $\hat{C}_p$ and $\check{C}_p$, for industrial buildings with roofs of 5-degree pitch (for any wind direction). (From Davenport, A.G., Surry, D. and Stathopoulos, T. 1977. Wind loads on low-rise buildings. Final report of phases I and II. University of Western Ontario, Boundary Layer Wind Tunnel Report BLWT-SS8-1977. With permission.)

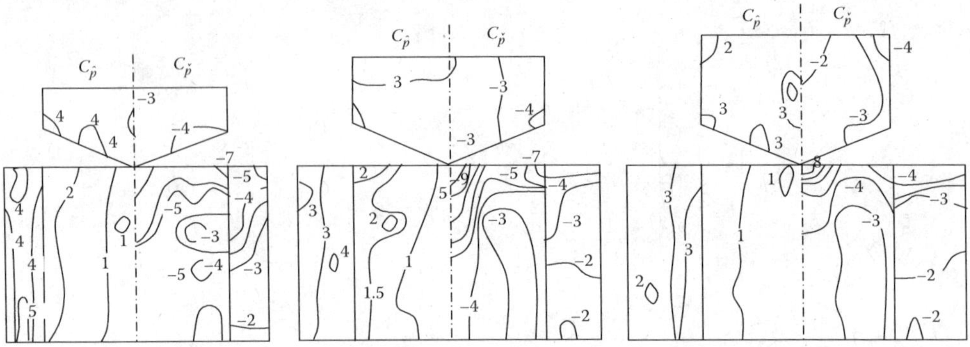

Figure 8.14 Largest maximum and minimum pressure coefficients, $\hat{C}_p$ and $\check{C}_p$, for industrial buildings with roofs of 18-degree pitch (for any wind direction). (From Davenport, A.G., Surry, D. and Stathopoulos, T. 1977. Wind loads on low-rise buildings. Final report of phases I and II. University of Western Ontario, Boundary Layer Wind Tunnel Report BLWT-SS8-1977. With permission.)

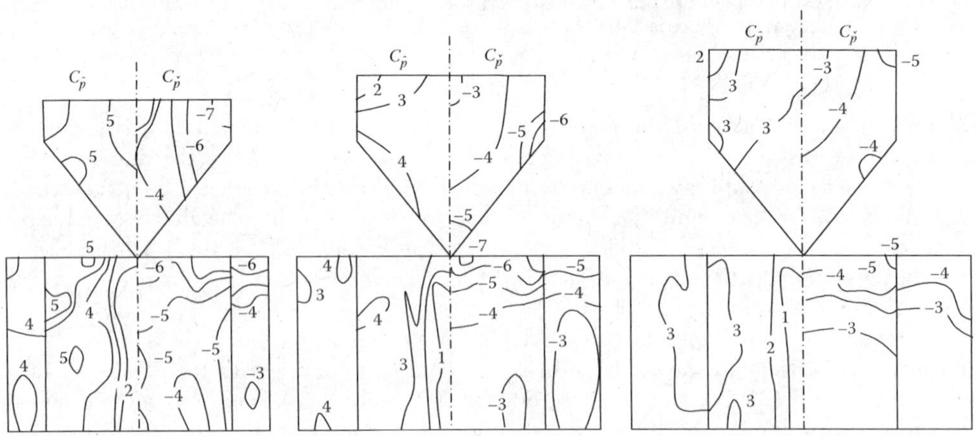

Figure 8.15 Largest maximum and minimum pressure coefficients, $\hat{C}_p$ and $\check{C}_p$, for industrial buildings with roofs of 45-degree pitch (for any wind direction). (From Davenport, A.G., Surry, D. and Stathopoulos, T. 1977. Wind loads on low-rise buildings. Final report of phases I and II. University of Western Ontario, Boundary Layer Wind Tunnel Report BLWT-SS8-1977. With permission.)

Plots such as those in Figures 8.11 through 8.15 can be used as a guide to the specification of wind loads for the design of cladding. However, it should be noted that if the design wind speeds are non-uniform with direction, as they normally will be, the contours of maximum and minimum *pressures* (as opposed to pressure coefficients) will be different, and will depend on the site and the building orientation.

8.4.2 Structural loads and equivalent static load distributions

The effective peak wind loads acting on a major structural element such as the portal frame of a low-rise building are dependent on two factors:

1. The correlation or statistical relationship between the fluctuating pressures on different parts of the tributary surface area 'seen' by the frame; this can be regarded as an area-averaging effect.

2. The influence coefficients which relate pressures at points or panels on the surface to particular load effects, such as bending moments or reactions.

Chapter 5 described methods for determining the effective static-loading distributions, which represent the wind loads which are equivalent in their structural effect to fluctuating (background) wind pressures, and to the resonant (inertial) loads when they are significant. For the low-rise buildings under discussion in this chapter, resonant effects can be ignored, but the fluctuating, or background, loading is quite significant because of the high turbulence intensities near the ground. Some examples of the application of the methods discussed in Chapter 5 will be given in this chapter.

To illustrate the problem, consider Figure 8.16. This shows instantaneous external pressure distributions occurring at three different times during a wind storm around a portal frame supporting a low-rise building. These pressure distributions are clearly different from each other in both shape and magnitude. The value of a load effect such as the bending moment at the knee of the frame will respond to these pressures in a way that might produce the time history of the bending moment versus time given in Figure 8.17. Over a given time period, a maximum bending moment will occur. A minimum bending moment will also occur. Depending on the sign of the bending moment produced by the dead loads acting on the structure, one of these extremes will be the critical one for the design of the structure. Methods for determination of the *expected* pressure distribution which corresponds to the maximum or minimum wind-induced bending moment were discussed in Chapter 5. The effective static pressure distribution so determined must lie between the extreme point pressure limits of the pressures around the frame, as shown in Figure 8.18.

It is of interest to consider the distributions of pressure coefficients given in wind codes and standards. Usually, an 'envelope' loading is specified with pressures uniformly distributed in length along the columns and rafters, as shown in Figure 8.16. These are usually, but not always, conservative loadings which will give over-estimates of load effects such as bending moments.

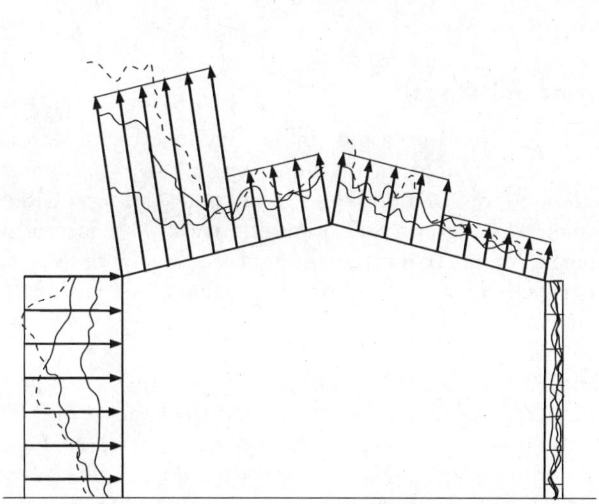

Figure 8.16 Instantaneous external pressure distributions on the frame of a low-rise building, and simplified code distributions. (From Holmes, J.D. and Syme, M.J. 1994. *Steel Construction (Australian Institute of Steel Construction)*, 28: 2–12.)

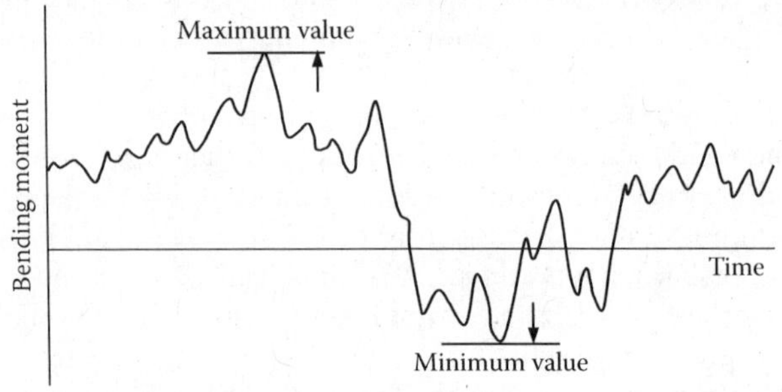

Figure 8.17 Time history of a bending moment. (From Holmes, J.D. and Syme, M.J. 1994. *Steel Construction (Australian Institute of Steel Construction)*, 28: 2–12.)

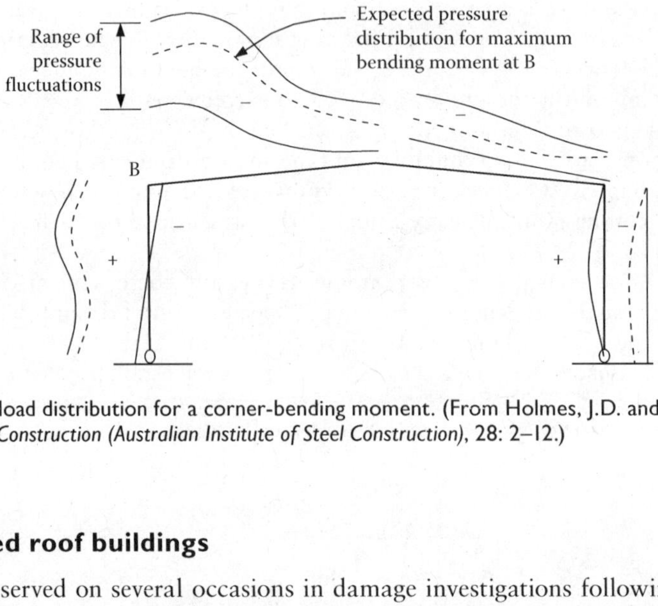

Figure 8.18 Peak load distribution for a corner-bending moment. (From Holmes, J.D. and Syme, M.J. 1994. *Steel Construction (Australian Institute of Steel Construction)*, 28: 2–12.)

8.4.3 Hipped roof buildings

It has been observed on several occasions in damage investigations following severe wind storms that hipped roof buildings have generally suffered lesser damage. Meecham et al. (1991) studied wind pressures on hipped and gable roof buildings of 18.4° pitch in a boundary-layer wind tunnel. Although there is little difference in the largest peak total lift force, or overturning moment, on the two roofs, the gable end region of the gable roof experiences around 50% greater peak negative local pressures, than does the corresponding region on the hipped roof. Furthermore, the largest area-averaged full-span truss load was about twice as high on the gable roof.

However, Xu and Reardon (1998), who studied pressures on hipped roofs with three different roof pitches (15°, 20° and 30°), found that the benefits of a hipped configuration compared with a gable roof type reduce as the roof pitch increases. Figure 8.19 shows contours of the worst minimum (negative) peak pressure coefficients, referenced to the mean dynamic pressure at eaves height, and can be compared with the equivalent gable roof values in Figures 8.11 and 8.12. Note that, at 30° pitch, the worst negative pressure coefficients of about –5.0 are similar for the two types of roof.

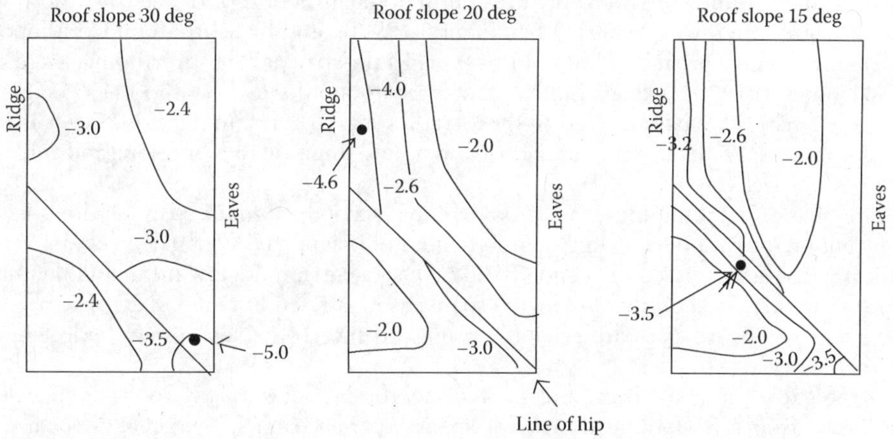

Figure 8.19 Largest minimum pressure coefficients for hipped roofs of 15-, 20- and 30-degree pitch, for any wind direction. (From Xu, Y.L. and Reardon, G.F. 1998. *Journal of Wind Engineering and Industrial Aerodynamics*, 73: 267–84.)

8.4.4 Effect of surrounding buildings: Shelter and interference

Most low-rise buildings are in an urban situation, and are often surrounded by buildings of similar size. The shelter and aerodynamic interference effect of upstream buildings can be very significant on the wind loads. This aspect was the motivation for the studies by Lee and Soliman (1977) and Hussain and Lee (1980) on grouped buildings, as discussed in Section 8.2.1. Three flow regimes were identified depending on the building spacing. The study on tropical houses, described by Holmes (1994), included a large number of grouped building situations for buildings with roofs of 10-degree pitch. This study showed that upstream buildings of the same height reduced the wall pressures and the pressures at the leading edge of the roof significantly, but had less effect on pressures on other parts of the roof. The building height/spacing ratio was the major parameter, with the number of shielding rows being of lesser importance.

A series of wind-tunnel pressure measurements, for both structural loads and local cladding loads, on a flat-roofed building, situated in a variety of 'random city' environments was carried out by Ho et al. (1990, 1991). It was found that the mean component of the wind loads decreased, and the fluctuating component increased, resulting in a less-distinct variation in peak wind load with direction. The expected peak loads in the urban environment were much lower than those on the isolated building. It was also found that a high coefficient of variation (60–80%) of wind loads occurred on the building in the urban environment due to the variation in *location* of the building. For the isolated building, similar coefficients of variation occurred, but in this case, they resulted from variation due to *wind direction*.

8.5 MULTI-SPAN BUILDINGS

The arrangement of industrial low-rise buildings as a series of connected spans is a common practice for reasons of structural efficiency, lighting and ventilation. Such configurations also allow for expansion in stages of a factory or warehouse.

Wind-tunnel studies of wind pressures on multi-span buildings of the 'saw-tooth' type with 20-degree pitch were reported by Holmes (1990b), and by Saathoff and Stathopoulos (1992) on 15-degree pitch buildings of this type. Multi-span gable roof buildings were studied by Holmes (1990b) (5-degree pitch), and by Stathopoulos and Saathoff (1994) (18- and 45-degree pitch). The main interest in these studies was to determine the difference in wind loads for multi-span buildings, and the corresponding single-span monoslope and gable roof buildings, respectively.

As for single-span buildings, the aerodynamic behaviour of multi-span buildings is quite dependent on the roof pitch. Multi-span buildings of low pitch (say <10 degrees) are aerodynamically flat, as discussed in Section 8.3.3. Consequently, quite low mean and fluctuating pressures are obtained on the downwind spans, as illustrated in Figure 8.20. The pressures on the first windward span are generally similar to those on a single-span building of the same geometry.

For the gable roof buildings, and for the saw-tooth roof with the roofs sloping downwards away from the wind, the downwind spans experience much lower magnitude negative mean pressures than the windward spans. For the opposite wind direction on the saw-tooth configuration, the highest magnitude mean pressure coefficients occur on the second span downwind, due to the separation bubble formed in the valley.

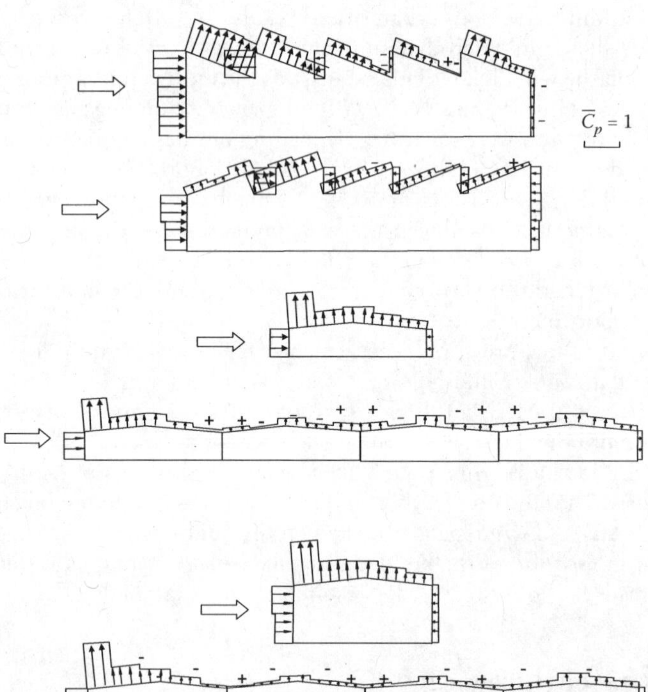

Figure 8.20 Mean pressure distributions on multi-span buildings and comparison with a single span. (From Holmes, J.D. 1990b. Wind loading of multi-span buildings. *Civil Engineering Transactions, Institution of Engineers, Australia*, CE32: 93–8.)

8.6 EFFECTS OF PARAPETS ON LOW-RISE BUILDINGS

A detailed wind-tunnel study of the wind effects of parapets on the roofs of low-rise buildings was carried out by Kopp et al. (2005a,b). Earlier work was reviewed by Stathopoulos and Baskaran (1988).

It was found that tall parapets ($h_p/(h + h_p) > 0.2$), where h_p is the parapet height, can reduce peak local negative pressures by up to 50%, in the corner regions of a roof, when they are installed around the complete perimeter of a roof. Lower parapets, ($h_p/(h + h_p) < 0.2$), increase the worst negative peak pressure coefficients, apparently by stabilising the corner conical vortices that occur on flat or near-flat roofs (see Figure 8.8). However, high parapets increase the positive (downwards) pressure peaks on the roof upwind of leeward parapets.

Isolated parapets – those installed adjacent to one wall of a building – always increase the corner roof loads, irrespective of their height (Kopp et al., 2005a).

Structural loads on roofs are affected somewhat differently than local loads. For a wind direction normal to a wall, a parapet will move the point of re-attachment (see Figure 8.5) further downwind, thus giving a larger region of separated flow. This increases the structural loads, presumably because of the increased correlations of roof pressures in the separated flow region. The loading on interior bays was increased by about 10% for low parapets ($h_p/(h + h_p) < 0.09$), with greater increases for higher parapets (Kopp et al., 2005b).

8.7 EFFECT OF BUILDING LENGTH

In codes and standards for wind loading, the effect of the horizontal aspect ratio (b/d) on the specified wind pressure coefficients for low-rise buildings is normally neglected. Coefficients are typically based on data obtained from wind-tunnel tests with b/d ratios of about 2. However, several studies have shown that horizontal aspect ratio (or building length) is a significant parameter when it becomes large, and particularly for long gable-roofed buildings with very high roof pitches. Such buildings are often used for bulk storage of solids, such as mineral ore or sugar.

The windward ends of these buildings can experience very high loads for oblique wind directions. Very high negative pressures have been observed on the leeward roof surfaces in these situations. The flow and resulting pressures are, in fact, similar to those occurring at the ends of free-standing walls for oblique wind directions (see Section 14.2.1 and Figure 14.3).

A wind-tunnel study of buildings with a 35-degree pitch (Ginger and Holmes, 2003) showed that some major codes and standards underestimated load effects such as frame-bending moments by up to 70%, for a building with a b/d ratio of 6. The Australian Standard of 1989 (Standards Australia, 1989) underestimated some bending moments for this case by 50%, but revisions included in the 2002 and 2011 editions (Standards Australia, 2011) have resulted in a much better agreement (Ginger and Holmes, 2003).

8.8 INTERNAL PRESSURES

In Chapter 6, the prediction of internal pressures in buildings in general is discussed. For low-rise buildings in particular, the internal pressure loading may form a high proportion of the total wind loading for both major structural elements and cladding. In severe wind storms, such as hurricanes or typhoons, failures of roofs often occur following window failure on the windward wall, which generates high positive internal pressures acting together with negative external pressures.

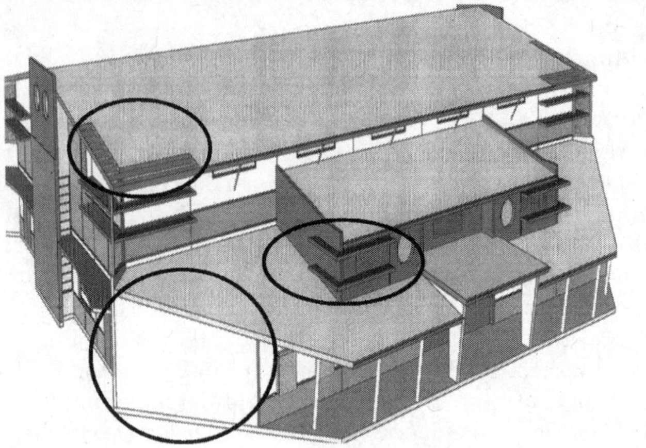

Figure 8.21 A low-rise building with design features to improve the aerodynamic characteristics in strong winds.

8.9 A CASE STUDY: OPTIMUM SHAPING OF A LOW-RISE BUILDING

Low-rise buildings are not generally shaped to optimise wind loading, unlike some larger structures such as long-span bridges, and some tall buildings. However, Figure 8.21 shows such an example – a standard design for public shelter buildings for towns threatened by tropical cyclones in the state of Queensland, Australia.

In order to satisfy onerous wind-loading guidelines with respect to both direct wind loading, and wind-borne debris (Department of Public Works, Queensland, 2006), several features to achieve this are incorporated. These include: chamfering of the corners on the walls on the lower wall corners, architectural 'fins' on the upper floor corners, venting of the corner eaves on the roof and heavy-duty debris-resistant screens (Figure 8.21).

8.10 WIND-TUNNEL DATABASES

A recent significant development for wind-engineering specialists and designers has been the availability of two very useful comprehensive sources of data on external shape factors and pressure coefficients on low-rise buildings on the Internet. These are:

- The National Institute of Standards and Technology (NIST) of the United States has provided data produced by the University of Western Ontario on 37 different configurations of low-rise gable roof buildings (no eaves). These include internal pressures as well as external pressures. Two configurations included roof parapets, and a shape representing the full-scale building of the Texas Tech Field Experiment (Figure 8.3) was included. The data are presented as time histories of local pressures which are available for further processing. These data are at: http://fris2.nist.gov/winddata/uwo-data/uwo-data.html
- The Wind Engineering Group of the Tokyo Polytechnic University has provided data on 116 configurations of low-rise buildings with gable hipped and flat roofs, both with and without eaves. The data are presented in a summarised form for local and area-averaged external pressures, as contour plots and graphs. In some cases,

time histories of fluctuating pressures are also provided as .MAT files. The uniform resource locator (URL) for this database is http://www.wind.arch.t-kougei.ac.jp/info_center/windpressure/lowrise/mainpage.html

8.11 SUMMARY

This chapter has discussed various aspects of the design of low buildings for wind loads. The long history of investigation into wind loads has been discussed, and the use of the modern boundary-layer wind tunnel for the determination of design-loading coefficients is covered. The characteristics of loads for major structural members and foundations, and for local cladding have been considered for buildings with flat and pitched roofs. The effect of shelter and interference from the surrounding buildings has been considered. Multi-span building configurations and parapets have also been discussed.

An example of a building in which architectural features have been used to minimise the aerodynamic effects of severe winds has been shown, and two web-based databases providing comprehensive information on aerodynamic shape factors and pressure coefficients for generic low-rise building shapes have been highlighted.

REFERENCES

American Society of Civil Engineers. 1936. Wind-bracing in steel buildings. Fifth Progress Report of Sub-Committee No. 31. *Proceedings ASCE*, March 1936, 397–412.

Bailey, A. and Vincent, N.D.G. 1943. Wind pressure on buildings including the effects of adjacent buildings. *Journal of the Institution of Civil Engineers*, 2: 243–75.

Baker, C.J. 1999. Aspects of the use of the technique of orthogonal decomposition of surface pressure fields. *10th International Conference on Wind Engineering*, Copenhagen, 21–24 June, 393–400, A.A. Balkema, Rotterdam.

Best, R.J. and Holmes, J.D. 1983. Use of eigenvalues in the covariance integration method for determination of wind load effects. *Journal of Wind Engineering and Industrial Aerodynamics*, 13: 359–70.

Bienkiewicz, B., Ham, H.J. and Sun, Y. 1993. Proper orthogonal decomposition of roof pressure. *Journal of Wind Engineering and Industrial Aerodynamics*, 50: 193–202.

Davenport, A.G., Surry, D. and Stathopoulos, T. 1977. Wind loads on low-rise buildings. Final report of phases I and II. University of Western Ontario, Boundary Layer Wind Tunnel Report BLWT-SS8-1977.

Department of Public Works, Queensland. 2006. Design guidelines for Queensland public cyclone shelters. Prepared by Mullins Consulting for the Queensland Government, September.

Eaton, K.J. and Mayne, J.R. 1975. The measurement of wind pressures on two-storey houses at Aylesbury. *Journal of Industrial Aerodynamics*, 1: 67–109.

Eaton, K.J., Mayne, J.R. and Cook, N.J. 1975. Wind loads on low-rise buildings – Effects of roof geometry. *Fourth International Conference on Wind Effects on Buildings and Structures*, London, September.

Flachsbart, O. 1932. Winddrucke auf geschlossene und offene Gebaude. in *Ergebnisse der Aerodynamischen Versuchanstalt zu Gottingen*, IV Lieferung, L. Prandtl, and A. Betz (eds.), Verlag von R. Oldenbourg, Munich and Berlin.

Ginger, J.D. and Holmes, J.D. 2003. Effect of building length on wind loads on low-rise buildings with a steep roof pitch. *Journal of Wind Engineering and Industrial Aerodynamics*, 91: 1377–400.

Ho, T.C.E., Davenport, A.G. and Surry, D. 1995. Characteristic pressure distribution shapes and load repetitions for the wind loading of low building roof panels. *Journal of Wind Engineering and Industrial Aerodynamics*, 57: 261–79.

Ho, T.C.E., Surry, D. and Davenport, A.G. 1990. The variability of low building wind loads due to surrounding obstructions. *Journal of Wind Engineering and Industrial Aerodynamics*, 36: 161–70.

Ho, T.C.E., Surry, D. and Davenport, A.G. 1991. Variability of low building wind loads due to surroundings. *Journal of Wind Engineering and Industrial Aerodynamics*, 38: 297–310.

Holmes, J.D. 1981. Non-Gaussian characteristics of wind pressure fluctuations. *Journal of Wind Engineering and Industrial Aerodynamics*, 7: 103–8.

Holmes, J.D. 1983. *Wind Loads on Low Rise Buildings – A Review*. CSIRO, Division of Building Research, Australia.

Holmes, J.D. 1990a. Analysis and synthesis of pressure fluctuations on bluff bodies using eigenvectors. *Journal of Wind Engineering and Industrial Aerodynamics*, 33: 219–30.

Holmes, J.D. 1990b. Wind loading of multi-span buildings. *Civil Engineering Transactions, Institution of Engineers, Australia*, CE32: 93–8.

Holmes, J.D. 1994. Wind pressures on tropical housing. *Journal of Wind Engineering and Industrial Aerodynamics*, 53: 105–23.

Holmes, J.D. and Carpenter, P. 1990. The effect of Jensen Number variations on the wind loads on a low-rise building. *Journal of Wind Engineering and Industrial Aerodynamics*, 36: 1279–88.

Holmes, J.D. and Syme, M.J. 1994. Wind loads on steel-framed low-rise buildings. *Steel Construction (Australian Institute of Steel Construction)*, 28: 2–12.

Holmes, J.D., Sankaran, R., Kwok, K.C.S. and Syme, M.J. 1997. Eigenvector modes of fluctuating pressures on low-rise building models. *Journal of Wind Engineering and Industrial Aerodynamics*, 69–71: 697–707.

Hussain, M. and Lee, B.E. 1980. A wind tunnel study of the mean pressures acting on large groups of low-rise buildings. *Journal of Wind Engineering and Industrial Aerodynamics*, 6: 207–25.

Irminger, J.O.V. 1894. Nogle forsog over trykforholdene paa planer og legemer paavirkede af luftstrominger. *Ingenioren*, 17.

Irminger, J.O.V. and Nokkentved, C. 1930. Wind pressures on buildings. *Ingeniorvidenskabelige Skrifter*, A23.

Jensen, M. 1958. The model law for phenomena in the natural wind. *Ingenioren*, 2: 121–8.

Jensen, M. and Franck, N. 1965. *Model-Scale Tests in Turbulent Wind. Part II*. Danish Technical Press, Copenhagen.

Kernot, W.C. 1893. Wind pressure. *Proceedings, Australasian Association for the Advancement of Science*, V: 573–81, and VI: 741–5, Adelaide, Australia.

Kopp, G.A., Mans, C. and Surry, D. 2005a. Wind effects of parapets on low buildings: Part 2. Structural loads. *Journal of Wind Engineering and Industrial Aerodynamics*, 93: 843–55.

Kopp, G.A., Surry, D. and Mans, C. 2005b. Wind effects of parapets on low buildings: Part 1. Basic aerodynamics and local loads. *Journal of Wind Engineering and Industrial Aerodynamics*, 93: 817–41.

Krishna, P. 1995. Wind loads on low rise buildings – A review. *Journal of Wind Engineering and Industrial Aerodynamics*, 55: 383–96.

Lee, B.E. and Soliman, B.F. 1977. An investigation of the forces on three-dimensional bluff bodies in rough wall turbulent boundary layers. *Journal of Fluids Engineering*, 99: 503–10.

Letchford, C.W. and Mehta, K.C. 1993. The distribution and correlation of fluctuating pressures on the Texas Tech Building. *Journal of Wind Engineering and Industrial Aerodynamics*, 50: 225–34.

Levitan, M.L. and Mehta, K.C. 1992a. Texas Tech field experiments for wind loads. Part I. Building and pressure measuring system. *Journal of Wind Engineering and Industrial Aerodynamics*, 43: 1565–76.

Levitan, M.L. and Mehta, K.C. 1992b. Texas Tech field experiments for wind loads. Part II. Meteorological instrumentation and terrain parameters. *Journal of Wind Engineering and Industrial Aerodynamics*, 43: 1577–88.

Meecham, D., Surry, D. and Davenport, A.G. 1991. The magnitude and distribution of wind-induced pressures on hip and gable roofs. *Journal of Wind Engineering and Industrial Aerodynamics*, 38: 257–72.

Mehta, K.C., Levitan, M.L., Iverson, R.E. and Macdonald, J.R. 1992. Roof corner pressures measured in the field on a low-rise building. *Journal of Wind Engineering and Industrial Aerodynamics*, 41: 181–92.

Richardson, E.B. and Miller, B.H. 1932. The experimental determination of the pressures and distribution of pressures of an airstream on model buildings. *Journal of the Institution of Engineers Australia*, 4: 277–82.

Robertson, A.P. 1992. The wind-induced response of a full-scale portal framed building. *Journal of Wind Engineering and Industrial Aerodynamics*, 43: 1677–88.

Saathoff, P. and Stathopoulos, T. 1992. Wind loads on buildings with sawtooth roofs. *ASCE Journal of Structural Engineering*, 118: 429–46.

Sill, B.L., Cook, N.J. and Blackmore, P.A. 1989. IAWE Aylesbury Comparative Experiment – Preliminary results of wind-tunnel comparisons. *Journal of Wind Engineering and Industrial Aerodynamics*, 32: 285–302.

Sill, B.L., Cook, N.J. and Fang, C. 1992. The Aylesbury Comparative Experiment – A final report. *Journal of Wind Engineering and Industrial Aerodynamics*, 43: 1553–64.

Simiu, E. and Scanlan, R.H. 1996. *Wind Effects on Structures – An Introduction to Wind Engineering*. 3rd edition. John Wiley, New York.

Standards Australia. 1989. *SAA Loading Code. Part 2: Wind Loads*. Standards Australia, North Sydney, NSW, Australia. Australian Standard, AS1170.2-1989.

Standards Australia. 2011. *Structural Design Actions. Part 2: Wind Actions*. Standards Australia, Sydney, NSW, Australia. Australian/New Zealand Standard, AS/NZS1170.2:2011.

Stathopoulos, T. 1984. Wind loads on low-rise buildings: A review of the state of the art. *Engineering Structures*, 6: 119–35.

Stathopoulos, T. 1995. Evaluation of wind loads on low buildings – A brief historical review. In P. Krishna (ed.), *A State of the Art in Wind Engineering*. Wiley Eastern Limited, New Delhi.

Stathopoulos, T. and Baskaran, A. 1988. Turbulent wind loading on roofs with parapet configurations. *Canadian Journal of Civil Engineering*, 29: 570–8.

Stathopoulos, T. and Saathoff, P. 1994. Codification of wind-pressure coefficients for multispan gable roofs. *ASCE Journal of Structural Engineering*, 120: 2495–519.

Surry, D. 1999. Wind loads on low-rise buildings: Past, present and future. *10th International Conference on Wind Engineering*, Copenhagen, 21–24 June, 105–4, A.A. Balkema, Rotterdam.

Xu, Y.L. and Reardon, G.F. 1998. Variations of wind pressure on hip roofs with roof pitch. *Journal of Wind Engineering and Industrial Aerodynamics*, 73: 267–84.

Chapter 9

Tall buildings

9.1 INTRODUCTION

Tall buildings, now up to 830 m in height, project well into the atmospheric boundary layer, and their upper levels may experience the highest winds of large-scale wind storms, such as tropical cyclones or the winter gales of the temperate regions. Resonant dynamic response in along-wind, cross-wind and torsional modes is a feature of the overall structural loads experienced by these structures. Extreme local cladding pressures may be experienced on their side walls.

The post-World War II generation of high-rise buildings was the stimulus for the development of the boundary-layer wind tunnel, which remains the most important tool for the establishment of design wind loads on major building projects in many countries.

In this chapter, the history of investigations into wind loading of tall buildings, the major response mechanisms and phenomena, and the available analytical and semi-analytical techniques will be discussed.

9.2 HISTORICAL

Tall buildings, or 'skyscrapers', are amongst the more wind-sensitive of structures. It was inevitable that their response to wind would be of concern to structural engineers, and attract the interest of early experimenters, both in the wind tunnel, and in full scale.

The Empire State Building, at 380 m, was the tallest building in the world for 40 years, and was the subject of three significant studies in the 1930s (Coyle, 1931; Dryden and Hill, 1933; Rathbun, 1940). These studies have been re-appraised in some detail by Davenport (1975).

Coyle (1931) used a portable horizontal pendulum to record the motion of the building. This clearly revealed resonant dynamic response with a period of around 8 s. Rathbun's (1940) extensive full-scale measurements were described by Davenport as 'a monumental piece of full-scale experimentation' (Davenport, 1975, p. 35). Wind pressures on three floors of the building were measured with 30 manometers and 28 flash cameras. The pressure coefficients showed considerable scatter, but were clearly much lower than those obtained by Dryden and Hill (1933) on a wind-tunnel model in a uniform flow some years earlier. Rathbun also performed deflection measurements on the Empire State Building using a plumb bob extending from the 86th floor to the 6th floor. These results (as re-analysed by Davenport) indicated the significantly different stiffness of the building in the east–west direction in comparison with the north–south direction (Figure 9.1).

In the 1960s and 1970s, a resurgence in the building of skyscrapers occurred – particularly in North America, Japan and Australia. There was great interest in wind loads on tall

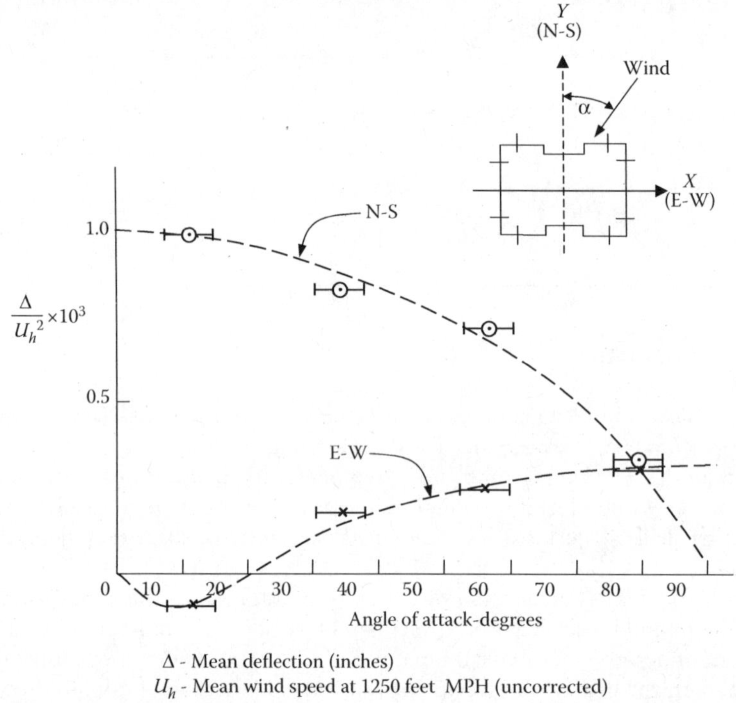

Figure 9.1 Full-scale measurements of mean deflection on the Empire State Building by Rathbun (1940) – reanalysed by Davenport (1981).

buildings at this time – this has continued to the end of the twentieth century. The two main problem areas to emerge were:

1. The vulnerability of glazed cladding to both direct wind pressures, and flying debris in wind storms
2. Serviceability problems arising from excessive motion near the top of tall buildings

From the early 1970s, many new building proposals were tested in the new boundary-layer wind tunnels (see Chapter 7), and quite a few full-scale monitoring programmes were commenced.

One of the most comprehensive and well-documented full-scale measurement studies, with several aspects to it, which lasted for most of the 1970s, was that on the 239-m tall Commerce Court building in Toronto, Canada (Dalgleish, 1975; Dalgleish et al., 1979, 1983). The full-scale studies were supplemented with wind-tunnel studies, both in the design stage (Davenport et al., 1969) and later on a pressure model (Dalgleish et al. 1979), and a multi-degree-of freedom aeroelastic model, in parallel with the full-scale studies (Templin and Cooper, 1981; Dalgleish et al., 1983).

The early full-scale pressure measurements on the Commerce Court building showed good agreement with the wind-tunnel study (at 1/400 scale) for mean pressure coefficients, and for the mean base shear and overturning moment coefficients. Not as good agreement with the 1/400 scale wind-tunnel tests, was found for the r.m.s. fluctuating pressure coefficients for some wind directions (Dalgleish, 1975). The later reported pressure measurements

(Dalgleish et al., 1979) showed better agreement for the fluctuating pressure and peak measurements on a larger (1/200) scale wind-tunnel model, with accurately calibrated tubing and pressure measurement system. The full-scale pressure study on Commerce Court highlighted the importance of short duration peak pressures in separated flow regions (at around this time similar observations were being made from the roof of the low-rise building at Aylesbury – Section 8.2.2). Subsequently, detailed statistical studies of these were carried out for application to glass loading (see Section 9.4.5). Although the Commerce Court pressure measurements were of a high quality, they suffered from the lack of an independent reference pressure for the pressure coefficients – an internal pressure reading from the building was used. For comparison of mean pressure coefficients with the wind-tunnel results, it was necessary to force agreement at one pressure tapping – usually in wake region.

The full-scale study of acceleration response (Dalgleish et al., 1983) showed the following features:

- The significance of the torsional (twisting) motions superimposed on the sway motions for one direction (E-W) – i.e. a 'coupled' mode. This was explained by an eccentricity in the north–south direction between the centre of mass, and the elastic axis.
- Generally good agreement between the final aeroelastic model, which included torsional motions, and the full-scale data, for winds from a range of directions.
- Reasonable agreement between the full-scale data and predictions of the National Building Code of Canada for along- and cross-wind accelerations.

The agreements observed occurred despite some uncertainties in the reference velocity measured at the top of the building, and in the dynamic properties (frequency and damping) of the building. An interesting observation, not clearly explained, was a clear decrease in observed building frequency as the mean speed increased.

Another important full-scale study, significant for its influence on the development of the British Code of Practice for Wind Loads, was that carried out on the 18-storey Royex House in London (Newberry et al., 1967). This study revealed aspects of the transient and fluctuating pressures on the windward and side walls.

The first major boundary-layer wind-tunnel study of a tall building was that carried out for the twin towers of the World Trade Centre, New York, in the mid-1960s, at Colorado State University. This was the first of many commercial studies, now numbering in the thousands, in boundary-layer wind tunnels.

9.3 FLOW AROUND TALL BUILDINGS

Tall buildings are bluff bodies of medium-to-high aspect ratio, and the basic characteristics of flow around this type of body were covered in some detail in Chapter 4.

Figure 9.2 shows the general characteristics of boundary-layer wind flow around a tall building. On the windward face there is a strong downward flow below the stagnation point, which occurs at a height of 70–80% of the overall building height. The down flow can often cause problems at the base, as high-velocity air from upper levels is brought down to street level. Separation and re-attachment at the side walls are associated with high local pressures. The rear face is a negative pressure region of lower magnitude mean pressures, and a low level of fluctuating pressures.

In a mixed extreme wind climate of thunderstorm downbursts (Section 1.3.5) and synoptic winds, the dominant wind for wind loading of tall buildings will normally be the latter, as the downburst profile has a maximum at a height of 50–100 m (Figure 3.3).

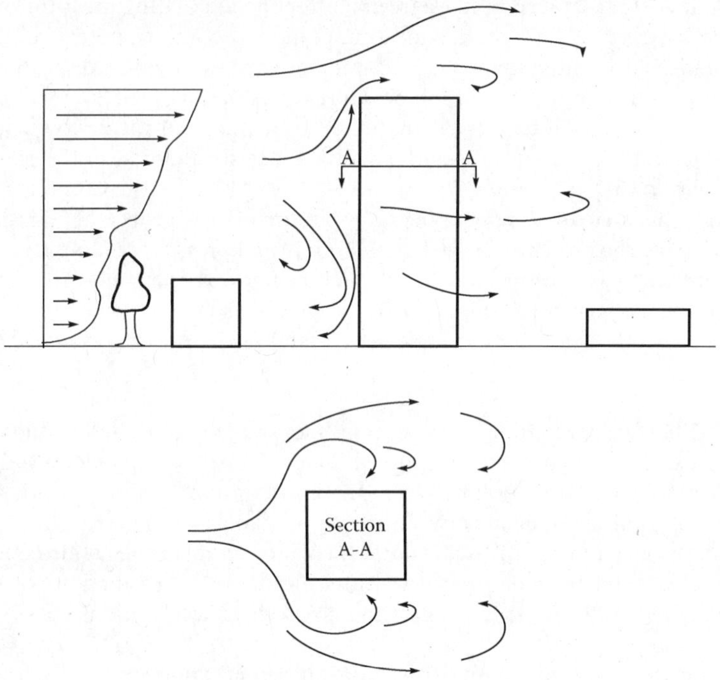

Figure 9.2 Wind flow around a tall building.

9.4 CLADDING PRESSURES

9.4.1 Pressure coefficients

As in previous chapters, pressure coefficients in this chapter will be defined with respect to a mean wind speed at the top of the building, denoted by $\bar{U}_h$. Thus, the mean, root-mean-square fluctuating (standard deviation), maximum and minimum pressure coefficients are defined according to Equations 9.1 through 9.4, respectively.

$$\bar{C}_p = \frac{\bar{p} - p_0}{(1/2)\rho_a \bar{U}_h^2} \tag{9.1}$$

$$C_p' = \sigma_{Cp} = \frac{\sqrt{\overline{p'^2}}}{(1/2)\rho_a \bar{U}_h^2} \tag{9.2}$$

$$\hat{C}_p = \frac{\hat{p} - p_0}{(1/2)\rho_a \bar{U}_h^2} \tag{9.3}$$

$$\check{C}_p = \frac{\check{p} - p_0}{(1/2)\rho_a \bar{U}_h^2} \tag{9.4}$$

In Equations 9.3 and 9.4, the maximum and minimum pressures, $\hat{p}$ and $\check{p}$, are normally defined as the average or expected peak pressure at a point in a given averaging time, which may be taken as a period between 10 min and 3 h in full scale. It is not usually convenient, or economical, to measure such average peaks directly in wind-tunnel tests, and various alternative statistical procedures have been proposed. These are discussed in Section 9.4.4.

9.4.2 Pressure distributions on buildings of rectangular cross-section

The local pressures on the wall of a tall building can be used directly for the design of cladding, which is generally supported over small tributary areas.

Figure 4.15 shows the distribution of mean pressure coefficient on the faces of tall prismatic shape, representative of a very tall building, with aspect ratio (height/width) of 8, in a boundary-layer flow.

Figures 9.3 to 9.5 show the variation in mean, maximum and minimum pressure coefficients on the windward, side and leeward faces, for a lower building of square cross-section, with aspect ratio equal to 2.1 (Cheung, 1984). The pressures were measured on a wind-tunnel model, which represented a building of 85 m height; the building is isolated, that is there is no shielding from buildings of comparable height, and the approaching flow was boundary-layer flow over suburban terrain. The value of Jensen number, h/z_0 (see Section 4.4.4) was then approximately 40.

Figure 9.3 shows a stagnation point on the windward face, where the value of $\bar{C}_p$ reaches a maximum, at about 0.8 h. The heights for largest maximum pressure coefficient are slightly lower than this.

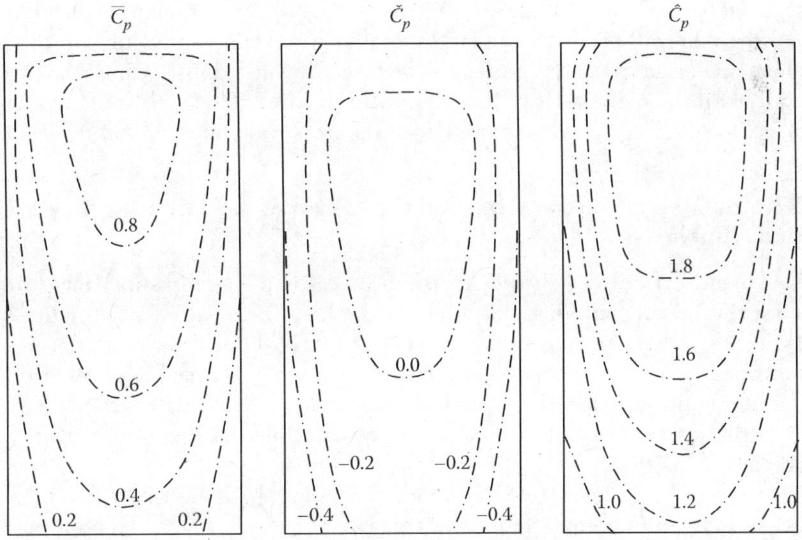

Figure 9.3 Mean, maximum and minimum pressure coefficients – windward wall of a building with square cross-section – height/width = 2.1. (From Cheung, J.C.K. 1984. Effect of tall building edge configurations on local surface wind pressures. *3rd International Conference on Tall Buildings*, Hong Kong and Guangzhou, 10–15 December. With permission.)

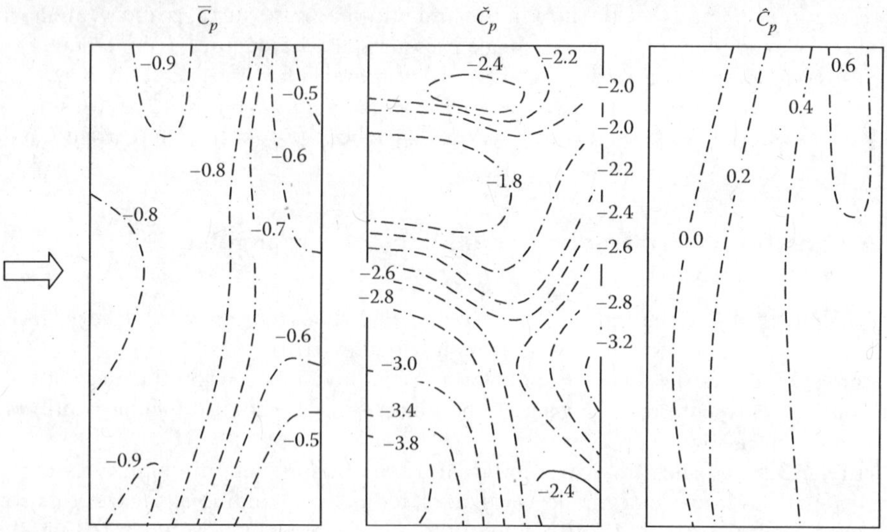

Figure 9.4 Mean, maximum and minimum pressure coefficients – side wall of a building with square cross-section – height/width = 2.1. (From Cheung, J.C.K. 1984. Effect of tall building edge configurations on local surface wind pressures. *3rd International Conference on Tall Buildings*, Hong Kong and Guangzhou, 10–15 December. With permission.)

The side walls (Figure 9.4) are adjacent to a flow which is separating from the front wall, and generating strong vortices (see Figures 4.1 and 9.2). The mean pressure coefficients are generally in the range from −0.6 to −0.8, and not dissimilar to the values on the much taller building in Figure 4.15. The largest magnitude minimum pressure coefficients of about −3.8 occur near the base of the buildings, unlike the windward wall pressures. A wind direction parallel to the side wall produces the largest magnitude negative pressures in this case.

The mean and largest peak pressures on the leeward wall (Figure 9.5) is also negative, but are typically half the magnitude of the side wall pressures. This wall is of course sheltered, and exposed to relatively slowly moving air in the near wake of the building.

9.4.3 The nature of fluctuating local pressures and probability distributions

As discussed in Section 9.2, in the 1970s, full-scale and wind-tunnel measurements of wind pressures on tall buildings, highlighted the local peak negative pressures, that can occur, for some wind directions, on the walls of tall buildings, particularly on side walls at locations near windward corners, and on leeward walls. These high pressures generally only occur for quite short periods of time, and may be very intermittent in nature. An example of the intermittent nature of these pressure fluctuations is shown in Figure 9.6 (from Dalgleish, 1971).

Several studies (e.g. Dalgleish, 1971; Peterka and Cermak, 1975) indicated that the probability densities of pressure fluctuations in separated flow regions on tall buildings were not well-fitted by the normal or Gaussian probability distribution (Appendix C3.1). This is the case, even though the latter is a good fit to the turbulent velocity fluctuations in the wind (see Section 3.3.2). The 'spiky' nature of local pressure fluctuations (Figure 9.6) results in probability densities of peaks of five standard deviations, or greater, below the mean

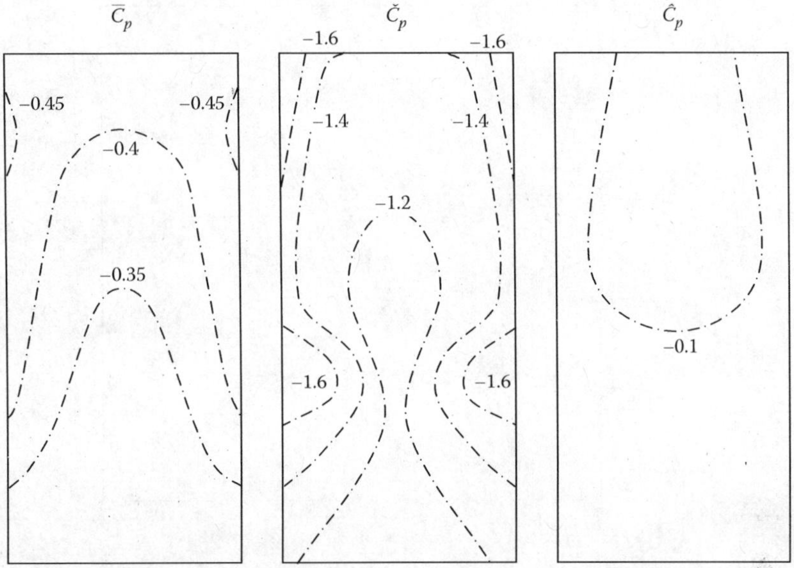

Figure 9.5 Mean, maximum and minimum pressure coefficients – leeward wall of a building with square cross-section – height/width = 2.1. (From Cheung, J.C.K. 1984. Effect of tall building edge configurations on local surface wind pressures. *3rd International Conference on Tall Buildings*, Hong Kong and Guangzhou, 10–15 December. With permission.)

pressure, being several times greater than that predicted by the Gaussian distribution. This is illustrated in Figure 9.7 derived from wind-tunnel tests of two tall buildings (Peterka and Cermak 1975).

A consequence of the intermittency and non-Gaussian nature of pressure fluctuations on tall buildings, is that the maximum pressure coefficient measured at a particular location on a building in a defined time period – say 10 min in full scale – may vary considerably from one time period to the next. Therefore, they cannot be predicted by knowing the mean and standard deviation, as is the case with a Gaussian random process (Davenport 1964). This

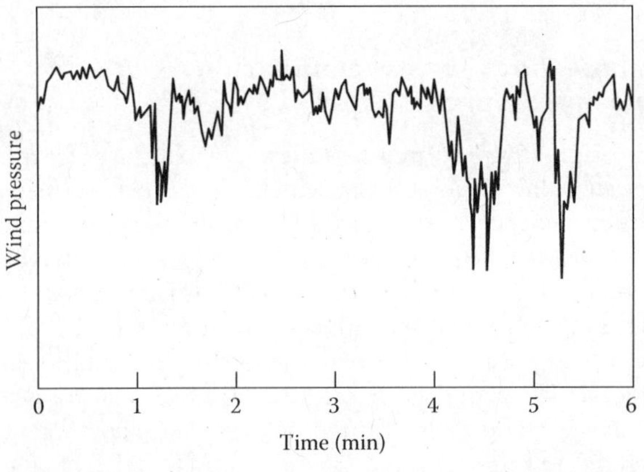

Figure 9.6 Record of fluctuating pressure from the leeward wall of a full-scale office building. (From Dalgleish, W.A. 1971. *ASCE Journal of the Structural Division*, 97: 2173–87. With permission.)

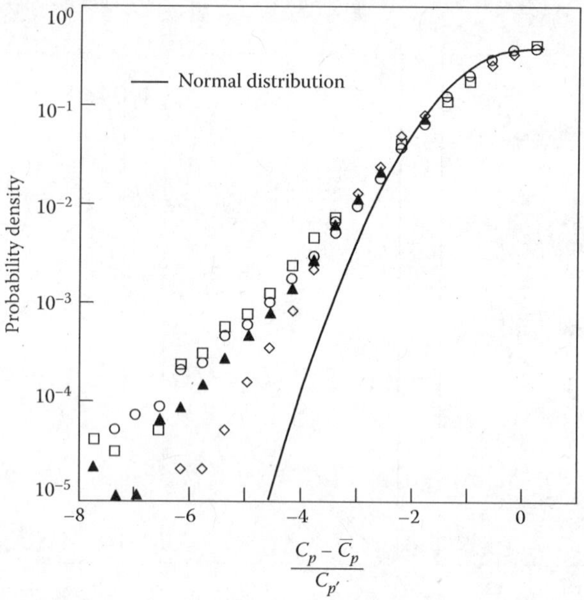

Figure 9.7 Probability densities of pressure fluctuations from regions in separated flow on tall buildings. (From Peterka, J.A. and Cermak, J.E. 1975. *ASCE Journal of the Structural Division,* 101: 1255–67. With permission.)

has led to a number of different statistical techniques being adopted to produce more consistent definitions of peak pressures for design – these are discussed in Section 9.4.4. A related matter is the response characteristics of glass cladding to short duration peak loads. The latter aspect is discussed in Section 9.4.5.

A detailed study (Surry and Djakovich, 1995) of local negative peak pressures on generic tall building models of constant cross-section, with four different corner geometries, indicated that the details of the corner geometry do not affect the general magnitude of the minimum pressure coefficients, but rather the wind direction at which they occur. The highest peaks were associated with vortex shedding.

9.4.4 Statistical methods for determination of peak local pressures

A simple approach, originally proposed by Lawson (1976), uses the parent probability distribution of the pressure fluctuations, from which a pressure coefficient, with a designated (low) probability of exceedence is extracted. The probability of exceedence is normally in the range 1×10^{-4} to 5×10^{-4}, with the latter being suggested by Lawson. This method can be programmed 'on the run' in wind-tunnel tests, relatively easily; sometimes a standard probability distribution, such as the Weibull type (see Appendix C3.4) is used to fit the measured data and interpolate, or extrapolate, to the desired probability level.

Cook and Mayne (1979) proposed a method in which the total averaging time, T, is divided into 16 equal parts and the measured peak pressure coefficient (maximum or minimum) within each reduced time period, t, is retained. A Type I Extreme Value (Gumbel) distribution (see Section 2.2.1 and Appendix C4) is fitted to the measured data, giving a mode, c_t, and slope ($1/a_t$). These can then be used to calculate the parameters of the Extreme Value

Type I distribution appropriate to the maxima (or minima) for the original time period, T, as follows:

$$c_T = c_t + a_t \log_e 16 \tag{9.5}$$

$$a_T = a_t \tag{9.6}$$

Knowing the distribution of the extreme pressure coefficients, the expected peak, or any other percentile, can then be easily determined. The method proposed by Cook and Mayne (1979), in fact, proposes an effective peak pressure coefficient C_p^* given by

$$C_p^* = c_T + 1.4a_T \tag{9.7}$$

Equation 9.7 can be rewritten in terms of the mean and standard deviation of the extremes (Kasperski, 2003). For the Extreme Value Type I distribution, the mean and standard deviation are related to the mode, c_T, and scale factor, a_T, by Equations 9.8 and 9.9.

$$\text{Mean} = m = u_T + 0.577a_T \tag{9.8}$$

$$\text{Standard deviation} = \sigma = (\pi/\sqrt{6})a_T = 1.282\ a_T \tag{9.9}$$

Hence, Equation 9.7 can be written as

$$\begin{aligned} C_p^* &= u + 0.577a + 0.823a = m + 0.64(1.282a) \\ C_p^* &= m \pm 0.64\sigma \end{aligned} \tag{9.10}$$

Equation 9.10 can be used to calculate C_p^* by estimating m and σ from the measured extremes without having to fit a distribution. In fact, it does not require an assumption about the distribution of the extremes. An alternative, more conservative, form is Equation 9.11; this corresponds to Equation 9.7 with the 1.4 replaced by 1.5.

$$C_p^* = m \pm 0.7\sigma \tag{9.11}$$

Peterka (1983) proposed the use of the probability distribution of 100 independent maxima within a time period equivalent to 1 h, to determine C_p^*.

Another approach is to make use of level crossing statistics. Melbourne (1977) proposed the use of a normalised rate of crossing of levels of pressure (or structural response). A nominal rate of crossing (e.g. 10^{-4} per h) is chosen to determine a nominal level of 'peak' pressure.

The parameters of the (Type I) extreme value distribution for the extreme pressure in a given time period can also be derived from level crossing rates as follows. The level crossings are assumed to be uncorrelated events that can be modelled by a Poisson distribution (Appendix C3.5).

The Poisson distribution gives the probability for the number of events, n, in a given time period, T, when the average rate of occurrence of the events is υ:

$$P(n,\upsilon) = \frac{(\upsilon T)^n}{n!}\exp(-\upsilon T) \tag{9.12}$$

The 'event' in this case can be taken as an upcrossing of a particular level, for example the exceedence at a particular pressure level. The probability of getting no crossings of a pressure level, p, during the time period, T, is also the probability that the largest value of the process $p(t)$, during the time period, is less than that level, that is the cumulative probability distribution of the largest value in the time period, T.

Thus,

$$F(p) = P(0,\upsilon) = \frac{(\upsilon T)^0}{0!}\exp(-\upsilon T) = \exp(-\upsilon T) \tag{9.13}$$

If we assume that the average number of crossings of level x in time T, is given by

$$\upsilon T = \exp\left[-\frac{1}{a}(p-u)\right] \tag{9.14}$$

where a and u are constants, then,

$$F(p) = \exp\left\{\exp\left[-\frac{1}{a}(p-u)\right]\right\} \tag{9.15}$$

This is the Type I (Gumbel) extreme value distribution with a mode of u and a scale factor of a.

From Equation 9.14, taking natural logarithms of both sides,

$$\log_e(\upsilon T) = -\frac{1}{a}(p-u) \tag{9.16}$$

The mode and scale factor of the Type I extreme value distribution of the process $p(t)$ can be estimated by the following procedure.

- Plot the natural logarithm of the rate of upcrossings against the level, p
- Fit a straight line. From Equation 9.16, the slope is $(-1/a)$, and the intercept ($p = 0$) is (u/a)
- From these values, estimate u and a, the mode and scale factor of the Type I extreme value distribution of p.

9.4.5 Strength characteristics of glass in relation to wind loads

Direct wind loading is a major design consideration in the design of glass and its fixing in tall buildings. However, the need to design for wind-generated flying debris (Section 1.5), particularly roof gravel, in some cities, also needs to be considered (Minor, 1994).

As has been discussed, wind pressures on the surfaces of buildings fluctuate greatly with time, and it is known that the strength of glass is quite dependent on the duration of the loading. The interaction of these two phenomena results in a complex design problem.

The surfaces of glass panels are covered with flaws of various sizes and orientations. When these are exposed to tensile stresses they grow at a rate dependent on the magnitude of the stress field, as well as relative humidity and temperature. The result is a strength reduction, which is dependent on the magnitude and duration of the tensile stress. Drawing on earlier studies of this phenomenon, known as 'static fatigue', Brown (1972) proposed a formula for damage accumulation, which has the form of Equation 9.17, at constant humidity and temperature.

$$D = \int_0^T [s(t)]^n \, dt \qquad (9.17)$$

where
 D is the accumulated damage
 $s(t)$ is the time-varying stress
 T is the time over which the glass is stressed
 n is a higher power (in the range of 12–20)

The expected damage, in time T, under a fluctuating wind pressure $p(t)$, in the vicinity of a critical flaw can be written as

$$E(D) = K\int_0^T E\{[p(t)]^m\}dt \qquad (9.18)$$

where K is a constant, and m is a different power, usually lower than n, but dependent on the size and aspect ratio of the glass, which allows for the non-linear relationship between load and stress for glass plates due to membrane stresses (Calderone and Melbourne 1993). $E\{\}$ is the expectation or averaging operation.

Calderone (1999), after extensive glass tests, found a power-law relationship between maximum stress anywhere in a plate, and the applied pressure, for any given plate; this may be used to determine the value of m for that plate. Values fall in the range of 5–20.

The integral on the right-hand side of Equation 9.18 is T times the mth moment of the pressure fluctuation, so that

$$E\{D\} = KT(\tfrac{1}{2}\rho\bar{U}^2)^m \int_0^\infty C_p{}^m f_{Cp}(C_p)dC_p \qquad (9.19)$$

where $C_p(t)$ is the time-varying pressure coefficient, and $f_{Cp}(C_p)$ is the probability density function for C_p.

The integral in Equation 9.19 is proportional to the rate at which damage is accumulated in the glass panel. It can be evaluated from known or expected probability distributions (e.g. Holmes 1985), or directly from wind-tunnel or full-scale pressure–time histories (Calderone and Melbourne 1993).

The high weighting given to the pressure coefficient by the power, m, in Equation 9.19 means that the main contribution to glass damage comes from isolated peak pressures, which typically occur intermittently on the walls of tall buildings (see Figure 9.6).

An equivalent static pressure coefficient, C_{ps}, which corresponds to a constant pressure, which gives the same rate of damage accumulation as a fluctuating pressure–time history, can be defined as

$$C_{p_s} = \left[\int_0^\infty C_p{}^m f_{Cp}(C_p)dC_p\right]^{1/m} \qquad (9.20)$$

For the structural design of glazing it is necessary to relate the computed damage caused by wind action, to failure loads obtained in laboratory tests of glass panels. The damage integral Equations 9.17 or 9.18, can be used to compute the damage sustained by a glass panel under the ramp loading (i.e. increasing linearly with time) commonly used in laboratory testing. In these tests, failure typically occurs in about 1 min.

An equivalent glass design coefficient, C_k, can be defined (Dalgleish, 1979) which, when multiplied by the reference dynamic pressure, $((1/2)\rho_a \bar{U}^2)$, gives a pressure, which produces the same damage in a 60 s ramp increase, as in a wind storm of specified duration.

Making use of Equations 9.19 and 9.20, it can be easily shown that for a statistically stationary (synoptic) wind storm of 1 h duration :

$$C_k = [60(1+m)]^{1/m} C_{p_s} \quad (9.21)$$

Using typical values of m and typical probability distributions, it can be shown (Dalgleish, 1979; Holmes, 1985) that C_k is approximately equal to the expected peak pressure coefficient occurring during the hour of storm wind. This fortuitous result, which is insensitive to both the value of m and the probability distribution, means that measured peak pressure coefficients from wind-tunnel tests are valid for use in calculation of design loads, for comparison with 1-min loads in glass design charts.

9.5 OVERALL LOADING AND DYNAMIC RESPONSE

In Chapter 5, the random, or spectral, approach to the along-wind response of tall structures was discussed. This approach is widely used for the prediction of the response of tall buildings in simplified forms in codes and standards (see Chapter 15). Dynamic response of a tall building in the along-wind direction is primarily produced by the turbulent velocity fluctuations in the natural wind (Section 3.3). In the cross-wind direction, loading and dynamic response is generated by random vortex shedding (Section 4.6.3) – that is it is a result of unsteady separating flow generated by the building itself, with a smaller contribution from cross-wind turbulence.

9.5.1 General response characteristics

In this section some general characteristics of the dynamic response of tall buildings to wind will be outlined.

By a dimensional analysis, or by application of the theory given in Section 5.3.1, it can be demonstrated (Davenport, 1966, 1971) that the root-mean-square fluctuating deflection at the top of a tall building of given geometry in a stationary (synoptic) wind, is given to a good approximation for the along-wind response by

$$\frac{\sigma_x}{h} = A_x \left(\frac{\rho_a}{\rho_b}\right)\left(\frac{\bar{U}_h}{n_1 b}\right)^{kx} \frac{1}{\sqrt{\eta}} \quad (9.22)$$

and for the cross-wind response:

$$\frac{\sigma_y}{h} = A_y \left(\frac{\rho_a}{\rho_b}\right)\left(\frac{\bar{U}_h}{n_1 b}\right)^{ky} \frac{1}{\sqrt{\eta}} \quad (9.23)$$

where,

h is the building height
A_x, A_y are constants for a particular building shape
ρ_a is the density of air
ρ_b is an average building density
$\bar{U}_h$ is the mean wind speed at the top of the building
b is the building breadth
k_x, k_y are exponents
n_1 is the first mode natural frequency
η is the critical damping ratio in the first mode of vibration

Equations 9.22 and 9.23 are based on the assumption that the responses are dominated by the resonant components. For along-wind response, the background component is independent of the natural frequency. In the case of the cross-wind response, there is no mean component but some background contribution due to cross-wind turbulence. The assumption of dominance of resonance is valid for slender tall buildings with first mode natural frequencies less than about 0.5 Hz, and damping ratios less than about 0.02.

The equations illustrate that the fluctuating building deflection can be reduced by either increasing the building density or the damping. The damping term, η, includes aerodynamic damping as well as structural damping; however, this is normally small for tall buildings.

The term $\left(\bar{U}_h/n_1 b\right)$ is a non-dimensional mean wind speed, known as the *reduced velocity*. The exponent, k_x, for the fluctuating along-wind deflection is greater than 2, since the spectral density of the wind speed near the natural frequency, n_1, increases at a greater power than 2, as does the aerodynamic admittance function (Section 5.3.1 and Figure 5.4) at that frequency. The exponent for cross-wind deflection, k_y, is typically about 3, but can be as high as 4.

Figure 9.8 shows the variation of (σ_x/h) and (σ_x/h) with reduced velocity for a building of circular cross-section (as well as the variation of $\bar{X}$).

9.5.2 Effect of building cross-section

In a study used to develop an optimum building shape for the U.S. Steel building, Pittsburgh, the response of six buildings of identical height and dynamic properties, but with different cross-sections were investigated in a boundary-layer wind tunnel (Davenport, 1971). The probability distributions of the extreme responses in a typical synoptic wind climate was determined, and are shown plotted in Figure 9.9. The figure shows a range of 3:1 in the responses with a circular cross-section producing the lowest response, and an equilateral triangular cross-section the highest. Deflection across the shortest (weakest) axis of a 2:1 rectangular cross-section was also large.

9.5.3 Corner modifications

Slotted and chamfered corners on rectangular building cross-sections have significant effects on both along- and cross-wind dynamic responses to wind (Kwok and Bailey, 1987; Kwok et al., 1988; Kwok, 1995). Chamfers of the order of 10% of the building width produce up to 40% reduction in the along-wind response and 30% reduction in the cross-wind response.

9.5.4 Prediction of cross-wind response

Along-wind response of isolated tall buildings can be predicted reasonably well from the turbulence properties in the approaching flow by applying the random vibration theory

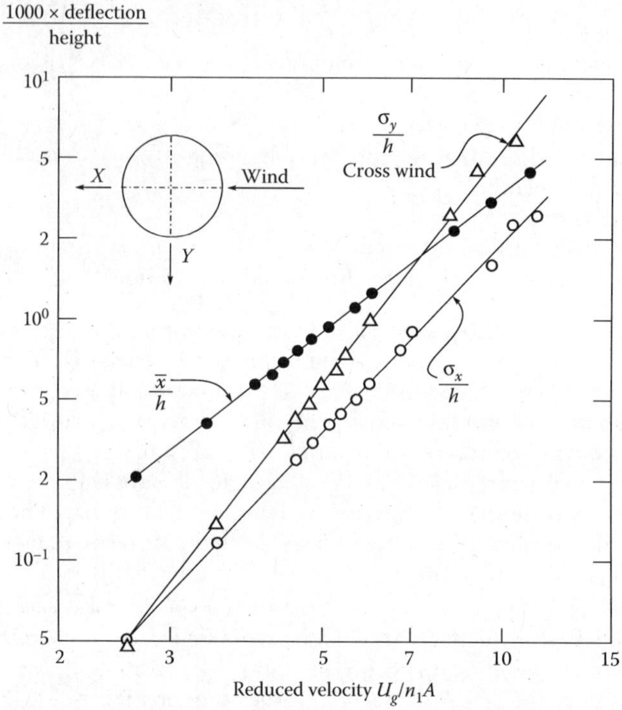

Figure 9.8 The mean and fluctuating response of a tall building of circular cross-section. (From Davenport A.G. 1971. *Philosophical Transactions, Royal Society, A*, 269: 385–94. With permission.)

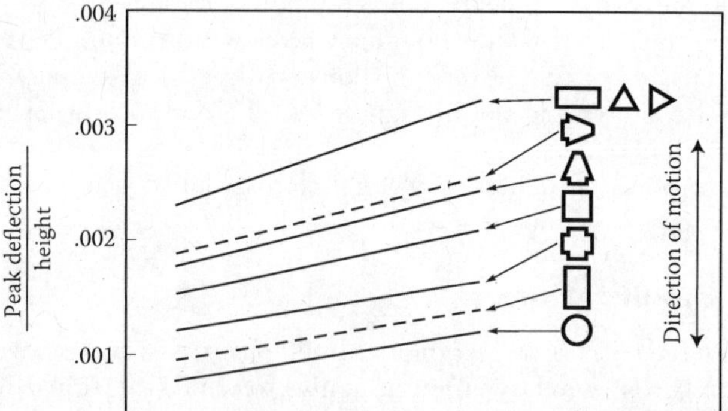

Figure 9.9 Effect of cross-sectional shape on maximum deflections of six buildings. (From Davenport A.G. 1971. *Philosophical Transactions, Royal Society, A*, 269: 385–94. With permission.)

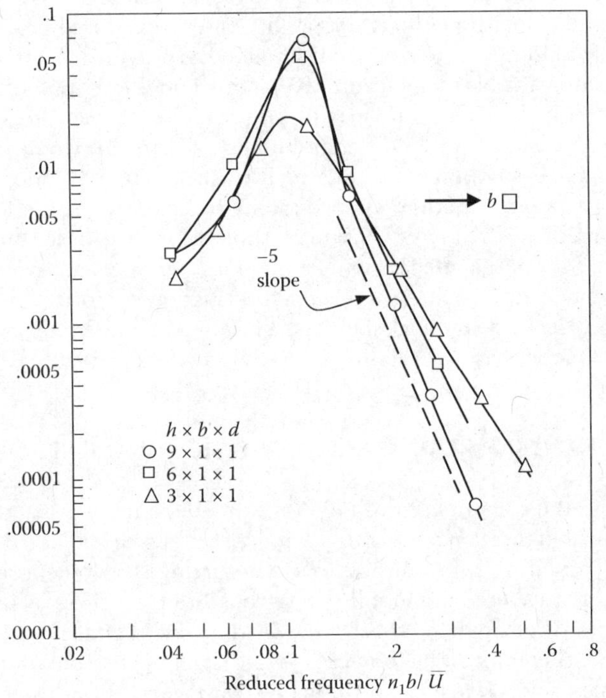

Figure 9.10 Cross-wind generalised force spectra for buildings of square cross-section. (From Saunders, J.W. 1974. Wind excitation of tall buildings. PhD thesis, Monash University. With permission.)

methods discussed in Section 5.3.1. Cross-wind response however is more difficult to predict, since vortex shedding plays a dominant role in the exciting forces in the cross-wind direction. However, an approach which has been quite successful, is the use of the high-frequency base balance technique to measure the spectral density of the generalised force in wind-tunnel tests (Section 7.6.2). Multiplication by the mechanical admittance and integration over frequency can then be performed to predict the building response.

Examples of generalised force spectra for buildings of square cross-section are shown in Figure 9.10. Non-dimensional spectra for three different height/breadth ratios are shown, and the approach flow is typical of suburban terrain. The mode shapes are assumed to be linear with height. The abscissa of this graph is reduced frequency – the reciprocal of reduced velocity.

For reduced velocities of practical importance (2–8), the non-dimensional spectra vary with reduced velocity to a power of 3–5, or with reduced frequency to a power of –3 to –5 (represented by the slope on the log–log plot). Such data have been incorporated in some standards and codes for design purposes (see Section 15.8).

9.5.5 Database for tall building loading and response

Most tall buildings of 100 m height, or greater, are the subject of special wind-tunnel tests at the later design stages; the techniques for this are well-developed and many of these are discussed in Section 7.6. However, the designer usually needs preliminary estimates of overall wind loading early in the design stage. As discussed in Chapter 15, several wind-loading codes and standards contain methods for prediction of along- and cross-wind response. A more comprehensive alternative to codes and standards is provided by Internet databases such as the one compiled by the Natural Hazards Modeling Laboratory of Notre Dame University (http://aerodata.ce.nd.edu/). The latter provides information on the spectral densities of three components of base moment for 27 different building shapes, in two different terrain types. This information has been obtained from a high-frequency base balance (Section 7.6.2). By application of the random vibration, or spectral approach, described in Section 5.3, reasonable preliminary predictions of basic building responses, such as base bending moments and accelerations at the top of the building can be obtained. Subsequently, more accurate predictions can be obtained from specific wind-tunnel tests, allowing for accurate reproduction of the building shape, the surroundings, including shielding and interference effects from other buildings, and the mode shapes, including sway-twist coupling effects.

9.6 COMBINATION OF ALONG- AND CROSS-WIND RESPONSE

When dealing with the response of tall buildings to wind loading, the question arises: how should the responses in the along- and cross-wind directions be combined statistically? Since clearly the along- and cross-wind responses are occurring simultaneously on a structure it would be unconservative (and potentially dangerous!) to treat these as separate load cases. The question arises when applying those wind-loading codes and standards, which provide methods for calculating both along- and cross-wind dynamic response for tall buildings (see Chapter 15). It also arises when wind-tunnel tests are carried out using either aeroelastic (Section 7.6.1), or base-balance methods (Section 7.6.2). In these cases, predictions are usually provided for each wind direction, with respect to *body*- or building-*axes* rather than *wind axes* (see Section 4.2.2 and Figure 4.2). These axes are usually the two principal axes for sway of the building.

Two cases can be identified:

1. 'Scalar' combination rules for load effects
2. 'Vector' combination of responses

The former case is the more relevant case for structural load effects being designed for strength, as in most cases structural elements will 'feel' internal forces and stresses from both response directions, and will be developed in the following. The second case is relevant when axisymmetric structures are under consideration, that is structures of circular cross-section such as chimneys.

Load effects (i.e. member forces and internal stresses) resulting from overall building response in two orthogonal directions (x- and y-) can very accurately be combined by the following formula :

$$\hat{\varepsilon}_t = \bar{\varepsilon}_x + \bar{\varepsilon}_y + \sqrt{(\hat{\varepsilon}_x - |\bar{\varepsilon}_x|)^2 + (\hat{\varepsilon}_y - |\bar{\varepsilon}_y|)^2} \tag{9.24}$$

where

$\hat{\varepsilon}_t$ is total combined maximum peak load effect (e.g. the axial load in a column)

$\bar{\varepsilon}_x$ is the load effect derived from the mean response in the x-direction (usually derived from the mean base bending moment in that direction)

$\bar{\varepsilon}_y$ is the load effect derived from the mean response in the y-direction

$\hat{\varepsilon}_x$ is the peak load effect derived from the response in the x-direction
$\hat{\varepsilon}_y$ is the peak load effect derived from the response in the x-direction

Equation 9.24 is quite an accurate one, as it is based on the combination of uncorrelated Gaussian random processes, for which it is exact. Most responses dominated by resonant contributions to wind, have been found to be very close to Gaussian, and if the two orthogonal sway frequencies are well-separated, the dynamic responses will be poorly correlated.

As an alternative approximation, the following load cases can be studied:

a. [Mean x-load +0.75(peak – mean)$_x$] with [mean y-load + 0.75(peak – mean)$_y$]
b. [Mean x-load + (peak – mean)$_x$] with [mean y-load]
c. [Mean x-load] with [mean y-load + (peak – mean)$_y$]

The case (a) corresponds to the following approximation to Equation 9.24 for peak load effect:

$$\varepsilon_t = \bar{\varepsilon}_x + \bar{\varepsilon}_y + 0.75((\hat{\varepsilon}_x - |\bar{\varepsilon}_x|) + (\hat{\varepsilon}_y - |\bar{\varepsilon}_y|)) \quad (9.25)$$

Equation 9.25 is a good approximation to Equation 9.24 for the range:

$$1/3 < (\hat{\varepsilon}_x - |\bar{\varepsilon}_x|)/(\hat{\varepsilon}_y - |\bar{\varepsilon}_y|) < 3$$

The other two cases (b) and (c) are intended to cover the cases outside this range, that is when $(\hat{\varepsilon}_x - |\bar{\varepsilon}_x|)$ is much larger than $(\hat{\varepsilon}_y - |\bar{\varepsilon}_y|)$, and vice versa.

9.7 TORSIONAL LOADING AND RESPONSE

The significance of torsional components in the dynamic response of tall buildings was highlighted by the Commerce Court study of the 1970s (Section 9.2), when a building of a uniform rectangular cross-section experienced significant and measurable dynamic twist due to an eccentricity between the elastic and mass centres. Such a possibility had been overlooked in the original wind-tunnel testing. Now, when considering accelerations at the top of tall buildings, the possibility of torsional motions increasing the perceptible motions at the periphery of the cross-section may need to be considered.

There are two mechanisms for producing dynamic torque and torsional motions in tall buildings:

1. Mean torque and torsional excitation resulting from non-uniform pressure distributions, or from non-symmetric cross-sectional geometries, and
2. Torsional response resulting from sway motions through coupled mode shapes and/or eccentricities between elastic (shear) and geometric centres.

The first aspect was studied by Isyumov and Poole (1983), Lythe and Surry (1990) and Cheung and Melbourne (1992). Torsional response of tall buildings has been investigated both computationally making use of experimentally obtained dynamic pressure or force data from wind-tunnel models (Tallin and Ellingwood, 1985; Kareem, 1985), and experimentally on aeroelastic models with torsional degrees of freedom (Xu et al., 1992a; Beneke and Kwok, 1993; Zhang et al., 1993).

A mean torque coefficient, $\bar{C}_{Mz}$, can be defined as

$$\bar{C}_{Mz} = \frac{\bar{M}_z}{(1/2)\rho_a \bar{U}_h^2 b_{max}^2 h} \qquad (9.26)$$

where
$\bar{M}_z$ is the mean torque
b_{max} is the maximum projected width of the cross-section
h is the height of the building

Lythe and Surry (1990), from wind-tunnel tests on 62 buildings, ranging from those with simple cross-sections to complex shapes, found an average value of $\bar{C}_{Mz}$, as defined above, of 0.085, with a standard deviation of 0.04. The highest values appear to be a function of the ratio of the minimum projected width, b_{min} to the maximum projected width, b_{max}, with a maximum value of $\bar{C}_{Mz}$ approaching 0.2, when (b_{min}/b_{max}) is equal to around 0.45 (Figure 9.11 from Cheung and Melbourne 1992). The highest value of $\bar{C}_{Mz}$ for any section generally occurs when the mean wind direction is about 60–80° from the normal to the widest building face.

Isyumov and Poole (1983) used simultaneous fluctuating pressures and pneumatic averaging (Section 7.5.2) on building models with a square or 2:1 rectangular cross-section in a wind tunnel, to determine the contribution to the fluctuating torque coefficient from various height levels on the buildings, and from the various building faces. The main contribution to the fluctuating torque on the square and rectangular section with the wind parallel to the long faces, came from pressures on the side faces, and could be predicted from the mean torque by quasi-steady assumptions (Section 4.6.2). On the other hand, for a mean wind

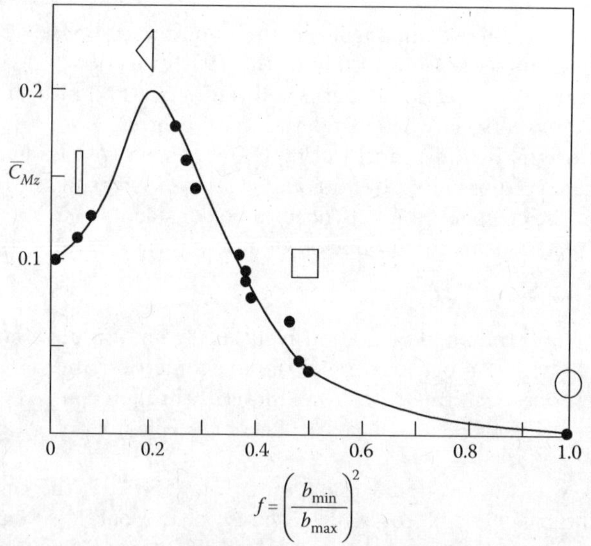

Figure 9.11 Mean torque coefficients on tall buildings of various cross-sections. (From Cheung, J.C.K. and Melbourne, W.H. 1992. *Journal of Wind Engineering and Industrial Aerodynamics*, 41–44: 1125–26. With permission.)

direction parallel to the short walls of the rectangular cross-section, the main contribution was pressure fluctuations on the rear face, induced by vortex shedding.

A double peak in the torque spectra for the wind direction parallel to the long face of a 2:1 building has been attributed to buffetting by lateral turbulence, and by reattaching flow on to the side faces (Xu et al., 1992a). Measurements on an aeroelastic wind-tunnel tall building model designed only to respond torsionally (Xu et al., 1992a), indicated that aerodynamic damping effects (Section 5.5.1) for torsional motion of cross-section shapes characteristic of tall buildings are quite small in the range of design reduced velocities, in contrast to bridge decks. However at higher reduced velocities, high torsional dynamic response and significant negative aerodynamic damping has been found for a triangular cross-section (Beneke and Kwok, 1993).

A small amount of eccentricity can increase both the mean twist angle and dynamic torsional response. For example for a building with square cross-section, a shift of the elastic centre from the geometric and mass centre by 10% of the breadth of the cross-section, is sufficient to double the mean angle of twist and increase the dynamic twist by 40–50% (Zhang et al., 1993).

9.8 INTERFERENCE EFFECTS

High-rise buildings are most commonly clustered together in groups – as office buildings grouped together in a city-centre business district, or in multiple building apartment developments, for example. The question of aerodynamic interference effects from other buildings of similar size on the structural loading and response of tall buildings arises.

9.8.1 Upwind building

A single similar upwind building on a building with square cross-section and height/width (aspect) ratio of six produces increases of up to 30% in peak along-wind base moment, and 70% in cross-wind moment, at reduced velocities representative of design wind conditions in suburban approach terrain (Melbourne and Sharp, 1976). The maximum increases occur when the upwind building is 2–3 building widths to one side of a line taken upwind, and about 8 building widths upstream. Contours of percentage increases in peak cross-wind loading for square-section buildings with an aspect ratio of 4, are shown in Figure 9.12. It can be seen that reductions, that is shielding, occurs when the upstream building is within 4 building heights upstream and ±2 building heights to one side of the downstream building. The effect of increasing turbulence in the approach flow, that is increasing roughness lengths in the approach terrain, is to reduce the increases produced by interference.

The effect of increasing aspect ratio is to further increase the interference effects of upstream buildings, with increases of up to 80% being obtained, although this was for buildings with an atypical aspect ratio of 9, and in relatively low turbulence conditions (Bailey and Kwok, 1985).

9.8.2 Downwind building

As shown in Figure 9.12, *downwind* buildings can also increase cross-wind loads on buildings if they are located in particular critical positions. In the case of the buildings of 4:1 aspect ratio of Figure 9.12, this is about one building width to the side, and two widths downwind.

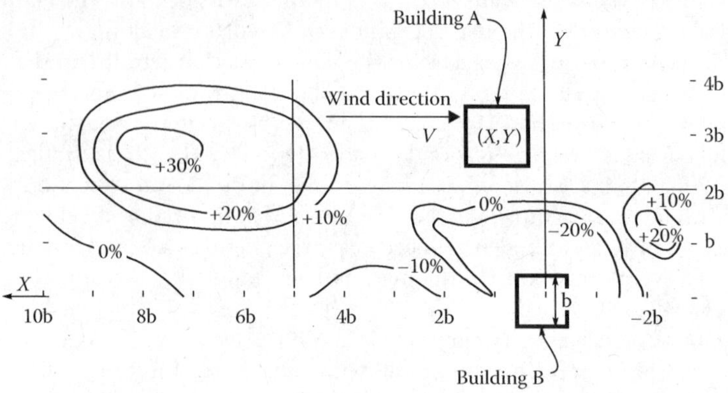

Figure 9.12 Percentage change in cross-wind response of a building (B) due to a similar building (A) at (*X*,*Y*). (Standards Australia. 1989. *Minimum Design Loads on Structures. Part 2: Wind Loads*. Standards Australia, North Sydney, Australian Standard AS1170.2-1989.)

More detailed reviews of interference effects on wind loads on tall buildings are given by Kwok (1995) and Khanduri et al. (1998). For a complex of tall buildings in the centres of large cities, wind-tunnel model tests (Chapter 7) will usually be carried out, and these should reveal any significant interference effects on *new* buildings such as those described in the previous paragraphs. Anticipated new construction should be included in the models when carrying out such tests. However, existing buildings may be subjected to unpredicted higher loads produced by new buildings of similar size at any time during their future life, and this should be considered by designers, when considering load factors.

9.8.3 Interference effects on local pressures

Adjacent buildings can also have dramatic effects on local cladding loads on tall buildings. An interesting, unusual example, based on wind-tunnel tests of a commercial high-rise development, was described by Surry and Mallais (1983). In this case, a *downwind* taller building resulted in an increase in the design local pressures by a factor of about 3. It was explained by the presence of the adjacent building inducing re-attachment of the separated shear layers (see Section 4.1) on to the wall of the upwind building (Figure 9.13).

9.9 DAMPING

The dynamic response of a tall building or other structures, to along- or cross-wind forces, depends on its ability to dissipate energy, known as 'damping'. Structural damping is derived from energy dissipation mechanisms within the material of the structure itself (i.e. steel, concrete etc.), or from friction at joints or from movement of partitions and so forth.

For some large structures, the structural damping alone has been insufficient to limit the resonant dynamic motions to acceptable levels for serviceability considerations, and auxiliary dampers have been added. Three types of auxiliary damping devices will be discussed in this chapter: viscoelastic dampers (Section 9.9.2), tuned mass dampers (TMDs) (Section 9.9.3) and tuned liquid dampers (TLDs) (Section 9.9.4).

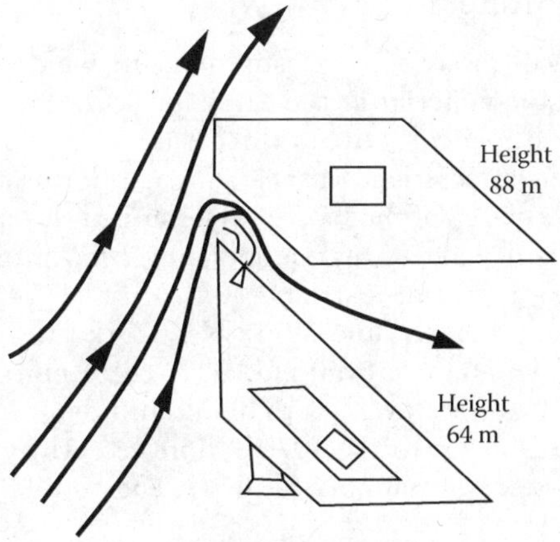

Figure 9.13 Effect of a downwind building on local cladding pressures. (From Surry, D. and Mallais, W. 1983. Adverse local wind loads induced by adjacent building. *ASCE Journal of Structural Engineering*, 109: 816–20. Reproduced by permission of ASCE.)

9.9.1 Structural damping

An extensive database of free vibration measurements from tall buildings in Japan has been collected (Tamura et al., 2000). This database includes data on frequency as well as damping. More than 200 buildings were studied, although there is a shortage of values at larger heights – the tallest (steel encased) reinforced concrete building was about 170 m in height, and the highest steel-framed building was 280 m.

For reinforced concrete buildings, the Japanese study proposed the following empirical formula for the critical damping ratio in the first mode of vibration, for buildings less than 100 m in height, and for low-amplitude vibrations (drift ratio, $(x_t/h) < 2 \times 10^{-5}$).

$$\eta_1 \cong 0.014 n_1 + 470\left(\frac{x_t}{h}\right) - 0.0018 \qquad (9.27)$$

where

n_1 is the first mode natural frequency
x_t is the amplitude of vibration at the top of the building $(z = h)$

The corresponding relationship for steel-framed buildings is

$$\eta_1 \cong 0.013 n_1 + 400\left(\frac{x_t}{h}\right) + 0.0029 \qquad (9.28)$$

The range of application for Equation 9.28 is stated to be: $h < 200$ m, and (x_t/h) less than 2×10^{-5}.

Equations 9.27 and 9.28 may be applied to tall buildings for serviceability limits states criteria (i.e. for the assessment of acceleration limits). Higher values are applicable for the high amplitudes appropriate to strength (ultimate) limits states, but unfortunately little, or no, measured data is available.

9.9.2 Visco-elastic dampers

Visco-elastic dampers incorporate visco-elastic material which dissipates energy as heat through shear stresses in the material. A typical damper, as shown in Figure 9.14, consists of two visco-elastic layers bonded between three parallel plates (Mahmoodi, 1969). The force versus displacement characteristic of such a damper forms a hysteresis loop as shown in Figure 9.15. The enclosed area of the loop is a measure of the energy dissipated per cycle, and for a given damper, is dependent on the operating temperature (Mahmoodi and Keel, 1986) and heat transfer to the adjacent structure.

The original World Trade Center buildings in New York City were the first major structures to utilise visco-elastic dampers (Mahmoodi, 1969). Approximately 10,000 dampers were installed in each 110-storey tower, with about 100 dampers at the ends of the floor trusses at each floor from the 7th to the 107th. More recently, visco-elastic dampers have been installed in the 76-storey Columbia Seafirst Center Building, in Seattle, USA. The

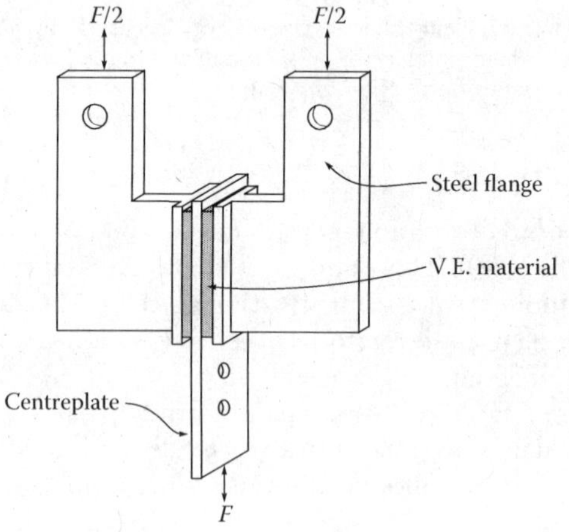

Figure 9.14 A viscoelastic damper. (From Mahmoodi, P. 1969. *ASCE, Journal of the Structural Division*, 95: 1661–72. With permission.)

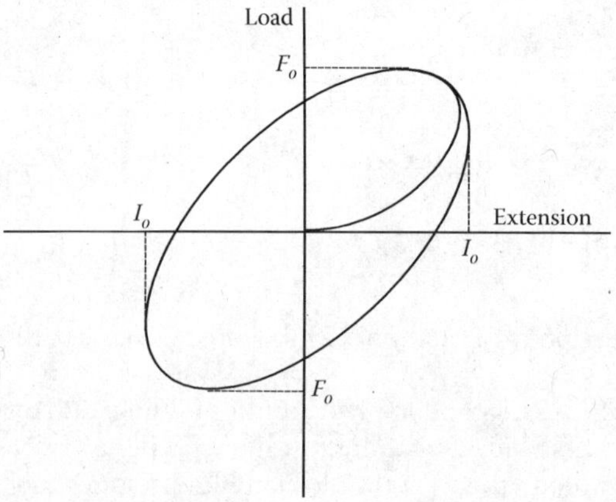

Figure 9.15 Hysteresis loop for viscoelastic dampers. (Mahmoodi, P. 1969. ASCE, *Journal of the Structural Division*, 95: 1661–72. With permission.)

dampers used in this building were significantly larger than those used at the World Trade Center, and only 260 were required to effectively reduce accelerations in the structure to acceptable levels, (Keel and Mahmoodi, 1986; Skilling et al., 1986).

A detailed review of the use of visco-elastic dampers in tall buildings has been given by Samali and Kwok (1995).

9.9.3 Tuned mass dampers

A relatively popular method of mitigating vibrations has been the *tuned mass damper* (TMD) or *vibration absorber*. Vibration energy is absorbed through the motion of an auxiliary or secondary mass connected to the main system by viscous dampers. The characteristics of a vibrating system with TMD can be investigated by studying the two-degree-of-freedom system shown in Figure 9.16 (e.g. den Hartog, 1956; Vickery and Davenport, 1970).

TMD systems have successfully been installed in the Sydney Tower in Australia, the Citycorp Center, New York (275 m), the John Hancock Building, Boston, USA (60 storeys), and in the Chiba Port Tower in Japan (125 m). In the first and last of these, extensive full-scale measurements have been made to verify the effectiveness of the systems.

For the Sydney Tower, a 180-tonne doughnut-shaped water tank, located near the top of the Tower, and required by law for fire protection, was incorporated into the design of the TMD. The tank is 2.1 m deep and 2.1 m from inner to outer radius, weighs about 200 tonnes, and is suspended from the top radial members of the turret. Energy is dissipated in 8 shock absorbers attached tangentially to the tank and anchored to the turret wall. A 40-tonne secondary damper is installed lower down on the tower to further increase the damping, particularly in the second mode of vibration (Vickery and Davenport, 1970; Kwok, 1984).

The system installed in the Citycorp Center, New York, (McNamara, 1977), consists of a 400-tonne concrete mass riding on a thin oil film. The damper stiffness is provided by pneumatic springs, whose rate can be adjusted to match the building frequency. The energy absorption is provided by pneumatic shock absorbers, as for the Sydney Tower. The building was extensively wind-tunnel tested (Isyumov et al., 1975). The aeroelastic model tests included the evaluation of the TMD. The TMD was found to significantly reduce the wind-induced dynamic accelerations to acceptable levels. The effective damping of the model damper was found to be consistent with theoretical estimates of effective viscous damping based on the two-degree-of-freedom model, (Vickery and Davenport, 1970).

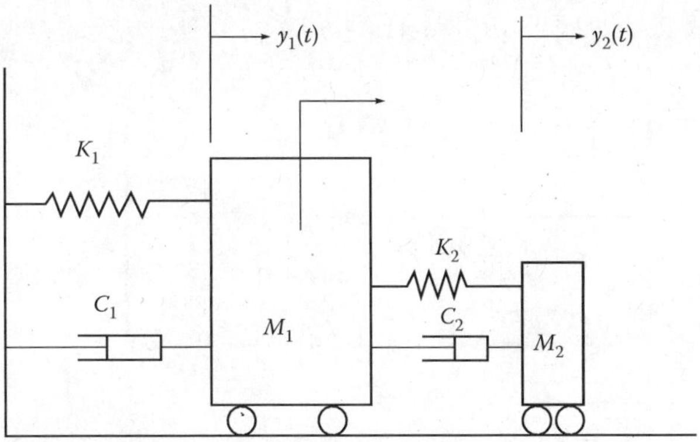

Figure 9.16 Two degree-of-freedom representation of a TMD.

TMD systems similar to those in the Citycorp Center have been installed in both the John Hancock Building, Boston, and in the Chiba Port Tower. In the case of the latter structure, the system has been installed to mitigate vibrations due to both wind (typhoon), and earthquake. Adjustable coil springs are used to restrain the moving mass, which is supported on frames sliding on rails in two orthogonal directions.

Taipei 101 in Taiwan, 508 m tall, has a pendulum-type TMD, consisting of a 660 tonne spherical steel mass, suspended at a length to tune it to the building frequency of 0.14 Hz. It is claimed to reduce the accelerations at the top of the building by 30–40%.

The performance of TMDs in tall buildings and towers under wind loading has been reviewed by Kwok and Samali (1995).

9.9.4 Tuned liquid dampers

TLDs are relatively new devices in building and structures applications, although similar devices have been used in marine and aerospace applications for many years. They are similar in principle to the TMD, in that they provide a heavily damped auxiliary vibrating system attached to the main system. However, the mass, stiffness and damping components of the auxiliary system are all provided by moving liquid. The stiffness is in fact gravitational; the energy absorption comes from mechanisms such as viscous boundary layers, turbulence or wave breaking, depending on the type of system. Two categories of TLD will be discussed briefly here: *tuned sloshing dampers* (TSD) and *tuned liquid column dampers* (TLCD).

The TSD type (Figure 9.17) relies on the motion of shallow liquid in a rigid container for absorbing and dissipating vibrational energy (Fujino et al. 1988; Sun et al. 1989). Devices of this type have already been installed in at least two structures in Japan (Fujii et al. 1990), and on a television broadcasting tower in Australia.

Although a very simple system in concept, the physical mechanisms behind this type of damper are in fact quite complicated. Parametric studies of dampers with circular containers were carried out by Fujino et al. (1988). Some of their conclusions can be summarised as follows:

- Wave breaking is a dominant mechanism for energy dissipation but not the only one.
- The additional damping produced by the damper is highly dependent on the amplitude of vibration.
- At small-to-moderate amplitudes, the damping achieved is sensitive to the frequency of sloshing of liquid in the container. For dampers with circular containers, the fundamental sloshing frequency is given by Equation 9.29.

$$n_s = (1/2\pi)\sqrt{[(1.84\, g/R)\tanh(1.84\, h/R)]} \tag{9.29}$$

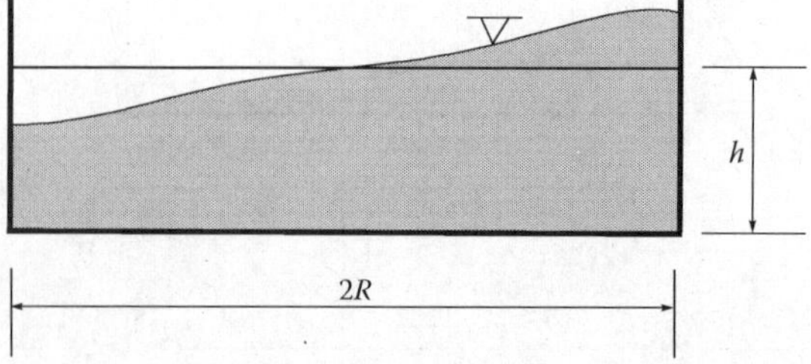

Figure 9.17 Tuned sloshing damper.

where g is the acceleration due to gravity, h is the height of the liquid and R is the radius of the container, as shown in Figure 9.17.

This formula is derived from linear potential theory of shallow waves.

- High viscosity sloshing liquid is not necessarily desirable at high amplitudes of vibration, as wave breaking is inhibited. However, at low amplitudes, at which energy is dissipated in the boundary layers on the bottom and side walls of the container, there is an optimum viscosity for maximum effectiveness (Sun et al., 1989).
- Roughening the container bottom does not improve the effectiveness because it has little effect on wave breaking.

The above conclusions were based on a limited number of free vibration tests with only two diameters of container. Further investigations are required, including the optimal size of TSD for a given mass of sloshing liquid. However, the simplicity and low cost of this type of damper makes them very suitable for many types of structure.

Variations in the geometrical form are possible, for example, Modi et al. (1990) has examined TSDs with torus ('doughnut')-shaped containers.

The TLCD damper (Figure 9.18) comprises an auxiliary vibrating system consisting of a column of liquid moving in a tube-like container. The restoring force is provided by gravity, and energy dissipation is achieved at orifices installed in the container, (Sakai et al. 1989, Hitchcock et al. 1997a, b). The same principle has been utilised in anti-rolling tanks used in ships.

The TLCD, like the TSD, is simple and cheap to implement. Unlike the TSD, the theory of its operation is relatively simple and accurate. Sakai et al. (1989) has designed a TLCD system for the Citycorp Center, New York as a feasibility study; he found that the resulting damper was simpler, lighter and presumably cheaper than the TMD system actually used in this building (Section 9.9.3). Xu et al. (1992b) have examined theoretically the along-wind response of tall, multi-degree-of-freedom structures, with TMDs, TLCDs and a hybrid damper – the tuned liquid column mass damper (TLCMD). They found that the TMD and TLCD, with the same amount of added mass, achieved similar response reductions. The TLCMD, in which the mass of the container, as well as the liquid, is used as part of the auxiliary vibrating system, is less effective when the liquid column frequency is tuned to the same frequency as the whole damper frequency (with the water assumed to remain still). The performance of the latter is improved when the liquid column frequency is set higher than the whole damper frequency.

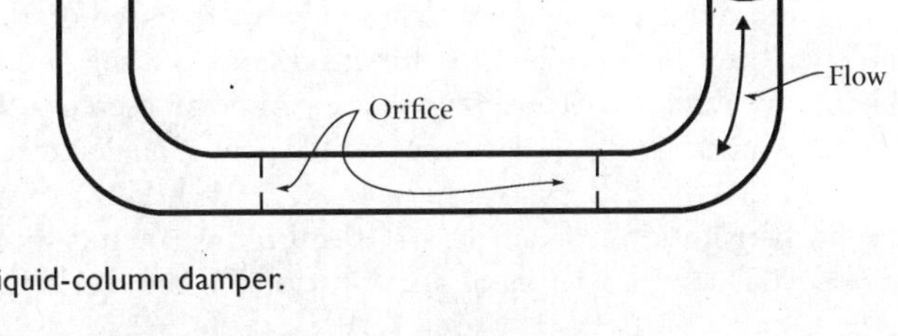

Figure 9.18 Tuned liquid-column damper.

The effectiveness of tuned liquid structures in several tall structures in Japan has been reviewed by Tamura et al. (1995).

9.10 MOTION PERCEPTION AND ACCELERATION CRITERIA

The limiting of the wind-induced sway and twist motion of tall buildings to acceptable levels is an important part of the assessment of a tall building response to dynamic wind forces. In some cases, this assessment may result in the need for installation of auxiliary damping systems such as those described in Section 9.9.

The response of occupants to building vibration is a complex mixture of physiological and psychological factors. Excessive vibrations may affect the ability of an individual to perform manual tasks (e.g. Burton et al., 2011). In extreme cases, some occupants may suffer motion sickness (Walton et al., 2011).

Studies have generally accepted that humans respond primarily to fluctuating and peak acceleration of a building, and the most generally accepted criterion is that given in Annex D of ISO 10137 (ISO, 2009). This provides acceptability limits of peak horizontal wind-induced accelerations for a 1-year return period. Separate limits are given for residences and offices, with the latter being higher by 50% – that is working occupants of a building are deemed to be accepting of higher motion than are those of a residential building. The criteria also reduce with frequency for the typical range of first-mode building frequencies of 0.06–1.0 Hz. For example for a frequency of 0.4 Hz, the limit on peak acceleration for office buildings is about 0.09 m/s^2 (~9 milli-gs), and for residences, about 0.06 m/s^2 (about 6 milli-gs). At 1 Hz, the limits are 0.06 m/s^2 for offices, and 0.04 m/s^2 for residences. Note that these limits are defined for buildings only, and higher limits might be expected to apply to other structures such as observation towers, in which the occupancy is transient.

9.11 DIRECTIONALITY

When, on occasions, the predicted responses of a tall building to wind action from two or more different wind consultants have been compared, the principal source of differences has often turned out to be the treatment of the wind climate, and its interaction with the measured response to wind of the building in wind-tunnel tests. This involves the directional variability of the wind climate at the building site and its probabilistic treatment. Several methods have been developed to account for the direction effects – however, these give results that are not necessarily consistent with each other.

This topic has been discussed in Section 2.2.6 and in detail elsewhere, particularly by Simiu and Miyata (2006), and Isyumov et al. (2013); hence, a brief summary will only be given here. The methods for dealing with directionality can be classified into four categories:

1. The up-crossing and out-crossing approaches. These are based on the *parent* distributions of wind speed (see Section 2.5) by direction, and assume that the extreme wind speeds, and hence extreme structural responses, can be predicted from probability distributions fitted to the parent population of wind speeds in each direction sector. This was designated as 'Method 3' by Isyumov et al. (2013). Since these approaches are based on parent distributions of hourly, or 10-min, mean wind speeds, it is not clear that they can be easily applied to the many locations in the world that are dominated

by thunderstorm-generated winds, such as downbursts (see Sections 1.3.5 and 2.4.1) that typically last only a few minutes.

2. Combined probability based on *extreme* wind speeds measured by direction sector. Two subsets of this approach are:
   a. The 'multi-sector' approach. Fitting of extreme value distributions to the responses in each direction sector and then computing a combined probability of non-exceedence by Equation 2.4 in this book (Holmes, 1990). That is Equation 2.4 is applied to a building response, instead of the wind speed itself. This method requires an assumption of independence between extreme wind speeds and responses in different direction sectors, and hence preferably angular increments between sectors that are not too small. However, use of smaller angular increments will lead to conservative predictions.
   b. Forming a time series of annual extreme responses from the directional wind-speed data, and directional response coefficients, and fitting a single extreme value probability distribution to the resulting 'one-dimensional' series of structural responses (e.g. Simiu and Filliben, 1981; Simiu and Miyata, 2006).
3. 'Sector-by-sector' approaches. These approaches attempt to produce directional wind speeds, or directional wind-speed multipliers, for a discrete number of defined wind directions. Directional wind speeds, or multipliers, derived by these methods, are discussed in more detail in Section 2.2.6. When the directional wind speeds are combined with the measured structural response coefficients, the largest resulting response from any direction is deemed to be an appropriate design value. This approach is particularly appropriate for codes and standards, due to the requirement for a reduced level of computation; however, variants of it have also been adopted by several wind-tunnel testing groups. Within this category, at least three variants have been developed (e.g. Cook, 1983; Melbourne, 1984; Simiu and Filliben, 2005). These are referred to as Methods (b), (c) and (a), respectively, in Section 2.2.6. The method by Melbourne (1984) has been used since 1989 in the Australian and Australian/New Zealand Standards for wind loading (Standards Australia, 1989 and 2011). The method of Cook (1983) was used in the British Standard, (British Standards Institution 1997).

A variant of this method was designated as 'Method 2' by Isyumov et al. and compared with predictions by other approaches. Despite its simplicity, this method gave reasonable results (although conservative in some cases), compared with the more computationally complex methods studied by Isyumov et al. (2013).

- The 'storm passage' approach. This approach simulates the passage of wind storms across a building site using methods discussed in Section 2.2.4. This allows extreme structural responses from each storm to be extracted and fitted with appropriate distributions to allow predictions of extreme building responses to be made for any defined risk level. This method was described by Isyumov et al. (2003), and designated as 'Method 4' in the comparisons by Isyumov et al. (2013). According to this reference, probabilistic methods, such as the up-crossing approach, under-predict peak load effects produced by tropical cyclones, such as hurricanes and typhoons. The storm passage approach appears to be an accurate one, but it is not yet available for less structured high-wind events such as extra-tropical cyclones or thunderstorms.

With such a variety of methods available, designers of tall buildings undergoing wind-tunnel testing, might be wise to request a sensitivity analysis of the results to the method used to account for directional variability when making extreme response predictions.

9.12 CASE STUDIES

Many tall buildings have been studied in wind tunnels over several decades. These studies include the determination of the overall loading and response, cladding pressures, and other wind effects, such as environmental wind conditions at ground level. However, these studies are usually proprietary in nature, and not generally available. However, Willford (1985) has described a response study for the Hong Kong and Shanghai Bank Building, Hong Kong. A detailed wind engineering study for a building of intermediate height, including wind-loading aspects, is presented by Surry et al. (1977).

Despite many hundreds of tall buildings that have been studied in boundary-layer wind tunnels over more than 40 years, very little 'benchmarking' has been carried out to promote confidence in structural designers that consistent and repeatable assessments of wind loads can be obtained for the same building by different groups in different facilities. An early attempt at this was the CAARC Standard Tall Building model that was studied in the 1970s – the results were summarised by Melbourne (1980). However, the measurements made, and techniques used, were relatively primitive compared to those currently available.

A more recent study, in which eight wind-tunnel laboratories participated, involved two 'benchmark' buildings (Holmes and Tse, 2014). This study was intended to compare the results of practitioners of the high-frequency base balance technique (Section 7.6.2), and base moments and accelerations. One building was 180 m tall and similar in dimensions to the CAARC building used in the 1970s, and mentioned in the previous paragraph. The second building, 240 m tall, was a modified version of an earlier building used as a benchmark for vibration control (Tse et al., 2007). Both buildings had simple rectangular cross-sections. The mean and peak base moments, including base torque, from the participating groups show quite good agreement with coefficients of variation about the averages of 8–20% for the lower building and 20–30% for the taller one. The peak corner accelerations from five groups show greater variation, especially for the lowest wind speed. The greater scatter in the acceleration predictions could be attributed to the dominance of the resonant components, which rely on accurate measurement of spectra of generalised forces at relatively high reduced frequencies requiring higher bandwidth in the instrumentation, a known difficulty with the high-frequency base balance technique.

Relatively few tall buildings have been studied in full scale for wind loads, although many have been studied for their basic dynamic properties (e.g. Tamura et al., 2000). Case studies of wind-induced accelerations on medium height buildings were described by Wyatt and Best (1984), and Snaebjornsson and Reed (1991).

Numerous auxiliary damping systems have been installed in tall structures. A listing of over 50 cases up to the mid-1990 (Holmes, 1995) included installations in chimneys, airport control towers and bridge pylons, as well as tall buildings.

'Super-tall' buildings, above 400 m in height, have particular problems from a wind-engineering perspective. Not the least of these is the lack of knowledge of wind and turbulence properties at these heights, especially as these structures are often being constructed in locations where 'well-behaved' large-scale boundary-layer winds are not necessarily dominant (Irwin, 2007). The 830 m Burj Khalifa (formerly 'Burj Dubai'), completed in 2009 (Figure 9.19), required extensive wind engineering in its design process with extensive wind-tunnel testing at two different scales (Irwin and Baker, 2005), including an aeroelastic model test using the 'spine' technique (Section 7.6.4). The tapering and multiple changes of cross-section over its height are entirely a result of the need to limit the cross-wind, vortex-induced vibrations of the building. This building had extensive monitoring of its behaviour during construction, and this has continued since. The instrumentation includes the permanent installation of accelerometers, a weather station and a GPS displacement

Figure 9.19 The 830-m tall Burj Khalifa building, Dubai, UAE.

measurement to continuously monitor the response of the structure to wind and seismic excitation (Abdelrazaq, 2011).

9.13 SUMMARY

This chapter has discussed various aspects of the design of tall buildings for wind loads. The general characteristics of wind pressures on tall buildings, and local cladding loads have been considered. The special response characteristics of glass have been discussed. The overall response of tall buildings in along- and cross-wind directions, and in twist (torsion) has been covered. Aerodynamic interference effects, and the application of auxiliary damping systems to mitigate wind-induced vibration have been discussed. Human perception of wind-induced tall building motions, and the several methods used to treat the varying response of building elements with wind direction, have also been discussed.

Discussion of several case studies includes the new generation of 'super-tall' buildings, whose design is dominated by wind-engineering considerations.

REFERENCES

Abdelrazaq, A. 2011. Validating the dynamics of the Burj Khalifa. CTBUH Technical Paper, Council of Tall Buildings and Urban Habitat, Chicago, Illinois, USA.

Bailey, P.A. and Kwok, K.C.S. 1985. Interference excitation of twin tall buildings. *Journal of Wind Engineering and Industrial Aerodynamics*, 21: 323–38.

Beneke, D.L. and Kwok, K.C.S. 1993. Aerodynamic effect of wind induced torsion on tall buildings. *Journal of Wind Engineering and Industrial Aerodynamics*, 50: 271–80.

British Standards Institution. 1997. *Loading for Buildings. Part 2: Code of Practice for Wind Loads*. BS 6399: Part 2:1997, BSI, London, UK.

Brown, W.G. 1972. A load duration theory for glass design. Division of Building Research. National Research Council of Canada. Research paper 508.

Burton, M.D., Kwok, K.C.S. and Hitchcock, P.A. 2011. Effect of low-frequency motion on the performance of a dynamic manual tracking task. *Wind and Structures*, 14: 517–36.

Calderone, I.J. 1999. The equivalent wind load for window glass design. PhD thesis, Monash University.

Calderone, I. and Melbourne, W.H. 1993. The behaviour of glass under wind loading. *Journal of Wind Engineering and Industrial Aerodynamics*, 48: 81–94.

Cheung, J.C.K. 1984. Effect of tall building edge configurations on local surface wind pressures. *3rd International Conference on Tall Buildings*, Hong Kong and Guangzhou, 10–15 December.

Cheung, J.C.K. and Melbourne, W.H. 1992. Torsional moments of tall buildings. *Journal of Wind Engineering and Industrial Aerodynamics*, 41–44: 1125–26.

Cook, N.J. 1983. Note on directional and seasonal assessment of extreme wind speeds for design. *Journal of Wind Engineering and Industrial Aerodynamics*, 12: 365–72.

Cook, N.J. and Mayne, J.R. 1979. A novel working approach to the assessment of wind loads for equivalent static design. *Journal of Industrial Aerodynamics*, 4: 149–64.

Coyle D.C. 1931. Measuring the behaviour of tall buildings. *Engineering News-Record*, 310–13.

Dalgleish, W.A. 1971. Statistical treatment of peak gusts on cladding. *ASCE Journal of the Structural Division*, 97: 2173–87.

Dalgleish, W.A. 1975. Comparison of model/full-scale wind pressures on a high-rise building. *Journal of Industrial Aerodynamics*, 1: 55–66.

Dalgleish, W.A. 1979. Assessment of wind loads for glazing design. *IAHR/IUTAM Symposium on Flow-induced Vibrations*, Karslruhe, September.

Dalgleish, W.A. Cooper, K.R. and Templin, J.T. 1983. Comparison of model and full-scale accelerations of a high-rise building. *Journal of Wind Engineering and Industrial Aerodynamics*, 13: 217–28.

Dalgleish, W.A., Templin, J.T. and Cooper, K.R. 1979. Comparison of wind tunnel and full-scale building surface pressures with emphasis on peaks. *Proceedings, 5th International Conference on Wind Engineering*, Fort Collins, Colorado, pp. 553–65, Pergamon Press, Oxford, U.K.

Davenport, A.G. 1964. Note on the distribution of the largest value of a random function with application to gust loading. *Proceedings, Institution of Civil Engineers*, 28: 187–96.

Davenport, A.G. 1966. The treatment of wind loading on tall buildings. *Proceedings Symposium on Tall Buildings*, Southampton, UK, April, 3–44.

Davenport, A.G. 1971. The response of six building shapes to turbulent wind. *Philosophical Transactions, Royal Society, A*, 269: 385–94.

Davenport, A.G. 1975. Perspectives on the full-scale measurements of wind effects. *Journal of Industrial Aerodynamics*, 1: 23–54.

Davenport A.G., Hogan, M. and Isyumov, N. 1969. A study of wind effects on the Commerce Court Tower, Part I. University of Western Ontario, Boundary layer Wind Tunnel Report, BLWT-7-69.

den Hartog, J.P. 1956. *Mechanical Vibrations*. McGraw-Hill, New York.

Dryden, H.L. and Hill, G.C. 1933. Wind pressure on a model of the Empire State Building. *Journal of Research of the National Bureau of Standards*, 10: 493–523.

Fujii, K., Tamura, Y., Sato, T. and Wakahara, T. 1990. Wind-induced vibration of tower and practical applications of tuned sloshing damper. *Journal of Wind Engineering and Industrial Aerodynamics*, 33: 263–72.

Fujino, Y., Pacheco, B.M., Chaiseri, P. and Sun, L.-M. 1988. Parametric studies on tuned liquid damper (TLD) using circular containers by free-oscillation experiments. *Structural Engineering/Earthquake Engineering, Japan Society of Civil Engineers*, 5: 381s–91s.

Hitchcock, P.A., Kwok, K.C.S., Watkins, R.D. and Samali, B. 1997a. Characteristics of liquid column vibration absorbers I. *Engineering Structures*, 19: 126–34.

Hitchcock, P.A., Kwok, K.C.S. Watkins, R.D. and Samali, B. 1997b. Characteristics of liquid column vibration absorbers II. *Engineering Structures*, 19: 135–44.

Holmes, J.D. 1985. Wind action on glass and Brown's integral. *Engineering Structures*, 7: 226–30.

Holmes, J.D. 1990. Directional effects on extreme wind loads. *Civil Engineering Transactions, Institution of Engineers, Australia*, 32: 45–50.
Holmes, J.D. 1995. Listing of installations. *Engineering Structures*, 17: 676–78.
Holmes, J.D. and Tse, T.K.T. 2014. International high-frequency base balance benchmark study. *Wind and Structures*, 18: 457–71.
International Organization for Standardization. 2009. Bases for design of structures – Serviceability of buildings and walkways against vibration. Annex D. Guidance for human response to wind-induced motions in buildings. International Standard ISO 10137:2007(E).
Irwin, P.A. 2007. Wind engineering challenges of the new generation of super-tall buildings. *12th International Conference on Wind Engineering*, Cairns, Queensland, Australia, 1–6 July.
Irwin, P.A. and Baker, W.F. 2005. The wind engineering of the Burj Dubai Tower. *7th CTBUH World Congress 'Renewing the urban landscape'*, New York, USA, 16–19 October .
Isyumov, N., Ho, T.C.E and, Case, P.C. 2013. Influence of wind directionality on wind loads and responses, *12th Americas Conference on Wind Engineering*, Seattle, Washington, 16–20 June.
Isyumov, N., Holmes, J.D., Surry, D. and Davenport, A.G. 1975. A study of wind effects for the First National City Corporation Project – New York, U.S.A. University of Western Ontario, Boundary Layer Wind Tunnel Laboratory, Special Study Report, BLWT-SS1-75.
Isyumov, N., Mikitiuk, M.J., Case, P.C., Lythe, G.R., and Welburn, A. 2003. Prediction of wind loads and responses from simulated tropical storm passages, *11th International Conference on Wind Engineering*, Lubbock, Texas, USA, 2–5 June.
Isyumov, N. and Poole, M. 1983. Wind induced torque on square and rectangular building shapes. *Journal of Wind Engineering and Industrial Aerodynamics*, 13: 183–96.
Kareem, A. 1985. Lateral-torsional motion of tall buildings to wind loads. *ASCE Journal of the Structural Division*, 111: 2479–96.
Kasperski, M. 2003. Specification of the design wind load based on wind-tunnel experiments. *Journal of Wind Engineering and Industrial Aerodynamics*, 91: 527–41.
Keel, C.J. and Mahmoodi, P. 1986. Design of viscoelastic dampers for Columbia Center Building. In *Building Motion in Wind*, ed. N. Isyumov and T. Tschanz, ASCE, New York.
Khanduri, A.C., Stathopoulos, T., and Bedard, C. 1998. Wind-induced interference effects on buildings – A review of the state-of-the-art. Engineering Structures, 20: 617–30.
Kwok, K.C.S. 1984. Damping increase in building with tuned mass damper. *ASCE, Journal of Engineering Mechanics*, 110: 1645–49.
Kwok, K.C.S. 1995. Aerodynamics of tall buildings. In *A State of the Art in Wind Engineering*, ed. P. Krishna. Wiley Eastern Limited, Delhi, India.
Kwok, K.C.S. and Bailey, P.A. 1987. Aerodynamic devices for tall buildings and structures. *ASCE, Journal of Engineering Mechanics*, 113: 349–65.
Kwok, K.C.S. and Samali, B. 1995. Performance of tuned mass dampers under wind loads. *Engineering Structures*, 17: 655–67.
Kwok, K.C.S., Wilhelm, P.A. and Wilkie, B.G. 1988. Effect of edge configuration on wind-induced response of tall buildings. *Engineering Structures*, 10: 135–40.
Lawson, T.V. 1976. The design of cladding. *Building and Environment*, 11: 37–38.
Lythe, G.R. and Surry, D. 1990. Wind induced torsional loads on tall buildings. *Journal of Wind Engineering and Industrial Aerodynamics*, 36: 225–34.
Mahmoodi, P. 1969. Structural dampers. *ASCE, Journal of the Structural Division*, 95: 1661–72.
Mahmoodi, P. and Keel, C.J. 1986. Performance of viscoelastic dampers for Columbia Center Building. In *Building Motion in Wind*, ed. N. Isyumov and T. Tschanz, ASCE, New York.
McNamara, R.J. 1977. Tuned mass dampers for buildings. *ASCE, Journal of the Structural Division*, 103: 1785–98.
Melbourne, W.H. 1977. Probability distributions associated with the wind loading of structures. *Civil Engineering Transactions, Institution of Engineers, Australia*, 19: 58–67.
Melbourne, W.H. 1980. Comparisons of measurements on the CAARC Standard tall building model in simulated model wind flows. *Journal of Wind Engineering and Industrial Aerodynamics*, 6: 73–88.

Melbourne, W.H. 1984. Designing for directionality. *1st Workshop on Wind Engineering and Industrial Aerodynamics*, Highett, Victoria, Australia, July.

Melbourne, W.H. and Sharp, D.B. 1976. Effects of upwind buildings on the response of tall buildings. *Proceedings Regional Conference on Tall Buildings*, Hong Kong, September, 174–191.

Minor, J.E. 1994. Windborne debris and the building envelope. *Journal of Wind Engineering and Industrial Aerodynamics*, 53: 207–27.

Modi, V.J., Welt, P. and Irani, P. 1990. On the suppression of vibrations using nutation dampers. *Journal of Wind Engineering and Industrial Aerodynamics*, 33: 273–82.

Newberry, C.W., Eaton, K.J. and Mayne, J.R. 1967. The nature of gust loading on tall buildings. *Proceedings, International Research Seminar on Wind effects on Buildings and Structures*, Ottawa, Canada, 11–15 September, pp. 399–428, University of Toronto Press, Toronto, Canada.

Peterka, J.A. 1983. Selection of local peak pressure coefficients for wind tunnel studies of buildings. *Journal of Wind Engineering and Industrial Aerodynamics*, 13: 477–88.

Peterka, J.A. and Cermak, J.E. 1975. Wind pressures on buildings – Probability densities. *ASCE Journal of the Structural Division*, 101: 1255–67.

Rathbun, J.C. 1940. Wind forces on a tall building. *Transactions, American Society of Civil Engineers*, 105: 1–41.

Sakai, F., Takeda, S. and Tamaki, T. 1989. Tuned liquid column damper – New type device for suppression of building vibrations. *International Conference on High-Rise Buildings*, Nanjing, China, 25–27 March.

Samali, B. and Kwok, K.C.S. 1995. Use of viscoelastic dampers in reducing wind- and earthquake-induced motion of building structures. *Engineering Structures*, 17: 639–54.

Saunders, J.W. 1974. Wind excitation of tall buildings. PhD thesis, Monash University.

Simiu, E. and Filliben, J.J. 1981. Wind direction effects on cladding and structural loads. *Engineering Structures*, 3: 181–6.

Simiu, E. and Filliben, J.J. 2005. Wind tunnel testing and the sector-by sector approach. *ASCE Journal of Structural Engineering*, 131: 1143–5.

Simiu, E. and Miyata, T. 2006. *Design of Buildings and Bridges for Wind*. John Wiley, New York.

Skilling, J.B., Tschanz, T., Isyumov, N., Loh, P. and Davenport, A.G. 1986. Experimental studies, structural design and full-scale measurements for the Columbia Seafirst Center. In *Building Motion in Wind*, ed. N. Isyumov and T. Tschanz, ASCE, New York.

Snaebjornsson, J. and Reed, D.A. 1991. Wind-induced accelerations of a building: A case study. *Engineering Structures*, 13: 268–80.

Standards Australia. 1989. *Minimum Design Loads on Structures. Part 2: Wind Loads*. Standards Australia, North Sydney, Australian Standard AS1170.2-1989.

Standards Australia. 2011. *Structural Design Actions. Part 2: Wind Actions*, Standards Australia, Sydney, New South Wales, Australia. Australian/New Zealand Standard, AS/NZS1170.2:2011.

Sun, L.-M., Chaiseri, P., Pacheco, B.M., Fujino, Y. and Isobe, M. 1989. Tuned liquid damper (TLD) for suppressing wind-induced vibration of structures. *2nd Asia-Pacific Symposium on Wind Engineering*, Beijing, 26–29 June.

Surry, D. and Djakovich, D. 1995. Fluctuating pressures on tall buildings. *Journal of Wind Engineering and Industrial Aerodynamics*, 58: 81–112.

Surry, D., Kitchen, R.B. and Davenport, A.G. 1977. Design effectiveness of wind tunnel studies for buildings of intermediate height. *Canadian Journal of Civil Engineering*, 4: 96–116.

Surry, D. and Mallais, W. 1983. Adverse local wind loads induced by adjacent building. *ASCE Journal of Structural Engineering*, 109: 816–20.

Tallin, A. and Ellingwood, B. 1985. Wind induced lateral–torsional motion of buildings. *ASCE Journal of the Structural Division*, 111: 2197–213.

Tamura, Y., Fujii, K, Ohtsuki, T., Wakahara, T. and Kohsaka, R. 1995. Effectiveness of tuned liquid dampers under wind excitation. *Engineering Structures*, 17: 609–21.

Tamura, Y., Suda, K and Sasaki, A. 2000. Damping in buildings for wind-resistant design. *First International Symposium on Wind and Structures for the 21st Century*, Cheju, Korea, 26–28 January.

Templin, J.T. and Cooper, K.R. 1981. Design and performance of a multi-degree-of-freedom aeroelastic building model. *Journal of Wind Engineering and Industrial Aerodynamics*, 8: 157–75.

Tse, K.T., Kwok, K.C.S., Hitchcock, P.A., Samali, B. and Huang, M.F. 2007. Vibration control of a wind-excited benchmark tall building with complex lateral-torsional modes of vibration, *Advances in Structural Engineering*, 10: 283–304.

Vickery, B.J. and Davenport, A.G. 1970. An investigation of the behaviour in wind of the proposed Centrepoint Tower, in Sydney, Australia. University of Western Ontario, Boundary Layer Wind Tunnel Laboratory, Research Report, BLWT-1-70.

Walton, D., Lamb, S. and Kwok, K.C.S. 2011. A review of two theories of motion sickness and their implications for tall building motion sway. *Wind and Structures*, 14: 499–515.

Willford, M.R. 1985. The prediction of wind-induced responses of the Hong Kong and Shanghai Banking Corporation headquarters, Hong Kong. *Engineering Structures*, 7: 35–45.

Wyatt, T.A. and Best, G. 1984. Case study of the dynamic response of a medium height building to wind-gust loading. *Engineering Structures*, 6: 256–61.

Xu Y.L., Kwok, K.C.S and Samali, B. 1992a. Torsion response and vibration suppression of wind-excited buildings. *Journal of Wind Engineering and Industrial Aerodynamics*, 43: 1997–2008.

Xu, Y.L., Samali, B. and Kwok, K.C.S. 1992b. Control of along-wind response of structures by mass and liquid dampers. *ASCE Journal of Engineering Mechanics*, 118: 20–39.

Zhang, W.J., Xu Y.L., and Kwok, K.C.S. 1993. Torsional vibration and stability of wind-excited tall buildings with eccentricity. *Journal of Wind Engineering and Industrial Aerodynamics*, 50: 299–309.

Chapter 10

Large roofs and sports stadiums

10.1 INTRODUCTION

Wind loading is usually the dominant structural loading on the roofs of large buildings, such as entertainment or exhibition centres, closed or partially closed sports buildings, aircraft hangars and so forth. The wind loads on these structures have some significant differences in comparison with those on the roofs of smaller low-rise buildings, that justifies separate treatment:

- The quasi-steady approach (Section 4.6.2), although appropriate for small buildings, is not applicable for large roofs.
- Resonant effects, although not dominant, can be significant.

These roofs are commonly of low pitch, and experience large areas of attached flow, with low correlations between the pressure fluctuations acting on different parts. Downward as well as upward external pressures can be significant. These roofs are often arched or domed structures, which are sensitive to the distributions of wind loads, and the possibility of critical 'unbalanced' pressure distributions should be considered.

This chapter will first consider the aerodynamic aspects of wind flow over large roofs, which will facilitate an understanding of the steady and fluctuating components of wind pressures acting on these structures. Then methods of obtaining design wind loads are described, with emphasis on the method of effective static wind-load distributions, introduced in earlier chapters. The incorporation of resonant contributions is also discussed.

10.2 WIND FLOW OVER LARGE ROOFS

Figure 8.5 in Chapter 8 shows the main features of the flow over a low-pitched roof, with the wind blowing normal to one wall. At the top of the windward wall, the flow 'separates' and 're-attaches' further along the roof, forming a separation 'bubble'. The turbulence in the wind flow plays an important role in determining the length of the separation bubble – high turbulence gives a shorter bubble length, low turbulence produces a longer bubble. Even in open country turbulence intensities in windstorms are equal to 10–20% of the mean or slowly varying wind speed, and in this situation mean separation bubble lengths are equal to 2–3 wall heights.

The separation bubble region is a very important one for large roofs because the upwards pressures are the greatest in this region. In the reattached flow region, the pressures are quite small. Thus for very large flat or near-flat roofs, only the edge regions within two to three wall heights from the edge will experience large pressures, and large areas of the roof will

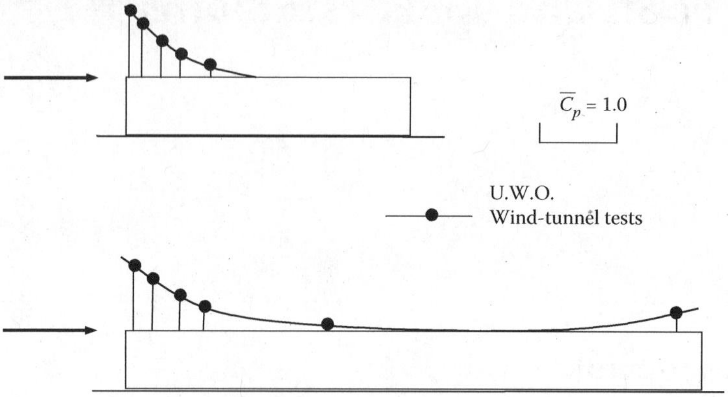

Figure 10.1 Mean pressure distributions on flat roofs.

experience quite low pressures. The variation of mean uplift pressures, measured in some wind-tunnel tests (Davenport and Surry, 1974) for flat roofs, shown in Figure 10.1, illustrates this point. It should be noted that fluctuations in pressures occur so that downwards as well as upwards pressures can occur for short time periods. Not all codes or standards on wind loads specify these downwards pressure coefficients.

As the roof pitch increases, the point of flow separation moves away from the leading edge of the roof and, in the case of a curved or arched roof, separation usually occurs downstream of the apex (Figure 10.2, from Blessmann, 1991). Upwind of the separation point, the pressures may be downwards (positive) or upwards (negative) near the leading edge, depending on the rise-to-span ratio, but are always upwards at the apex. Downwind of the separation point, they are upwards with small magnitudes.

The form of the net mean pressure coefficient distribution (i.e. the top surface pressure minus the bottom surface pressure) on a large cantilevered stadium roof is shown in Figure 10.3. Negative values indicate net upwards pressure differences. The largest uplift occurs at the leading edge, and reduces to a small pressure difference at the rear. The top surface experiences flow separation, so that the characteristic pressure distribution peaks at the leading edge, and reduces quite rapidly downstream. Underneath the flow stagnates at the back of the grandstand, if there is no gap present, to reach a pressure approaching the dynamic pressure of the freestream. However, the underside pressure will reduce in magnitude with increasing vent gaps at the back of the grandstand.

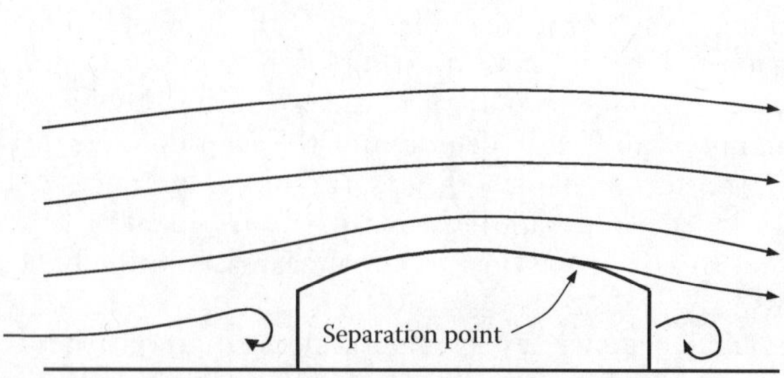

Figure 10.2 Flow separations over arched roofs. (From Blessmann, J. 1991. *Acao do vento em telhados.* SAGRA, Porto Alegre, Brazil. With permission.)

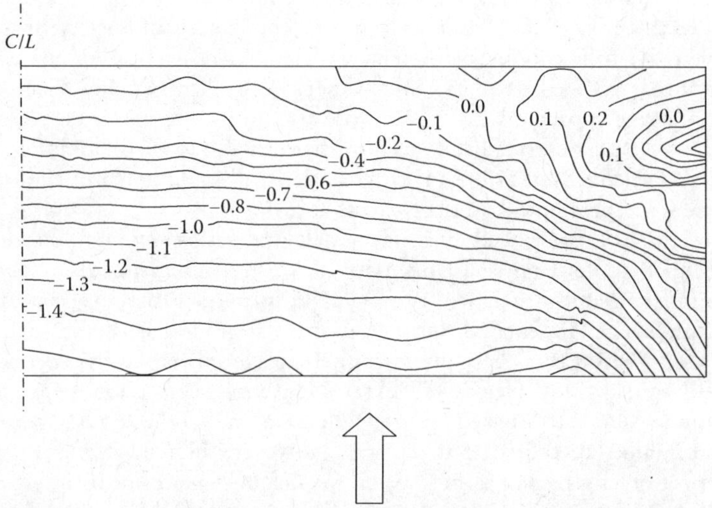

Figure 10.3 Mean net pressure distributions on a cantilevered stadium roof. (From Lam, K.M. and To, A.P. 1995. *Journal of Wind Engineering and Industrial Aerodynamics*, 54/55: 345–57. With permission.)

10.3 ARCHED AND DOMED ROOFS

10.3.1 Arched roofs

Arched roofs are structurally efficient, and are popular for structures like aircraft hangars, and enclosed sports arenas, which require large clear spans. Figure 10.4 shows the geometric variables that are relevant to the wind loading of arched-roof buildings. The variables are: the span, S; the length, L; the rise, R; the height of the walls to the eaves level, h_e.

Some very early studies on arched roofs were carried out in an aeronautical wind tunnel in the Soviet Union in the 1920s (Bounkin and Tcheremoukhin, 1928). These data found

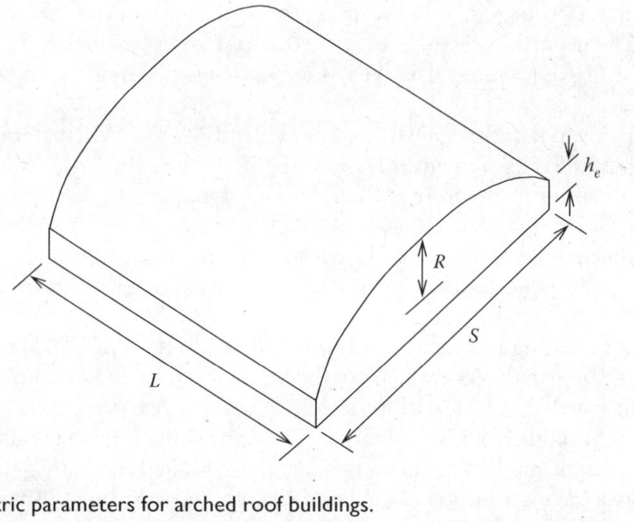

Figure 10.4 Geometric parameters for arched roof buildings.

their way into a number of national codes and standards on wind loading, and are still widely used to the present day, after reference by the American Society of Civil Engineers (1936). Some early full-scale measurements on the Akron Airship Hangar, which had an arched roof of high rise-to-span ratio, were described by Arnstein and Klemperer (1936).

Arched roofs were apparently given very little attention by researchers after 1936 until the 1980s. Grillaud (1981) described full-scale studies of wind loads on an inflatable structure, and Hoxey and Richardson (1983) also measured full-scale loads on film plastic greenhouses. These structures both had rise/span ratios of 0.5.

Johnson et al. (1985) reviewed existing models and full-scale data, and described some new wind-tunnel results from the University of Western Ontario. They found significant Reynolds number effects in their wind-tunnel data for models with a rise/span ratio of 0.5.

Toy and Tahouri (1988) carried out measurements on models of semi-cylindrical structures ($R/S = 0.5$; $h_e/R = 0$). These wind-tunnel measurements were carried out with a smooth-wall boundary flow (very high Jensen Number – see Section 4.4.5), as well as low Reynolds number (6.6×10^4, based on model height), and so the results are questionable in terms of applicability to full-scale structures. However, the data are useful in illustrating the strong effect of length/span ratio (L/S) on the mean pressures near the crest of the roofs. In this study the effects of lengthening the cross-section to produce a 'flat-top', and shortening it to produce a 'ridge', were also investigated. The latter modification has a particularly strong effect in modifying the mean pressure distribution over the roof.

Cook (1990), as well as describing the measurements of Toy and Tahouri in some detail, also discusses some work carried out by Blessmann (1987a,b), on arched roofs mounted on flat vertical walls. It is suggested that flow separations occur at the eave when the roof pitch angle there is less than about 30°.

In a computational study of mean wind pressures on arched roofs (Paterson and Holmes, 1993), 11 separate geometrical configurations were examined. Figures 10.5 and 10.6 show the computed mean external pressure coefficients on a building with a rise/span ratio (R/S) of 0.2, a length/span ratio (L/S) of 1.0, and a height to eaves/rise ratio (h_e/R) 0.45, for wind directions of 0° and 45° from the normal to the axis of the arch. Because of symmetry, values on one-half only are shown for the 0° case.

For the 0° direction, positive pressure coefficients occur on the windward wall and the windward edge of the roof, with negative values over the rest of the structure. The highest magnitude negative values occur just upwind of the apex to the roof.

At a wind direction of 45°, positive pressures only occur near the windward corner of the building. The negative pressures on the roof and walls are generally higher than those obtained for the 0° case, with particularly high suctions occurring along the windward end of the arch roof.

The effect of rise/span ratio is illustrated in Figure 10.7. The rise/span ratio of the building in this figure is 0.50, compared with the building in Figures 10.5 and 10.6, which has a rise/span of 0.20. It should be noted that the reference dynamic pressure ($(1/2)\rho_a \bar{U}_h^2$) is taken as the apex height of the structure in both cases, so that, for a fixed span and wall height, the reference dynamic pressure will increase with increasing rise/span ratio.

As for high pitch gable roofs (Figure 8.6), there is a region of positive pressure on the windward side of the roof.

The effect of increasing length/span ratio is to increase the magnitude of both the positive and negative pressures in the central part of the building as the flow becomes more two dimensional. Increasing wall height-to-rise ratio (h_e/R) produces more negative values of external pressure coefficient on the roof, side walls and leeward wall (Paterson and Holmes, 1993).

For wind directions parallel to the axis of the arch, arched roofs are aerodynamically flat, with similar pressure distributions as gable roofs for the same wind direction.

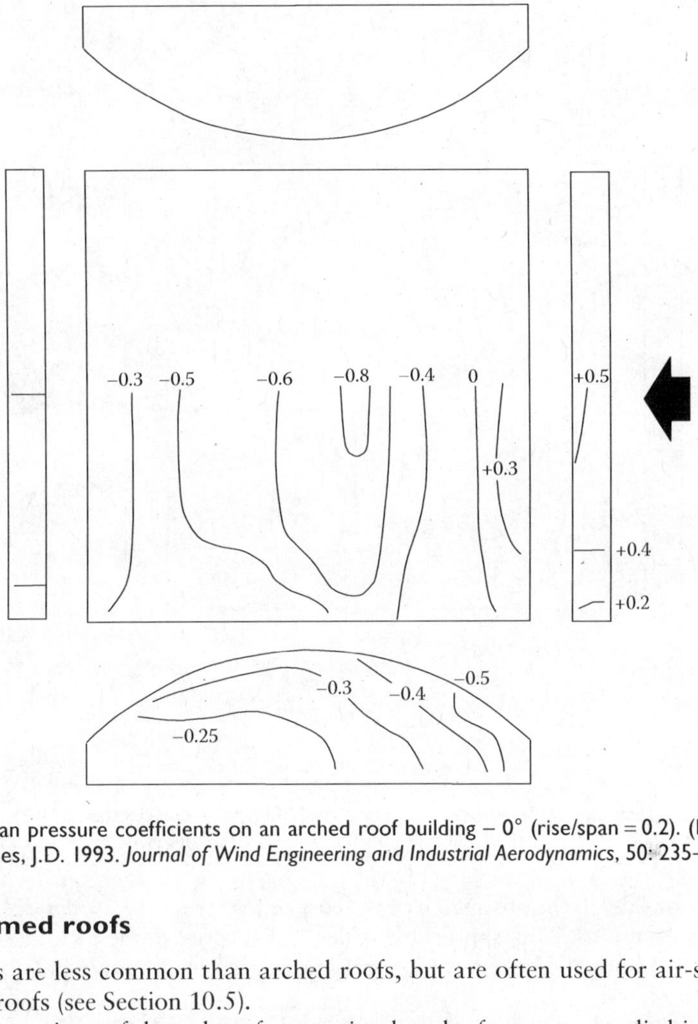

Figure 10.5 Mean pressure coefficients on an arched roof building – 0° (rise/span = 0.2). (From Paterson, D.A. and Holmes, J.D. 1993. *Journal of Wind Engineering and Industrial Aerodynamics*, 50: 235–43.)

10.3.2 Domed roofs

Domed roofs are less common than arched roofs, but are often used for air-supported, or 'pneumatic' roofs (see Section 10.5).

Two configurations of domed roofs on a circular planform were studied in a wind tunnel, with boundary-layer flow, by Kawamura and Kiuchi (1986). These both had a rise/span ratio of 0.15; eaves height-to-span ratios were 0.1 and 0.2, giving eaves height/rise ratios of 0.67 and 1.33, respectively. The mean pressure coefficients were negative over most of the roof, in both cases, with only a small positive pressure on the windward side. At the apex, the mean pressure coefficients (based on the mean wind speed upwind at the same height) were about –0.7 and –0.8 for the eaves height-to-span ratios of 0.1 and 0.2, respectively.

For design purposes, domed roofs can conservatively be assumed to have similar pressure distributions as those on arched roofs, of the same rise/span ratio, for a wind direction normal to the axis.

10.4 EFFECTIVE STATIC LOAD DISTRIBUTIONS

Because of the large fluctuating component in the wind loading on large roofs, the statistical correlation between pressures separated by large distances can be quite small. Designers

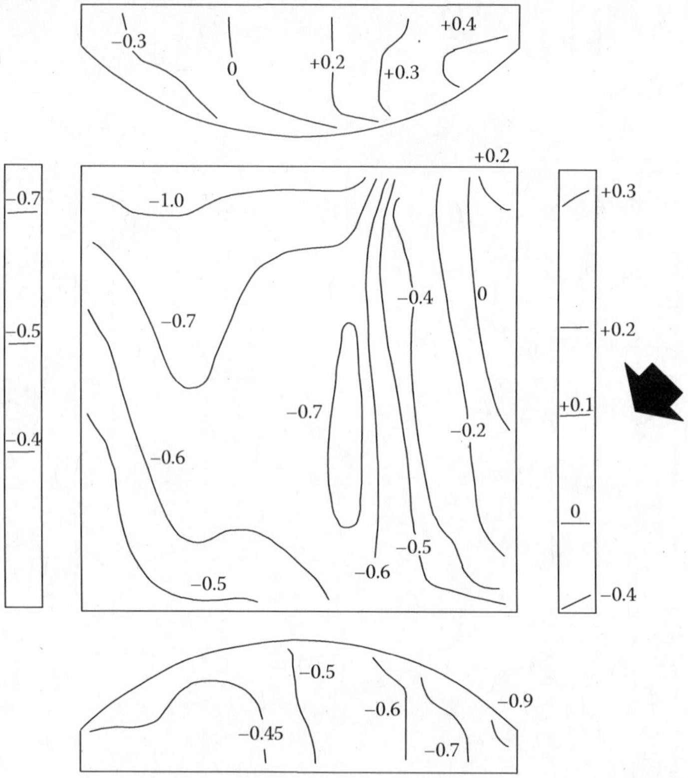

Figure 10.6 Mean pressure coefficients on an arched roof building – 45° (rise/span = 0.2). (From Paterson, D.A. and Holmes, J.D. 1993. *Journal of Wind Engineering and Industrial Aerodynamics*, 50: 235–43.)

can make use of this, to the advantage of the cost of the structure, by determining effective static load distributions. This approach enables realistic and economical design wind-load distributions to be obtained using wind-tunnel tests. Two methods are possible:

1. A direct approach in which simultaneous time histories of fluctuating pressures from the whole roof are recorded and stored. These are subsequently weighted with structural influence coefficients to obtain time histories of load effects. The instantaneous pressure distributions coinciding with peak-load effects are then identified and averaged.
2. In the other approach, correlations between pressure fluctuations at different parts of the roof are obtained, and expected pressure distributions corresponding to peak-load effects are obtained, using methods discussed in Chapter 5.

The effective static load distribution method, discussed in Section 5.4, tries to simplify the complex time and space variation of wind pressures on structures (produced by upwind turbulence and local building-induced effects) into a number of effective static pressure distributions for structural design. It is a particularly appropriate method for large roofs, over which the pressure fluctuations are not strongly correlated (or statistically related to each other). Significant reductions in design load effects, such as axial forces and bending moments in major structural members, can be obtained by this method, although normally wind-tunnel tests are required to obtain the necessary statistical data.

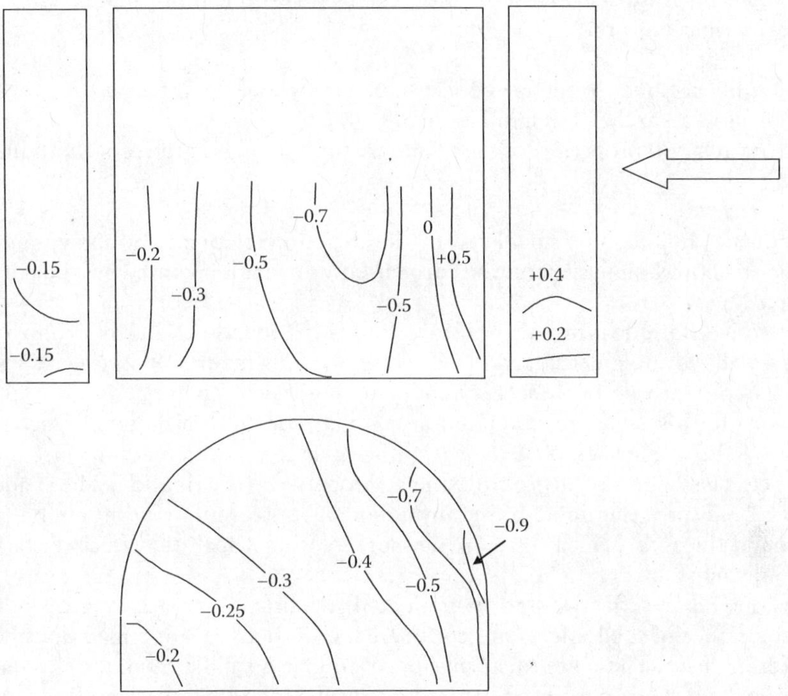

Figure 10.7 Mean pressure coefficients on an arched roof building (rise/span = 0.5). (From Paterson, D.A. and Holmes, J.D. 1993. *Journal of Wind Engineering and Industrial Aerodynamics*, 50: 235–43.)

The principles behind the method as applied to large roofs are illustrated in Figure 10.8. In this figure, a section through a large arched roof is shown and the instantaneous external pressure distributions at three different points in time are shown. Clearly, there are considerable variations from time to time in these loadings. The variations are due to turbulence in the approaching wind flow, and local effects such as vortex shedding at the leading edge of the roof. The mean pressure distribution indicates only the average pressure at each point, but this distribution usually forms the basis for the design distributions of pressure found in codes. However, the instantaneous pressure distributions producing the largest load effects may be quite different in shape to the mean.

The question of interest to the structural engineer is: what are the critical instantaneous distributions which produce the largest structural load effects in the structure? The maximum and minimum values of each load effect will be produced by two particular expected

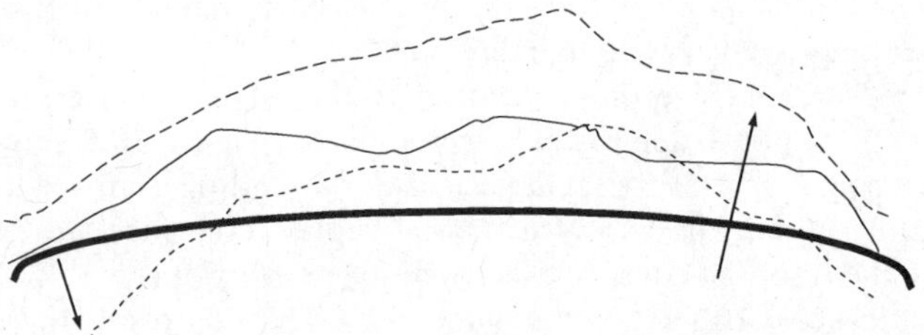

Figure 10.8 Instantaneous pressure distributions at three different times.

instantaneous pressure distributions, which can be determined. The main factors determining these distributions are:

- The influence line for the load effect – an example of the influence line for a bending moment in an arch is given in Figure 5.8.
- The correlation properties of the wind pressures acting on the roof (both internal and external).

The influence lines can be calculated by the structural engineer, by applying point loads in a static structural analysis, and the correlation information can be obtained easily from wind-tunnel tests.

The effective static loading distributions for the various load effects of interest can be obtained by the formula developed by Kasperski and Niemann (1992) (see also Kasperski, 1992). Examples of two of these distributions are given in Figure 5.10 in Chapter 5. The distributions for a support reaction and a bending moment are shown. Clearly these two distributions differ considerably from each other, due to the different influence lines for the two load effects. They also differ from the mean pressure distribution. The shaded area in Figure 5.10 indicates the limits of the instantaneous maximum and minimum peak pressures around the arch, which form an 'envelope' within which the effective static loading distributions must fall.

When applying the effective static wind-load distribution approach to large roofs, usually a limited number of load effects are considered and effective static load distributions are computed for them. These are then 'enveloped' to give a smaller number of wind-pressure distributions, which are then used by the structural engineers to design all the members of the structure. If required by structural designers, the peak values of critical load effects, such as forces in main members, or deflections can be directly computed.

10.4.1 Contributions of resonant components

When considering dynamic response of any structure to wind, it is necessary to distinguish between the resonant response at or near the natural frequencies of the structure, and the fluctuating response at frequencies below the first or lowest frequency, or 'background response', which is usually the largest contributor. As for all structures, the significance of resonant dynamic response to wind for large roofs depends on the natural frequencies of vibration, which are in turn dependent on the mass (inertia) and stiffness properties, and the damping. For roofs which are supported on two or four sides in the case of a rectangular plan, or all the way round in the case of a circular plan, the stiffness is usually high enough that resonant response is very small and can be ignored. For totally enclosed buildings, additional stiffness may be provided by compression of the air inside the building. Also there may be additional positive aerodynamic damping which further acts to mitigate any resonant dynamic response.

Extra-large stadium roofs may have several natural frequencies below 1 Hz, although these can be expected to have quite high damping.

Roofs supported on one side only – cantilevered roofs – however, are more prone to significant resonant response due to the lower stiffness. Figure 10.9 shows some resonant response in the time history of vertical deflection at the leading edge of a model cantilevered roof in wind-tunnel tests. The use of stiffening cables often increases the stiffness sufficiently to reduce the resonant contribution to minor proportions.

Most codes and standards do not include the effects of resonant response on large roofs – the Australian Standard AS/NZS1170.2 (Standards Australia, 2011) is an exception – it contains a design load distribution which is dependent on natural frequency.

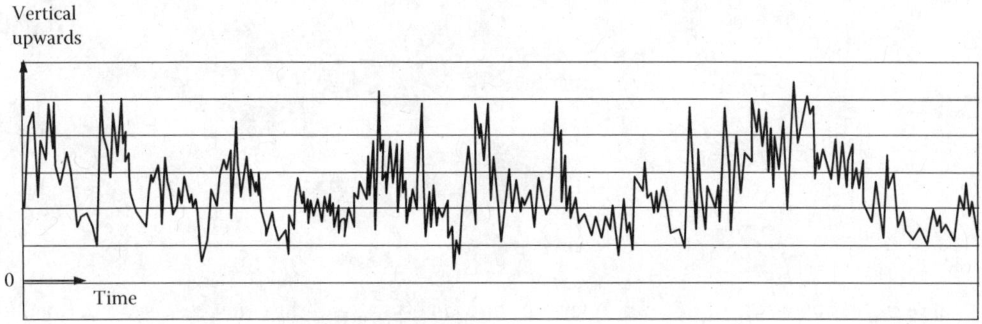

Figure 10.9 Vertical displacement of the leading edge of a cantilevered roof showing some resonant contributions to the response. (From Melbourne, W.H. 1995. *Journal of Wind Engineering and Industrial Aerodynamics,* 54/55: 325–35. With permission.)

If resonant response is anticipated to be substantial on a large roof, wind-tunnel testing with an aeroelastic model is often recommended. These can be very useful but have limitations, in that accurate effective load distributions cannot be determined from them, and the structural stiffness cannot be altered to accommodate design changes once the model has been built. For important structures, a rigid pressure model test is also advisable to obtain the distributions in pressure for the mean and background components, as discussed earlier. The resonant response can also be computed from the spectra and cross-spectra of the fluctuating pressures at the natural frequency, or from the time histories of the generalised forces in the contributing modes of vibration. Either method is computationally complex and requires simultaneous pressure measurement over the entire roof (including the underside pressure for an open stadium roof), but this is certainly feasible and has been used for large projects at wind-tunnel laboratories in Australia and elsewhere.

Usually, the resonant response will comprise no more than 10–20% of the peak values of critical load effects (Holmes et al., 1997), and this contribution can be calculated separately and added to the fluctuating background response using a 'root-sum-of squares approach'. The effective static load distribution corresponding to each peak-load effect can then be scaled up to match the recalculated peak-load effect.

For very large roofs several resonant modes can contribute, and the evaluation of effective static loads becomes more difficult. In general, it is necessary to adopt the approach of Section 5.3.7 in which the background response is separated from the resonant components, as these components all have different loading distributions. The magnitude of the contribution from each resonant mode depends on the load effect through its influence line. Section 12.3.4 describes the application of the equivalent static load approach to long-span bridges, when more than one resonant mode contributes. This approach can also be applied to very large roofs; in this case the background contribution is treated as an additional 'mode', for which the effective load distribution is calculated separately.

Thus, the effective static load distribution for the combined background and resonant contributions is

$$p'_{\text{eff}}(x) = W'_B \cdot p_{\text{eff,back}}(x) + \sum_{j}^{N} W_j{}'m(x)\phi_j(x) \tag{10.1}$$

where the weighting factors are given by

$$W'_B = \frac{\sigma_{r,B}}{\left\{\sigma_{r,B}^2 + \sum_{j=1}^{N} \alpha_j^2 \omega_j^4 \overline{a_j^2}\right\}^{1/2}} \tag{10.2}$$

$$W'_j = \frac{\alpha_j \omega_j^4 \overline{a_j^2}}{\left\{\sigma_{r,B}^2 + \sum_{j=1}^{N} \alpha_j^2 \omega_j^4 \overline{a_j^2}\right\}^{1/2}} \tag{10.3}$$

where $\sigma_{r,B}$ is the background component of the load effect, and the other terms are defined in Section 12.3.4. The derivation of the background effective static load distribution, $p_{\text{eff,back}}(x)$, is described in Chapter 5.

10.5 AIR-SUPPORTED ROOFS

Roofs made from light-weight fabric, and supported by air pressure, may be an attractive option economically for sporting venues that require a large column-free space. These roofs form a special class that require special treatment when considering the action of wind pressures (e.g. Tryggvason, 1979; Kind, 1984; Kawamura and Kiuchi, 1986).

The resistance to wind forces is provided by the internal pressurisation and by the geometric and elastic stiffness of the membrane. The latter component is usually small in comparison to the first two, and has usually been neglected in the scaling of aeroelastic wind-tunnel models (Tryggvason, 1979, and Section 7.6.5).

Air changes are required within the volume of a building of this type for ventilation purposes, and must be provided by fans and controlled leakage. Kind (1984) clearly showed that fluctuations in this air flow provided large pneumatic damping of the roof, particularly in large structures. This damping effectively suppresses any possibility of significant dynamic response to wind of these roofs. Designers should rather concentrate on providing resistance to overall uplift at the supports of the roof, and on maintaining sufficient internal air pressure to avoid local distortions, or total collapse. For these purposes, consideration should be given to the fluctuations in wind pressure, and spatial variations in instantaneous distributions in wind pressure, as discussed earlier in this chapter.

10.6 WIND-TUNNEL METHODS

As discussed in previous sections, large roofs are usually dominated by the mean wind pressures and the background fluctuating components. Resonant contributions to the wind-induced structural load effects are usually small, even though natural frequencies as low as 0.5 Hz can occur for the largest roofs. The main reason for this behaviour is the nature of the separating–reattaching flow over large roofs of low pitch, and the consequent very low correlations between fluctuating pressures acting on different parts of the roof. Excitation of a dynamic mode requires pressure 'modes', which are coincident with the mode shape, at the modal frequency in question. Usually, the excitation energy satisfying these conditions is small. Another reason for low resonant response is high damping with significant positive contributions from aerodynamic damping (Section 5.5.1).

For the reasons given above, modern wind-tunnel testing of large roofs for sports stadiums or arenas is usually carried out with rigid models on which detailed pressure measurements are made. The techniques used are described in Section 7.6.8. Using recorded time histories of fluctuating pressures, computations can be made of the resonant contributions, and added to the mean and background fluctuating contributions.

Full aeroelastic models of large roofs, although used quite frequently in the past, are now much less common. They are quite expensive to design and build, are structure dependent, and do not lend themselves to changes in the underlying structure during the design process. Also they can only normally be used for deflection measurements. However, for very flexible cantilevered roofs, the use of aeroelastic models may be required in conjunction with tests on rigid models.

10.7 CASE STUDIES

Holmes (1984) carried out wind-tunnel model measurements for an arched-roof aircraft hangar building with a rise/span ratio (R/S) of 0.20. Although the tests were carried out at low Reynolds numbers, the curved roof surface was roughened. The effect of a ridge ventilator on the apex of the roof was also investigated and found to be significant. A significant aspect of this work was an early attempt to establish effective static load distributions for load effects such as axial forces and bending moments in the arch, and structural influence lines for the arch with both pinned and fixed supports were incorporated into the processing of the pressure data.

Wind-tunnel studies for a large cable-supported roof system with an elliptical planform, of the Stadio delle Alpi, in Turin, Italy, were described by Vickery and Majowiecki (1992). These consisted of pressure measurements and deflection measurements on an aeroelastic model. Due to the lack of multiple pressure-measuring instrumentation available at that time, unfortunately the pressure measurements covered only about 10% of the total roof area. However, the spectra of the fluctuating pressures indicated a dominant frequency which was attributed to vortex shedding from the upwind roof. This frequency was about one-quarter of the first mode natural frequency of the roof, and the resonant response was found to be relatively small, as it often is for large stadium roofs supported circumferentially (see Section 10.4.1). Cross-spectra (i.e. coherence functions) for panels separated on the roof indicated low correlations between fluctuating pressures on different parts of the roof. Interestingly, Vickery and Majowwiecki identified the need to provide designers of structures such as these with a number of effective wind-pressure distributions, and suggested reducing the large amount of data generated from pressure–time histories by weighting with orthogonal functions (this suggests the use of 'proper orthogonal decomposition' as discussed in Section 8.3.4). Techniques for achieving these objectives were subsequently developed, and have been described earlier in this chapter.

A study for the re-roofing of the Olympic Stadium in Rome was described by Borri et al. (1992). This roof has a similar planform and cable-supported structure to the Turin stadium discussed in the previous paragraph. The limited fluctuating pressure measurements from a wind-tunnel study were supplemented with computer-generated time histories to cover the whole roof; these were applied as inputs to a non-linear finite element structural analysis program to compute the response of the structure to wind action. As for the roof of the Turin stadium, the response was primarily quasi-static with only small resonant contributions.

Sykes (1994) described wind-tunnel measurements of mean net pressure distributions on two large tension roofs for EXPO'92 in Seville, Spain. The pressure coefficients and overall vertical and horizontal force coefficients were similar to those expected for 'rigid' free roofs with a similar slope. The effect of 30% porosity of one roof resulted in an approximate halving of the overall force coefficients. However, these studies did not include measurement of wind-pressure fluctuations, and the structural loads in large roofs of this type are sensitive to spatially varying wind-load distributions as are large roofs generally, as discussed in Section 10.4. A difference with tensioned fabric roofs is their non-linear structural behaviour; this results in structural influence coefficients that vary with applied load. This can be handled by determining the influence coefficients by superimposing unit point or 'patch loads' on the design mean wind loading.

The effective static wind-load distribution method (Section 10.4), based on measurement of correlations between fluctuating pressures on panels on different parts of the roof, as applied in conjunction with wind-tunnel tests, to two large stadium roofs in Australia is described by Holmes et al. (1997). This reference also discusses the effects of resonant load components. Some results from that study is given in Figure 10.10.

The alternative approach, based on the direct weighting of the recorded fluctuating pressures by influence coefficients, is described by Xie (2000). This is a case study of a stadium roof consisting of two large cantilevered panels with a complex curvature.

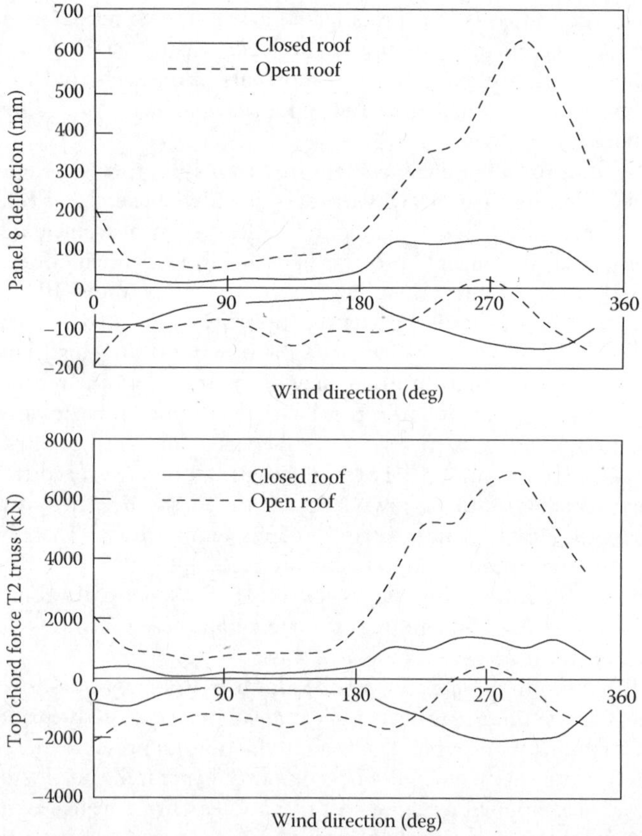

Figure 10.10 Variation of a deflection and a main truss force for a large stadium roof computed from a wind-tunnel pressure model test. (From Holmes, J.D. et al. 1997. Wind loading and response of large stadium roofs. *Proceedings, IASS International Symposium '97 on Shell and Spatial Structures*, Singapore, 10–14 November.)

There have been many studies of the response to wind of large stadium roofs, which are unpublished, and hence cannot be discussed in detail here. However, one example worth mentioning briefly is the roof of Wembley Stadium in London. This is a very large roof with a number of resonant modes below 1 Hz that have the potential to be excited by wind action. However, it was found that the percentage of dynamic amplification of the internal forces in individual members of the roof structures varied greatly – between about 2% and 40%. This can be attributed to the similarity, or lack of, between the influence lines of the particular structural load effect associated with a member, and the mode shapes for any of the resonant modes excited by the wind.

10.8 SUMMARY

This chapter has attempted to cover the main aspects of wind loads on large roofs, including those used increasingly for sports stadiums. The characteristics of airflow and mean pressure distributions on flat, arched and domed roofs are discussed. There is some overlap with Chapter 8, 'Low-rise Buildings', but there are some significant differences – namely the large effects of the reduced correlations between fluctuating pressures over large expanses of low-pitch roofs, and the possibility of some resonant response contributions.

The application of wind-tunnel methods, using pressure measurements on rigid models, to the determination of effective static wind-load distributions is discussed. Several case studies documented in the public domain have been discussed.

REFERENCES

American Society of Civil Engineers. 1936. Wind-bracing in steel buildings. Fifth Progress Report of Sub-Committee No. 31. *Proceedings ASCE*, March: 397–412.

Arnstein, K. and Klemperer, W. 1936. Wind pressures on the Akron Airship-dock. *Journal of the Aeronautical Sciences*, 3: 88–90.

Blessmann, J. 1987a. Acao do vento em coberturas curvas, la Parte. Caderno Tecnico CT-86. Universidade Federale do Rio Grande do Sul.

Blessmann, J. 1987b. Vento em coberturas curvas – pavilhoes vizinhos. Caderno Tecnico CT-88. Universidade Federale do Rio Grande do Sul.

Blessmann, J. 1991. *Acao do vento em telhados*. SAGRA, Porto Alegre, Brazil.

Borri, C., Majowiecki, M. and Spinelli, P. 1992. Wind response of a large tensile structure: The new roof of the Olympic Stadium in Rome. *Journal of Wind Engineering and Industrial Aerodynamics*, 41–44: 1435–46.

Bounkin, A. and Tcheremoukhin, A. 1928. Wind pressures on roofs of buildings. Transactions, Central Aero- and Hydrodynamical Institute, Moscow, No. 35.

Cook, N.J. 1990. *The Designer's Guide to Wind Loading of Building Structures. Part 2. Static Structures.* Building Research Establishment, Watford, UK.

Davenport, A.G. and Surry, D. 1974. The pressures on low-rise structures in turbulent wind. *Canadian Structural Engineering Conference*, Toronto.

Grillaud, G. 1981. Effet du vent sur une structure gonflable. *Colloque, 'Construire avec le vent'*, Nantes, France, June.

Holmes, J.D. 1984. Determination of wind loads for an arch roof. *Civil Engineering Transactions, Institution of Engineers, Australia*, CE26: 247–53.

Holmes, J.D., Denoon, R.O., Kwok, K.C.S. and Glanville, M.J. 1997. Wind loading and response of large stadium roofs. *Proceedings, IASS International Symposium '97 on Shell and Spatial Structures*, Singapore, 10–14 November.

Hoxey, R. and Richardson, G.M. 1983. Wind loads on film plastic greenhouses. *Journal of Wind Engineering and Industrial Aerodynamics*, 11: 225–37.

Johnson, G.L., Surry, D. and Ng, W.K. 1985. Turbulent wind loads on arch-roof structures: A review of model and full-scale results and the effect of Reynolds number. *5th U.S. National Conference on Wind Engineering*, Lubbock, TX, 6–8 November.

Kasperski, M. 1992. Extreme wind load distributions for linear and nonlinear design. *Engineering Structures*, 14: 27–34.

Kasperski, M. and Niemann, H.-J. 1992. The L.R.C. (load-response-correlation) method: A general method of estimating unfavourable wind load distributions for linear and non-linear structural behaviour. *Journal of Wind Engineering & Industrial Aerodynamics*, 43: 1753–63.

Kawamura, S. and Kiuchi, T. 1986. An experimental study of one-membrane type pneumatic structure – Wind load and response. *Journal of Wind Engineering and Industrial Aerodynamics*, 23: 127–40.

Kind, R.J. 1984. Pneumatic stiffness and damping in air-supported structures. *Journal of Wind Engineering and Industrial Aerodynamics*, 17: 295–304.

Lam, K.M. and To, A.P. 1995. Generation of wind loads on a horizontal grandstand roof of large aspect ratio. *Journal of Wind Engineering and Industrial Aerodynamics*, 54/55: 345–57.

Melbourne, W.H. 1995. The response of large roofs to wind action. *Journal of Wind Engineering and Industrial Aerodynamics*, 54/55: 325–35.

Paterson, D.A. and Holmes, J.D. 1993. Mean wind pressures on arched-roof buildings by computation. *Journal of Wind Engineering and Industrial Aerodynamics*, 50: 235–43.

Standards Australia. 2011. *Structural Design Actions. Part 2: Wind Actions*. Standards Australia, Sydney, Australian/New Zealand Standard AS/NZS1170.2:2011.

Sykes, D.M. 1994. Wind loading tests on models of two tension structures for EXPO'92, Seville. *Journal of Wind Engineering and Industrial Aerodynamics*, 52: 371–83.

Toy, N. and Tahouri, B. 1988. Pressure distributions on semi-cylindrical structures of different geometrical cross-sections. *Journal of Wind Engineering and Industrial Aerodynamics*, 29: 263–72.

Tryggvason, B.V. 1979. Aeroelastic modelling of pneumatic and tensioned fabric structures. *Proceedings, 5th Internal Conference on Wind Engineering*, Fort Collins, Colorado, pp. 1061–1072. Pergamon Press, Oxford, UK.

Vickery, B.J. and Majowiecki, M. 1992. Wind induced response of a cable supported stadium roof. *Journal of Wind Engineering and Industrial Aerodynamics*, 41–44: 1447–58.

Xie, J. 2000. Gust factors for wind loads on large roofs. *First International Symposium on Wind and Structures*, Cheju, Korea, January.

Chapter 11

Towers, chimneys and masts

11.1 INTRODUCTION

In this chapter the wind-loading and wind-induced response of a variety of slender vertical structures will be considered: chimneys of circular cross section, free-standing lattice towers, observation towers of varying cross section, poles carrying lighting arrays or mobile telephone antennas, and guyed masts. Natural draft cooling towers, although not slender, are large wind-sensitive structures; the loading and response under wind action of these structures will be considered briefly in Section 11.6.

The methodology for determination of the loading and response of slender structures will first be described (making use of the general principles outlined in Chapters 1–7), and then followed by descriptions of several test case examples.

The dynamic response to wind of slender structures is quite similar in nature to that of tall buildings (described in Chapter 9). There are some significant differences, however:

- Fundamental mode shapes are generally non-linear
- Higher modes are more likely to be significant in the resonant dynamic response
- Since the aspect ratio is higher – the width is much less than the height – aerodynamic 'strip' theory can be applied. That is, total aerodynamic coefficients for the cross section can be used with the wind properties upstream, at the same height
- If the mass per unit height is low, aerodynamic damping (Section 5.5.1) will be significant
- As for tall buildings, cross-wind response can be significant (except for lattice structures). However because of the smaller cross-wind breadth, the velocity at which the vortex-shedding frequency (or the maximum frequency of the cross-wind force spectrum) coincides with the first mode vibration frequency is usually much lower than for tall buildings, and within the range of frequently occurring mean wind speeds

11.2 HISTORICAL

11.2.1 Lattice towers

When the Eiffel Tower in Paris was completed in 1889, it was easily the tallest structure in the world at 300 m and one of the first major towers of lattice construction. The designer Gustav Eiffel described the wind-loading assumptions used in the design in an address to the Societe' des Ingenieurs Civils (Eiffel, 1885). He assumed a static horizontal pressure of 2 kPa at the base increasing to 4 kPa at the top. Over a large part of the top and base of the tower, he replaced the area of members in the lattice with solid surfaces with the same enclosed area. In the middle section where the tower solidity is lower, he assumed a frontal area equal to 'four times the actual area of iron'. These very conservative assumptions, of course, resulted in a very stiff structure with no serviceability problems in strong winds.

Eiffel constructed a laboratory at the top of the Tower, and carried out various scientific experiments, including measurements of the deflection of the tower, using a telescope aimed vertically at the target at the top. Some of these measurements were later analysed by Davenport (1975). These indicated that the effective drag coefficient used in the design was approximately 3.5 times that required to produce the measured deflections, and that currently used in design for a tower with a solidity of about 0.3 (see Figure 11.1).

Later on the tower, Eiffel, perhaps concerned with the over-conservatism of his designs, carried out some experiments on wind forces on simple plates.

The development of high-voltage power transmission, and radio and television broadcasting, from the 1920s onwards promoted the efficient use of steel for lattice tower construction.

11.2.2 Tall chimneys

In the nineteenth and early twentieth centuries, most factory and power station chimneys were of masonry construction. With the known weakness of masonry joints to resist tension, these structures would have relied on dead load to resist the overturning effect of wind loads. Although undoubtedly many of these failed in severe windstorms, Kernot commented in 1893 that: '... there are thousands of such chimneys in existence, many in very open and exposed situations, which, apart from the adhesion of the mortar, would infallibly overturn with a pressure of not more than 15 pounds per square foot' (Kernot, 1893). Kernot concluded that the currently used design wind pressures were over-conservative (perhaps an early recognition of the effect of correlation), and proceeded to carry out some important early research in wind loads (see Section 7.2.1 and Figure 7.1).

The first full-scale wind-pressure measurements on a cylindrical chimney were performed by Dryden and Hill on the newly erected masonry chimney of the power plant of the Bureau of Standards near Washington, DC (Dryden and Hill, 1930). These measurements were carried out together with full-scale measurements on another shorter cylinder (aspect ratio of 3) mounted on a roof, and wind-tunnel measurements on circular cylinders. Through comparison of the resulting pressure distributions, this important study recognised, at an early stage, the effects of Reynolds number, surface roughness (Section 4.5.1) and aspect

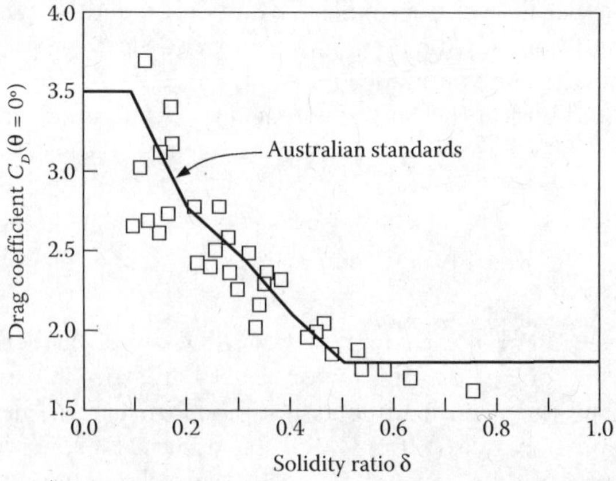

Figure 11.1 Drag coefficients for square towers with flat-sided members.

ratio (Section 4.5.2) on the pressure distribution and drag coefficients of slender circular cylinders.

In the 1950s, extensive work on the cross-wind vibration of steel chimneys was carried out at the National Physical Laboratory (NPL) in the United Kingdom under the direction of C. ('Kit') Scruton. This work (e.g. Scruton and Flint, 1964) included some important measurements on circular cylinders obtained in a compressed air wind tunnel, and the development of the now-ubiquitous helical strakes for the mitigation of vibration due to vortex shedding on tall chimneys (Section 4.6.3 and Figure 4.26).

11.3 BASIC DRAG COEFFICIENTS FOR TOWER SECTIONS

11.3.1 Drag coefficients for solid cross-sections

Many observation towers, communication towers and chimneys have cross sections that are circular or square. Drag coefficients for these cross sections were discussed in Chapter 4. The effect of aspect ratios less than 20 is significant on the effective total drag coefficient (see Figures 4.10 and 4.19). Other cross sections may require wind-tunnel tests to determine drag coefficients.

The mean or time-averaged drag force per unit height, and hence bending moments, can be calculated using an appropriate sectional drag coefficient with a wind speed appropriate to the height, using an appropriate expression for mean wind-speed profile (see Equation 5.32).

11.3.2 Drag coefficients for lattice towers

A basic formula for drag force for winds blowing at any angle to a face of a rectangular lattice tower is

$$D = C_D \cdot A_z \cdot q_z \tag{11.1}$$

where

D is the drag force on a complete tower panel section (i.e. all four sides of a square section tower)

C_D is the drag coefficient for the complete tower section – it depends on the solidity of a face, and the wind direction

A_z is the projected area of tower members in one face of the tower

$q_z(=(1/2)\rho_a U_z^2)$ is the dynamic wind pressure at the average height, z, of the panel under consideration

Figure 11.1 shows the values of C_D specified in the Australian Standard for steel lattice Towers, AS3995 (Standards Australia, 1994) for square sections with flat-sided members, as a function of the solidity, compared with experimental values obtained from wind-tunnel tests for wind blowing normally to a face. For the range of solidity from 0.1 to 0.5, the following equations are appropriate (from Bayar, 1986).

$$C_D = 4.2 - 7\delta \text{ (for } \delta < 0.2) \tag{11.2}$$

$$C_D = 3.5 - 3.5\delta \text{ (for } 0.2 < \delta < 0.5) \tag{11.3}$$

The ASCE Guidelines (2010) and CSIR Recommendations (1990) for transmission line structures give equations for the wind drag force on a section of a lattice tower for any arbitrary wind direction, θ, with respect to the face of the tower. The CSIR equation may be written as follows:

$$D = q_z\ [C_{dn1}\, A_{n1}\, \cos^2\theta + C_{dn2}\cdot A_{n2}\, \sin^2\theta]\, K_\theta \tag{11.4}$$

where

C_{dn1}, C_{dn2} are drag coefficients for wind normal to adjacent faces, 1 and 2, of the tower
A_{n1}, A_{n2} are the total projected areas of faces 1 and 2, respectively
θ is the angle of incidence of the wind with respect to the normal to face 1 of the tower
K_θ is a wind incidence factor (derived empirically), given by

$$K_\theta = 1 + 0.55\cdot\delta\cdot\sin^2(2\theta) \tag{11.5}$$

where δ is the solidity ratio (for $0.2 \le \delta \le 0.5$).

The ASCE Guidelines (2010) give a similar form to Equation 11.4, with a slightly different form for K_θ.

The drag of a lattice tower can also be computed by summing the contributions from every member. However, this is a complex calculation, as the effect of varying pitch and yaw angles on the various members, must be considered. This method also cannot easily account for interference and shielding effects between members and faces.

11.4 DYNAMIC ALONG-WIND RESPONSE OF TALL SLENDER TOWERS

The application of random vibration theory to the along-wind response of structures with distributed mass is discussed in Sections 5.3.6 and 5.3.7. The application of the equivalent static load distribution method to the along-wind response of tall structures is described in Section 5.4. These methods are applicable to all the structures covered in this chapter. However, a simple gust response factor (Section 5.3.2), in which a single multiplier, G, is applied to the mean pressure distribution, or a structural response derived from it, is generally not applicable in its simplest form to slender structures. Modifications are required to allow for a varying gust response factor, depending on the height, s, at which the load effect is required. A similar argument applies when a dynamic response factor approach is used (Section 5.3.4).

Two effects produce an increase in the gust response factor with height of load effect:

1. The curved mode shape which gives an increasing contribution from the resonant component as the height, s, increases.
2. Since wind gusts of size equal to, or greater than, the distance (h–s) between the height s and height of the top of the structure, h, are fully effective in producing stresses at the level s, the background contribution also increases as the height s increases.

An analysis for slender towers (Holmes, 1994) gives the following expressions for the gust response factors for shearing force, G_q, and bending moment, G_m, at any arbitrary height level, s, on a tower.

$$G_q = 1 + \frac{r\sqrt{g_B^2 B_s F_2 + g_R^2 (SE/\eta_1) F_3 F_4 F_5}}{F_1} \tag{11.6}$$

$$G_m = 1 + \frac{r\sqrt{g_B^2 B_s F_7 + g_R^2 (SE/\eta_1) F_3 F_4 F_8}}{F_6} \tag{11.7}$$

where

r is a roughness factor ($=2I_u$), that is twice the longitudinal turbulence intensity at the top of the tower (Section 3.3.1)

B_s is a background factor reflecting the reduction in correlation of the fluctuating loads between the height level s and the top of the tower (Section 4.6.6)

g_B and g_R are peak factors (Section 5.3.3) separately calculated for the background and resonant components

S is a size factor representing the aerodynamic admittance (Section 5.3.1) evaluated at the natural frequency of the tower

$E = (\pi n_1 S_u(n_1)/4\sigma_u^2)$, is a non-dimensional form of the spectral density of longitudinal turbulence (Section 3.3.4) evaluated at the natural frequency of the tower

η_1 is the critical damping ratio for the first mode of vibration (this should also include aerodynamic damping contributions)

$F_1 \ldots F_8$ are non-dimensional parameters depending on properties of the approaching wind and geometrical and dynamic properties of the tower, such as mean velocity profile, taper ratio, mode shape and mass distribution. They also depend on the ratio (s/h), that is the ratio of the height level, s, at which the shearing force and bending moments are required, and the height of the top of the tower.

By evaluation of Equations 11.6 and 11.7 for a typical lattice tower (Holmes, 1994), it was shown that the increase in the value of gust response factor over the height of a structure will typically be in the range of 5–15%.

A similar analysis for the deflection at the top of the tower, x, gives a similar expression to Equations 11.6 and 11.7 for the gust response factor for deflection, G_x (Holmes, 1996a):

$$G_x = 1 + \frac{r\sqrt{g_B^2 B_o F_{11} + g_R^2 (SE/\eta_1) F_3 F_4 F_{12}}}{F_{10}} \tag{11.8}$$

where B_o is B_s evaluated at s equal to 0 (the reduction due to correlation over the whole height of the tower is important. F_{10}, F_{11} and F_{12} are additional non-dimensional parameters; F_{12} is a non-dimensional stiffness for the tower.

It can be seen from Equations 11.6, 11.7 and 11.8 that the gust response factor depends on the type of load effect under consideration, as well as the height on the tower at which it is evaluated.

An alternative approach, for the along-wind loading and response of slender towers and chimneys is the equivalent (or effective) static load distribution approach discussed in Section 5.4 (see also Holmes, 1996b). This approach allows variations in dimension shape and mass over the height of a tower of complex shape to be easily incorporated. Examples of effective static wind-load distributions derived for a 160-m tower are given in Figures 5.11 and 5.12.

11.5 CROSS-WIND RESPONSE OF TALL SLENDER TOWERS

The strength of regular vortex shedding from a tower of uniform or slightly tapered cross section, is often strong enough to produce significant dynamic forces in the cross-wind direction. If the damping of a slender tower of a solid cross section is low, high-amplitude vibrations can occur if the frequency of vortex shedding coincides with a natural frequency of the structure. The velocity at which this coincidence occurs is known as the *critical velocity*. If the critical velocity is very high, that is outside the design range, no problems should arise, as the resonant condition will not occur. Conversely, if the critical velocity is very low, there will also not be a problem as the aerodynamic excitation forces will be low. However, significant vibration could occur if a critical velocity falls in the range 10–40 m/s.

Because of the higher rate of vortex shedding for a circular cross section compared with that for a square or rectangular section of the same cross-wind breadth, the critical velocity is significantly lower.

Methods of calculation of cross-wind response of slender towers or chimneys fall into two classes:

1. Those based on sinusoidal excitation
2. Those based on random excitation

In the following sections, methods developed mainly for structures of circular cross section, are described. However, in principle they can be applied to structures of any (constant) cross section.

11.5.1 Sinusoidal excitation models

The assumption that the vortex-shedding phenomenon generates near-sinusoidal cross-wind forces on circular cylinders can be linked to the work of Scruton and co-workers in the 1950s and 1960s (summarised in Scruton, 1981). In the original formulation, the excitation forces were treated solely as a form of negative aerodynamic damping, but this is equivalent to sinusoidal excitation by applied forces. Such models are good ones for situations in which large oscillations occur, and the shedding has 'locked-in' to the cross-wind motion of the structure (Section 5.5.4).

Sinusoidal excitation models were also proposed by Rumman (1970) and Ruscheweyh (1990).

Unlike other loading models in wind engineering, sinusoidal excitation models are *deterministic*, rather than random. The assumption of sinusoidal excitation leads to responses that are also sinusoidal.

To derive a simple formula for the maximum amplitude of vibration of a structure undergoing cross-wind vibration due to vortex shedding, the following assumptions will be made:

- Sinusoidal cross-wind force variation with time
- Full correlation of the forces over the height over which they act
- Constant amplitude of fluctuating cross-wind force coefficient

None of these assumptions are very accurate for structures vibrating in the turbulent natural wind. However they are useful for simple initial calculations to determine whether vortex-induced vibrations are a potential problem.

The structure is assumed to vibrate in the jth mode of vibration (in practice j will be equal to 1 or 2), so that Equation 5.17 applies:

$$G_j\ddot{a}_j + C_j\dot{a}_j + K_j a_j = Q_j(t) \tag{5.17}$$

where

G_j is the generalised mass equal to $\int_0^h m(z)\phi_j^2(z)\,dz$
$m(z)$ is the mass per unit length along the structure
h is the height of the structure
C_j is the modal damping
K_j is the modal stiffness
ω_j is the natural undamped circular frequency for the jth mode $\left(=2\pi n_j = \sqrt{K_j/G_j}\right)$
$Q_j(t)$ is the generalised force, equal to $\int_{z1}^{z2} f(z,t)\phi_j(z)\,dz$, where $f(z,t)$ is the fluctuating force per unit height
z_1 and z_2 are the lower and upper limits of the height range over which the vortex-shedding forces act

In this case, the applied force is assumed to be harmonic (sinusoidal) with a frequency equal to the vortex-shedding frequency, n_s, The maximum amplitude of vibration will occur at resonance, when n_s is equal to the natural frequency of the structure, n_j.

Thus, the generalised force (Section 5.3.6) is given by

$$Q_j(t) = \int_{z1}^{z2} f(z,t)\phi_j(z)\,dz = \left(\frac{1}{2}\right)\rho_a C_\ell b \sin(2\pi n_j t + \psi)\int_{z1}^{z2} \bar{U}^2(z)\phi_j(z)\,dz$$
$$= Q_{j,\max}\sin(2\pi n_j t + \psi)$$

where $Q_{j,\max}$ is the amplitude of the applied generalised force, given by

$$Q_{j,\max} = \left(\frac{1}{2}\right)\rho_a C_\ell\, b\int_{z1}^{z2} \bar{U}^2(z)\phi_j(z)\,dz \tag{11.9}$$

where

C_ℓ is the amplitude of the sinusoidal lift (cross-wind force) per unit length coefficient
ρ_a is the density of air.

The result for the maximum amplitude at resonance for a single-degree-of-freedom system can be applied:

$$a_{\max} = \frac{Q_{j,\max}}{2K_j\eta_j} = \frac{Q_{j,\max}}{8\pi^2 n_j^2 G_j \eta_j} \tag{11.10}$$

where η_j is the critical damping ratio for the jth mode, equal to $\left(C_j/2\sqrt{G_j K_j}\right)$

Substituting for $Q_{j,\max}$ from Equation 11.9 into Equation 11.10,

$$a_{\max} = \frac{(1/2)\rho_a C_\ell\, b\int_{z1}^{z2} \bar{U}^2(z)\phi_j(z)\,dz}{8\pi^2 n_j^2 G_j \eta_j}$$
$$= \frac{\rho_a C_\ell\, b^3\int_{z1}^{z2} \phi_j(z)\,dz}{16\pi^2 G_j \eta_j St^2} \tag{11.11}$$

where St is the Strouhal number for vortex shedding (Section 4.6.3), which in this case can be written as

$$St = \frac{n_s b}{\bar{U}(z_e)} = \frac{n_j b}{\bar{U}(z_e)}$$

where z_e is an average or effective height for the vortex-shedding frequency.

The maximum amplitude of deflection at any height on the structure is given by:

$$y_{max}(z) = a_{max} \cdot \phi_j(z)$$
$$= \frac{\rho_a C_\ell b^3 \phi_j(z) \int_{z1}^{z2} \phi_j(z)\,dz}{16\pi^2 G_j \eta_j St^2} \tag{11.12}$$

For a tower with a uniform mass per unit height, the maximum deflection at the tip ($z = h$), and where $\varphi(h)$ is chosen as 1.0, is given by

$$\frac{y_{max}(h)}{b} = \frac{\rho_a C_\ell b^2 \int_{z1}^{z2} \phi_j(z)\,dz}{16\pi^2 G_j \eta_j St^2} = \frac{C_\ell \int_{z1}^{z2} \phi_j(z)\,dz}{4\pi\, Sc\, St^2 \int_0^h \phi_j^2(z)\,dz} \tag{11.13}$$

where Sc is the *Scruton number*, or 'mass-damping parameter', previously introduced in Equation 5.48 and, in this case, defined as

$$Sc = \frac{4\pi m \eta_j}{\rho_a b^2} \tag{11.14}$$

where m is the average mass per unit length along the structure.

The ratio of vibration amplitude at the tip of a uniform cantilevered tower, to the tower breadth, can thus be evaluated as

$$\frac{y_{max}}{b} = \frac{k \cdot C_\ell}{4\pi \cdot Sc \cdot St^2} \tag{11.15}$$

where $k\left(= \left(\int_{z1}^{z2} \phi_j(z)\,dz \Big/ \int_0^h \phi_j^2(z)\,dz\right)\right)$ is a parameter dependent weakly on the mode shape of vibration.

Ruscheweyh (1990) has modified the basic sinusoidal model by the use of a 'correlation length'. The term 'correlation length' is one that is normally applied to random processes or excitation (Section 4.6.5), and a better term would be 'excitation length'. The vortex-shedding forces are applied over a height range less than the total height of the structure in this model.

A simple formula, based on Equation 11.13 can be derived to estimate the maximum amplitude of vibration as a fraction of the diameter. The version in the Eurocode (BSI, 2005) is written as follows:

$$\frac{y_{max}}{b} = \frac{1}{St^2} \cdot \frac{1}{Sc} \cdot K \cdot K_w \cdot C_{lat} \tag{11.16}$$

where

y_{max} is the maximum amplitude of vibration at the critical wind speed
K_w is an effective correlation length factor
K is a mode shape factor
C_{lat} is a lateral (cross-wind) force coefficient ($=C_\ell$)

11.5.2 Random excitation model: Vickery–Basu model

A random excitation model, for vortex-shedding response prediction, was developed by Vickery and Basu (1983). With some approximations, the peak deflection at the tip, as a ratio of diameter, can be written in the following form for a uniform cantilever:

$$\frac{\hat{y}}{b} = g\frac{[n_1 S_{C\ell}(n_1)]^{1/2}(\rho_a b^2/m)}{16\pi^{3/2}\eta^{1/2}St^2}f(\phi) \tag{11.17}$$

where

$S_{C\ell}(n)$ is the spectral density of the generalised cross-wind force coefficient
$f(\phi)$ is a function of mode shape
g is a peak factor, which depends on the resonant frequency, but is usually taken as 3.5–4
η is the critical damping ratio, comprising both structural and aerodynamic components

Equation 11.17 has some similarities with Equation 11.13, but it should be noted that in the case of random vibration, the response is inversely proportional to the *square root* of the damping, whereas in the case of sinusoidal excitation, the peak response is inversely proportional to the damping. The peak factor (ratio between peak and r.m.s. response) is also much greater than the value of $\sqrt{2}$ in the sinusoidal model. The spectral density includes the effect of correlation length on the fluctuating forces.

In Vickery and Basu's procedure, the spectral density of the local lift force per unit length, is represented by a Gaussian function, as follows:

$$\frac{n \cdot S_\ell(n)}{\sigma_\ell^2} = \frac{(n/n_s)}{B\sqrt{\pi}}\exp\left[-\left(\frac{1-n/n_s}{B}\right)^2\right] \tag{11.18}$$

where B is a bandwidth parameter.

This function is based on the assumption of a constant Strouhal number and the shedding frequency varying with wind speed, as the large-scale turbulence generates a Gaussian variation in wind speed about the mean value (Vickery and Basu, 1983).

Lock-in (Sections 4.6.3 and 5.5.4), in which the vortex-shedding frequency 'locks-in' to the natural frequency of the structure, results in an increase in the magnitude of the fluctuating cross-wind forces, and an increase in their correlation along the length of the structure. It is dealt in the Vickery and Basu model with a non-linear, amplitude-dependent, aerodynamic damping, within the random excitation model.

Equation 11.17 can be written in the form:

$$\frac{\hat{y}}{b} = \frac{A}{\left[(Sc/4\pi) - K_{ao}\left(1 - y^2/y_L^2\right)\right]^{1/2}} \tag{11.19}$$

where

A incorporates all parameters not associated with damping
y is the root-mean-square fluctuating amplitude
y_L is a limiting r.m.s. amplitude
K_{ao} is a non-dimensional parameter associated with the negative aerodynamic damping.

Equation 11.19 can be used to define three response regimes:

1. A randomly 'forced' vibration regime, at high values of Scruton number
2. A 'lock-in' regime for low values of Scruton number, in which the response is driven by the negative aerodynamic damping, and is largely independent of A
3. A transition regime between the above two regimes

These three regimes, with an empirical fit based on Equation 11.19, are shown in Figure 11.2 (from Vickery and Basu, 1983), and compared with experimental data from a model chimney (Wooton, 1969).

With appropriate input parameters, the Vickery/Basu method is applicable to any full-scale structure of constant, or slightly tapered cross section, but it has been calibrated to the vortex-induced response of large concrete chimneys.

When making predictions on real towers, or chimneys, in atmospheric turbulence it is necessary to include the effect of lateral turbulence. Referring to Figure 11.3, the effect of lateral (horizontal) turbulence is for the instantaneous flow direction to be at angle to the mean flow direction of θ, where,

$$\sin\theta \cong \frac{v}{\bar{U}}$$

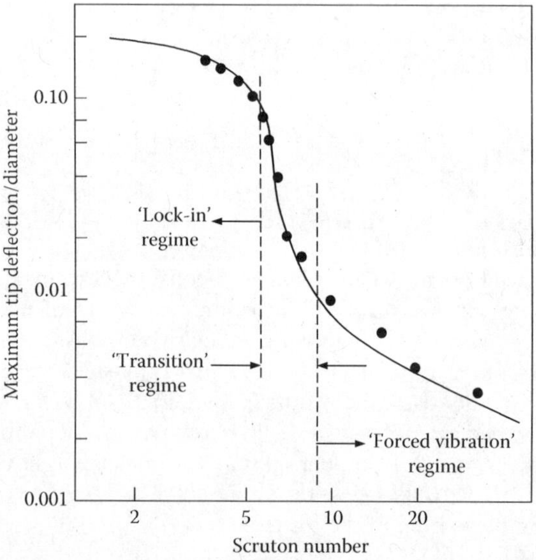

Figure 11.2 Response regimes for cross-wind vibration of circular towers and chimneys. (From Vickery, B.J. and Basu, R.I. 1983. *Journal of Wind Engineering and Industrial Aerodynamics*, 12: 49–73. With permission.)

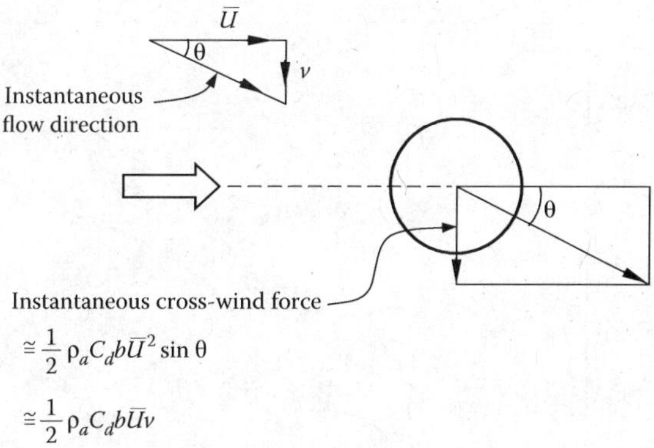

Figure 11.3 Cross-wind force due to lateral turbulence.

Thus, for a circular cross section, the instantaneous lateral force per unit length based on quasi-steady assumptions can be written as

$$f_t(z,t) = \frac{1}{2}\rho_a b C_d \bar{U}^2 \sin\theta = \frac{1}{2}\rho_a b C_d \bar{U} v(z,t) \tag{11.20}$$

Basu and Vickery (1983), in developing a method sui for prediction of the combined cross-wind response of real structures in the atmospheric boundary layer, used the following expression for the mean square modal coordinate in the jth mode:

$$\overline{a_j^2} = \frac{\pi n_j [S_{\ell,v}(n_j) + S_{\ell,t}(n_j)]}{4K_j^2(\eta_s + \eta_a)} = \frac{S_{\ell,v}(n_j) + S_{\ell,t}(n_j)}{(4\pi n_j)^3 G_j^2(\eta_s + \eta_a)} \tag{11.21}$$

where $S_{\ell,v}(n_j), S_{\ell,t}(n_j)$ are respectively the spectral densities, evaluated at the natural frequency, n_j, of the cross-wind forces due to vortex shedding and lateral turbulence. Equation 11.21 is based on the assumption that the spectral density is constant over the resonant peak, as previously used to derive Equation 5.13.

A comparison of the peak-to-peak cross-wind deflection at the top of the 330-m high Emley Moor television tower computed by the random vibration approach of Vickery and Basu, and compared with measurements, is shown in Figure 11.4. Calculations were made for the first four modes of vibration. There was some uncertainty in the appropriate structural damping for this tower, but generally a good agreement was obtained.

Comparisons were also made with full-scale response measurements from several reinforced-concrete chimneys (Vickery and Basu, 1984). The average agreement was quite good, but some scatter was shown.

11.5.3 Random excitation model: Hansen model

The Vickery amd Basu model for vortex-induced cross-wind vibration was developed into a more usable closed-form model by Hansen (2007). This version has been adopted in the design codes of the International Committee on Industrial Chimneys (CICIND, 2002), and by the Eurocode on wind actions (BSI, 2005).

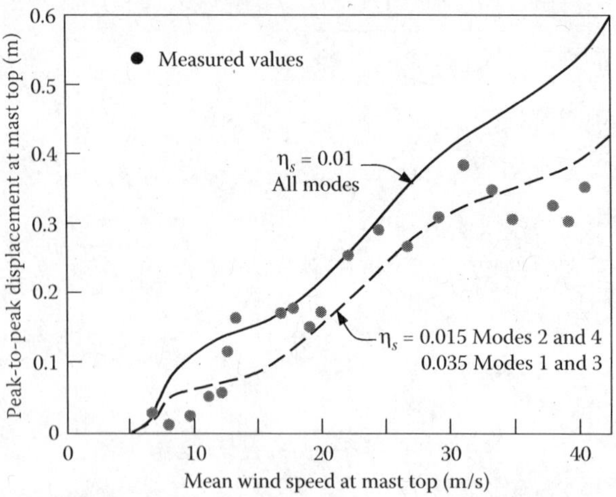

Figure 11.4 Comparison of measured and computed cross-wind response on the Emley Moor television tower. (From Basu, R.I. and Vickery, B.J. 1983. *Journal of Wind Engineering and Industrial Aerodynamics*, 12: 75–97. With permission.)

The equations for the ratio of the maximum standard deviation of the cross-wind deflection to the diameter are as follows (Hansen, 2007):

$$\left(\frac{\sigma_{max}}{b}\right)^2 = c_1 + \sqrt{c_1^2 + c_2} \tag{11.22}$$

$$c_1 = \frac{a_L^2}{2}\left(1 - \frac{Sc}{4\pi K_a}\right); \tag{11.23}$$

$$c_2 = \frac{a_L^2}{K_a}\frac{\rho b^2}{m}\frac{C_c^2}{St^4}\frac{b}{h}. \tag{11.24}$$

where

$a_L = \gamma_L \cdot a_{L,ref}$ is a limiting amplitude
$C_c = \gamma_C \cdot C_{c,ref}$ is an aerodynamic excitation parameter
K_a = aerodynamic damping parameter
H = length of structure exposed to vortex-shedding forces (normally the height of a vertical structure
ρ is the density of air
b is the breadth (diameter)
m is the mass per unit length
Sc is the Scruton number (Section 11.5.1)
St is the Strouhal number (Section 4.6.3)

Equation 11.22 is the solution of a quadratic equation for $(\sigma_{max}/b)^2$ similar to Equation 11.19 (Hansen, 2007). As a closed-form solution, iterations are avoided.

Table 11.1 Aerodynamic parameter for cross-wind response of circular sections

Parameter	$Re \leq 10^5$	$Re = 5 \times 10^5$	$Re \geq 10^6$
$a_{L,ref}$	0.4	0.4	0.4
$C_{c,ref}$	0.02	0.005	0.01
K_a	2	0.5	1

Source: Hansen, S.O. 2007. Vortex-induced vibrations of structures. *Proceedings, Third Structural Engineers World Congress*, Bangalore, India, 2–7 November. With permission.

Table 11.2 Correction factors for mode shape

Mode shape	γ_L	γ_C
Uniform, $\xi(z) = 1$	1	1
Linear, $\xi(z) = z/h$	1.73	1.29
Parabolic, $\xi(z) = (z/h)^2$	2.24	1.34
Cantilever	1.41	1.16

Source: Hansen, S.O. 2007. Vortex-induced vibrations of structures. *Proceedings, Third Structural Engineers World Congress*, Bangalore, India, 2–7 November. With permission.

For a circular cylinder, values of $a_{L,ref}$, $C_{c,ref}$ and K_a suggested by Hansen (2007) for various ranges of Reynolds number are given in Table 11.1. The correction factors for mode shape, γ_L and γ_C, are given for various shapes (normalised to a maximum value of 1.0) in Table 11.2.

The standard deviation of deflection $\sigma_{y,max}$ given by Equation 5.29 must be multiplied by a peak factor to give the peak deflection y_{max}.

Hansen (2007) recommended the following expression for peak factor (from Ruscheweyh and Sedlacek, 1988):

$$g = \sqrt{2}\left[1 + 1.2 arctan\left(0.75\left(\frac{Sc}{4\pi K_a}\right)^4\right)\right] \tag{11.25}$$

For large amplitudes of vibration (i.e. 'lock-in' conditions), Equation 11.25 gives a peak factor which approaches √2, which is the value for sinusoidal vibrations. At small amplitudes the value approaches 3.5–4.

11.5.4 Hybrid model of ESDU

Item 96030 of the Engineering Sciences Data Unit (ESDU, 1996) covers the response of structures of circular and polygonal cross section to vortex shedding. A computer program and spreadsheet is provided to implement the methods. ESDU 96030 covers uniform, tapered and stepped cylindrical or polygonal structures, and also yawed flow situations.

The method used in ESDU 96030 appears to be a hybrid of the two previously described approaches. For low amplitudes of vibration, a random excitation model similar to that of Vickery and Basu, has been adopted. At high amplitudes, that is in lock-in situations, a sinusoidal excitation model has been adopted, with a cross-wind force coefficient that is non-linearly dependent on the vibration amplitude. The response is postulated to switch intermittently between a random wide-band response and a constant amplitude sinusoidal type, as lock-in occurs.

The effect of cross-wind turbulence excitation is also included in this method. This contribution becomes more significant with increasing wind speed, and thus is more important for larger cylinders (e.g. large diameter reinforced-concrete chimneys with high critical wind speeds).

The ESDU method has the disadvantage of a discontinuity between the two response regimes.

11.5.5 Comparison of predictions of cross-wind response

In this section, a comparison of the computed response to vortex shedding for three representative slender structures with circular cross section is made:

1. A 100-m steel chimney
2. A 250-m reinforced-concrete chimney
3. A 25-m thin-walled, steel lighting pole

The relevant details of the three structures are given in Table 11.3.

These represent a wide range of structural types for which the cross-wind response needs to be assessed. In all three cases, the structures were assumed to be located in open country terrain, with relevant velocity profile and turbulence properties. In this comparison, only the first mode of vibration was considered.

The maximum r.m.s. ratio of tip deflection/mean diameter, for the three structures have been calculated by the following methods and tabulated in Table 11.4: (a) the sinusoidal excitation method given in the European pre-standard (BSI, 2005); (b) Vickery and Basu's random excitation approach (Structures 1,2 only); (c) the hybrid approach of ESDU (ESDU, 1996).

The three methods compared in Table 11.4 clearly give significant variations in estimated response to vortex shedding, for all three structures. In the case of structure (1), all methods predict large amplitudes characteristic of lock-in, although methods (b) and (c) predict

Table 11.3 Structural properties

Property	*Structure 1*	*Structure 2*	*Structure 3*
Height (m)	100	250	25
Diameter (m)	4.9	20	0.55–0.20 (tapered)
Surface roughness (mm)	0.1	1	0.15
Natural frequency (Hz)	0.5	0.3	0.5
Mode shape exponent	2	1.6	2
Mass/unit height (kg/m) (top third)	1700	50,000	30
Critical damping ratio	0.005	0.01	0.005

Table 11.4 Calculated values of maximum r.m.s. tip deflection/diameter (at or near critical velocity)

Method	*Structure 1*	*Structure 2*	*Structure 3*
a.	0.080	0.032	0.016
b.	0.214	0.0045	n.a.
c.	0.308	0.0054	0.014

higher amplitudes. Method (a), based on sinusoidal excitation, overestimates the response of structure 2 (a large reinforced-concrete chimney), which is subject to wide-band excitation with low amplitudes. Methods (b) and (c) predict similar maximum response for structure 2.

Vickery and Basu's model has generally been used for high Reynolds numbers only, and has not been applied to structure 3, which is clearly in the sub-critical regime. The other methods predict a low response amplitude for structure 3, which has a very low critical velocity in the first mode, although this type of low-mass pole, or mast, has a history of occasional large vortex-shedding responses, sometimes in higher modes, and often producing fatigue problems. One of the main problems in predicting their behaviour is in predicting the structural damping ratio, which is often very amplitude dependent.

Verboom and van Koten (2009) applied three methods for prediction of the cross-wind response to vortex shedding of 13 steel chimneys in Europe, the operational history of which was known in some detail. The three methods used were:

- *Approach* 1 in Eurocode 1. This is a variant of the sinusoidal excitation model discussed in Section 11.5.1.
- *Approach* 2 in Eurocode 1. This is the Hansen model (Section 11.5.3), and essentially the same as that in the CICIND (2002) code for steel chimneys.
- *Approach* 3 was described as 'an accurate implementation' of the Vickery–Basu model. In particular this included the effect of atmospheric turbulence on the bandwidth parameter, B, and on the aerodynamic damping parameter, K_{ao} (Section 11.5.2).

Using the expected fatigue life as a criterion, Verboom and van Koten concluded that *Approach* 1 'seriously underestimated' the stresses caused by cross-wind vibrations, and hence would have overestimated the fatigue life for 5 of the 13 chimneys. *Approach* 2 mostly overestimated the stresses, and hence unjustly rejected 6 out of the 13 chimneys as being 'unsafe'.

Approach 3, which accurately took account of the effects of turbulence was stated as giving a 'good indication' of the stresses due to cross-wind vibrations, and accurately predicted the fatigue life for three out of the four chimneys, for which the operational life is known.

Hence, it appears that neither approach currently given in Eurocode 1, for the prediction of the cross-wind response of steel chimneys, is completely satisfactory, and that the effects of turbulence intensity on the aerodynamic parameters are significant. However, it will be difficult to accurately incorporate the latter into a code-based approach, as the turbulence intensity can often vary considerably at the same site, depending on the wind speed, and other factors such the atmospheric temperature, and the occurrence of temperature inversions, as well as the roughness of the surrounding terrain, for various directions.

11.6 COOLING TOWERS

The vulnerability of large hyperbolic natural draught cooling towers to wind action was emphasised in the 1960s by the collapse of the Ferrybridge towers in the United Kingdom (Figure 1.11). This event provoked research work on the wind loading and response of these large structures, especially in Europe. The sensitivity of wind pressures on circular cross sections to Reynolds number, means that like chimneys, there are some questions about the validity of wind-tunnel tests to produce reliable results.

The main factors affecting wind loading of large cooling towers are:

- The partially correlated nature of fluctuating wind pressures acting on such large bluff structures, which means that quasi-steady design wind pressures are inadequate.

- The non-linear nature of the thin reinforced concrete.
- Aerodynamic interference effects from adjacent similar structures (as illustrated by the Ferrybridge failures).

Since the lowest natural frequency in the uncracked state usually exceeds 1 Hz, these structures are not particularly dynamically sensitive to wind, although after cracking of the concrete, the frequencies can apparently reduce significantly, with significant resonant contributions to the response (Zahlten and Borri, 1998).

A detailed discussion on the wind loading of these special structures will not be given in this chapter, although they are covered in some detail by Simiu and Scanlan (1996). There are a number of specialist design codes for cooling towers, which include specification of wind loads (e.g. VGB, 1990; BSI, 1992).

Other useful references are by Shu and Wenda (1991) for soil interaction effects, Niemann and Köpper (1998) for aerodynamic interference, Zahlten and Borri (1998) for resonant amplification effects and Niemann and Ruhwedel (1980) for wind-tunnel modelling.

11.7 GUYED MASTS

Since most guyed masts are lattice structures (usually with triangular cross sections), wind-tunnel testing is neither appropriate nor required for this type of structure. Analytical methods are usually used for tall guyed masts.

However, guyed masts are complex structures to analyse for wind loading for various reasons.

- Their structural behaviour is non-linear.
- The influence lines for load effects such as bending moments and guy tensions are complex.
- When resonant dynamic response is important (for masts greater than about 150 m in height) many modes participate, and they are often coupled.

Generally, the dynamic response to wind may be analysed using the methods of random vibration outlined in Chapter 5. However, simple gust response factor approaches are not appropriate, because of the complex influence lines, with alternating positive and negative portions. The non-linear nature of the structure may be readily dealt with by computing the free-vibration frequencies and mode shapes, about the deflected position under the mean wind loading, rather than the 'no wind' condition. The effective static-load methods outlined in Section 5.4 are very useful to derive effective static-load distributions for both the background and resonant response of these structures.

A simplified approach to the dynamic response of tall guyed masts, in which the responses due to 'patch loads' are scaled to match the response calculated more rigorously from random vibration theory, is described by Davenport and Sparling (1992), and Sparling et al. (1996). The patch loads are applied on each span of the mast between adjacent guy levels, and from midpoint to midpoint of adjacent spans. The magnitude of the patch loads is taken as equal to the r.m.s. fluctuating drag force per unit height, at each height level, z:

$$d(z) = \rho_a C_d(z)\, b(z)\bar{U}(z)\, \sigma_u(z) \tag{11.26}$$

To simulate the lack of correlation of the fluctuating wind loads, the responses (bending moments, shear, deflections) due to the individual patch loads are combined by a root-sum-of squares as in Equation 11.27.

$$\tilde{r}_{Pl} = \sqrt{\sum_{i=1}^{N} r_i^2} \tag{11.27}$$

where $\tilde{r}_{Pl}$ is the resultant patch-load response, r_i is the response due to the i th patch load, and N is the total number of patch loads.

The design peak response is then determined from Equation 11.28.

$$\hat{r}_{Pl} = \tilde{r}_{Pl} \cdot \lambda_B \cdot \lambda_R \cdot \lambda_{TL} \cdot g \tag{11.28}$$

where g is a peak factor, $\lambda_B, \lambda_R, \lambda_{TL}$ are a 'background scaling factor', a 'resonant magnification factor' and a 'turbulent length scale factor', respectively. These factors were determined by calibrating the method against the results of a full dynamic (random vibration) analysis for eight guyed masts ranging in height from 123 to 622 m. Expressions for these factors resulting from this calibration are given by Sparling et al. (1996).

This patch method has been adopted by the British Standard for lattice towers and masts (BSI, 1994). The results from the analysis of a 295-m guyed mast are shown in Figure 11.5. This shows that a good agreement is achieved between the patch-load method, and the full dynamic analysis. The results from a conventional gust response factor approach (Section 5.3.2) are also shown. In this method, the mast is analysed under the mean wind loading, and the resulting responses are factored up by a constant factor (in this case 2.0 was used). Clearly, this method grossly underestimates the peak bending moments between the guy levels.

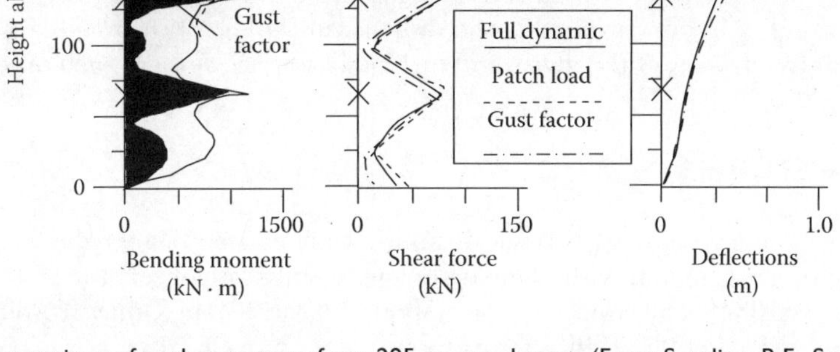

Figure 11.5 Comparison of peak responses for a 295-m guyed mast. (From Sparling, B.F., Smith, B.W. and Davenport, A.G. 1996. *Journal of the International Association for Shell and Spatial Structures*, 37: 89–106. With permission.)

Figure 11.6 A failed wind turbine tower following a typhoon. (Picture by Dr. M. Matsui, Tokyo Polytechnic University. With permission.)

11.8 WIND TURBINE TOWERS

With the development of wind energy farms in many parts of the world, during the last 20 years, there are now many large wind turbine towers in existence, with heights approaching 100 m. These towers are relatively flexible and carry a large mass at the top, due to the nacelles and turbine blades. Hence, their frequencies can be quite low (i.e. <0.5 Hz) and the towers, which are normally in exposed locations, may be subjected to significant turbulent buffeting (Section 5.3) at high wind speeds. There have been a number of failures of these towers in extreme winds (Figure 11.6).

The along-wind response of a typical wind turbine tower was investigated numerically by Murtagh et al. (2005) using simulated turbulent wind forces. The tower and the blades were modelled dynamically as multi-degree-of-freedom systems. The responses of the blades themselves were calculated using drag force time histories derived from rotationally sampled wind spectra. The responses of the blades and the tower were coupled using compatibility of displacement at the top of the tower. It was found that neglect of the blade–tower interaction can significantly underestimate the response at the top of the tower, especially if the fundamental frequencies of the tower and turbine blades are close to each other.

11.9 CASE STUDIES

An overview of the comprehensive wind-tunnel study carried out for the 555-m high CN Tower in Toronto, Canada, with comparisons with full-scale observations is presented by Isyumov et al. (1984). The wind-induced response of the 309-m Sydney Tower is described by Kwok and Macdonald (1990); the response was found to decrease markedly after a tuned mass–damper system (Section 9.9.3) was installed. Numerical and wind-tunnel simulations of the wind-induced response of the 310-m Nanjing Tower are described by Kareem et al. (1998).

Figure 11.7 Aeroelastic wind-tunnel model of a large free-standing tower.

A case study of the wind loading and response study of the 338-m tall Macau Tower, which incorporates both wind-tunnel studies, and calculations, is described by Holmes (2000). The full aeroelastic model (1/150 scale) of the Macau Tower, used for the wind-tunnel testing, is shown in Figure 11.7.

There have also been a number of full-scale studies on the dynamic response of large reinforced-concrete chimneys. Notable amongst these are studies by Muller and Nieser (1975), Hansen (1981), Melbourne et al. (1983) and Waldeck (1992). Ruscheweyh (1990) reported on some measurements on a number of steel stacks of cross-wind vibration, and makes comparisons with predictions based on the sinusoidal model (Section 11.5.1).

Measurements on two tall guyed masts have been made by Peil et al. (1996) for comparison with theoretical predictions. One of these studies entailed the detailed measurement of turbulent wind speed at 17 height levels up to 340 m height (Peil and Nölle 1992).

The wind-induced acceleration response of an air-traffic control tower was investigated by Park et al. (2006), using both a high-frequency base balance (Section 7.6.2), and an aeroelastic (Section 7.6.4) test in a wind tunnel. The excessive acceleration response at the top of the tower was controlled using a hybrid active passive tuned mass damper (Section 9.9.3). The effectiveness of the damper system was demonstrated by free vibration tests.

11.10 SUMMARY

In this chapter, the wind loading of slender towers, chimneys and masts of various types has been discussed. These structures are usually dynamically sensitive to wind, and response in both along-wind and cross-wind directions may need to be considered. Theoretical methods for calculating dynamic response, in both directions, are discussed.

The wind loading of hyperbolic cooling towers, guyed masts and wind turbine towers, are complex due to their structural behaviour. The main features of the wind loading and response of these structures are discussed.

REFERENCES

American Society of Civil Engineers. 2010. *Guidelines for Electrical Transmission Line Structural Loading*. 3rd edition. American Society of Civil Engineers, Reston, VA, USA.

Basu, R.I. and Vickery, B.J. 1983. Across-wind vibrations of structures of circular cross-section. Part II. Development of a mathematical model for full-scale applications. *Journal of Wind Engineering and Industrial Aerodynamics*, 12: 75–97.

Bayar, D.C. 1986. Drag coefficients of latticed towers. *A.S.C.E., Journal of Structurql Engineering*, 112: 417–30.

British Standards Institution. 1992. Water cooling towers. Part 4. Code of practice for structural design and construction. British Standard, BS 4485: Part 4: 1992.

British Standards Institution. 1994. Lattice towers and masts. Part 4. Code of practice for lattice masts. British Standard, BS 8100: Part 4: 1994.

British Standards Institution. 2005. Eurocode 1: Actions on structures – Part 1–4: General actions – Wind actions. BS EN 1991-1-4.6, BSI, London, UK.

CICIND (International Committee on Industrial Chimneys). 2002. Model code for steel chimneys. Revision 1 – December 1999, Amendment A, March 2002, CICIND, Zurich, Switzerland.

CSIR. 1990. Transmission Line Loading. Part I: Recommendations and Commentary. Part II: Appendices. Engineering Structures Programme, CSIR Building Technology, South Africa.

Davenport, A.G. 1975. Perspectives on the full-scale measurement of wind effects. *Journal of Industrial Aerodynamics*, 1: 23–54.

Davenport, A.G. and Sparling, B.F. 1992. Dynamic gust response factors for guyed masts. *Journal of Wind Engineering and Industrial Aerodynamics*, 44: 2237–48.

Dryden, H.L. and Hill, G.C. 1930. Wind pressure on circular cylinders and chimneys. *Journal of Research of the National Bureau of Standards*, 5, 653–93.

Eiffel, G. 1885. Projet d'une tour en fer de 300 m de hauteur. *Memoires de la Societe' des Ingenieurs Civils I*, 345–70.

ESDU. 1996. Response of structures to vortex shedding: Structures of circular or polygonal cross-section. Engineering Sciences Data Unit (ESDU International, London, UK), ESDU Data Item 96030.

Hansen, S.O. 1981. Cross-wind vibrations of a 130 metre tapered concrete chimney. *Journal of Wind Engineering and Industrial Aerodynamics*, 8: 145–56.

Hansen, S.O. 2007. Vortex-induced vibrations of structures. *Proceedings, Third Structural Engineers World Congress*, Bangalore, India, 2–7 November.

Holmes, J.D. 1994. Along-wind response of lattice towers: Part I – Derivation of expressions for gust response factors. *Engineering Structures*, 16: 287–92.

Holmes, J.D. 1996a. Along-wind response of lattice towers: Part II – Aerodynamic damping and deflections. *Engineering Structures*, 18: 483–8.

Holmes, J.D. 1996b. Along-wind response of lattice towers: Part III – Effective load distributions. *Engineering Structures*, 18: 489–94.

Holmes, J.D. 2000. Wind loading of the Macau Tower – Application of the effective static load approach. *Proceedings, First International Symposium on Wind and Structures for the 21st Century*, Cheju, Korea, 26–28 January, pp. 81–90.

Isyumov, N., Davenport, A.G., and Monbaliu, J. 1984. CN Tower, Toronto: Model and full-scale response to wind. *Proceedings, 12th Congress, International Association for Bridge and Structural Engineering*, Vancouver, Canada, 3–7 September, pp. 737–46.

Kareem, A., Kabat, S. and Haan, F.L. 1998. Aerodynamics of Nanjing Tower: A case study. *Journal of Wind Engineering and Industrial Aerodynamics*, 77–78: 725–39.

Kernot, W.C. 1893. Wind pressure. *Proceedings, Australasian Society for the Advancement of Science*, V: 573–81.

Kwok, K.C.S. and Macdonald, P.A. 1990. Full-scale measurements of wind-induced acceleration response of Sydney Tower. *Engineering Structures*, 12: 153–62.

Melbourne, W.H., Cheung, J.C.K. and Goddard, C. 1983. Response to wind action of 265-m Mount Isa stack. *ASCE, Journal of Structural Engineering*, 109: 2561–77.

Muller, F.P. and Nieser, H. 1975. Measurements of wind-induced vibrations on a concrete chimney. *Journal of Industrial Aerodynamics*, 1: 239–48.

Murtagh, P.J., Basu, B. and Broderick, B.M. 2005. Along-wind response of a wind turbine tower with blade coupling subjected to rotationally sampled wind loading. *Engineering Structures*, 27: 1209–19.

Niemann, H.-J. and Köpper, H.-D. 1998. Influence of adjacent buildings on wind effects on cooling towers. *Engineering Structures*, 20: 874–80.

Niemann, H.-J. and Ruhwedel, J. 1980. Full-scale and model tests on wind-induced, static and dynamic stresses in cooling tower shells. *Engineering Structures*, 2: 81–9.

Park, W., Park, K.-S., Koh, H.-M. and Ha, D.-H. 2006. Wind-induced response control and serviceability improvement of an air traffic control tower. *Engineering Structures*, 28: 1060–70.

Peil, U. and Nölle, H. 1992. Guyed masts under wind load. *Journal of Wind Engineering and Industrial Aerodynamics*, 41–44: 2129–40.

Peil, U., Nölle, H., and Wang, Z.H. 1996. Nonlinear dynamic behaviour of guys and guyed masts under turbulent wind load. *Journal of the International Association for Shell and Spatial Structures*, 37: 77–88.

Rumman, W.S. 1970. Basic structural design of concrete chimneys. *ASCE, Journal of the Power Division*, 96: 309–18.

Ruscheweyh, H. 1990. Practical experiences with wind-induced vibrations. *Journal of Wind Engineering and Industrial Aerodynamics*, 33: 211–18.

Ruscheweyh, H. and Sedlacek, G. 1988. Cross-wind vibrations of steel stacks – Critical comparisons between some recently-proposed codes. *Journal of Wind Engineering and Industrial Aerodynamics*, 33: 173–83.

Scruton, C. 1981. *An Introduction to Wind Effects on Structures*. Oxford University Press, Oxford, UK.

Scruton, C. and Flint, A.R. 1964. Wind-excited oscillations of structures. *Proceedings, Institution of Civil Engineers (UK)*, 27: 673–702.

Shu W. and Wenda L. 1991. Gust factors for hyperbolic cooling towers on soils. *Engineering Structures*, 13: 21–6.

Simiu, E. and Scanlan, R.H. 1996. *Wind Effects on Structures – Fundamentals and Applications to Design*. 3rd edition. John Wiley, New York.

Sparling, B.F., Smith, B.W. and Davenport, A.G. 1996. Simplified dynamic analysis methods for guyed masts in turbulent winds. *Journal of the International Association for Shell and Spatial Structures*, 37: 89–106.

Standards Australia. 1994. *Design of Steel Lattice Towers and Masts*. Standards Australia, North Sydney, AS3995-1994.

VGB. 1990. *VGB: BTR Bautechnik bei Kültürmen*. (Construction guidelines for cooling towers.) VGB Association of Large Powerplant Operators, Essen, Germany.

Verboom, V.K. and van Koten, H. 2009. Vortex excitation: Three design rules tested on 13 industrial chimneys. *Journal of Wind Engineering and Industrial Aerodynamics*, 98: 145–54.

Vickery, B.J. and Basu, R.I. 1983. Across-wind vibrations of structures of circular cross-section. Part I. Development of a mathematical model for two-dimensional conditions. *Journal of Wind Engineering and Industrial Aerodynamics*, 12: 49–73.

Vickery, B.J. and Basu, R I. 1984. The response of reinforced concrete chimneys to vortex shedding. *Engineering Structures*, 6: 324–33.

Waldeck, J.L. 1992. The measured and predicted response of a 300 m concrete chimney. *Journal of Wind Engineering and Industrial Aerodynamics*, 41: 229–40.

Wooton, L.R. 1969. The oscillations of large circular stacks in wind. *Proceedings of the Institution of Civil Engineers* (UK), 43: 573–98.

Zahlten, W. and Borri, C. 1998. Time-domain simulation of the non-linear response of cooling tower shells subjected to stochastic wind loading *Engineering Structures*, 20: 881–9.

Chapter 12

Bridges

12.1 INTRODUCTION

As discussed in Chapter 1, bridges have featured in some spectacular failures during wind storms (Figure 1.10). The history of the dynamically wind-sensitive suspension bridge from the nineteenth century onwards, including the periodic failures that have occurred, has been well documented (e.g. Steinman and Watson, 1957; Billington, 1977; Petroski, 1996).

Most of the early interest was in the drag, or along-wind, forces, and Baker (1884), Kernot (1893) and others noted that peak wind forces acting on large areas, such as a complete bridge girder, were considerably less than those on a small plate or board. However, the great American builder of suspension bridges, John Roebling, was aware of the dynamic effects of wind as early as 1855. In commenting on the failure of the Wheeling Bridge, Ohio, in the previous year, he wrote: 'That bridge was destroyed by the momentum acquired by its own dead weight, when swayed up and down by the force of the wind.... A high wind, acting upon a suspended floor, devoid of inherent stiffness, will produce a series of undulations, which will be corresponding from the center each way' (Steinman and Watson, 1957, p. 221).

However it took more than 80 years for the dramatic failure of the first Tacoma Narrows suspension bridge in 1940 (Section 1.4), to direct serious attention to the dynamic actions of the wind, and other wind actions on bridge decks: vertical cross-wind forces and torsional moments.

The cable-stayed bridge emerged in the 1950s in Germany, as an efficient method of spanning intermediate length crossings. Gimsing (1983) and Virlogeux (1999) have reviewed recent developments in the design of bridges of this type.

At the start of the twenty-first century, the spans of the long-span suspension and cable-stayed bridges were extended to new limits. The longest bridge in the world in 2014 is the suspension bridge across the Akashi-Kaikyo Straits in Japan, which has an overall length of nearly 4 km, with a main span of 1990 m (Figure 12.1). The design of this bridge was dominated by its aerodynamic characteristics.

The cable-stayed bridge with the longest span in 2014 is the Russky Bridge, near Vladivostok, Russia, with an overall length of 3100 m, and a main span of 1104 m.

As the spans increase, wind actions become more critical in bridge design, and for the longest suspension or cable-stayed bridges, extensive wind studies are normally undertaken. The dynamic wind forces will excite resonant response, often in several modes, and *aero-elastic* forces, in which the motion of the structure itself generates forces, are important. Long-span bridges are usually crossings of large expanses of water, and may be exposed to relatively low-turbulence flow, at least at low wind speeds. This has contributed to a number of cases of vibrations of bridge decks induced by vortex shedding (Section 4.6.3). Recently, the spans of cable-stayed bridges have been limited by problems with cable vibrations, sometimes involving rain, as well as wind (Section 12.5).

Figure 12.1 Akashi-Kaikyo Bridge, Japan.

In the following sections, a review of the main aspects of wind forces and the wind-induced excitation of long-span bridges, and their supporting cables is given. The aerodynamics of bridges is a large and specialised topic, and an in-depth treatment will not be given in this book. The reader is referred to other sources (e.g. Larsen, 1992; Simiu and Scanlan, 1996) which treat the subject in more detail.

12.2 BASIC FORCE COEFFICIENTS FOR BRIDGES

As for other structures, all bridges are subjected to mean and fluctuating wind forces. These may be estimated by the use of mean, or steady state, force coefficients, usually determined from wind-tunnel tests. Such coefficients are also required to determine dynamic response from turbulent buffeting.

Many wind-tunnel section tests of decks for long-span bridges (Section 7.6.3) have been carried out, primarily to determine their aerodynamic stability (Section 12.3.2). Determination of the basic section force coefficients, as a function of wind angle of attack, is also routinely done during the tests.

Most nineteenth century suspension bridges were built with open lattice truss sections. This use has continued, as this type of section has some benefits from the point of view of dynamic response. The open structure prevents the formation of vortices, and dynamic excitation from vortex shedding (Section 4.6.3) is not usually a problem. Provided the torsional stiffness can be made high enough, the critical speed for flutter instability (Sections 5.5.3 and 12.3.2) will be high. However, the drag coefficients for open truss sections are high in comparison with other sections. For example, the drag coefficients for two cross-sections considered for the Little Belt suspension bridge completed in the 1960s in Denmark are shown in Figure 12.2 (Ostenfeld and Larsen, 1992). The drag coefficient for the trussed cross-section is more than three times that of the streamlined box girder section; the latter was eventually used for the bridge. However, after extensive aerodynamic testing (Miyata et al., 1992), a truss girder, 11 m deep, was chosen for the Akashi-Kaikyo suspension bridge – the world's longest (Figure 12.1).

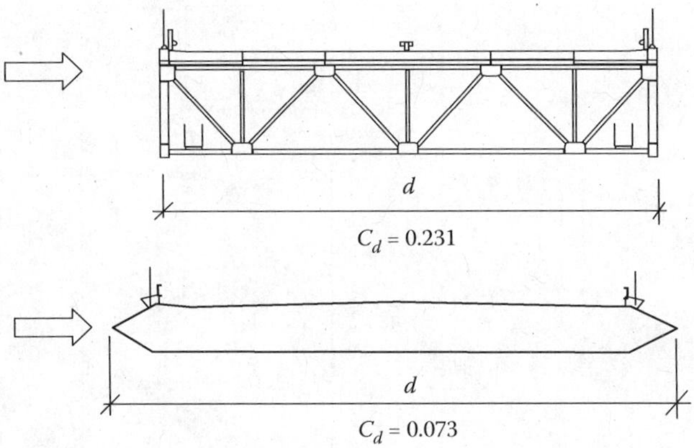

Figure 12.2 Comparison of drag coefficients for two bridge deck cross-sections. (From Ostenfeld, K.H. and Larsen, A. 1992. Bridge engineering and aerodynamics. *Proceedings of the First International Symposium on Aerodynamics of Large Bridges*, Copenhagen, Denmark, 19–21 February 1992, pp. 3–22. A.A. Balkema, Rotterdam. Reproduced with permission.)

Note that the along-wind *chord* dimension, d, rather than the cross-wind dimension, b, has been used to define the drag coefficients. This is usually the convention for bridges.

Very slender deck cross-sections, such as the box girder section shown in Figure 12.2, although having very low drag coefficients, will have high lift (cross-wind) force coefficients (Section 4.2.2) when the wind has a significant angle of attack, in a similar way to an airfoil. This situation will occur instantaneously in turbulent flow. This characteristic makes deck sections of this type prone to buffeting by vertical turbulence (Section 12.3.3).

Examples of the variation of static horizontal and vertical force coefficients, and moment coefficient about the mass centre of a bridge deck section, with angle of attack, are given in Figure 12.3.

The conventional definition of section force and pitching moments for bridges is as follows:

$$C_X = \frac{F_x}{1/2\,\rho_a U^2 d} \quad C_Z = \frac{F_z}{1/2\,\rho_a U^2 d} \quad C_M = \frac{M}{1/2\,\rho_a U^2 d^2} \tag{12.1}$$

12.3 THE NATURE OF DYNAMIC RESPONSE OF LONG-SPAN BRIDGES

There are several mechanisms, in various wind speed ranges, which can excite resonant dynamic response in the decks of long-span bridges:

- Vortex shedding excitation (Section 4.6.3) which usually occurs in low wind speeds and low turbulence conditions (e.g. Frandsen, 2001).
- Flutter instabilities (Section 5.5.3) of several types, which occur at very high wind speeds for *aerodynamically stable* decks, as a result of the dominance of self-excited aerodynamic forces (Sabzevari and Scanlan, 1968). These always involve torsional (rotational) motions, and may also involve vertical bending motions.

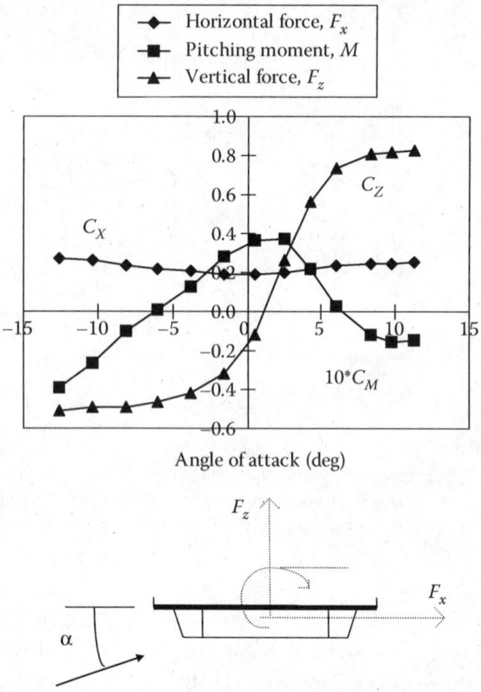

Figure 12.3 Static force coefficients for a typical bridge deck section.

- Buffeting excitation (Section 4.6.1) caused by the fluctuating forces induced by turbulence (Davenport, 1962; Scanlan and Gade, 1977). This occurs over a wide range of wind speeds, and normally increases monotonically with increasing wind speed.

The nature of these mechanisms is discussed in the following sections.

12.3.1 Vortex-shedding excitation

Under certain conditions, vortex-shedding excitation can induce significant, but limited, amplitudes of vibration. The conditions required for this to occur are most, or all, of the following:

- Wind direction normal to the longitudinal axis of the bridge
- Low turbulence conditions (typically I_u <0.05)
- A wind speed in a narrow critical range (5–12 m/s)
- Low damping (1% of critical or less)

The above conditions can be satisfied for both short-to-medium span cable-stayed bridges crossing water, and longer span suspension bridges. With Strouhal numbers in the range of 0.1–0.2 (based on the depth of the deck cross section), and natural frequencies in the range of 0.1–0.6 Hz, critical velocities of 6–15 m/s can produce significant amplitudes. Low turbulence conditions can occur in 'stable' atmospheric conditions, often in the early morning or evening. Recorded examples of this behaviour are listed in Table 12.1.

Table 12.1 Some recorded cases of vortex-shedding-induced vibrations of bridges

Name	*Natural frequency (Hz)*	*Critical velocity (m/s)*	*Max. amplitude (mm)*	*Reference*
Long's Creek Bridge	0.6	12	100–170	Wardlaw (1971)
Wye Bridge	0.46	7.5	35	Smith (1980)
Waal River	0.44	9–12	50	van Nunen and Persoon (1982)
Great Belt East	0.13–0.21	4.5–9	320	Larsen et al. (1999) Frandsen (2001)

Section tests carried out in smooth flow in wind tunnels can provide reasonably good predictions of the full-scale behaviour (Wardlaw, 1971; van Nunen and Persoon, 1982). In the case of the Long's Creek Bridge, Canada, where the vibrations were large enough to require remedial action, triangular fairings on the ends, and a soffit plate underneath the deck were added to the prototype structure, with satisfactory results (Wardlaw, 1971). Guide vanes were used at the lower corners of the box girder of the Great Belt East suspension bridge, Denmark, a method known to be successful in suppressing vortex shedding vibrations, which occurred at four different frequencies and a corresponding wide range of wind speeds. Lock-in effects (Sections 4.6.3 and 5.5.4) were also observed in the vortex-induced vibration on this bridge (Frandsen, 2001).

12.3.2 Flutter instabilities and prediction of flutter speeds

The coupled motion (rotation and vertical displacement) of a suspended bluff body was discussed in Section 5.5.3. Equations 5.49 and 5.50 – the coupled equations of motion – are repeated as follows:

$$\ddot{z} + 2\eta_z\omega_z\dot{z} + \omega_z^2 z = \frac{F_z(t)}{m} + H_1\dot{z} + H_2\dot{\theta} + H_3\theta \quad (5.49)$$

$$\ddot{\theta} + 2\eta_\theta\omega_\theta\dot{\theta} + \omega_\theta^2\theta = \frac{M(t)}{I} + A_1\dot{z} + A_2\dot{\theta} + A_3\theta \quad (5.50)$$

Equations 5.49 and 5.50 are simplified forms of the full equations of motion, which include the horizontal motions of the deck, and as many as 18 different aeroelastic derivatives, corresponding to all possible motion-induced forces. Many of these terms are small, however. The propensity of a bridge deck to flutter instability depends on the magnitudes and signs of some of the aeroelastic derivatives, or *flutter derivatives*, of the particular deck cross-section as a function of the wind speed. For example, a positive value of the derivative A_2 is an indication of flutter in a pure rotational motion – sometimes known as 'stall flutter' in aeronautical terminology. This can be seen from Equation 5.50 when the term $A_2\dot{\theta}$ is transposed to the left-hand-side of the equation – it then has the form of a negative damping term, with the ability to *extract* energy from the flow. If the magnitude of the negative aerodynamic is greater than the structural damping, then vibrations will grow in amplitude – i.e. an aeroelastic instability will occur.

The most commonly understood use of the term 'flutter' however is to describe the coupled translational–rotational form of instability, which is largely governed by the signs of the derivatives H_2 and A_1 (see Table 5.1).

Data on the flutter derivatives A_i to H_i are usually obtained experimentally from section tests in wind tunnels (see Section 7.6.3). Tests are usually done in smooth (low turbulence) flow – it

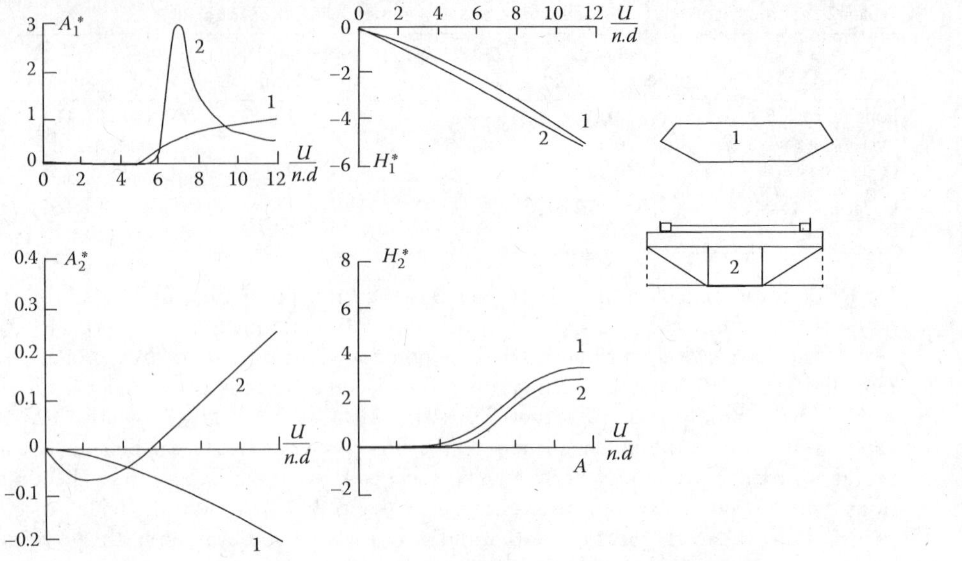

Figure 12.4 Aeroelastic derivatives for two types of bridge deck. (From Scanlan, R.H. and Tomko, J.J. 1971. *ASCE Journal of the Engineering Mechanics Division*, 97: 1717–37. With permission.)

has been found that the effects of turbulence on the derivatives are generally small (Scanlan and Lin, 1978). The derivatives are a function of reduced velocity, (*U*/*nd*) which incorporates the variation with frequency of vibration, *n*, as well as the wind speed, *U*. The following non-dimensional forms are usually used for the derivatives (Scanlan and Tomko, 1971).

$$\begin{aligned} H_1^* &= \frac{mH_1}{\rho_a d^2 \omega}; A_1^* = \frac{IA_1}{\rho_a d^3 \omega} \\ H_2^* &= \frac{mH_2}{\rho_a d^3 \omega}; A_2^* = \frac{IA_2}{\rho_a d^4 \omega} \\ H_3^* &= \frac{mH_3}{\rho_a d^3 \omega^2}; A_3^* = \frac{IA_3}{\rho_a d^4 \omega^2} \end{aligned} \tag{12.2}$$

where *m* and *I* are the mass and moment of inertia per unit length (spanwise), *d* is the width (chord) of the deck, ρ_a is the air density and ω is the circular frequency ($=2\pi n$).

Examples of aeroelastic derivatives determined for two common types of bridge deck – an open truss and a box-girder – are shown in Figure 12.4.

Although the magnitude and sign of the derivatives give some indication of the tendency of a particular section, in the design stage of important long-span bridges to aerodynamic instability, it is usual to attempt to determine a 'critical flutter speed' for the deck cross-section. If this wind speed does not exceed, by a substantial margin, the design wind speed of the site at the deck height (suitably factored for ultimate limit states), then modifications to the deck cross-section are usually made.

Several methods may be used to determine the critical flutter speed:

- Empirical formulae (e.g. Selberg, 1963)
- Experimental determination by use of section model testing

- Theoretical stability analysis of the equations of motion (Equations 5.46 and 5.47), with values of A_i and H_i obtained experimentally (e.g. Simiu and Scanlan, 1996)

Selberg (1961, 1963) proposed an empirical equation for critical flutter speed, U_F, which, in its simplest form, can be written as

$$U_F = 0.44d\sqrt{(\omega_T^2 - \omega_V^2)\frac{\sqrt{\upsilon}}{\mu}} \tag{12.3}$$

where $\upsilon = 8(r/d)^2$ and $\mu = \pi\rho_a d^2/2\,m$; r is the radius of gyration of the cross-section ($I = mr^2$); m is the mass per unit length.

ω_T ($=2\pi n_T$) and ω_V ($=2\pi n_V$) are the circular frequencies in the first torsional mode and first vertical bending modes, respectively.

Alternative ways of expressing the Selberg formula are as follows:

$$U_F = 0.417\omega_T\left(\frac{d}{2}\right)\sqrt{\left(1 - \frac{\omega_V^2}{\omega_T^2}\right)\frac{mr}{\rho_a (d/2)^3}} \tag{12.4}$$

where the half chord of the bridge deck ($d/2$) has been used

$$\frac{U_F}{n_T d} = 3.72\sqrt{\left(\frac{mr}{\rho_a d^3}\right)\left(1 - \frac{n_V{}^2}{n_T{}^2}\right)} \tag{12.5}$$

Figure 12.5 shows measured flutter speeds for several bridge deck sections compared with predictions from the Selberg formula. Reasonable agreement is obtained although there is an overestimation at low angles of attack. It would appear to be unwise to rely on a prediction based on an empirical formula alone.

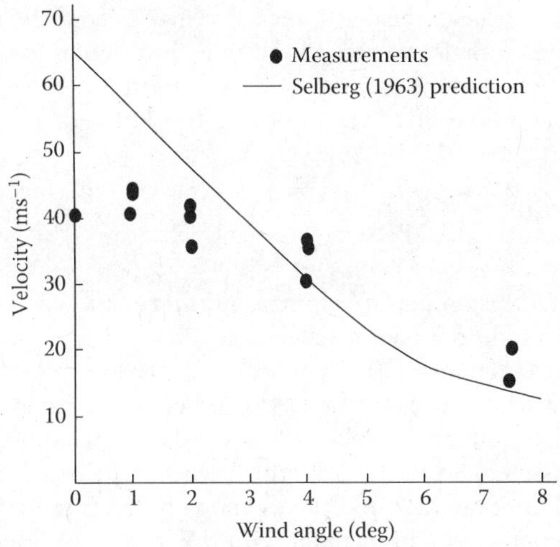

Figure 12.5 Measured critical flutter speeds and comparisons with the Selberg prediction formula. (From Wardlaw, R.L. 1971. Some approaches for improving the aerodynamic stability of bridge road decks. *3rd International Conference on Wind Effects on Buildings and Structures*, Tokyo, Japan, 6–9 September. Saikon Shuppan Co., Tokyo. With permission.)

The analytical estimation of flutter speeds is a specialist function of bridge aerodynamicists, but Ge and Tanaka (2000) have given a useful summary of the techniques available.

12.3.3 Buffeting of long-span bridges

A bridge that is otherwise stable in flutter up to a high wind speed, and does not suffer from vortex-induced vibrations at low wind speeds, will still experience dynamic response to atmospheric turbulence, known as *buffeting* over a wide range of wind speeds. This response will normally determine the size of the structural members and require evaluation at the design stage.

Davenport (1962) was the first to apply random vibration methods to the buffeting of a long-span suspension bridge. These methods were later validated by comparison with model studies in turbulent boundary layer flow in the 1970s (e.g. Holmes, 1975, 1979; Irwin, 1977).

The methodology described in Section 5.3.6 for the along-wind response of distributed mass structures can be adapted to the *cross*-wind response of bridge decks excited by vertical turbulence components.

The sectional cross-wind force per unit span can be written applying a 'strip' assumption:

$$f'(z,t) = \rho_a \bar{U} d\left[C_{Zo}u'(t) + \tfrac{1}{2}\frac{dC_Z}{d\alpha}w'(t)\right] \tag{12.6}$$

where

C_{Zo} is the vertical force coefficient at zero angle of attack
$dC_Z/d\alpha$ is the slope of the vertical force coefficient versus angle of attack, α
$u'(t)$ and $w'(t)$ are the horizontal and vertical velocity fluctuations upstream of the deck section in question

C_{Zo} and $dC_Z/d\alpha$ can be obtained from static section tests of the deck cross-section (Section 12.2). If there is significant angular rotation of the bridge deck under the mean wind load (as is often the case with suspension bridges), then C_{Zo} may need to be replaced by the value of C_Z at the mean (non-zero) angle of attack under the mean wind loading.

Following an argument similar to that used in Section 5.3.6, the spectral density of the generalised force for the jth mode of vibration can be obtained.

$$S_{Qj}(n) = (\rho_a \bar{U} d)^2\left[C_{Zo}^2 S_u(n) + \frac{1}{4}\left(\frac{dC_Z}{d\alpha}\right)^2 S_w(n)\right]X^2(n)\int_0^L\int_0^L \rho(y_1,y_2,n)\phi_j(y_1)\phi_j(y_2)dy_1dy_2 \tag{12.7}$$

In Equation 12.7, $X^2(n)$ is an *aerodynamic admittance*, allowing for the fact that smaller gusts (higher frequencies) do not completely envelope the bridge cross-section. Konishi et al. (1975), Shiraishi and Matsumoto (1977), Jancauśkas (1986) and others have directly measured this function for bridge deck sections and other bluff shapes. Note that this aerodynamic admittance which applies to vertical (cross-wind) aerodynamic forces is similar, but not identical, to that discussed in Section 5.3.1, which relates to along-wind forces and response.

Analysis based on Equation 12.7 and the methods of random vibration analysis outlined in Section 5.3.6 have given good agreement with the measured response on full aeroelastic wind-tunnel models (e.g. Holmes, 1975) and full-scale measurements (Melbourne, 1979). However, for large-span bridges, the towers and cables play important parts in the overall bridge response, and it is the practice to carry out full aeroelastic model studies in simulated turbulent boundary-layer flow, as described in Section 12.4.

12.3.4 Effective static load distributions

The method of equivalent static load distributions discussed in Section 5.4 can be applied to the response of bridges. In many cases of long-span bridges, the background response can be neglected in comparison to the resonant contributions. However, it is often the case that several modes are significant. The following approach (Holmes, 1999) gives the correct method of combining inertial force distributions from more than one resonant mode of vibration. This approach is consistent with the weighting factor method discussed in Section 5.4.5.

The mean square fluctuating value of a load effect, r, resulting from the resonant response in mode j, can be written as

$$\sigma_{rj}^2 = \alpha_j^2 \omega_j^4 \overline{a_j^2} \tag{12.8}$$

where the displacement response of the deck is written as

$$y\,(x, t) = \phi_j(x) \cdot a_j(t)$$

where

$\phi_j(x)$ is the mode shape
$a_j(t)$ is the modal coordinate for the jth mode
ω_j is the circular frequency in mode j ($=2\pi n_j$)
α_j is the integral

$$\int_0^L m(x) \cdot \phi_j(x) \cdot I(x) dx \tag{12.9}$$

where

$m(x)$ is the mass per unit length
$I(x)$ is the influence line for the load effect in question

The contribution of the load at each spanwise position to the load effect is the product of the inertial load on a small increment of span, centred at that position, $m(x) \cdot \omega_j^2 \cdot y(x,t) \cdot \delta x$, multiplied by the influence function, $I(x)$. Equation 12.8 is then obtained by integration of the contributions over the span, L, squaring and taking the mean value.

The *total* mean square fluctuating response is then obtained by summing the contributions from the N contributing modes.

$$\sigma_r^2 = \sum_j^N \alpha_j^2 \omega_j^4 \overline{a_j^2} \tag{12.10}$$

To obtain Equation 12.10, we have assumed that the modes are well separated, and hence the resulting responses can be assumed to be uncorrelated with each other.

The *envelope* of the combined dynamic loadings at each point along the span of a bridge can be obtained by taking the root sum of squares of the inertial loads from the contributing modes along the span, and adding to the mean loading. Thus,

$$f_{\text{env}}(x) = \bar{f}(x) \pm \left[\sum_j^N \left(m(x) \cdot \omega_j^2 \phi_j(x) \right)^2 \overline{a_j^2} \right]^{1/2} \tag{12.11}$$

where $\bar{f}(x)$ is the mean wind loading at x.

Note that the envelope is independent of the influence line $I(x)$ of the load effect. It represents the limits within which the effective static load distributions for all load effects must lie.

The contribution of each mode to the total static equivalent load corresponding to a peak load effect (e.g. a bending moment at any point along the span) depends on the shape of the influence line for that load effect. Thus, there is not a single static equivalent load. The weighting factor to be applied to obtain the contribution from mode j to the combined inertial load for a *root-mean-square* value of a given load effect, when a total of N modes contribute, is given by

$$W_j' = \frac{\alpha_j \omega_j^4 \overline{a_j^2}}{\left\{\sum_j^N \alpha_j^2 \omega_j^4 \overline{a_j^2}\right\}^{1/2}} \tag{12.12}$$

It can be demonstrated that Equation 12.12 will result in the correct mean square fluctuating response, as given by Equation 12.10.

The effective loading distribution for the root mean square fluctuating response, σ_r, obtained by summing over all modes is

$$p'_{\text{eff}}(x) = m(x)\sum_j^N W_j'\phi_j(x) \tag{12.13}$$

The total root mean square fluctuating response is then,

$$\sigma_r = \int_0^L p'_{\text{eff}}(x)\cdot I(x)\cdot dx = \int_0^L I(x)\cdot m(x)\sum_j^N W_j'\phi_j(x)\cdot dx$$

$$\frac{\sum_j^N \alpha_j^2 \omega_j^4 \overline{a_j^2}}{\left\{\sum_j^N \alpha_j^2 \omega_j^4 \overline{a_j^2}\right\}^{1/2}} = \left\{\sum_j^N \alpha_j^2 \omega_j^4 \overline{a_j^2}\right\}^{1/2}$$

which agrees with Equation 12.10.

The weighting factor for the contribution from mode j to the effective static loading for the *peak* (maximum or minimum) load effect, r, in a specified time period, T, can be written to a good approximation as

$$W_j = \frac{\left\{\sum_j^N \alpha_j^2 g_j^2 \omega_j^4 \overline{a_j^2}\right\}^{1/2} \alpha_j \omega_j^4 \overline{a_j^2}}{\sum_j^N \alpha_j^2 \omega_j^4 \overline{a_j^2}} \tag{12.14}$$

where g_j is an expected peak factor for the response in mode j.

Equation 12.14 can be obtained from Equation 12.12, as follows:

$$W_j = g_r W_j'$$

where g_r is the peak factor for the response, which can be approximated quite accurately by

$$g_r \cong \frac{\left\{\sum_j^N \alpha_j^2 g_j^2 \omega_j^4 \overline{a_j^2}\right\}^{1/2}}{\left\{\sum_j^N \alpha_j^2 \omega_j^4 \overline{a_j^2}\right\}^{1/2}} \tag{12.15}$$

This is a weighted average of the peak factors for the various modes.

When only one mode is significant, Equation 12.14 reduces to

$$W_j = g_j \omega_j^2 (\overline{a_j^2})^{1/2} \tag{12.16}$$

that is, simply the peak inertial force in the mode, j.

Note that Equation 12.16 is independent of α_j, and hence of the influence line $I(x)$.

The contribution to the total inertial loading from mode j at a given spanwise position is then given by the product of W_j with the mass/unit length, $m(x)$, and the mode shape at that position. The total effective static loading for the peak load effect, r, is then

$$f_{\text{eff}}(x) = \bar{f}(x) + m(x)\sum_j^N W_j \varphi_j(x) \tag{12.17}$$

The effective static loading depends on the influence line for r through the parameter α_j. Thus, the effective static loading will be different for load effects, for example, bending moments at different spanwise positions. If the influence line is symmetrical about the centre of the bridge, as for example that for the bending moment at centre span, then α_j will be zero for anti-symmetrical modes, that is, only symmetrical modes will contribute.

It should also be noted that since g_r from Equation 12.15 can be either positive or negative, the second term on the right-hand side of Equation 12.17 can also be either positive or negative, that is, it may add or subtract from the mean loading.

12.4 WIND-TUNNEL TECHNIQUES

The verification of aerodynamic stability and determination of response to wind of long-span bridges, for structural design, is still largely an experimental process, making use of modern wind-tunnel techniques. Some of the experimental techniques were discussed in Chapter 7 (Sections 7.6.3 and 7.6.4).

A full wind-tunnel test programme for a major long-span bridge might consist of all or some of the following phases.

- Section model tests to determine basic static aerodynamic force and moment coefficients (Section 12.2) for the deck section.
- Section model free or forced vibration tests to determine the aerodynamic or flutter derivatives (Sections 5.5.3 and 12.3.2).
- Section model tests in which the natural frequencies in vertical translation and rotation are scaled to match those of the prototype bridge, and critical flutter speeds are thence determined by slowly increasing the wind-tunnel speed (Section 7.6.3). This

Figure 12.6 A full aeroelastic model of a cable-stayed bridge in a construction stage.

may be done in both smooth (low turbulence) and turbulent flow. (An alternative method which better reproduces the mode shapes of the prototype bridge is the 'taut strip' method described in Section 7.6.3.)

- Scaled aeroelastic models of the completed bridge, that is, deck, towers, cables, tested in turbulent boundary-layer flow (Section 7.6.4). The multi-mode aeroelastic modelling scales the various parts of the bridge for elastic properties, mass (inertial), as well as geometric properties. Such tests are quite expensive, with much of the cost in the model design and manufacture.
- Scaled aeroelastic partial models of the bridge in various stages of erection. In most cases, the erection stages find a bridge in its most vulnerable state with respect to wind loading, with lower frequencies making them more prone to turbulent buffeting (Section 12.3.3) and lower flutter speeds, since flutter instabilities tend to occur at constant *reduced* velocity. The erection stage tests may include separate aeroelastic tests of the bridge towers as free-standing structures.

A complete series of tests as outlined above may require two or three different wind tunnels. The wind-tunnel testing of bridges tends to be a specialist activity for wind-tunnel laboratories, with few facilities being capable of carrying out all the above-listed tests. Some facilities restrict their involvement to section testing for bridge decks; others only carry out boundary-layer wind-tunnel tests. However, it should be noted that, to satisfactorily carry out aeroelastic tests on full models of the largest suspension bridges, a test section of at least 10 m width (e.g. Figure 12.6) is required. Few boundary layer wind tunnels are of this size.

12.5 VIBRATION OF BRIDGE CABLES

As the spans of cable-stayed bridges have increased and the cables themselves have become longer, cable vibration has become more of a problem. One of the more interesting excitation mechanisms, and until recently, least-understood ones, is the so-called 'rain–wind' vibration. In the following sections, the history of occurrences of this phenomenon, suggested excitation mechanisms, and methods of mitigating the vibrations are reviewed.

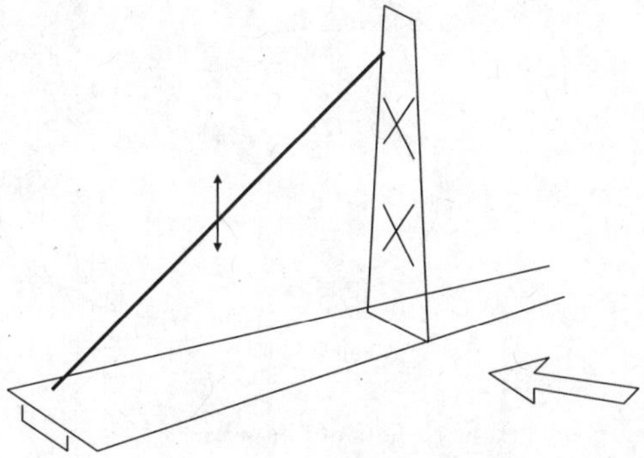

Figure 12.7 Typical cable/wind orientation for rain–wind vibration. (From Matsumoto, M. et al. 1993. Response characteristics of rain–wind induced vibration of stay cables of cable-stayed bridges. *Ist. European-African Regional Congress on Wind Engineering*, Guernsey, 20–24 September. With permission.)

12.5.1 Rain–wind vibration

The first, clearly defined, occurrence of wind-induced cable vibration, during which the presence of rain was an essential feature, was observed during the construction of Meiko-Nishi Bridge at Nagoya Harbour, Japan, in 1984. Low-frequency (1–3 Hz) vibrations of some cables, with double amplitudes up to 300 mm, were observed, over a five-month period. This bridge has a main span of 405 m with cables up to 165 mm in diameter, and lengths varying from 65 to 200 m. The vibrations occurred in wind speeds between 7 and 14 m/s; these speeds greatly exceeded the critical wind speeds for vortex shedding at the low frequencies observed. Using a section of polyethylene pipe casing from the prototype structure, wind-tunnel tests were conducted, with and without simulated rain, and it was clearly established that the rain was necessary to induce vibration over a defined range of wind speeds (Hikami and Shiraishi, 1988).

Later, it was found that rain–wind-induced vibration had occurred on six bridges in Japan. A common feature was that the vibrating cables were usually sloping downwards in a downwind direction, with the wind approaching obliquely to the plane of the cable (Figure 12.7). Vibrations were apparently observed only for cables encased in polyethylene.

Outside Japan, rain–wind vibration of bridge cables have been observed on the Faroe Bridge (Denmark), Bretonne Bridge (France), Koehlbrand Bridge (Germany), Normandie Bridge (France), Fred Hartman Bridge (USA) and the Anzac Bridge (Australia). Many other bridges have experienced cable vibrations – some from different mechanisms such as high-frequency vortex-shedding excitation, or from unknown or undefined mechanisms.

12.5.2 Excitation mechanisms

The wind-tunnel studies carried out following the vibrations observed on the Meiko-Nishi Bridge indicated that the motion was induced by the presence of two water 'rivulets', that oscillated in circumferential position with the cable motion. At low wind speeds, a single rivulet formed on the underside. Motion commenced at higher wind speeds when a second rivulet formed on the upper surface. The rivulets act as trigger points to promote flow separation on the vibrating cable, as shown in Figure 12.8. In this figure, the effective cross-wind shape is postulated to be elliptical. Other observations have suggested that the

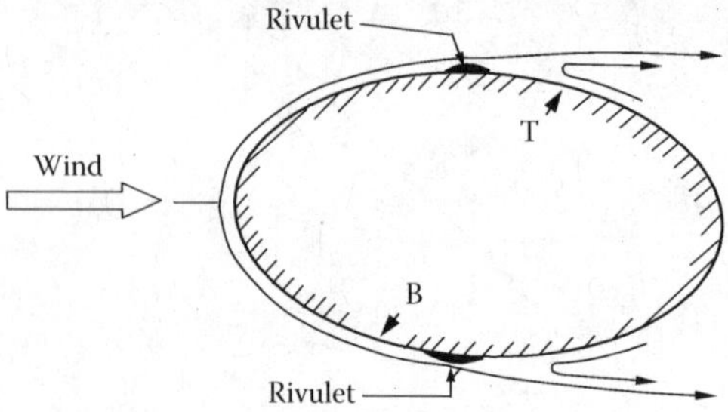

Figure 12.8 Flow separations produced by rivulets of rain water.

circumferential motion was not two-dimensional, and that the width and depth of the rivulet on the upper surface was less than that on the lower surface.

Wind-tunnel tests in France for the Normandie Bridge (Flamand, 1994) showed that carbon combustion products deposited on the surface of the casing were necessary for aerodynamic instability to occur, indicating the role played by surface tension in allowing the water rivulet to be maintained.

Fundamental wind-tunnel model studies of inclined cable aerodynamics, with and without rain, have been made at various angles of pitch (inclination), yaw, and rivulet position (Matsumoto et al., 1993). Severe vibrations usually commenced at a reduced wind velocity of ($U/n_c\, b$), (where U is the wind velocity, n_c is the cable frequency and b is the diameter) of about 40.

Matsumoto (2012) summarised more than 25 years of Japanese research into the excitation mechanisms for stay cable vibration. He concluded that inclined cable vibration is a form of galloping instability, but not one that can be explained by the 'quasi-steady' theory of Section 5.5.2. He suggested that rivulet formation, axial flow along the cable in the wake, and Reynolds number variations were all contributing factors to the 'unsteady galloping' mechanism.

The criterion of the Federal Highway Administration (FHWA, 2007) suggests that if a cable has sufficient damping so that it has a Scruton number (defined according to Equation 11.14) of 125 or greater, then rain–wind vibrations will not occur. If this is not possible to achieve, then other mitigation options may be adopted, as discussed in the following section.

12.5.3 Solutions

The solutions that have been successful in eliminating, or mitigating rain–wind-induced vibration of bridge cables, can be divided into the following categories:

- Aerodynamic treatments – that is, geometrical modifications of the outer cable casing
- Auxiliary cable ties
- Auxiliary dampers

Model measurements were carried out by Miyata et al. (1994), on sections of cable models with the same diameter as full-size cables, with a variety of roughened surface treatments (Figure 12.9). Discrete roughness, of about 1% of the diameter, was found to be effective in suppressing rain–wind-induced vibration. The explanation was that supercritical flow was promoted at lower Reynolds numbers than would occur on cables with smooth surface finish.

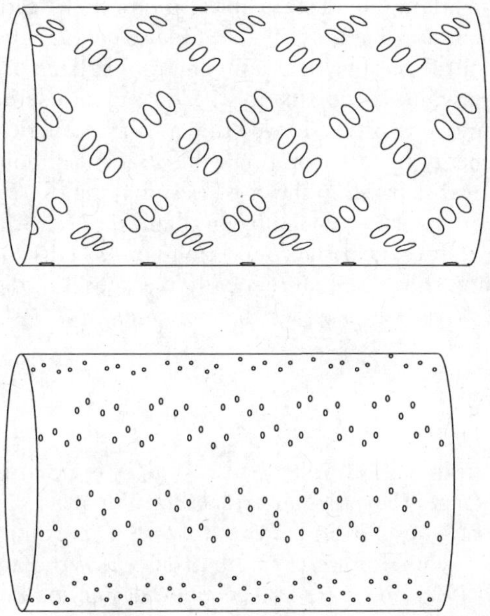

Figure 12.9 Surface roughness treatments for cable vibration mitigation. (From Miyata, T., Yamada, H. and Hojo, T. 1994. Aerodynamic response of PE stay cables with pattern indented surface. *International Conference on Cable-Stayed and Suspension Bridges*, Deauville, France, 12–15 October. With permission)

Wind-tunnel tests in France (Flamand, 1994) found that parallel surface projections did stabilise a cable model, but produced a high drag coefficient in the super-critical Reynolds number range. An alternative solution that minimised the drag increase was adopted, namely the use of a double-helix spiral, 1.3 mm high, 2 mm wide, and with a pitch of 0.6 m. This configuration was adopted for the Normandie Bridge.

Usually only one- or two-stay cables from a harp or fan array will experience rain–wind vibration in particular atmospheric conditions. This observation led to a solution that has been used on several bridges – cable cross-ties. They have also been used on the Normandie Bridge, where they are known as 'aiguilles'. They have been adopted for the Dane Point Bridge, Florida, (USA), the Fred Hartman Bridge, Texas, (USA), and the Tatara Bridge, Japan (Figure 12.10).

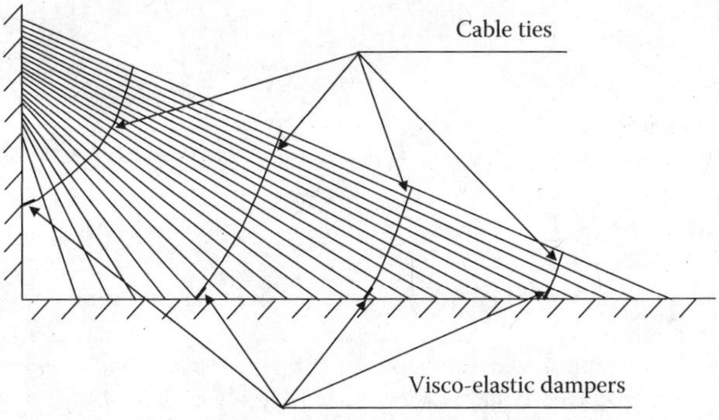

Figure 12.10 Cable ties for vibration mitigation used on the Tatara Bridge.

A fundamental study of damping in stay cables, and of the effectiveness of cross ties, was carried out by Yamaguchi and Fujino (1994). Measurements on cables of a typical cable-stayed bridge indicated a range of critical damping ratios, from about 0.001 to 0.003, for the first mode, with lower values occurring for the low sag ratios, that is, a higher pre-stress. A laboratory experiment on cross ties showed that a 'stiff' cross tie performed a function of transferring vibration energy from a vibrating cable to its neighbours. By use of 'soft' cross ties, energy could also be dissipated in the cross ties, making this system more effective.

Energy dissipation can also be provided by auxiliary damping devices mounted between the cable and the bridge girder, near the connection points. This solution is more expensive than the cross-tie method, but more aesthetically pleasing. Oil dampers and visco-elastic dampers (Section 9.9.2) have been used for this purpose.

12.6 CASE STUDIES

The literature on the aerodynamics of long-span bridges is extensive, and many papers on the subject contain references to particular bridges for illustration purposes. Sections 12.3.1 and 12.5 contain several examples in relation to vortex-shedding-induced vibrations and cable vibrations, respectively. Holmes (1999) has described the application of the equivalent static load method (Section 12.3.4) to generate design loadings for the Baram River (Malaysia) cable-stayed bridge.

The extensive wind engineering studies carried out for two of the longest bridges in the world are described by Miyata et al. (1992) for the Akashi Kaikyo Bridge, and by Reinhold et al. (1992) and Larsen and Jacobsen (1992) for the Great Belt East Bridge (Denmark). The wind design of the Normandie Bridge from the designer's point of view is well covered by Virlogeux (1992).

12.7 SUMMARY

In this chapter, the aerodynamics of bridges have been presented in a summary form. Long-span bridges are probably the most 'wind-sensitive' of all structures, and their aerodynamics are complex and the sphere of specialists. The main phenomena of vortex shedding, flutter and buffeting have been discussed.

The vibration of the cables on cable-stayed bridges has become the limiting factor on their ultimate spans, and this topic, with alleviation measures, has been discussed in some detail.

REFERENCES

Baker, B. 1884. The Forth Bridge. *Engineering*, 38: 213.

Billington, D.P. 1977. History and esthetics in suspension bridges. *ASCE Journal of the Structural Division*, 103: 1655–72.

Davenport, A.G. 1962. Buffeting of a suspension bridge by storm winds. *ASCE Journal of the Structural Division*, 88: 233–68.

Federal Highway Administration (FHWA). 2007. Wind-induced vibration of stay cables. TechBrief. FHWA-HRT-05-084, July.

Flamand, O. 1994. Rain–wind induced vibrations of cables. *International Conference on Cable-Stayed and Suspension Bridges*, Deauville, France, 12–15 October.

Frandsen, J.B. 2001. Simultaneous pressures and accelerations measured full-scale on the Great Belt East suspension bridge. *Journal of Wind Engineering and Industrial Aerodynamics*, 89: 95–129.

Ge, Y.J. and Tanaka, H. 2000. Aerodynamic flutter analysis of cable-supported bridges by multi-mode and full-mode approaches. *Journal of Wind Engineering and Industrial Aerodynamics*, 86: 123–53.

Gimsing, N.J. 1983. *Cable-Supported Bridges*. John Wiley, New York, USA.

Hikami, Y. and Shiraishi, N. 1988. Rain–wind induced vibrations of cables in cable-stayed bridges. *Journal of Wind Engineering and Industrial Aerodynamics*, 29: 409–18.

Holmes, J.D. 1975. Prediction of the response of a cable-stayed bridge to turbulence. *Proceedings, 4th International Conference on Wind Effects on Buildings and Structures*, London, 8–12 September, pp. 187–197. Cambridge University Press, Cambridge, UK.

Holmes, J.D. 1979. Monte Carlo simulation of the wind-induced response of a cable-stayed bridge. *Proceedings, 3rd International Conference on Applications of Statistics and Probability in Soil and Structural Engineering (ICASP-3)*, University of New South Wales, Sydney, pp. 551–565.

Holmes, J.D. 1999. Equivalent static load distributions for resonant dynamic response of bridges. *Proceedings, 10th International Conference on Wind Engineering*, Copenhagen, 21–24 June, pp. 907–911. A.A. Balkema, Rotterdam.

Irwin, H.P.A.H. 1977. Wind tunnel and analytical investigations of the response of Lions' Gate Bridge to a turbulent wind. National Aeronautical Establishment, Canada. Laboratory Technical Report, LTR-LA-210, June 1977.

Jancauskas, E.D. 1986. The aerodynamic admittance of two-dimensional rectangular section cylinders in smooth flow. *Journal of Wind Engineering & Industrial Aerodynamics*, 23: 395 – 408.

Kernot, W.C. 1893. Wind pressure. *Proceedings, Australasian Society for the Advancement of Science*, V: 573–81.

Konishi, I., Shiraishi, N. and Matsumoto, M. 1975. Aerodynamic response characteristics of bridge structures. *Proceedings, 4th Internal Conference on Wind Effects on Buildings and Structures*, London, 8–12 September, pp. 199–208. Cambridge University Press, Cambridge, UK.

Larsen, A. (ed.) 1992. Aerodynamics of large bridges. *Proceedings of the First International Symposium on Aerodynamics of Large Bridges*, Copenhagen, Denmark, 19–21 February. A.A. Balkema, Rotterdam.

Larsen, A. and Jacobsen, A.S. 1992. Aerodynamic design of the Great Belt East Bridge. *Proceedings of the First International Symposium on Aerodynamics of Large Bridges*, Copenhagen, Denmark, 19–21 February, pp. 269–283. A.A. Balkema, Rotterdam.

Larsen, A., Esdahl, S., Andersen, J.E. and Vejrum, T. 1999. Vortex shedding excitation of the Great Belt suspension bridge. *Proceedings, 10th International Conference on Wind Engineering*, Copenhagen, 21–24 June, pp. 947–954. A.A. Balkema, Rotterdam.

Matsumoto, M., Shirato, H., Saito, H., Kitizawa, H. and Nishizaki, T. 1993. Response characteristics of rain–wind induced vibration of stay cables of cable-stayed bridges. *1st European–African Regional Congress on Wind Engineering*, Guernsey, 20–24 September.

Matsumoto, M. 2012. On galloping instability of stay cables of cable-stayed bridges. *2012 World Congress on Advances in Civil, Environmental and Materials Research*, Seoul, Korea, 26–30 August.

Melbourne, W.H. 1979. Model and full-scale response to wind action of the cable-stayed, box-girder, West Gate Bridge. *IAHR/IUTAM Symposium on Flow-Induced Vibrations*, Karlsruhe, Germany, 3–8 September.

Miyata, T., Yamada, H. and Hojo, T. 1994. Aerodynamic response of PE stay cables with pattern indented surface. *International Conference on Cable-stayed and Suspension Bridges*, Deauville, France, 12–15 October.

Miyata, T., Yokoyama, Y., Yasuda, M. and Hikami, Y. 1992. Akashi Kaikyo Bridge: Wind effects and full model tests. *Proceedings of the First International Symposium on Aerodynamics of Large Bridges*, Copenhagen, Denmark, 19–21 February 1992, pp. 217–236. A.A. Balkema, Rotterdam.

Ostenfeld, K.H. and Larsen, A. 1992. Bridge engineering and aerodynamics. *Proceedings of the First International Symposium on Aerodynamics of Large Bridges*, Copenhagen, Denmark, 19–21 February 1992, pp. 3–22. A.A. Balkema, Rotterdam.

Petroski, H. 1996. *Engineers of Dreams*. Vintage Books, New York.

Reinhold, T.A., Brinch, M. and Damsgaard, A. 1992. Wind-tunnel tests for the Great Belt link. *Proceedings of the First International Symposium on Aerodynamics of Large Bridges*, Copenhagen, Denmark, 19–21 February 1992, pp. 255–267. A.A. Balkema, Rotterdam.

Sabzevari, A. and Scanlan, R.H. 1968. Aerodynamic instability of suspension bridges. *ASCE Journal of the Engineering Mechanics Division*, 94: 489–519.

Selberg, A. 1961. Aerodynamic effects on suspension bridges. *Acta Polytechnica Scandinavica*, Engineering and Building Construction Series No. 13, Trondheim, Norway.

Selberg, A. 1963. Aerodynamic effects on suspension bridges. *Proceedings, International Conference on Wind Effects on Buildings and Structures*, Teddington, UK, 26–28 June, pp. 462–486.

Scanlan, R.H. and Gade, R.H. 1977. Motion of suspended bridge spans under gusty winds. *ASCE Journal of the Structural Division*, 103: 1867–83.

Scanlan, R.H. and Lin, W.-H. 1978. Effects of turbulence on bridge flutter derivatives. *ASCE Journal of the Engineering Mechanics Division*, 104: 719–33.

Scanlan, R.H. and Tomko, J.J. 1971. Airfoil and bridge flutter derivatives. *ASCE Journal of the Engineering Mechanics Division*, 97: 1717–37.

Shiraishi, N. and Matsumoto, M. 1977. Aerodynamic responses of bridge structures subjected to strong winds. *Symposium on Engineering for Natural Hazards*, Manila, September.

Simiu, E. and Scanlan, R.H. 1996. *Wind Effects on Structures – Fundamentals and Applications to Design*. 3rd edition. John Wiley, New York.

Smith, I.J. 1980. Wind induced dynamic response of the Wye Bridge. *Engineering Structures*, 2: 202–8.

Steinman, D.B. and Watson, S.R. 1957. *Bridges and Their Builders*. Dover Publications Inc., New York.

van Nunen, J.W.G. and Persoon, A.J. 1982. Investigation of the vibrational behaviour of a cable-stayed bridge under wind loads. *Engineering Structures*, 4: 99–105.

Virlogeux, M. 1992. Wind design and analysis for the Normandy Bridge. *Proceedings of the First International Symposium on Aerodynamics of Large Bridges*, Copenhagen, Denmark, 19–21 February, 1992, pp. 183–216. A.A. Balkema, Rotterdam.

Virlogeux, M. 1999. Recent evolution of cable-stayed bridges. *Engineering Structures*, 21: 737–55.

Wardlaw, R.L. 1971. Some approaches for improving the aerodynamic stability of bridge road decks. *3rd International Conference on Wind Effects on Buildings and Structures*, Tokyo, Japan, 6–9 September. Saikon Shuppan Co., Tokyo.

Yamaguchi, H. and Fujino, T. 1994. Damping in cables in cable-stayed bridges with and without damping control measures. *International Conference on Cable-Stayed and Suspension Bridges*, Deauville, France, 12–15 October.

Chapter 13

Transmission lines

13.1 INTRODUCTION

Electrical transmission lines and their supporting towers are, like other structures, subjected to severe wind storms of various types, and their safe and economic design for wind loading is of concern to the power utilities. There are significant differences between the response of high-voltage transmission towers and other structures to wind:

- They are structurally designed with generally lower safety margins against collapse than other structures.
- The overall length of a transmission line system is relevant when considering probability and risk of receiving strong winds from localised wind storms such as thunderstorm downbursts and tornadoes.

This chapter deals with the wind loading of the transmission lines themselves, and risk issues associated with a long transmission line as a system. The wind loading of the supporting towers and poles is covered elsewhere in the book, in particular, Chapter 11.

13.2 STRUCTURAL RESPONSE AND CALCULATION OF WIND LOADS

The basic design data for wind loads on transmission line conductors in temperate synoptic winds has been compiled by the American Society for Civil Engineers (ASCE, 2010), and CSIR in South Africa (CSIR, 1990).

13.2.1 Nature of the response

Fortunately, resonant dynamic response does not appear to be a major problem with transmission line systems. Although the suspended lines themselves usually have natural frequencies less than 1 Hz, the resonant response is largely damped out because of the very large aerodynamic damping (Section 5.5.1) (e.g. Matheson and Holmes, 1981).

The natural frequencies of supporting towers up to 50 m in height are normally greater than 1 Hz, and hence the resonant response is also negligible. Thus, except for extremely tall supporting towers and long line spans, we can safely compute the peak response of a transmission line system, neglecting the resonant dynamic response. Then the peak response is directly related to the instantaneous gusts upwind, and hence transmission line structures can be designed using gust wind speeds. However, because of the non-uniform spatial gust structure, assumption of the same peak gust along the full span is conservative; this leads to the concept of a *span reduction factor.*

For those cases where resonant response is significant, such as very high supporting towers, and very long spans, a simplified random response model of the tower-line combination, based on the gust response factor concept is available (Davenport, 1979).

13.2.2 Wind forces on conductors

The nominal wind force acting on a single conductor perpendicular to the span can be taken to be

$$F_c = q_{zc} \cdot C_d \cdot A_c \sin^2\theta \cdot \alpha \tag{13.1}$$

where q_{zc} is the free-stream dynamic wind pressure (= $(1/2)\rho_a \hat{U}_{zc}^2$) at a suitable mean conductor height, z_c. A suitable value for z_c is shown in Figure 13.1, taken from the South African recommendations for transmission line loading (CSIR, 1990).

C_d is the drag force coefficient for the conductor
A_c is the reference area, which may be taken as $s \times b$, where s is the wind span (see Figure 13.1), and b is the conductor diameter
θ is the horizontal angle of incidence of the wind in relation to the direction of the line
α is a span reduction factor

The ASCE Guidelines show experimental data for the drag force coefficient as a function of Reynolds number, Re (Section 4.2.4) for several conductor types, based on wind tunnel tests. These data are reproduced in Figure 13.2. The Reynolds number can be calculated by

$$\text{Re} = \frac{U_{zc} b}{15 \times 10^{-6}} \tag{13.2}$$

where U_{zc} is the design gust wind speed in metres per second at the mean conductor height, z_c. The conductor diameter, b, is in metres.

The South African design recommendations (CSIR, 1990) have simplified the data to give the design line shown in Figure 13.3.

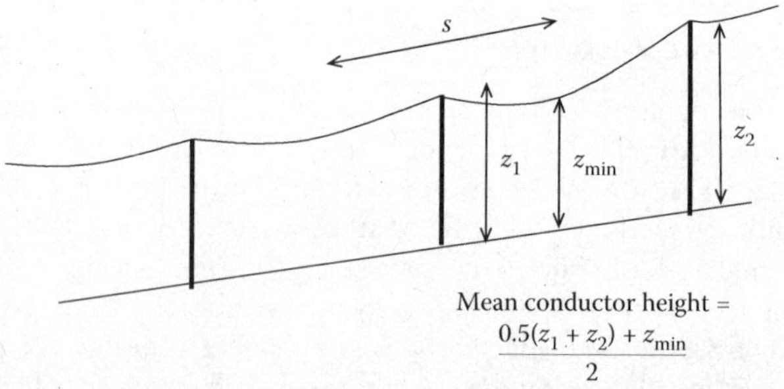

Figure 13.1 Mean conductor height for calculation of wind loads. (From CSIR. 1990. Transmission Line Loading. Part I: Recommendations and Commentary. Part II: Appendices. Engineering Structures Programme. CSIR Building Technology, South Africa. With permission.)

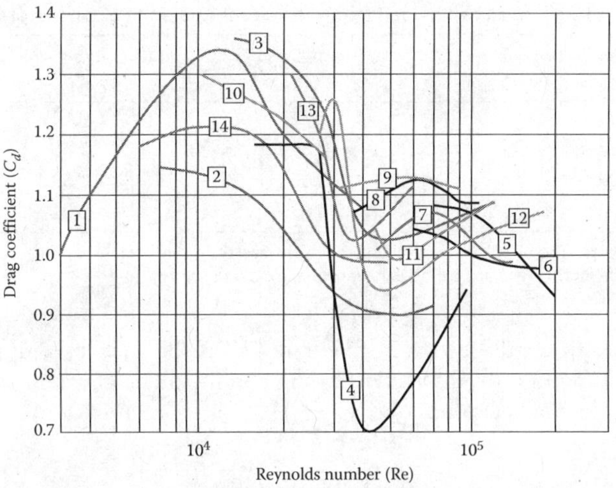

Figure 13.2 Drag force coefficients on conductors from wind-tunnel tests. (From American Society of Civil Engineers 2010. *Guidelines for Electrical Transmission Line Structural Loading*. 3rd edition. American Society of Civil Engineers, ASCE, Reston, VA, USA; CSIR. 1990. Transmission Line Loading. Part I: Recommendations and Commentary. Part II: Appendices. Engineering Structures Programme. CSIR Building Technology, South Africa. With permission.)

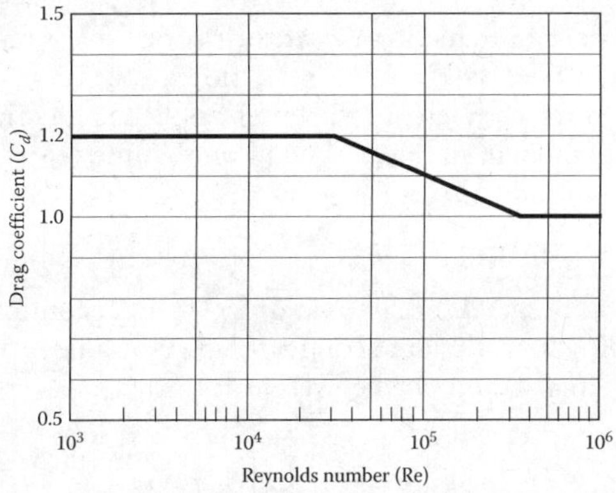

Figure 13.3 Design values of conductor drag coefficient. (Recommended by CSIR 1990. Transmission Line Loading. Part I: Recommendations and Commentary. Part II: Appendices. Engineering Structures Programme. CSIR Building Technology, South Africa.)

13.2.3 Span reduction factor

The span reduction factor, α, allows for the reduction in peak wind along the span of a conductor, due to the non-simultaneous action of the gusts. Since it is determined by the structure of turbulence in the approaching wind flow, the span reduction factor is a function of the approach terrain, the mean conductor height and the span. This factor has a direct relationship with the gust response factor, G (Section 5.3.2). The relationship is as follows:

$$\alpha = G\left(\frac{\bar{U}_z}{\hat{U}_z}\right)^2 \tag{13.3}$$

Table 13.1 Span reduction factors for trransmission line conductors

Conductor span (m)	*Rural terrain ($z_0 \cong 0.02$ m)*	*Urban terrain ($z_0 \cong 0.2$ m)*
200	0.72	0.62
300	0.66	0.56
400	0.63	0.53
500	0.61	0.51

Note: The above values are applicable to large-scale synoptic winds. Span reduction factors in thunderstorm downbursts are much greater.

where $\bar{U}_z$ is the mean wind speed at height z, and $\hat{U}_z$ is the gust speed at the same height.

Span reduction factors are insensitive to the conductor height, and the following equations can be used to predict values of α.

$$\alpha = 0.58 + 0.42\exp\left(\frac{-s}{180}\right) \quad \text{for rural terrain} \tag{13.4}$$

$$\alpha = 0.50 + 0.50\exp\left(\frac{-s}{140}\right) \quad \text{for urban terrain} \tag{13.5}$$

where s is the span in metres

In Table 13.1, values of span reduction factor for various spans have been calculated using Equations 13.4 and 13.5. Clearly, the span reduction factor reduces with increasing span, and with increasing terrain roughness. In the latter case, the reduction occurs because of the increased fluctuating component in the peak load on the line.

13.2.4 Conductor shielding

In both the ASCE Guidelines (ASCE, 2010) and the CSIR Recommendations (CSIR, 1990), no allowance for shielding for individual conductors in a bundle is permitted. Such shielding effects would be small, and would not be present for every angle of attack of the instantaneous wind to the line.

13.2.5 Wind forces on lattice supporting towers

The calculation of wind forces on lattice towers typical of those used in high-voltage transmission line system is discussed in Section 11.3.2. The overall drag coefficients for lattice towers depends upon the solidity of the towers. Higher solidity results in greater mutual interference and shielding, and a reduction in drag coefficient, based on the projected area of members.

13.3 RISK MODELS FOR TRANSMISSION LINE SYSTEMS

Transmission line systems often extend for several hundred kilometres, and are prone to impact by small intense local windstorms, such as tornadoes (Section 1.3.4) and downbursts (Section 1.3.5). There has been a history of failures of transmission line systems from these events – especially in large continental countries such as Australia, Brazil and Argentina

Figure 13.4 Failure of a high-voltage transmission tower following a local downburst event.

(e.g. Hawes and Dempsey, 1993). Figure 13.4 shows the result of one such event. The risk of failure of *any one tower* along a line is much greater than that for a single isolated structure. Design of the supporting structures requires knowledge of the total risk of the complete line to these small intense windstorms. Knowledge of the risk of failures enables a balance to be made between the cost of failures, and the cost of replacement towers. This may vary from country to country, as in some cases there are alternative routes for power transmission.

13.3.1 Tornado risk model

Twisdale and Dunn (1983) describe several tornado risk models for point and 'lifeline' targets. Milford and Goliger (1997) developed a tornado risk model for transmission line design which considered normal intersection of a tornado with the line direction. A simplified probabilistic model is discussed in this section, considering both normal and oblique intersections of tornado tracks with the line target, representing a transmission line.

Since the width of tornado tracks (usually <100 m) is almost always less than the span length between towers, the critical factor in line failure is intersection of a tornado with a tower. Thus, the rate of intersection with a tower is required, rather than with the conductors.

Consider a region specified by its area, A (square kilometres), in which there is an average tornado occurrence of n events per year, so that the per square kilometre rate for the region is

$$\upsilon = n/A \tag{13.6}$$

Normal intersection of a tornado path of length ℓ, and width, w, with a line of overall length L occurs only for those tracks whose centre falls within the zone of area, $L \times \ell$, adjacent to the line (see Figure 13.5), giving a rate of intersection, r,

$$r = \upsilon\, L\, \ell \tag{13.7}$$

This model can be extended to variable intersection angle as follows.

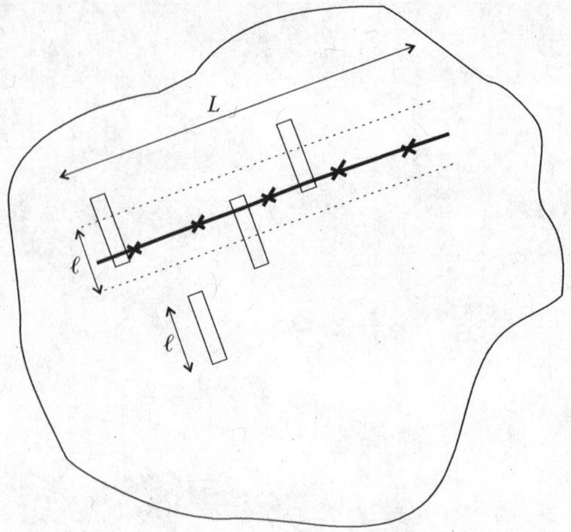

Figure 13.5 Normal intersection of a tornado with a transmission line system.

For a tornado path intersecting the transmission line at an angle β to normal (Figure 13.6), the width of the zone of intersection reduces to $\ell \cos \beta$ and the rate of intersection (with the line) per annum is now given by

$$r = \upsilon L \ell \cos \beta \tag{13.8}$$

Now, the width of the intersection zone along the line is given by $w/\cos \beta$ and the probability of a *given single point* on the line falling within this zone is $w/(L \cos \beta)$, which may represent a single tower. Thus, the number of intersections of tornadoes with this tower per year is given by

$$\begin{aligned} r &= \upsilon L \ell \cos \beta \cdot w/(L \cos \beta) \\ &= \upsilon w \ell \end{aligned} \tag{13.9}$$

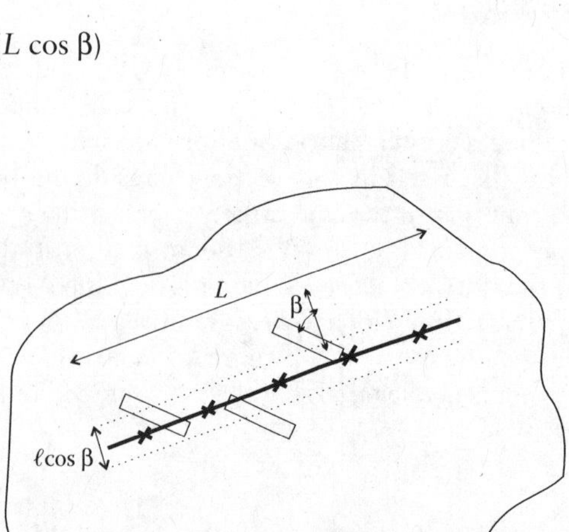

Figure 13.6 Oblique intersection of a tornado with a transmission line system.

If the span length between towers is s, the number of towers along a line of length, L, is equal to L/s, and assuming that intersections are independent, (i.e. only one tower is intersected by any tornado), then the total number of intersections with *any* tower along the line per year is given by

$$r_t = (\upsilon\, w\, \ell)\,(L/s) \tag{13.10}$$

It should be noted that the rate of intersection is independent of the intersection angle, β. Equation 13.10 may also be written as

$$r_t = n\,(a/A)\,N = \upsilon\, a\, N \tag{13.11}$$

where n is the number of events per year in an area A, a is the area of tornado path and N is the number of towers in the area.

EXAMPLE

Assume

$L = 500$ km; $s = 0.5$ km; $\ell = 5$ km; $w = 0.1$ km; $\upsilon = 10^{-4}$ per km^2 per year

Then, from Equation 13.10, the number of intersections with this line per year = $10^{-4} \times 0.1 \times 5 \times (500/0.5) = 0.05$, i.e. average of 1 intersection every 20 years.

13.3.2 Downburst risk model

Damage 'footprints' produced by severe thunderstorm downbursts (Section 1.3.5) are usually wider than those produced by tornadoes. The lengths of the damaged areas produced by downbursts are generally shorter than those of tornadoes, however. The increased width usually results in several transmission line spans being enveloped by damaging winds, and several adjacent towers often fail as a group. The direct wind load on the conductors themselves is therefore a significant component of the overall wind load in downburst events. This must be incorporated into a risk model.

Oliver et al. (2000) describe a downburst risk model for transmission lines, which allows the prediction of an event frequency, where an event is the intersection of a region of wind above a given or design wind speed with a line of some defined length. The probability of such an event is dependent on:

- The overall length of the line, L
- The relative angle, $\theta-\varphi$, between the direction of the downburst path, θ, and the line orientation, φ
- The probability of exceedence of the threshold wind speed of interest, U, at any point in the surrounding region, derived from the anemometer records
- The width of the path of winds above the threshold, w_u

The return period, $R_{U,L}$, of the event was shown (Oliver et al., 2000) to be given by

$$R_{U,L} = (w_u/L)\Bigg/\left\{\sum_{i=1}^{N} Pr(u > U/|\sin(\theta_i - \varphi)|)\cdot Pr(\theta_i)|\sin(\theta_i - \varphi)|\right\} \tag{13.12}$$

where it is assumed that:

- There is an average or characteristic downburst damage footprint width associated with each wind speed U, given by w_u.
- For each direction, all downburst tracks can be represented in discrete directional ranges, centred on a characteristic direction θ_i and the summation is over each of these directions.
- The relative probability that the downburst should lie along each of these directions is directly related to the directional frequency of measured gusts.
- The distribution of wind speed, given a direction, is independent of the directional sector.

The presence of the overall line length, L, in the denominator of Equation 13.12 indicates that, as the overall transmission line length increases, the return period for damaging intersections decreases. Thus, for very long lines orientated at right angles to the prevailing directions of severe thunderstorm winds, the risk of failure may be very high, if these parameters have not been taken into account of in design. This is the experience in large continental countries such as Australia and Argentina, where many failures have occurred (e.g. Hawes and Dempsey, 1993).

An alternative model of downburst risk for transmission line systems has been developed for Argentina by de Schwarzkopf and Rosso (2001).

13.4 WIND-INDUCED VIBRATIONS OF TRANSMISSION LINES

Wind can induce vibrations in transmission lines through several different mechanisms, and over a range of frequencies and wind speeds. This topic is covered in specialist publications, usually originating within the electrical transmission line industry, and coverage in this book is restricted to summaries in the following sections.

13.4.1 Vortex-induced 'aeolian' vibration

High-frequency cross-wind vibrations occur on transmission lines under certain atmospheric conditions, and were first observed as early as the 1920s. Early experiments (e.g. Varney, 1926) identified the exciting mechanism to be the regular shedding of vortices (see Section 4.6.3). It was observed that the vibrations consisted of travelling waves that are reflected at the points of support. Fatigue failures usually occurred at or near the point of support, as a result of local flexural stresses. It was also observed that vibrations occurred in relatively low wind speeds, and in conditions of low turbulence. Hence, significant vibrations are only observed on lines located in flat, open country, where the turbulence intensities (Section 3.3.1) are low. Several types of damping devices to mitigate this vibration have been devised over the years, and are widely used in the industry.

'Aelioan' vibration of transmission lines is covered in detail elsewhere (e.g. EPRI, 1979 and 2009). However, the main features of the phenomenon, and mitigation measures, will be discussed here.

The main conditions for significant aeolian vibrations of transmission lines are as follows:

- Wind direction near normal to the line.
- Mean wind speeds less than about 7 m/s.
- Low turbulence intensities – this may occur in thermally stable atmospheric conditions in which turbulence in the wind flow is likely to be suppressed.

Figure 13.7 Spiral damper for mitigating aeolian (vortex-induced) vibrations of transmission line conductors.

Some characteristics of the vibration are:

- Excitation over short lengths of span by vortex shedding at high frequency. The Strouhal number for the vortex shedding (Section 4.6.3) for typical transmission line cross sections is in the range of 0.18–0.19.
- 'Lock-in' (Section 5.5.4) may occur resulting in the vortex-shedding synchronising with the vibration frequency.
- Amplitudes of vibrations are usually <1 diameter.
- Frequencies of vibrations are typically in the range of 10–150 Hz.
- Fatigue damage is typically initiated near supports by fretting or rubbing between strands of a conductor.
- Vibration amplitudes are enhanced by increasing line temperatures and reduced temperatures.

Various types of damper have been devised to mitigate aeolian vibrations of transmission lines. These can broadly be categorised into impact dampers, and tuned mass dampers. An example of the former is the spiral type, which consists of a helically wound spiral bar of length 1.2–1.5 m. The damper is attached to a conductor by a short 'gripping' section close to one end of a span. The cross-wind motion of the line is mitigated by impacts on the loosely wound part of the damper, which has a diameter of about twice the diameter of the conductor (Figure 13.7).

Mass dampers for transmission lines come in several forms – but notably the 'Stockbridge' or 'dog-bone' type (Figure 13.8). These are nominally 'tuned' mass dampers (Section 9.9.3), but in fact operate over a wide range of frequencies. Their performance is generally determined by experiment rather than by theoretical means.

13.4.2 'Galloping' vibrations

A common phenomenon in cold climates is the 'galloping' of iced-up conductors – low frequency, generally cross-wind, vibrations of large amplitude, produced by the 'galloping' instability mechanism discussed in Section 5.5.2. The presence of ice may change the shape of the cross section to an asymmetric one with aerodynamic properties that make it prone to galloping instability. Galloping vibrations are in the lower modes of vibration – with a typical frequency range of 0.1–1 Hz. Large deflection amplitudes of up to 10 m can occur. This can result in line failure with a few hours from the onset of vibration.

Several devices designed to reduce or eliminate large-amplitude galloping vibrations were described by Richardson (1991). These generally act to increase the drag coefficient, C_D,

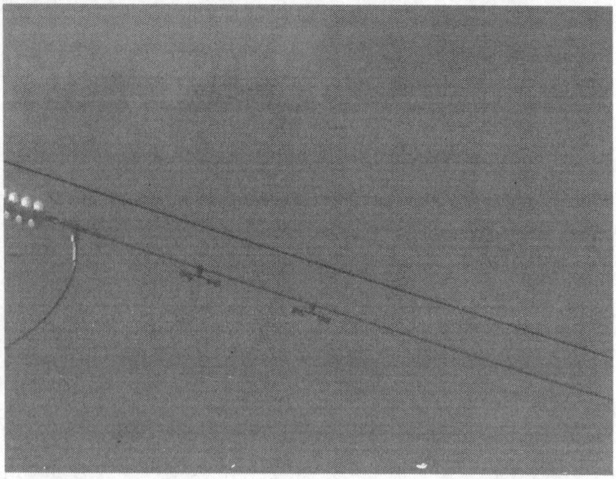

Figure 13.8 A pair of 'dogbone' dampers.

and/or the lift curve slope ($dC_L/d\alpha$) so increasing the critical wind speed for galloping in Equation 5.46.

13.4.3 Wake-induced vibrations of bundled conductors

Wake-induced vibrations can occur when two or more circular cross sections are grouped together and have been commonly observed in 'bundled' power conductors. The motion in these vibrations involves rotation of the bundle as a group, as well as translation motion. Wake-induced vibrations typically occur in a low-frequency range (up to 10 Hz), with displacement amplitudes of up to 1 m.

A quasi-steady linear theory has been developed for a pair of circular cylinders, representing a pair of conductors, with one located in the wake for the other (e.g. Blevins, 1977). When the wind direction is such that the downwind cylinder is in the wake of the upwind one, but offset from the centreline, the downwind cylinder will develop lift as well as drag forces. For certain along-wind and cross-wind separations, this can lead to instability. No closed-form solution of the coupled equations for the deflections of the downwind cylinder is available, and numerical solutions of a fourth-order stability polynomial are required. Given the dynamic properties of the cylinders, or conductors (i.e. mass distribution and frequencies), and the variation of lift and drag coefficients with spacing between the cylinders, stability boundaries can be determined. The system is stable – i.e. no vibrations, when either the downwind cylinder is immediately downwind of the upwind one, or it is outside the wake of the upwind cylinder (Blevins, 1977). However, if the downwind cylinder falls within certain regions in the wake, either side of the centreline, unstable oscillations can occur; the extent of these regions increases with increasing wind speed. Also, like galloping, there is a critical wind speed below which no oscillations will theoretically occur.

13.4.4 Turbulent buffeting

Like any other structure, or system, buffeting by atmospheric turbulence will lead to fluctuating response in sway of overhead lines, particularly in urban areas where the turbulence intensities are high. However, because of the high aerodynamic damping (see Section

5.5.1) on overhead lines in the along-wind direction – at high wind speeds this may exceed 20% of critical damping (Matheson and Holmes, 1981), the resonant contributions in low-frequency sway modes are generally small. However, quasi-static fluctuating stresses (i.e. 'background' response below the resonant frequencies) will still exist, and over a long period of time may induce fatigue damage (see Section 5.6).

13.5 SUMMARY

The available data for the specification of wind loads on transmission line structures has been critically reviewed. Risk models which consider the risk of intersection of small intense storms such as tornadoes and downbursts with long transmission line systems are discussed. Wind-induced vibrations of transmission and other overhead lines by various mechanisms, and their mitigation, are also described.

REFERENCES

American Society of Civil Engineers. 2010. *Guidelines for Electrical Transmission Line Structural Loading*. 3rd edition. American Society of Civil Engineers, ASCE, Reston, VA, USA.

Blevins, R.D. 1977. *Flow-Induced Vibrations*. Van Nostrand Reinhold, New York.

CSIR. 1990. Transmission Line Loading. Part I: Recommendations and Commentary. Part II: Appendices. Engineering Structures Programme. CSIR Building Technology, South Africa.

Davenport, A.G. 1979. Gust response factors for transmission line loading. *Proceedings, 5th International Conference on Wind Engineering*, Fort Collins, Colorado, pp. 899–909. Pergamon Press, Oxford, UK.

EPRI. 1979. Transmission line reference book – Wind-induced conductor motion. Electric Power Research Institute, Palo Alto, CA, USA.

EPRI. 2009. Transmission line reference book – Wind-induced conductor motion – Revised edition. Electric Power Research Institute, Palo Alto, CA, USA.

Hawes, H. and Dempsey, D. 1993. Review of recent Australian transmission line failures due to high intensity winds. *Report to Task Force on High Intensity Winds on Transmission Lines*, Buenos Aires, 19–23 April 1993.

Matheson, M.J. and Holmes, J.D. 1981. Simulation of the dynamic response of transmission lines in strong winds. *Engineering Structures*, 3: 105–10.

Milford, R.V. and Goliger, A.M. 1997. Tornado risk model for transmission line design. *Journal of Wind Engineering and Industrial Aerodynamics*, 72: 469–78.

Oliver, S.E., Moriarty, W.W. and Holmes, J.D. 2000. A risk model for design of transmission line systems against thunderstorm downburst winds. *Engineering Structures*, 22: 1173–9.

Richardson, A.S. 1991. A study of galloping conductors on a 230 kV transmission line. *Electric Power Systems Research*, 21: 43–55.

de Schwarzkopf, M.L.A. and Rosso, L.C. 2001. *A Method to Evaluate Downdraft Risk*. University of Buenos Aires, Argentina.

Twisdale, L.A. and Dunn, W.L. 1983. Probabilistic analysis of tornado wind risks. *Journal of Structural Engineering (ASCE)*, 109: 468–88.

Varney, T. 1926. Notes on the vibration of transmission line conductors. *AIEE Transactions*, XLV: 791–5.

Chapter 14

Other structures

14.1 INTRODUCTION

In this chapter the wind loads on some structures not covered in Chapters 8–13, and appendages attached to buildings, will be considered. Some of these structures may be of lesser economic importance, but are often sensitive to wind loads, fail early during a severe wind storm, and provide a source of flying debris.

In the following sections, wind loads on free-standing walls (including noise barriers along freeways or motorways) and hoardings are discussed (Section 14.2). Wind loading of free-standing roofs or canopies, including tensioned fabric roofs (Section 14.3), solar panels attached to the roofs of buildings, as well as other appendages attached to buildings such as canopies, awnings and balconies, are also discussed (Section 14.4). Free-standing paraboloidal antennas for radio telescopes, and antennas of various geometries attached to towers or buildings (Section 14.5), and lighting frames and luminaires (Section 14.6) are considered.

Petrochemical, mining and other industrial structures are often located in high-wind regions of the world. Wind codes and standards are generally written primarily for buildings and hence industrial structures are often a secondary consideration in those documents. Section 14.7 discusses some approaches, including other sources of shape coefficients and force coefficients, for such structures, including offshore platforms.

14.2 WALLS AND HOARDINGS

14.2.1 Single walls under normal and oblique winds

In Sections 4.3.1 and 4.3.2, the mean drag coefficients on walls on the ground are discussed in the context of bluff-body aerodynamics. Discussion of wind loads on free-standing walls under normal and oblique winds will be expanded in this chapter.

In Figure 14.1, mean and maximum net pressure difference coefficients acting on complete walls of various breadth/height ratios are shown plotted. These values are based on boundary-layer wind-tunnel measurements (Letchford and Holmes, 1994) in simulated open country terrain (Jensen numbers h/z_o in the range 50–160). The net pressure coefficient, C_{pn}, is defined in Equation 14.1 and, in this case, is equivalent to a drag coefficient.

$$C_{pn} = \frac{p_w - p_L}{\frac{1}{2}\rho_a \bar{U}_h^2} \quad (14.1)$$

where

p_w is the area-averaged pressure coefficient on the windward face of the wall

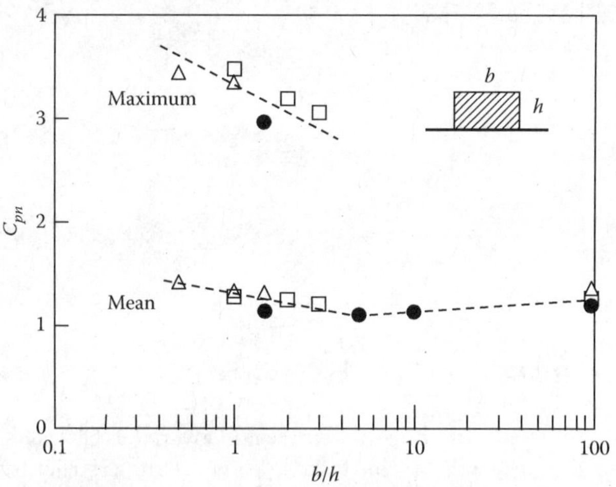

Figure 14.1 Mean and maximum pressure difference coefficients for free-standing walls (normal wind). (From Letchford, C.W. and Holmes, J.D. 1994. *Journal of Wind Engineering and Industrial Aerodynamics*, 51: 1–27.)

p_L is the area-averaged pressure coefficient on the leeward face of the wall
$\bar{U}_h$ is the mean wind speed at the top of the wall

The maximum values were expected values for periods equivalent to 10 min in full scale. The mean net pressure coefficients show a small reduction in the range of *b/h* from 0.5 to 5, as previously shown in Figure 4.5. A larger reduction occurs for the maximum pressure coefficients – this is due to the reduction in spatial correlation for longer lengths of wall. About a 20% reduction in peak net load occurs as the wall length increases from 1 to 4 wall heights.

For a wind direction at 45° to the plane of the walls, the average net pressure coefficients are shown in Figure 14.2. In this case, the net mean pressure coefficient reaches a maximum for a *b/h* ratio of about 3 with lower values for longer walls. For this wind direction, there is a strong separation on the leeward face of the walls of this length ratio. For longer walls, re-attachment occurs and generates lower magnitude pressures on the leeward face.

For mean wind directions normal to the wall, the net pressures do not vary much along the length of the wall. However, this is not the case for the oblique wind direction. Figure 14.3 shows how the mean net pressure coefficient varies along the wall length. The flow separation behind the windward edge generates very high pressures for the first 1–2 wall heights from the windward edge. This also occurs for elevated hoardings (Figure 4.8 and Section 14.2.4), and is usually the critical design case for wind loads.

14.2.2 Walls with corners

The effect of a right-angled corner at a free end of a wall for various wind directions on mean pressure coefficients averaged over a vertical line, at a distance *y* from the corner, is shown in Figures 14.4 and 14.5.

For a wind direction of 0°, with the corner running downwind, the effect is small; however, for 180° there is an increase in mean pressure coefficient of up to 30% (Figure 14.4). However, for the 45° wind direction (i.e. blowing from outside the corner), there is a

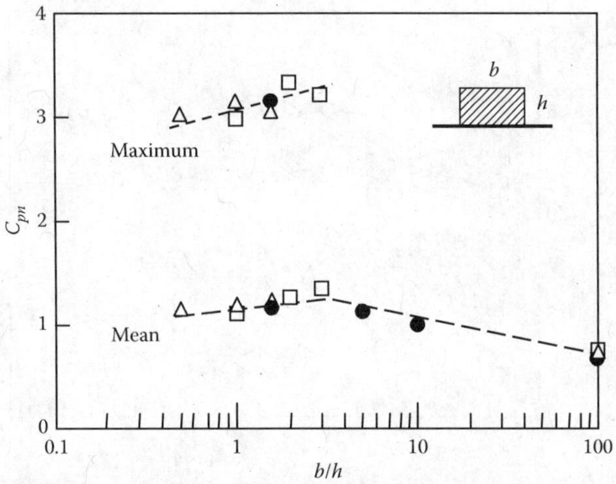

Figure 14.2 Mean and maximum pressure difference coefficients for free-standing walls (oblique wind). (From Letchford, C.W. and Holmes, J.D. 1994. *Journal of Wind Engineering and Industrial Aerodynamics*, 51: 1–27.)

significant reduction in mean pressure coefficients, for the region immediately adjacent to the corner (Figure 14.5).

14.2.3 Parallel two-dimensional walls

There is an increasing tendency to provide noise barriers along freeways and motorways when they pass through urban areas. These are generally parallel walls spaced at the width of the roadway, so that shielding effects from the opposite wall may be important for certain wind directions. The wind loads on these walls are also affected by other disturbances to the wind flow, such as topographic features and elevated bridges.

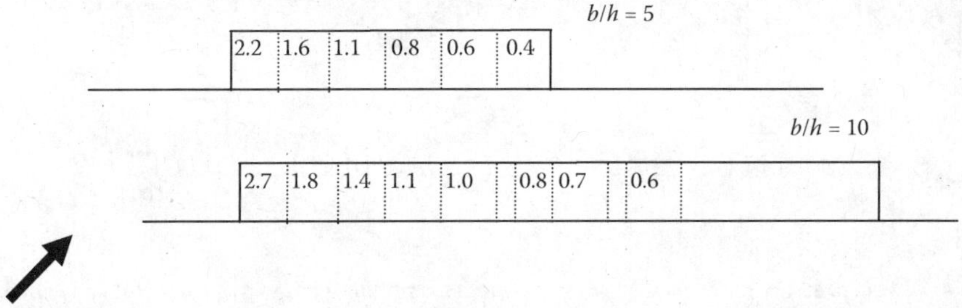

Figure 14.3 Mean pressure difference coefficients for free-standing walls (oblique wind).

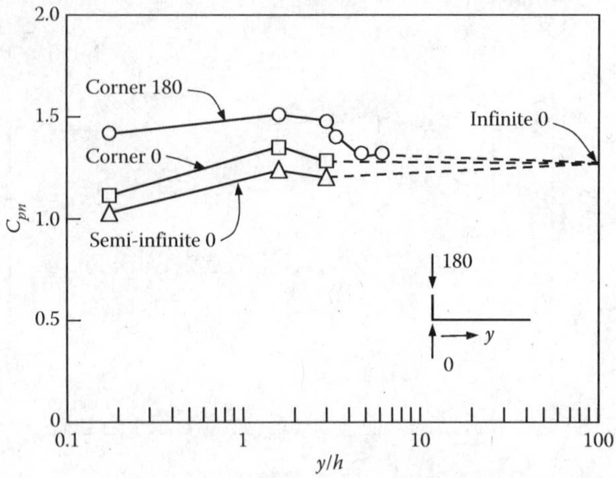

Figure 14.4 Mean pressure difference coefficients for free-standing walls with corners (normal winds). (From Letchford, C.W. and Holmes, J.D. 1994. *Journal of Wind Engineering and Industrial Aerodynamics*, 51: 1–27.)

Figure 14.6 shows the variation of mean, r.m.s., maximum and minimum net pressure coefficients on one wall of a pair of parallel ones, for various spacings. The pressure coefficients are based on the mean wind speed at wall height in the undisturbed flow. A negative value of wall spacing/wall height means the second wall is *downwind*. These measurements were carried out in simulated atmospheric boundary-layer flow in a wind tunnel. The values of Jensen number, h/z_o (Section 4.4.5) for the wall heights used in the tests were about 10–20.

The pressure tappings were arranged in vertical rows, with spacings chosen that the pressures averaged together as a group of four (Section 7.5.2) gave a measure of the bending moment at the base of the wall. Thus, the measurements are of base moment coefficients defined as

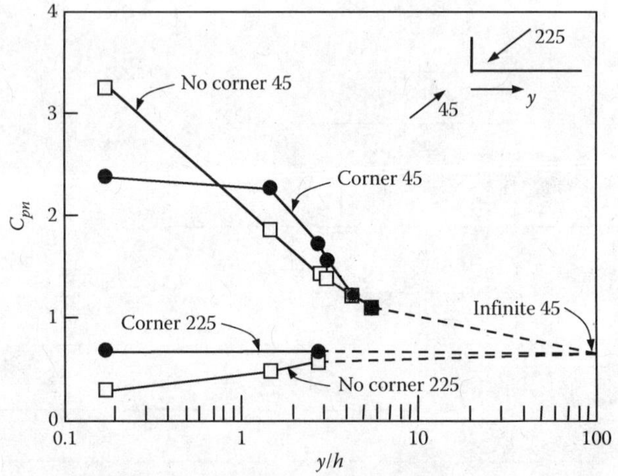

Figure 14.5 Mean pressure difference coefficients for free-standing walls with corners (oblique winds). (From Letchford, C.W. and Holmes, J.D. 1994. *Journal of Wind Engineering and Industrial Aerodynamics*, 51: 1–27.)

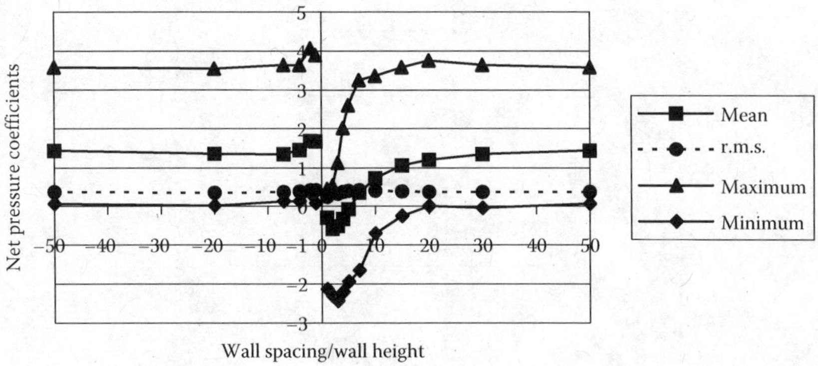

Figure 14.6 Parallel walls on flat level ground – effect of wall spacing. (From Holmes, J.D. 2001. *Journal of Wind Engineering and Industrial Aerodynamics*, 89: 1397–1407.)

$$C_M = \frac{M}{(1/4)\rho_a \bar{U}^2 h^2} \qquad (14.2)$$

where M is the moment about the base, per unit length of wall. This is also an effective net pressure coefficient which, when applied uniformly over the well height, will give the correct base moment. Averaging of peak and fluctuating pressures was carried out over one wall height horizontally along the wall axis.

The mean pressure difference is negative when the upwind wall is about two wall heights away from the shielded wall; i.e. it acts *upwind*. Small shielding effects are felt when the upwind wall is as much as 20 wall heights upwind.

Figure 14.7 compares the mean, maximum and r.m.s. net effective pressure coefficients for the windward wall of the pair of parallel walls on a bridge with two different values of clear space underneath (Holmes, 2001). The thickness of the bridge deck was equal to the wall height. Values are for s/h equal to 0, 2 and 4, where s is the clear spacing under the

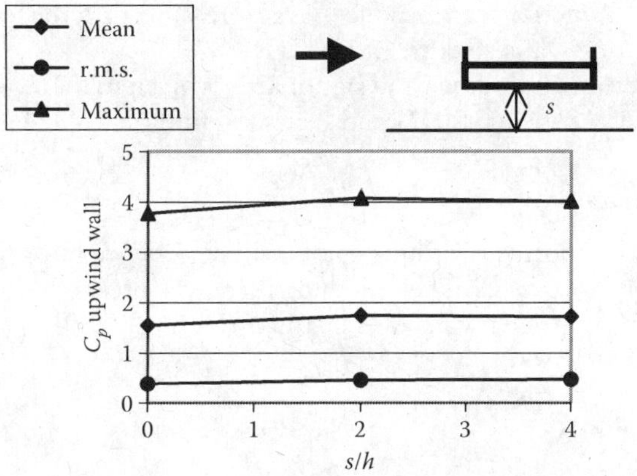

Figure 14.7 Effect of clear space, s, for parallel walls on bridges. (From Holmes, J.D. 2001. *Journal of Wind Engineering and Industrial Aerodynamics*, 89: 1397–1407.)

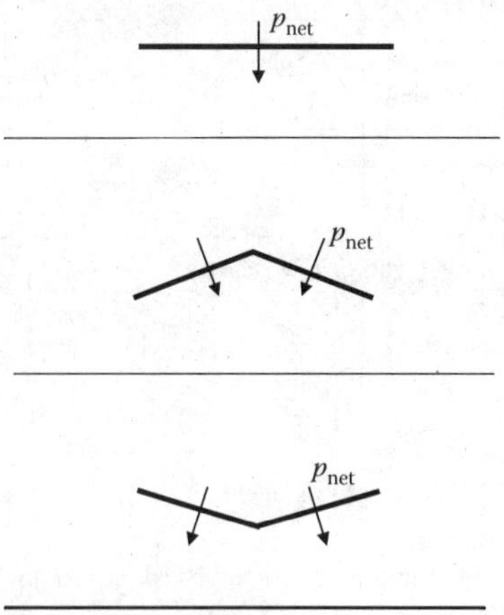

Figure 14.8 Types of free-standing roof, and sign convention for net pressures.

bridge. All pressure coefficients are calculated with respect to the mean wind speed at the height of the top of the wall ($s + 2\,h$) in the undisturbed flow.

Figure 14.7 shows that there is little difference between the net pressure coefficients for the cases of s/h equal to 2 or 4, when there is airflow beneath the bridge. However, when s/h is equal to 0 – that is the 'bridge' forms a flat-topped cliff, the mean and maximum net pressure coefficients are about 90% of the values on the elevated bridges; the r.m.s. pressures are about 80% of those in the elevated case.

14.2.4 Elevated hoardings

The net wind pressure coefficients on elevated hoardings have generally similar characteristics to those on free-standing walls. The effect of elevation is to increase the magnitude of the net pressure coefficient for winds normal to the surface. The average mean pressure coefficient depends on the spacing to the ground beneath the hoarding. For a spacing equal to the depth of the hoarding, a mean net pressure coefficient (with reference to the mean velocity at the top of the hoarding) of about 1.5 occurs.

The oblique wind direction can produce large pressure differences near the windward end, as for free-standing walls (Figure 4.8).

Design data for elevated hoardings and signboards is given in the American (ASCE, 2010), Australian (Standards Australia, 2011) and British Standards (BSI, 1997).

14.2.5 Spanwise averaging

Walls, or hoardings, supported over long spans will experience lower peak wind loads than those supported over short spans. The following form was proposed for the reduction factor for peak loads on free-standing walls over spans, s (Holmes, 2001).

$$\alpha = 0.5 + 1.35\exp\left[-\left(\frac{s}{h}\right)^{0.15}\right] \tag{14.3}$$

Equation 14.3 gives the ratio between peak net pressures for a span s (greater than the wall height, h) and the peak net pressure on a wall with a span, s, equal to the height, h.

Thus, for $s = h$, α is equal to 1.0. The equation was derived for both unshielded walls and shielded parallel walls.

14.3 FREE-STANDING ROOFS AND CANOPIES

14.3.1 Pitched-free roofs

Free-standing, or 'canopy', roofs, without walls, are often used for basic shelter structures – such as those at motor vehicle service stations and railway stations, or for coverage of industrial, mineral or agricultural products. The wind loads on roofs of this type attached to buildings are discussed in Section 14.4.1. Wind loads on tensioned fabric-free roofs and canopies are covered in Section 14.3.3.

Free-standing roofs which are completely free of stored material underneath allow air to flow freely underneath; this generally results in negative, or near-zero, underside pressures with respect to atmospheric pressure. The addition of stored material underneath the roof in sufficient quantity will cause full or partial stagnation of the airflow, and positive pressures underneath. The nature of the upper surface pressures depends on the roof pitch and the wind direction.

Wind-pressure coefficients on free-standing roofs are usually quoted in the form of net pressure coefficients, as defined in Equation 14.1. The pressures can normally be assumed to act normal to the roof surface. The usual sign convention is that positive net pressures act downwards. This sign convention and the most common three types of free-standing roof geometry are shown in Figure 14.8.

Although the pressures normal to the roof surface are the dominant ones, frictional forces acting parallel to the roof surfaces can also be significant, and it may be necessary to consider them, when designing the bracing required to resist horizontal forces.

Free-standing roofs have been studied in both wind-tunnel tests (Gumley, 1984; Letchford and Ginger, 1992; Ginger and Letchford, 1994) and full-scale experiments (Robertson et al., 1985).

Net pressure coefficients along the centreline of a free-standing 'Dutch barn' with 15° roof pitch, measured in full scale (Robertson et al., 1985), are shown in Figure 14.9. The roof is completely empty underneath. Positive (downwards) pressure differences exist over the windward quarter of the roof for all wind directions. The largest negative loads occur near the middle of the roof for a wind direction normal to the ridge.

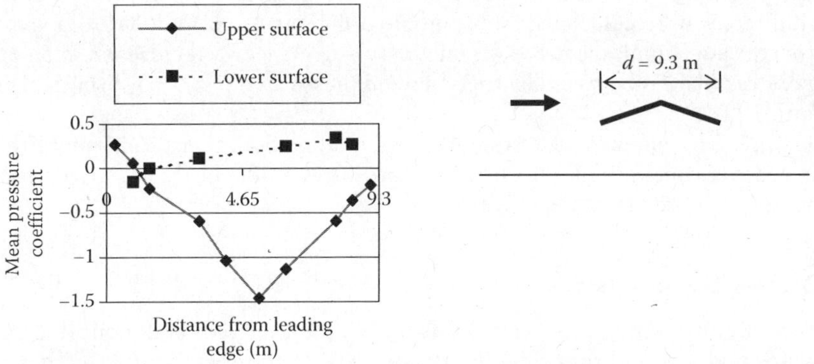

Figure 14.9 Mean pressure difference coefficients along the centre line of a free-standing roof with 15° pitch. (From Robertson, A.P., Hoxey, R.P. and Moran, P. 1985. *Journal of Wind Engineering and Industrial Aerodynamics*, 21: 113–25. With permission.)

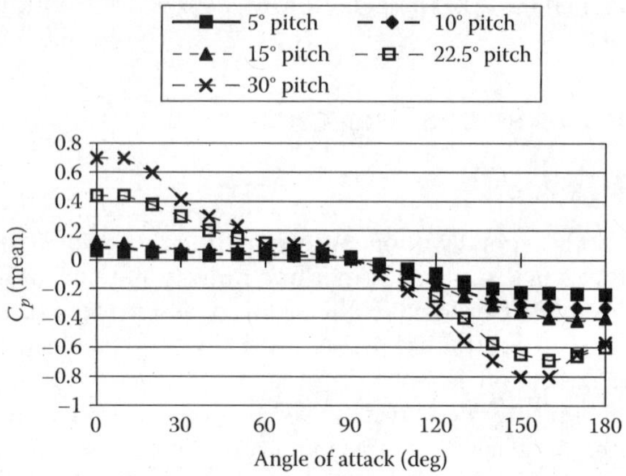

Figure 14.10 Mean pressure difference coefficients for pitched free roofs, averaged over a half roof. (From Letchford, C.W. and Ginger, J.D. 1992. *Journal of Wind Engineering and Industrial Aerodynamics*, 45: 25–45.)

Letchford and Ginger (1992), and Ginger and Letchford (1994), carried out extensive wind-tunnel measurements on pitched free roofs (empty under) of approximately square plan, with a range of pitches up to 30°. Mean and fluctuating pressure measurements from single points and area-averaged (Section 7.5.2) over six panels were made. In addition, correlation coefficients (Section 3.3.5) were measured for the six panel pressures enabling fluctuating total forces (Section 4.6.6), and equivalent static loading distributions to be derived (Section 5.4.3).

Mean area-averaged net pressure coefficients for half the pitched roof are shown in Figure 14.10. For the 0° wind direction, the half roof is on the windward side. Figure 14.10 shows that significant positive pressures (for wind directions of 0–30°) and negative pressures (for wind directions of 120–180°) occur for roof pitches of 22.5° and 30°. For roof pitches of 15° or less, the net pressure difference coefficients are not large for any wind direction.

The peak (maximum and minimum) area-averaged pressure difference coefficients generally showed a similar behaviour to the mean coefficients shown in Figure 14.10, with the 22.5° and 30° pitch roofs clearly showing larger magnitudes. When peak total uplift and horizontal forces were calculated, substantial reductions of up to 50% from values calculated from the non-simultaneous peaks on windward and leeward halves were obtained, due to the poor correlation between fluctuating wind pressures on the two surfaces (Ginger and Letchford, 1994).

Appendix F gives an example of the calculation of maximum and minimum lift and drag on a pitched-free roof, and the effective static pressures producing them, based on data from Ginger and Letchford (1994).

14.3.2 Effect of porosity

The effect of porosity on mean wind loads on pitched-, hipped- and monoslope-free roofs was investigated, in wind-tunnel model studies, by Letchford et al. (2000). Two porosities of 11% and 23% were studied for comparison with the equivalent values for solid roofs (i.e. 0% porosity) for three different roof pitches. Figure 14.11 shows some of the results of Letchford et al. The coefficients shown are the average mean net pressure coefficients for

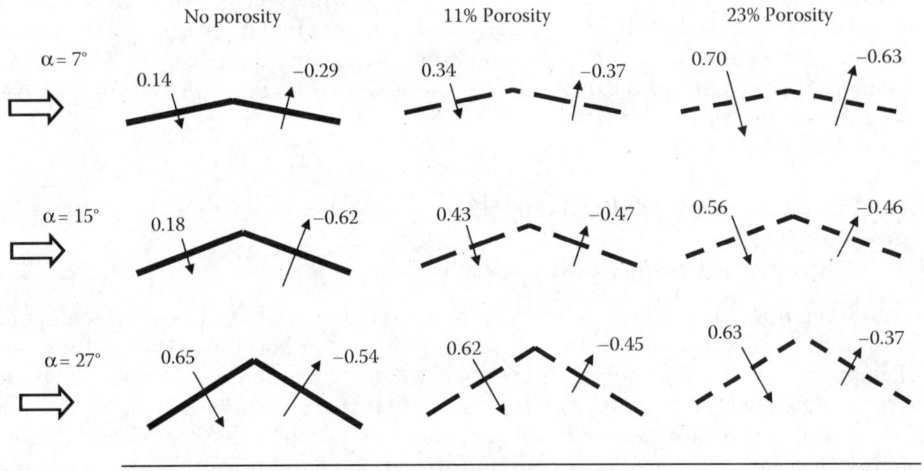

Figure 14.11 Effect of porosity on mean net pressure coefficients for pitched-free roofs – referenced to mean velocity at average roof height. (Adapted from data presented by Letchford, C.W. et al. 2000. *Journal of Wind Engineering and Industrial Aerodynamics*, 84: 197–213.)

each roof half. At the higher pitches, the main effect of porosity is on the net pressure on the leeward side. However, for the 7° pitch case, the porosity affects the net pressure on both halves.

The data in Figure 14.11 can also be interpreted in terms of mean drag and lift coefficients. For the 7° and 15° roof pitches, the mean drag force reduces with increasing porosity, whereas for the 27° pitch roof the opposite is the case. The effect of increasing porosity is to reduce the positive (upward) lift force for all roof pitches. For the 27° pitch case the average lift force is negative (i.e. downwards) for all porosities, and this downward force increases with increasing porosity.

14.3.3 Tensioned fabric roofs and shade sails

In hot climates, relatively cheap tensioned fabric roofs are often used to provide shading to school playgrounds, house yards and outdoor shopping malls and restaurants and so forth. These are sometimes known as 'shade sails' or 'membrane roofs'. Larger prestigious architect-designed tensioned fabric roofs are sometimes used for pavilions at international exhibitions and 'EXPOS' (e.g. Sykes, 1994).

Knitted or woven fabrics, with some porosity to air flow, are often used for shade sails in hot climates, such as northern Australia. The results of Letchford et al. (2000), discussed in the previous section, can be applied to determine the effects of porosity on simple pitched and monoslope roof shapes made from such fabrics. By comparing pressure loss coefficients, Letchford et al. found that the porous plates with 11% and 23% solidity were equivalent to shade fabrics with about 90% ultra-violet (UV) radiation reduction.

Wind loads on tensioned fabric structures have been rarely investigated experimentally, partly due to the difficulty in installing pressure tappings in thin wind-tunnel models, and partly due to the fact that fluctuating wind pressures will deform a roof and change its geometry over short time intervals.

Although roofs of this type with long spans may have quite low natural frequencies, significant resonant response to wind is unlikely, as they have high positive aerodynamic damping (see Section 5.5.1) for vertical motion at high wind speeds. For example, Sun and Gu (2014) in a numerical study of a 'saddle' membrane roof found aerodynamic damping of up to 10% of critical – about five times the assumed structural damping.

Another class of tensioned fabric roofs comprises air-supported or 'pneumatic' roofs. These are discussed in Chapter 10.

14.4 ATTACHMENTS TO BUILDINGS

14.4.1 Canopies, awnings and parapets

Several configurations of horizontal canopy attached to one wall of a low-rise building have been investigated (Jancauskas and Holmes, 1985). The width of the canopy and the height of the canopy position on the wall were the variables that were investigated. A narrow canopy mounted at the top of the wall behaves similarly to eaves on the roof.

For wind directions normal to the adjacent wall, the peak net force across the canopy is strongly dependent on the non-dimensional ratios, h_c/h, and h_c/w_c. h_c is the height of the canopy above the ground, h is the total height of the adjacent wall and w_c is the width of the canopy.

For the peak vertical uplift force coefficient, $\hat{C}_Z$, based on the mean wind speed at the height of the canopy, the following conservative relationships were proposed, based on the wind-tunnel measurements:

$$
\begin{aligned}
&\text{for } \frac{h}{h_c} = 1.0, \quad \hat{C}_Z = 1.0 + 1.3\left(\frac{h_c}{w_c}\right) \text{or 4.0, whichever is the lesser} \\
&\text{for } \frac{h}{h_c} = 1.0, \quad \hat{C}_Z = 1.0 + 0.4\left(\frac{h_c}{w_c}\right) \text{or 4.0, whichever is the lesser} \\
&\text{for } \frac{h}{h_c} = 0.5, \quad \hat{C}_Z = 1.0
\end{aligned}
\tag{14.4}
$$

where

$$C_Z = \frac{F_z}{(1/2)\rho_a \bar{U}_c^2 A}$$

F_z is the net vertical force on the canopy (positive upwards)
$\bar{U}_c$ is the mean wind speed at the canopy height
A is the plan area of the canopy

Equation 14.4 can be applied to canopies with pitch angles within 5° of the horizontal. Appropriate adjustment is required if it is applied with gust wind speeds; such adjustment has been made for the rule incorporated in the Australian wind-loading standard (Standards Australia, 2011).

The relationships of Equation 14.4 are compared with the experimental data in Figure 14.12.

The higher values obtained for canopies or awnings near the top of the wall (or eaves) can be explained by the high flow velocities occurring on the upper side of the canopy producing significant negative pressures; on the underside of the canopy, stagnation and hence positive pressures occur. When the canopy is mounted part-way up the wall, stagnation of the flow occurs on the wall, both above and below the canopy. In this situation, the mean net force coefficients are low, but turbulence produces finite peak loads in both directions.

Parapets, and their effect on roof pressures on flat roofs, have been the subject of several wind-tunnel studies. In the early work there were some conflicting conclusions drawn by

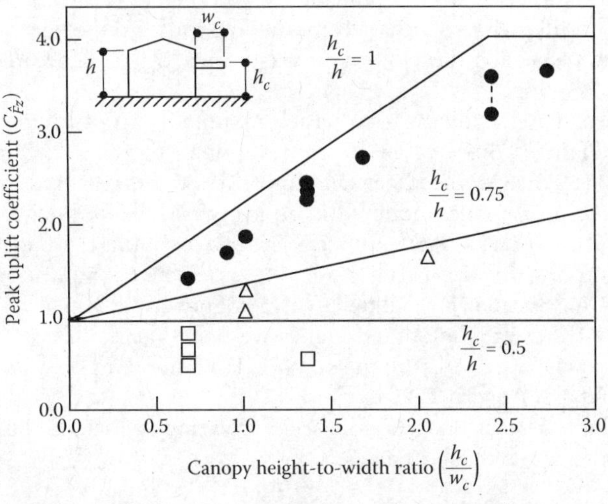

Figure 14.12 Peak uplift force coefficients for attached canopies. (From Jancauskas, E.D. and Holmes, J.D. 1985. Wind loads on attached canopies. *Fifth U.S. National Conference on Wind Engineering*, Lubbock, TX, USA, 6–8 November. With permission)

different laboratories, but the issue was largely resolved using large models and a high density of pressure tappings (Kind, 1988). With or without parapets, the worst suction peaks occur in small zones near the upwind corner of the roof, for wind directions nearly bisecting the corner. The worst suction coefficients decrease monotonically with increasing relative parapet height. The amount of the reduction depends also on the height/width ratio of the building to which the parapet is attached (see also Section 8.6).

14.4.2 Solar panels on roofs

The wind loads on solar panels attached to the roofs of a building are closely related to the flow over the roofs of the building itself, since the latter is a much larger bluff body. Figure 14.13 shows the various geometric variables that are significant in determining the wind loads on solar collector panels on a pitched roof building (Tieleman et al., 1980).

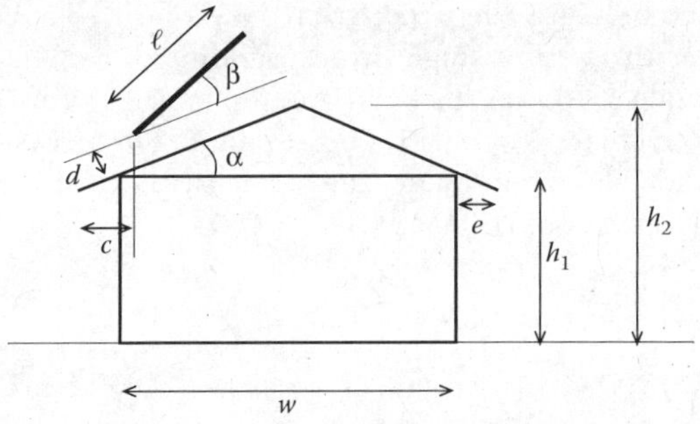

Figure 14.13 The variables affecting wind loads on solar panels. (From Tieleman, H.W., Akins, R.E. and Sparks, P.R. 1980. An investigation on wind loads on solar collectors. Virginia Polytechnic Institute and State University, College of Engineering, Report VPI-E-80-1. With permission.)

Wind-tunnel pressure tests were conducted by Wood et al. (2001) on a 1:100 scale model of an industrial building with solar panels mounted parallel to the flat roof. The orientation of the panels with respect to the wind flow was found to have a greater effect than panel height and lateral spacing.

A comprehensive model study for solar panels mounted on roofs of low-rise buildings was undertaken by the James Cook Cyclone Testing Station in Townsville, Queensland, Australia (Ginger et al., 2011). Area-averaged net wind pressures were measured on panels, of dimensions 1.7 m × 1.0 m in full scale, mounted in an array, parallel to pitched (gable) roofs with pitches of 7.5°, 15° and 22.5°. The gap between the panel and the underlying roof surface was either 100 mm or 200 mm in full scale. Measurements were also made by the James Cook Cyclone Testing Station (JCCTS) on an array of panels mounted at 15° and 30° to a flat roof. Results from the parallel-mounted panels were used as the basis of net pressure coefficients on solar panels in the Australian/New Zealand Standard on Wind Actions AS/NZS 1170.2:2011 (Standards Australia, 2011).

The following summarises the general effects of various panel and building variables on wind loads on roof-mounted solar panels:

- 'Stand-off' spacing from the roof – increasing stand-off appears to reduce net uplift load (normal to roof), but increases the wind force acting on the panel parallel to the roof (Newton, 1983).
- The effect of the width of the gap between panel and roof, for parallel-mounted panels, on the net pressures on the panels is small (Ginger et al., 2011).
- Module shape and size – the combined peak load on a row of panels is significantly less than that on a single panel, due to area reduction effects on the fluctuating pressures.
- Roof pitch – higher roof pitch produces lower uplift loads, but increasing downwards wind loads (as for the loads on the roofs of low-rise buildings generally).
- Proximity to eaves or gable ends – the end panel (adjacent to the eaves) experiences considerably higher loads than the interior panel in a row.
- Net wind pressures on panels near the windward gable end of a pitched roof are generally similar in magnitude to the external pressures on the roof itself without the presence of panels (Ginger et al., 2011).
- Net pressures on panels in the central part of the roof can be *greater than* the local external pressures on the bare roof at the same location (Ginger et al., 2011). However, these net pressures are lower in magnitude than those on panels located near the edges of a roof.
- Wind direction – the worst uplifts occur for oblique wind directions to a row of collector panels.
- Roof height – the pressure *coefficients* for panels on two-storey buildings are lower than the equivalent values for single-storey buildings.
- Increasing the angle β, so that the inclination of panels is greater than that of the roof pitch, appears to increase wind loads on the panels. For arrays of inclined solar panels on flat roofs, 'guide plates', or 'guide vanes' are useful for reducing net uplift forces on the panels (e.g. Chung et al., 2013a,b).

14.5 ANTENNAS

14.5.1 Radio telescopes

Wind loads on the antennas of large steerable radio telescopes – usually with dish reflectors of paraboloidal shape – are of critical importance for several design criteria (Wyatt, 1964):

1. Overall strength for safety in extreme winds
2. Loads on drive system
3. Freedom from oscillations
4. Pointing accuracy
5. Distortion of the reflector

Conditions 2 to 5 are serviceability criteria. Very small tolerances are required for the operation of these antennas.

The main source of wind loads is the paraboloidal dish itself. If the dish is impermeable, the pressures acting on it may be assumed to act normal to the surface, with negligible contributions from skin friction. For a paraboloid, the normal to any point on the surface passes through the generating axis, at a point $2f$ measured along the axis from that point, where f is the focal length. Therefore, it may be assumed that the resultant aerodynamic force will act through a point on the axis, distant from the vertex by $2f$ plus half the depth of the dish, d (Wyatt, 1964).

Considering first the case with the wind direction normal to the altitude axis of rotation of the dish is shown in Figure 14.14. Resolving the aerodynamic forces in body axes (Section 4.2.2), the force coefficients are given by

$$C_X = \frac{F_x}{(1/2)\rho_a \bar{U}_b^2 A} \tag{14.5}$$

$$C_Y = \frac{F_Y}{(1/2)\rho_a \bar{U}_b^2 A} \tag{14.6}$$

where A is the projected area normal to the dish, given by $\pi(b^2/4)$.

Following the arguments in the previous paragraph, the eccentricity, e, of the aerodynamic force can be closely approximated by (Wyatt, 1964)

$$e = 2f + \frac{d}{2} = 2f\left[1 + \left(\frac{b}{8f}\right)^2\right] \tag{14.7}$$

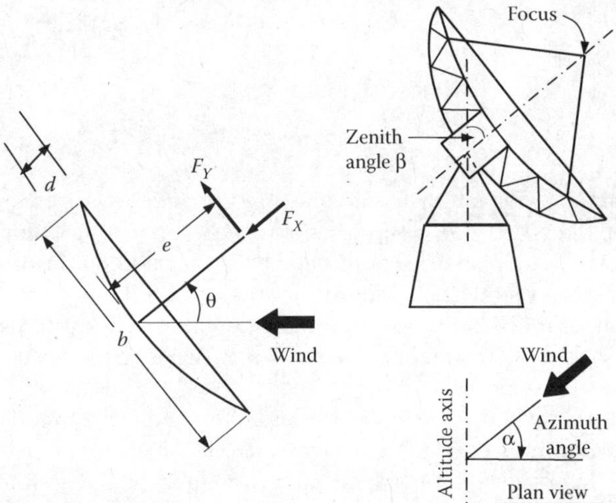

Figure 14.14 Resultant aerodynamic forces on the dish antenna of a radio telescope. (From Wyatt, T.A. 1964. *Annals, New York Academy of Sciences*, 116: 222–38. With permission.)

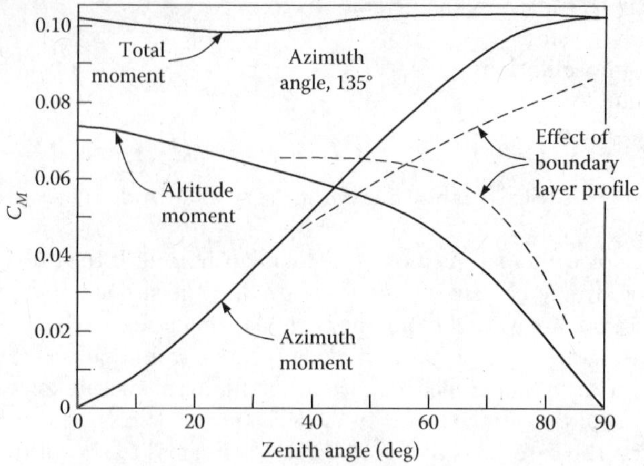

Figure 14.15 The effect of velocity profile on the aerodynamic moments on a radio telescope. (From Wyatt, T.A. 1964. *Annals, New York Academy of Sciences*, 116: 222–38. With permission.)

Tests in smooth uniform flow (Wyatt, 1964) indicate maximum values of C_x of about 1.7 when the angle of attack, α, is about 45°. The transverse force coefficient C_Y is approximately constant with f/d when expressed in the form $(f/d)C_Y$, with a maximum value of about 0.05, for α equal to about 135°. The transverse force F_Y generates a moment about the vertex equal to $F_Y \cdot e$.

It is found that the effect of a boundary-layer mean wind profile has a relatively small effect for wind directions facing the wind. However, the effect is greater when the wind is blowing obliquely on to the rear of the paraboloid. As shown in Figure 14.15, the effect is to increase the moment about the altitude axis and decrease it about the azimuth axis (Wyatt, 1964).

In Figure 14.15, the moment coefficients are defined as follows:

$$C_M = \frac{M}{(1/2)\rho_a \bar{U}_h^2 A d} \tag{14.8}$$

14.5.2 Microwave dish antennas

The drag forces acting on small dish antennas used for microwave frequency transmission are of interest for the structural design of the towers supporting them. In the past, total drag forces for tower design have been obtained by simply adding the drag measured on the antennas in isolation to that determined for the tower without antennas. This will overestimate the total drag in many cases, as usually the antennas will shield part of the tower, or vice versa; also the drag on an antenna itself in the presence of the tower will be different to that on the antenna in isolation.

Figure 14.16 shows the drag coefficient for an impermeable unshrouded dish obtained as a function of the wind incidence angle measured from the normal to the plane of the dish, measured in both smooth (~1% turbulence intensity), and turbulent flow (10% turbulence intensity) (Holmes et al., 1993). The reference area is the projected area of the dish, $\pi(b^2/4)$.

The drag coefficient for the isolated dish is maximum with a wind direction normal to the plane of the dish, but does not reduce much in an angular window within 30° to the

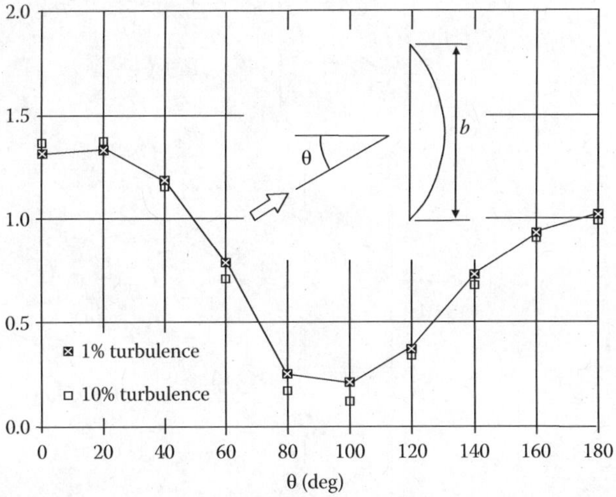

Figure 14.16 Drag coefficient as a function of angle of attack for an isolated dish antenna. (From Holmes, J.D., Banks, R.W. and Roberts, G. 1993. *Journal of Wind Engineering and Industrial Aerodynamics*, 50: 263–9.)

normal. The maximum drag coefficient based on the disc area is about 1.4. A large reduction occurs for wind directions from 40° to 80° to the normal. The effect of turbulence intensity is small.

The concept of *interference factor* is illustrated in Figure 14.17. The drag of an isolated antenna should be multiplied by this factor to give the measured incremental contribution to the total tower drag. The sum of the drag on the tower segment, D_t, and the incremental contribution from the antenna, $K_i \cdot D_a$, gives a total effective drag, D_e.

The interference factor for a single dish attached to a face of a lattice tower, with square cross-section and a solidity ratio of 0.3, is shown graphically, as a function of wind direction, θ, relative to the tower face in Figure 14.18 (Holmes et al., 1993). The maximum interference factor of about 1.3 occurs at wind directions for which the dish accelerated the airflow over the tower, that is for wind directions of 90° and 270°. For wind directions of 0° and 180°, where mutual shielding occurs, interference factors as low as 0.5 can occur.

An empirical form for the interference factor, K_i, based only on the solidity and drag coefficient of the tower, which fits the experimental data in Figure 14.17, and data from other cases, takes the form:

$$K_i = \exp\left[-k(C_D\,\delta)^2\right] \cdot \left[(1+t) + t\cos 2(\theta - \theta_d - 90°)\right] \tag{14.9}$$

where

C_D is the drag coefficient for the tower or mast section alone, based on the projected area of members in one face, measured normal to the face
δ is the solidity of a face of the tower
k is a parameter equal to 1.2 for a square tower (ESDU, 1981)
t is an adjustable parameter (equal to 0.5 in Figure 14.18)
θ_d is the angle of the normal to the dish antenna relative to the tower

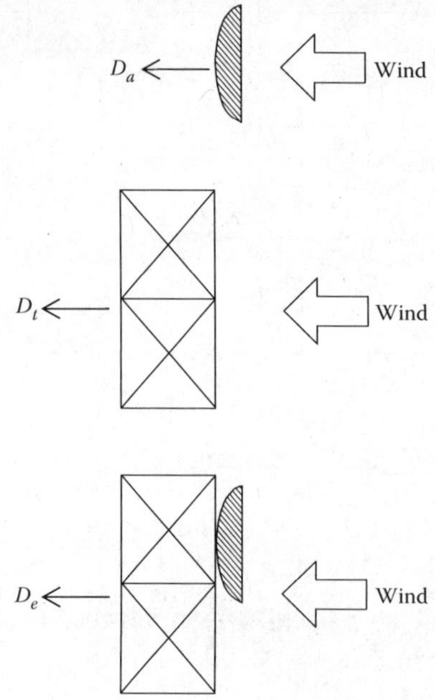

Interference factor

$$K_i = \frac{D_e - D_t}{D_a}$$

$$D_e = D_t + K_i \cdot D_a$$

Figure 14.17 Concept of interference factor for incremental antenna drag.

As well as drag (along-wind) forces, there may be significant cross-wind forces acting for wind directions parallel, or nearly parallel, to the plane of a solid dish. These should be taken account of when designing support attachments for the dish. Basic aerodynamic force coefficients are often obtainable from the antenna manufacturers, although these would not generally include interference effects.

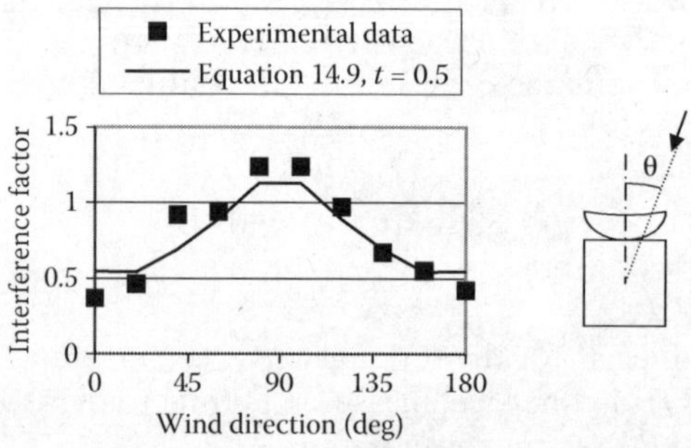

Figure 14.18 Interference factor as a function of wind direction for a single microwave dish added to a square lattice tower. (From Holmes, J.D., Banks, R.W. and Roberts, G. 1993. *Journal of Wind Engineering and Industrial Aerodynamics*, 50: 263–9.)

14.5.3 Rotating radar antennas

Aerodynamic loads on large rotating radar antennas, such as those used at large airports, pose a particular serviceability problem due to the variations in torque that arise. The operation of the antennas imposes strict limits on variations in angular velocity, and this in turn limits the variations in torque that the drive motor must overcome. Wind-induced variations in torque arise from two sources:

1. Variations in the azimuth angle between the wind direction and the antenna
2. Horizontal wind turbulence

At the rates of rotation used in practice, the first source of aerodynamic torque variation appears to be dominant.

The effect of rotation of the antenna can be treated by a quasi-steady approach. This results in the predicted variation of torque being obtained from static tests in a wind tunnel, in which the azimuth angle is varied. The effect of rotation is assumed to result in a static shift in the fluctuating torque curve obtained from such tests (Sachs, 1978; Lombardi, 1989). However, the quasi-steady theory has been found to be only approximately correct at high rotational speeds (Lombardi, 1991).

The use of small fins on the back of the antenna has been found to be effective in reducing the aerodynamic torque. These are small lifting surfaces that produce a counter-acting torque. Figure 14.19 shows measured torque coefficients obtained from a rotating wind-tunnel model with and without fins (Lombardi, 1991).

Unfortunately, all the wind-tunnel measurements on rotating radar antenna have been carried out in smooth uniform flow. The effect of turbulent boundary-layer flow is uncertain, but the most likely effect is to smooth out the torque versus yaw angle graphs, such as those shown in Figure 14.19.

14.5.4 Mobile telephone antennas

Antennas for mobile telephone cells typically consist of several radiating antennas within fibreglass or plastic radomes, mounted on poles or towers, which may in turn be mounted above buildings or other structures. By their nature they are in exposed positions, and thus

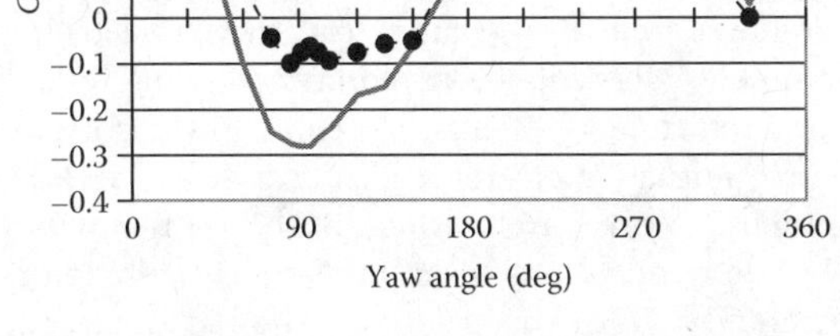

Figure 14.19 Aerodynamic torque coefficient versus yaw angle for rotating radar antennas. (From Lombardi, G. 1991. *Engineering Structures*, 13: 345–50. With permission.)

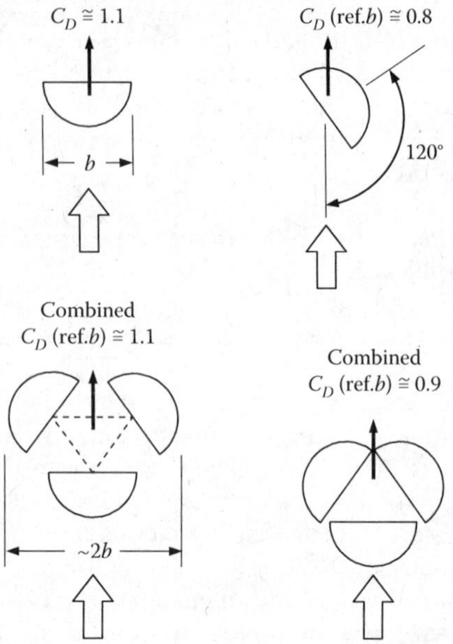

Figure 14.20 Typical drag force coefficients for mobile telephone antenna elements.

the interference, or shelter, effects from other structures are usually small. However, the mutual aerodynamic interference between radomes can be considerable.

Many of these antennas have been tested at full scale in large wind tunnels, for aerodynamic force coefficients but the data are usually proprietary in nature, and not freely available. The force coefficients have been found to be dependent on Reynolds number, so that model testing at small scales will produce unreliable results. However, drag coefficients at high Reynolds number from full-size measurements, which illustrate the mutual interference effects, are shown in Figure 14.20.

The drag coefficient for wind normal to the curved face of an antenna is around 1.1 based on the projected frontal area. This value is reduced for wind directions in which the frontal area presented to the wind is reduced, as illustrated for the value shown for a wind direction 120° from the normal.

When the antenna elements are grouped in threes, the combined drag coefficient (based on the frontal area of *one* radome element) is greatly reduced. As shown in Figure 14.20, the effect of the two downwind elements in the widely spaced (left side) cluster is neutral – that is the drag of the upwind element in isolation is the same as the combined drag of the group of three. For the closely spaced cluster, which resembles a single bluff body with curved surfaces, the overall drag is more than 20% less than that of the upwind radome in isolation.

Sometimes up to nine antennas are grouped together on a triangular frame, as shown in Figure 14.21. For the case shown, the antennas on a single face are well separated to avoid large aerodynamic interference effects, but those at the corners may experience slight increases in drag due to mutual interference (Section 4.3.1, and Marchman and Werme, 1982). The six downwind antennas are shielded both by the upwind antennas, and by the supporting pole. Full-scale wind-tunnel tests on complete antenna headframes indicate an overall reduction of about 30% in the combined drag is obtained, when comparing the combined drag of the group, with that obtained by the summation of contributions from individual elements.

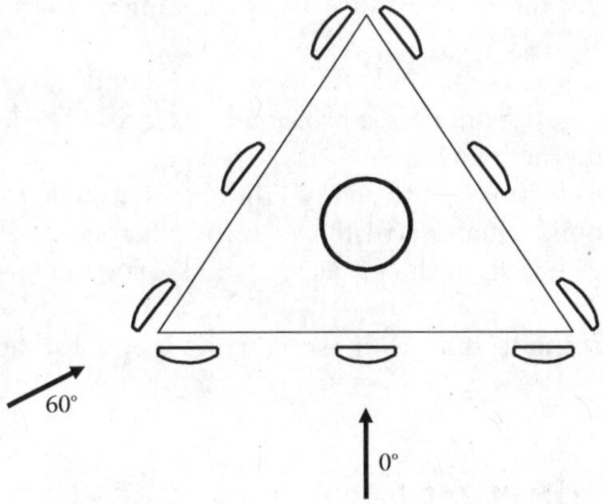

Figure 14.21 A group of mobile telephone antenna elements on a triangular frame.

Isolated radome elements will also experience cross-wind forces at oblique wind directions. However these will be largely cancelled by opposite forces from other elements, when they are part of a group.

14.5.5 UHF television antennas

Antennas for the transmission of ultra-high-frequency television-broadcasting signals (including digital television), consist of fibreglass or plastic radomes mounted on four- or five-sided masts, and are usually quite 'solid' cross-sections. They are of the order of 1 m in breadth, and about 20 m long. They are usually mounted at the top of free-standing or guyed towers.

The drag coefficient for these cross-sections depends on the porosity of the cross-section – that is the ability of the wake to be vented to the windward side. Measurements on full-size antenna sections have yielded drag coefficients in the range of 1.1–1.6, with some dependency on wind direction.

UHF antennas have experienced significant cross-wind response due to vortex shedding (Sections 4.6.3 and 11.4). This has often occurred for antennas on guyed masts, which have a lower damping than free-standing lattice towers. Such responses occur at a critical velocity, that is quite low, and in atmospheric conditions with low turbulence intensities. The prediction of cross-wind response due to vortex shedding for *circular* cross-sections was discussed in Section 11.5. Methods also exist for *non-circular* cross-sections (e.g. ESDU, 1990). These methods require information on the Strouhal number (rate of vortex shedding – see Section 4.6.3), and fluctuating cross-wind force coefficients (Section 4.6.4). These would not be well defined for the complex cross-sections of UHF antennas.

If vibrations occur, they can be mitigated by the use of simple damping devices, such as liquid dampers (Section 9.9.4), or *hanging chain* dampers (Koss and Melbourne, 1995).

14.6 LIGHTING FRAMES AND LUMINAIRES

Street lighting, flood lighting for railway yards, sporting grounds and industrial areas are bluff bodies of a variety of shapes and porosities. There are considerable interference effects when luminaires are arranged in groups. As for antennas, the drag of many types has been measured in wind tunnels, but such tests are usually commercially sponsored, and the results are not readily obtainable.

The largest drag coefficients of single lights for any wind direction fall in the range of 1.0–1.5, based on the largest frontal area projected vertically. The lower value applies to the more rounded types, and the higher value to sharp-edged lights.

Large rectangular headframes, with many luminaires attached, such as large floodlight systems for sporting grounds, may be treated as porous flat plates (see Section 4.3.1). A value of drag coefficient of 1.5 based on the projected 'solid' or 'wind' area is an appropriate one for solidities of 0.3–0.7.

The wind loads on, and response of, supporting poles for lighting are discussed in Chapter 11.

14.7 INDUSTRIAL COMPLEXES AND OFFSHORE PLATFORMS

Estimation of wind loads for elements of industrial complexes such as power stations, petroleum refineries, or mineral processing plants is an extremely difficult problem. Such complexes consist of a large number of closely spaced bluff bodies, with considerable aerodynamic interference between them. It would normally be extremely conservative to estimate the total wind drag force by summing up the contributions from individual elements, as if they were isolated bluff bodies, although this is often done. The complexity and unique layouts of these plants mean that it is difficult or impossible to give general rules for estimating wind forces, except for some relatively common situations such as closely spaced circular cylinders. A useful source of shape factors for petrochemical, and other industrial, structures is a guide published by a committee of the American Society of Civil Engineers (ASCE, 2011).

A useful approach for densely packed industrial structures, which avoids gross overestimation of drag forces, is to treat a closely spaced complex of bodies in a 'global' way as a single 'porous' bluff body. This approach has been adopted for high-solidity open industrial structures by the ASCE guidelines (ASCE, 2011, Appendix 5B). The following formula for the maximum drag coefficient of such a structure, based on the projected frontal 'solid' area, is suggested:

$$C_D = \frac{1.4\delta \cdot \sqrt{b^2 + d^2}}{2b\left(1 + (b^2/d^2) - (b/d^2)\sqrt{b^2 + d^2}\right)} \qquad (14.10)$$

where b and d are the cross-wind breadth, and along-wind depth respectively and δ is the solidity ratio (viewed from the front elevation, and including all equipment).

Mining equipment and equipment for loading mined ore on to ships at ports are also in need of source data to enable wind-loading calculations. Drag coefficients for open-lattice conveyor support structures can be derived from those specified for open lattice towers (see Section 11.3.2). However, when one of these is clad for environmental reasons (e.g. to prevent airborne dust), it becomes a slender bluff body, and may be subjected to cross-wind response caused by vortex shedding (see Section 11.5). Offshore platforms, used for oil exploration and production, are similar in complexity, with the topsides often exposed to severe wind storms (in many cases tropical cyclones), as well as wave action. In these cases the overall wind forces on the above water-exposed structure is of interest in the design of the underwater foundations and supporting structure.

Flare towers, on offshore platforms, or onshore in refineries or liquid natural gas plants, are usually open lattice towers, and can be treated as other structures of this type, such as communication towers. However, the flares and risers attached to these structures often have significant frontal area, and the appropriate interference and shielding factors may not be available.

The low frequencies of 'compliant' offshore structures, such as tension leg or guyed structures in deep water locations, are of special concern because of the need to consider resonant excitation by dynamic wind forces. The frequencies of some structures of this type can be so low that they are near the peak of the spectrum of wind forces in synoptic winds (Section 3.3.4). However, it appears that hydrodynamic damping, resulting from the underwater motion of the structure (Cook et al., 1986), largely mitigates resonant effects. The special problems of wind effects on compliant offshore structures are discussed in a number of specialist publications (e.g. Smith and Simiu, 1986).

14.8 SUMMARY

In this chapter, wind loads on structures not covered in Chapters 8–13 have been discussed. This category includes free-standing walls and hoardings, attachments to buildings such as canopies and awnings and solar collectors.

Communications and broadcasting antennas of various types, particularly those impermeable enough to attract substantial wind loading, are considered in some detail. Some discussion on wind loads on elements of complex industrial structures, such as petrochemical plant, and offshore oil platforms has also been given.

REFERENCES

American Society of Civil Engineers. 2010. *Minimum Design Loads for Buildings and Other Structures*. ASCE/SEI 7-10. ASCE, Reston, Virginia.

American Society of Civil Engineers. 2011. Wind loads for petrochemical and other industrial facilities. Prepared by the Task Committee on Wind-induced Forces of the Petrochemical Energy Committee. ASCE, Reston, Virginia.

British Standards Institution. 1997. *Loading for Buildings. Part 2. Code of Practice for Wind Loads*. BS 6399: Part 2: 1997.

Chung, K.-M., Chou, C.-C., Chang, K.-C. and Chen, Y.-J. 2013a. Effect of a vertical guide plate on the wind loading of an inclined flat plate. *Wind and Structures*, 17: 537–52.

Chung, K.-M., Chang, K.-C., Chen, C.-K. and Chou, C.-C. 2013b. Guide plates on wind uplift on a solar collector model. *Wind and Structures*, 16: 213–24.

Cook, G.R., Kumarasena, T. and Simiu, E. 1986. Amplification of wind effects on compliant platforms. *Structures Congress '86*, New Orleans, September 15–18, (Proceedings of session: 'Wind effects on compliant offshore structures'). ASCE, New York.

ESDU. 1981. Lattice structures Part 2 – Mean forces on tower-like space frames. Engineering Sciences Data Unit (ESDU International PLC, London, UK). ESDU Data Items 81028.

ESDU. 1990. Structures of non-circular cross section. Engineering Sciences Data Unit (ESDU International PLC, London, UK). ESDU Data Items 90036.

Ginger, J.D. and Letchford, C.W. 1994. Wind loads on planar canopy roofs – Part 2: Fluctuating pressure distributions and correlations. *Journal of Wind Engineering and Industrial Aerodynamics*, 51: 353–70.

Ginger, J.D., Payne, M., Stark, G., Sumant, B. and Leitch, C. 2011. Investigation on wind loads applied to solar panels mounted on roofs. Cyclone Testing Station Report, TS821, James Cook University, Townsville, Queensland, Australia.

Gumley, S.J. 1984. A parametric study of extreme pressures for the static design of canopy structures. *Journal of Wind Engineering and Industrial Aerodynamics*, 16: 43–56.

Holmes, J.D. 2001. Wind loading of parallel free-standing walls on bridges, cliffs, embankments and ridges. *Journal of Wind Engineering and Industrial Aerodynamics*, 89: 1397–1407.

Holmes, J.D., Banks, R.W. and Roberts, G. 1993. Drag and aerodynamic interference on microwave dish antennas and their supporting towers. *Journal of Wind Engineering and Industrial Aerodynamics*, 50: 263–9.

Jancauskas, E.D. and Holmes, J.D. 1985. Wind loads on attached canopies. *Fifth U.S. National Conference on Wind Engineering*, Lubbock, Texas, USA, 6–8 November.

Kind, R.J. 1988. Worst suctions near edges of flat rooftops with parapets. *Journal of Wind Engineering and Industrial Aerodynamics*, 31: 251–64.

Koss, L.L. and Melbourne, W.H. 1995. Chain dampers for control of wind-induced vibration of tower and mast structures. *Engineering Structures*, 17: 622–5.

Letchford, C.W. and Ginger, J.D. 1992. Wind loads on planar canopy roofs – Part 1: Mean pressure distributions. *Journal of Wind Engineering and Industrial Aerodynamics*, 45: 25–45.

Letchford, C.W. and Holmes, J.D. 1994. Wind loads on free-standing walls in turbulent boundary layers. *Journal of Wind Engineering and Industrial Aerodynamics*, 51: 1–27.

Letchford, C.W., Row, A., Vitale, A. and Wolbers, J. 2000. Mean wind loads on porous canopy roofs. *Journal of Wind Engineering and Industrial Aerodynamics*, 84: 197–213.

Lombardi, G. 1989. Wind-tunnel tests on a model antenna with different fin configurations. *Engineering Structures*, 11: 134–8.

Lombardi, G. 1991. Wind-tunnel tests on a model antenna rotating in a cross flow. *Engineering Structures*, 13: 345–50.

Marchman, J.F. and Werme, T.D. 1982. Mutual interference drag on signs and luminaires. *ASCE Journal of the Structural Division*, 108: 2235–44.

Newton, J.R.H. 1983. Wind effects on buildings – Recent studies at Redland wind tunnel. *Journal of Wind Engineering and Industrial Aerodynamics*, 11: 175–86.

Robertson, A.P., Hoxey, R.P. and Moran, P. 1985. A full-scale study of wind loads on agricultural canopy roof ridged structures and proposals for design. *Journal of Wind Engineering and Industrial Aerodynamics*, 21: 113–25.

Sachs, P. 1978. *Wind Forces in Engineering*. 2nd edition. Pergamon Press, Oxford, UK.

Smith, C.E. and Simiu, E. (eds.) 1986. Wind effects on compliant offshore structures. *Proceedings of a Session at Structures Congress '86*, New Orleans, 15–18 September, ASCE, New York.

Standards Australia. 2011. *Structural Design Actions. Part 2: Wind Actions*. Standards Australia, Sydney, NSW Australian/New Zealand Standard AS/NZS1170.2:2011.

Sun, F.-J. and Gu, M. 2014. A numerical solution to fluid–structure interaction of membrane structures under wind action. *Wind and Structures*, 19: 35–58.

Sykes, D.M. 1994. Wind loading tests on models of two tension structures for EXPO92, Seville. *Journal of Wind Engineering and Industrial Aerodynamics*, 52: 371–83.

Tieleman, H.W., Akins, R.E. and Sparks, P.R. 1980. An investigation on wind loads on solar collectors. Virginia Polytechnic Institute and State University, College of Engineering, Report VPI-E-80-1.

Wood, G.S., Denoon, R.O. and Kwok, K.C.S. 2001. Wind loads on industrial solar panel arrays and supporting roof structure. *Wind and Structures*, 4: 481–94.

Wyatt, T.A. 1964. The aerodynamics of shallow paraboloid antennas. *Annals, New York Academy of Sciences*, 116: 222–38.

Chapter 15

Wind-loading codes and standards

15.1 INTRODUCTION

Wind-loading codes and standards, emerged in the second half of the twentieth century, have achieved wide acceptance, and are often the practising structural engineer's only contact with information for wind-loading calculations. Although they may be based on extensive research, they are, by necessity, simplified models of wind loading. Thus, great accuracy cannot be expected from them. Often this is consistent with the knowledge of the structure of the wind storms themselves in their country of use.

Advanced wind-loading codes and standards invariably contain the following features:

- A specification of a basic or reference wind speed for various locations, or zones, within a jurisdiction. Almost always a reference height of 10 m in flat, open country terrain is chosen (this is the standard location for wind measurements, specified by the World Meteorological Organization).
- Modification factors for the effects of height and terrain type, and sometimes for: change of terrain, wind direction, topography and shelter.
- Shape factors (pressure or force coefficients) for structures of various shapes.
- Some account of possible resonant dynamic effects of wind on flexible structures.

The above elements represent the first four links in the wind loading 'chain' proposed by A.G. Davenport (1977, 1982), and shown in Figure 15.1. Each of these links contributes to the overall strength of the design process for wind loads, and weakness or inaccuracy in any one reduces the reliability of the overall design process (see also Section 2.6).

This chapter reviews the wind-loading provisions of several prominent national, multinational and international documents, and highlights their similarities and differences. As codes and standards are continually being revised and updated, the overview is, by necessity, time dependent.

Other comparisons between major wind-loading codes and standards have been made by Cook (1990), Mehta (1998) and by Kijewski and Kareem (1998) for dynamic effects. A special issue of the journal *Wind and Structures* in 2005 comprised five papers in which all aspects of codification for wind loads were reviewed (Holmes et al. 2005a,b; Kasperski and Geurts 2005; Letchford et al. 2005; Tamura et al. 2005).

Holmes et al. (2009) described a comparison of wind load and response calculations for three buildings (low-rise, medium-rise and high-rise) from 15 wind codes and standards in the Asia-Pacific region. Holmes (2014) described a comparison of the calculated response of a generic tall building from three codes and standards, with the consensus along-wind response determined by several wind-tunnel groups using the high-frequency base balance technique (see Section 7.6.2).

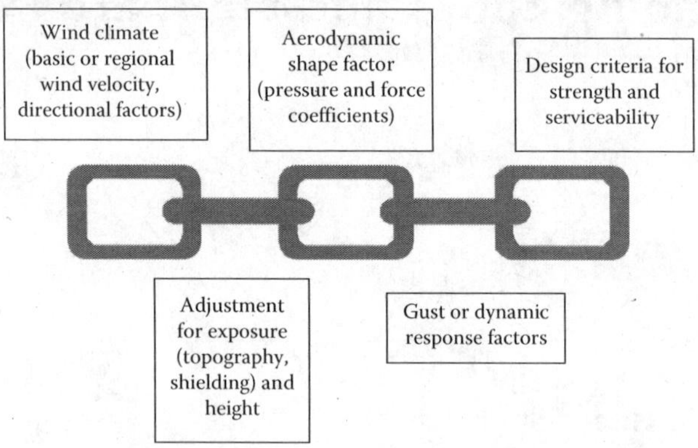

Figure 15.1 The wind loading 'chain'. (Adapted from Davenport A.G. 1977. The prediction of risk under wind loading. *Proceedings, 2nd International Conference on Structural Safety and Reliability*, Munich, Germany, 19–21 September, pp. 511–96; Reprinted from *Engineering Meteorology*, Chapter 12, Davenport, A.G., The interaction of wind and structures, pp. 527–572, Copyright 1982, with permission from Elsevier.)

15.2 GENERAL DESCRIPTIONS

The following five standards will be described, and critically reviewed, in this chapter:

1. ISO 4354:2009, Wind actions on structures – published in 2009.
2. EN 1991-1-4.6:2005, Eurocode 1: Actions on Structures – Part 1.4: General Actions – Wind Actions – published in 2005.
3. ASCE Standard ASCE/SEI 7-10. Minimum Design Loads for Buildings and Other Structures – published in 2010.
4. AIJ Recommendations for Loads on Buildings – published in 2004.
5. Australian/New Zealand Standard. Structural design actions. Part 2: Wind actions. AS/NZS1170.2:2011 – published in 2011 with amendments in 2012 and 2013.

The documents reviewed are those current at the time of writing. Although there are many other wind codes and standards in the world, many of these have been derived from, or are closely related to, one or other of the above documents, so that the comments following have wide and general application.

15.2.1 ISO/DIS 4354: Wind Actions on Structures

The current version of ISO International Standard 4354 – Wind Actions on Structures, published by the International Organization for Standardization (ISO), was issued in 2009 (International Standards Organization, 2009). The current version has been completely re-formulated and bears little resemblance to the earlier (1997) edition. As described in the introduction to ISO 4354, the document is intended 'for use by countries without an adequate wind-loading standard and to bridge between existing standards'. Most of the technical information in ISO 4354 is provided in a series of informative annexes. Two general methods of calculation of wind forces are given – one based on a peak velocity (nominally with a 3-s duration), and the other on a mean velocity (nominally with a 10-min averaging time). However, for structures for which dynamic response effects are not important, the peak velocity method should be used.

The main part of the document is quite short, and consists largely of definitions of the terms in the expression used to calculate wind pressure:

$$p = (q_{site})(C_p)(C_{dyn}) \tag{15.1}$$

The site peak dynamic pressure, q_{site}, is derived from the peak site wind speed, which, in turn, is derived from a reference peak wind speed (in the standard exposure of 10 m height in open country terrain):

$$q_{site} = 0.5\rho(V_{site})^2 \tag{15.2}$$

$$V_{site} = V_{ref} \cdot C_{exp} \tag{15.3}$$

C_{dyn} in Equation 15.1 allows for the effects of fluctuating pressures due to upwind turbulence, in the wake of a structure, as well as forces resulting from resonant dynamic response.

Section 9 of ISO 4354 allows one of the several methods to be used to determine pressure and force coefficients, including the use of wind-tunnel tests. The basic requirements for suitable wind-tunnel test procedures are given in Annex H in ISO 4354. Interestingly, Section 9 also apparently allows the use of 'computational based data'. However, this is somewhat contradicted by Annex I in ISO 4354, which cautions against the use of computational fluid dynamics techniques (CFD) for this purpose, at the present stage of their development. The comments in Annex I on the applicability of CFD techniques are generally compatible with those in Section 7.8 of this book.

Some basic aerodynamic pressure and force coefficients for simple shapes are provided in an informative Annex D. This Annex contains some interesting and innovative features:

a. A clear specification of the target probability level associated with the specified pressure and force coefficients. This has been set at the 80% fractile of the extremes, and can be estimated as follows:

   $$C_{p,80\%} = \bar{C}_p \pm 0.7C_{p,rms} \tag{15.4}$$

   when a reference time of 1 h (or equivalent in a wind-tunnel test) is adopted,

   $$\text{or, } C_{p,80\%} = \bar{C}_p \pm 2.1C_{p,rms} \tag{15.5}$$

   when a reference time of 10 min, or equivalent, is used.

   $C_{p,rms}$ is the root-mean-square value determined from the set of the *extreme* pressure coefficients determined from multiple samples with the same reference period. (Note this is not the same as the r.m.s. fluctuating pressure coefficient, C_p', for a complete pressure–time history, as defined in Section 8.3.1 of this book.)
b. The use of the load–response correlation (LRC) method (see Section 5.4.3) to determine some effective pressure coefficients for overall lift and drag forces for some simple shapes of low-rise buildings. These are given in Section D10 of ISO 4354.

In Sections D6 and D7 of ISO 4354, external and internal pressure coefficients, that originated in the Australian/New Zealand Standard, AS/NZS 1170.2, have been specified. However, these should be used with caution in the conjunction with ISO 4354, as they have not been adjusted for the different gust durations in the two Standards (0.2 s in AS/NZS 1170.2, versus 3 s in ISO 4354). This could be achieved through the dynamic response factor, C_{dyn}, which should be greater than 1.0 for small structures. The suggested value of C_{dyn} of 0.85 in Section D5 of ISO 4354 is clearly un-conservative for small low-rise buildings, when used with 3-s gust velocities and quasi-steady pressure coefficients (see Section 15.3.2 of this chapter, and Holmes et al., 2014).

Annex E in ISO 4354 provides detailed informative advice on the dynamic response factor for dynamically sensitive structures such as tall buildings and towers. Methods are given for estimating along-wind, cross-wind and torsional response, and these are closely related to those given the Recommendations of the Architectural Institute of Japan (AIJ).

15.2.2 EN 1991-1-4.6 Eurocode 1. Parts 1–4 wind actions

Eurocode 1 on wind loads, issued in 2005 (British Standards Institution, 2005), is a European Standard (EN), which is intended for use in most European countries. The version in each member country contains a 'National Annex', applicable only to that particular country. EN 1991-1-4.6 represents several years of work by representatives from many countries of the European Union and two separate committees, and is the nearest document to a truly multi-national wind-loading standard currently in existence.

This is a lengthy document with comprehensive methods of static and dynamic design for wind loads. The code is applicable to buildings and other structures, with heights up to 200 m, and to bridges with spans less than 200 m. No basic wind speeds are provided in the standard – these are provided separately in each National Annex. However, the basic wind velocity in each country is a 10-min mean velocity at 10 m height in open country terrain, with an annual probability of exceedence of 0.02 (50-year return period).

The mean wind velocity $v_m(z)$ at a height, z, above the terrain is given by

$$v_m(z) = c_r(z) \cdot c_o(z) \cdot v_b \tag{15.6}$$

where v_b is the basic wind velocity at 10 m height over open country terrain; $c_r(z)$ is a roughness factor, which varies with both height and terrain type; $c_o(z)$ is the orography (i.e. topography) factor. For the roughness factor, five different terrain types are defined with roughness lengths, z_o (see Section 3.2.1) varying from 0.003 to 1.0 m.

The peak velocity pressure, $q_p(z)$, is given by

$$q_p(z) = c_e(z) \cdot q_b \tag{15.7}$$

where $q_b = (1/2)\rho\, v_b^2 \cdot$ is the density of air, given in the National Annexes, and $c_e(z)$ is an 'exposure factor' given by

$$c_e(z) = [1 + 7 \cdot \mathrm{I}_v(z)] \cdot [c_r(z) \cdot c_o(z)]^2 \tag{15.8}$$

$I_v(z)$ is the turbulence intensity at height, z. The term $[1 + 7 \cdot I_v(z)]$ is effectively a gust factor for the velocity pressure, and is an approximation to $[1 + g_v \cdot I_v(z)]^2$, with the peak factor g_v, taking a value of 3.5.

Thus, the exposure factor, $c_e(z)$, combines gusting effects, terrain-height, and topographic effects into a single height- and terrain-dependent factor, and enables the 10-min mean wind velocity to be effectively converted into a gust wind velocity, and velocity pressure. The use of a peak factor of 3.5 with a 10-min reference period corresponds to an equivalent gust duration of about 0.1 s (see Table 3.4 in Chapter 3).

The number of shape factors presented in Section 7 of EN 1991-1-4.6 is extensive, with the number of cases covered exceeding those in most other codes and standards. External pressure coefficients on buildings are given for loaded areas of 1 m^2, and 10 m^2, denoted by $C_{p.e1}$ and $C_{p.e10}$, respectively. For areas between 1 and 10 m^2, linear interpolation is applied.

A 'structural factor' which allows for reductions in effective loading due to correlation effects on large structures, and for possible increases due to resonant dynamic effects, is defined in Section 6. This is discussed in more detail in Section 15.8.

Two alternative and independent procedures for calculating parameters for along-wind dynamic response (B^2 and R^2) are given in separate Annexes (Annex B and Annex C, respectively). A separate Annex D provides graphical information on the structural factor as a function of height and width of the structure, with separate graphs given for concrete and steel buildings, and for chimneys, with or without liners.

The several methods given for calculation of the structural factor are potentially confusing for the user, and may have legal implications. However, the multiple alternative options may be resolved within the National Annexes for each participating country.

Annex E contains detailed information on vortex-induced response of slender structures, such as chimneys. Two different and independent approaches for calculation of the cross-wind amplitude are given in Sections E.1.5.2 ('Approach 1') and E.1.5.3, respectively ('Approach 2'). Approach 1 is based on a sinusoidal excitation model (see Section 11.5.1 of this book); Approach 2 is derived from a random excitation model (see Section 11.5.2).

15.2.3 ASCE Standard ASCE 7-10. Minimum design loads for buildings and other structures

ASCE/SEI 7-10 (American Society of Civil Engineers, 2010) is a complete loading standard covering all types of loads. The wind-loading part consists of Chapters 26–31.

From 1995 onwards, ASCE-7 incorporated a number of significant changes in the wind-load provisions from the 1993 and earlier editions. This includes the use of a 3-s gust wind speed instead of the 'fastest-mile-of wind' as used in the past, a new zoning system for basic wind speeds, the use of high average recurrence intervals (300, 700 and 1700 years) for ultimate limits states design, the incorporation of topographic factors, some new data on pressure coefficients, a simplified procedure for buildings less than 9 m in height, and a revised method for along-wind dynamic response calculation.

The 2010 version includes a significant re-formatting of the previous (2005) edition for ease of use, with the previous single chapter broken up into five chapters that each cover an alternative method of assessing wind loads:

- Chapter 26 gives 'General Requirements' including wind hazard maps, exposure categories, topographic multipliers, gust effects factors and internal pressure coefficients.
- Chapter 27 provides a 'Directional Procedure' for the 'main wind-force resisting system' (MWFRS) for buildings of all heights.
- Chapter 28 gives an 'Envelope Procedure' (non-directional) for low-rise buildings (defined as having a mean roof height less than 18 m).
- Chapter 29 describes a directional procedure for building appurtenances (such as rooftop structures and equipment), and other structures (such as freestanding walls and signage, chimney tanks, lattice frameworks and towers).
- Chapter 30 provides procedures for assessing wind loads on 'components and cladding'.

- Chapter 31 describes the main requirements for the 'Wind Tunnel Procedure'. This may be required in earlier chapters for certain wind-sensitive structures, and is available as an option for all structures or parts of structures covered by Chapters 27–30.

The ASCE Standard has no legal standing of its own, but its provisions are cited by many of the regional, city and county building codes. The three major regional building codes in the United States have merged to form a single 'International Building Code'. This draws on the ASCE Standard for wind-load provisions.

15.2.4 AIJ Recommendations for loads on buildings

The Recommendations of the Architectural Institute of Japan were revised in 2004 (Architectural Institute of Japan, 2004), and form a comprehensive loading code including the effects of dead, live, snow, seismic, temperature, earth and hydraulic pressure, as well as wind loads. Chapter 6 on wind loads comprises 54 pages, with 114 pages of Commentary. The derivation of the wind-loading section of the 2004 edition of the AIJ, and revisions from the 1993 version were described in detail by Tamura et al. (2004).

Similar to the ASCE Standard, this is a comprehensive and advanced wind-loading document, although the Recommendations have no legally binding standing in Japan. The Building Law of Japan has a separate set of wind-loading rules – BSLJ-2000 (Ministry of Land, Infrastructure and Transport, 2000). Since the latter does not have a comprehensive set of rules for cross-wind and torsional dynamic response, the AIJ is commonly used by structural designers for buildings greater than 60 m in height.

15.2.5 Australian/New Zealand Standard AS/NZS 1170.2

The current edition of the joint Australian and New Zealand Standard for Wind Loads was issued in 2011 (Standards Australia and Standards New Zealand, 2011) as the second edition of a combined Standard, replacing separate documents from the two countries. A significant amendment was made in 2012. It is a comprehensive document of 96 A4 pages, and is supported by a separate handbook (Australasian Wind Engineering Society 2012), and a guide for designers is also available (Holmes and King 2005).

AS/NZS1170.2 has an indirect legal status in Australia by being called up in the Building Code of Australia (Australian Building Codes Board 2004). This document, or Part 0 of the joint Standards on Structural design actions, AS/NZS 1170.0 (Standards Australia 2002) must be consulted to obtain the appropriate annual probability of exceedence for the importance and use of the building, before use of AS/NZS 1170.2, Wind actions.

The nominal basic wind speed in AS/NZS1170.2 is a 0.2-s gust measured at 10 m height in open country terrain, and values are specified for a range of annual recurrence intervals from 1 to 10,000 years, for four regions. The gust duration was re-defined in 2012 from the value of 3 s given in earlier editions of the Standard. The justification for this was given by Holmes and Ginger (2012). For most buildings, excluding those with large numbers of occupants, and important post-disaster facilities, the annual risk of exceedence for ultimate limit states wind speeds is specified, in the Building Code of Australia, as 1/500. Tall buildings generally are assessed to have a higher 'importance level' and an annual risk of exceedence of 1/1000 is adopted. Other structures such as temporary ones, and cyclone shelters, may have lower or higher values, respectively.

15.3 BASIC WIND SPEEDS OR PRESSURES

15.3.1 Averaging times

Codes and standards for wind loading are currently based on extreme wind speeds with a variety of nominal averaging times. These variations have occurred for a number of reasons, such as the type of recording systems used by meteorological services to record winds, or the type of extreme wind event that dominates designs for wind in a particular jurisdiction.

Some codes are based on wind speeds averaged over relatively long periods, such as 10 min or 1 h. However, the wind speed is effectively converted into a gust speed within the format of the code, before calculating building pressures or forces. Gust factors, being the ratios between the expected maximum gust and the mean value, in an averaging time such as 1 h, are therefore important for these conversions (see Section 3.3.3).

Many national wind codes and standards are based on a maximum gust wind speed, with a defined gust duration; it is most common to find it stated as a '3-second' gust. The reason for this value is partly historical, with a perception of this being a typical averaging time of anemometers used to record historical data on gust wind speeds. However, as discussed by Holmes et al. (2014), the effective frontal area associated with a gust of this duration, at typical design wind speeds, is equivalent to that of a tall building. For smaller structures, a code or standard based on a 3-s gust should therefore incorporate a 'gust effect factor', or amplification factor, somewhat greater than 1.0, to allow for the lack of reduction due to correlation effects over small frontal areas.

15.3.2 Basic wind speeds in major codes and standards

Table 15.1 summarises the basic wind-speed characteristics used, or recommended, in the five documents surveyed in this chapter. In all cases the standard meteorological reference position of 10-m height in flat, open country is used.

The ISO Standard, as previously discussed, does not give basic wind speeds or dynamic pressures. Annex B in ISO 4354 provides peak factors and gust factors for conversion of wind data with various other averaging times, to the 3-s gust and 10-min velocities used as a basis for calculation of wind loads in ISO 4354. For synoptic winds, more comprehensive terrain- and wind-speed-dependent conversions are given in ESDU 83054 (ESDU International 1983), and by Holmes et al. (2014).

The Eurocode EN 1991-1-4.6 also does not give basic wind speeds, although a previous 1994 draft (C.E.N. 1994) gave 'reference wind velocities', for 18 countries in Europe in an informative Annex. National Annexes provide basic wind-speed information for individual countries in Europe.

Table 15.1 Definitions of basic wind speeds

Code	*Averaging time*	*Basic return periods*[a]
ISO 4354:2009	3 s (10 min)	Not specified
EN 1991-1-4.6	10 min	50 years
ASCE 7-10	3 s	350–700–1700 years
AIJ	10 min	100 years
AS/NZS1170.2:2011	0.2 s	500–1000 years[b]

[a] For ultimate limit states design.

[b] A wide range of return periods are provided in AS/NZS 1170.2 for various limits states, and important levels.

The American Standard (ASCE-7) contains maps with two zones in the majority of the country, and closely specified contours for Alaska and the coastal regions adjacent to the Gulf of Mexico and the Atlantic Ocean. In the latter case, the effects of hurricanes are of particular concern. The values of basic wind speed given on these maps are peak gust wind speeds, with mean recurrence intervals of 300, 700 and 1700 years. The methodology for the derivation of the basic wind-speed maps for the non-hurricane regions of the United States has been described by Peterka and Shahid (1998).

The Recommendations of the Architectural Institute of Japan (AIJ) gives a detailed map showing contours of the basic wind speed (10-min mean with 100-year return period). Single values are given for outlying territories such as Okinawa. A map of 500-year return period values is also given to enable users to interpolate for intermediate return periods.

In the Australian/New Zealand Standard, the 500–1000 years return periods shown in Table 15.1 apply to the majority of buildings in Australia (Importance Levels 2 or 3 in the Building Code of Australia) for assessment of ultimate limit states design criteria. Basic wind speeds are given in the form of maps with five regions, denoted by A, B, C, D and W. Two of these regions (C and D) comprise a coastal strip exposed to the effects of tropical cyclones (Section 1.3.2). Regional wind speeds are specified for each Region as a function of annual probability of exceedence. The analysis of extreme wind speeds for Region A, covering most of Australia, in the 2002 Australian Standard was described by Holmes (2002). The analysis of wind speeds for Regions C and D was discussed by Dorman (1984).

15.4 MODIFICATION FACTORS ON WIND VELOCITY

All the documents include modifiers for the effect of terrain/height and topography, although in the case of ASCE 7, these act on the dynamic *pressure*, rather than wind *speed*.

In the Eurocode, the mean wind speed is modified for terrain and height (roughness factor c_r), and for topography (described as 'orography'), c_o, then converted into a *gust* dynamic pressure at the height of interest, by a factor involving turbulence intensity (i.e. a gust factor acting on the dynamic pressure). The exposure coefficient $c_e(z)$ includes terrain/height and topographic (orographic) effects within Equation 15.8.

EN 1991-1-4.6 and AS/NZS1170.2 (for regions not affected by tropical cyclones) use a logarithmic law (or a modification for gust speeds) to define the terrain/height variation, ASCE 7 and AIJ use a power-law variation, and ISO 4354 gives parameters for both. AS/NZS1170.2 allows for averaging of terrain roughness upwind of the site, with an interpolation of terrain/height multipliers.

All five documents provide factors or multipliers for topographic, or orographic, speed-up effects on wind speeds. None of the documents allow for any shielding effects of topography. However, there are significant differences in the magnitudes of the speed ups predicted by the various documents for the same topographic geometry, as discussed by Holmes et al. (2005a).

The Australian/New Zealand Standard, AS/NZS1170.2, is unique in having a 'Shielding Multiplier', which allows for reductions in velocity when there are buildings upwind of greater or similar height.

Table 15.2 summarises the formats for calculation of design wind velocities and dynamic pressures in various documents.

15.5 BUILDING EXTERNAL PRESSURES

Table 15.2 also shows the general format for calculation of external pressures on wall or roof surfaces of enclosed buildings.

Table 15.2 Calculation formats for velocity, dynamic pressures and building pressure

Code	*Velocity*	*Dynamic pressure*	*Building pressure/force*
ISO4354	$V_{site} = V_{ref} \cdot C_{exp}$	$q_{site} = (1/2)\ \rho\ (V_{site})^2$	$p = q_{site.} C_p \cdot C_{dyn}$
EN 1991-1-4.6	$v_b = c_{dir}\ c_{season} v_{b,o}$	$q_p(z) = c_e(z)\ (1/2)\ \rho\ v_b^2$	$w_e = q_p(z)\ c_{pe}$
ASCE 7	V	$q_z = (1/2)\ \rho\ K_z K_{zt}\ K_d V^2\ I$	$p = q\ (GC_p)$
AIJ	$U_H = U_o K_D E_H k_{rW}$	$q_H = (1/2)\ \rho\ U_H^2$	$W_f = q_H\ C_f\ G_f\ A^a$
AS/NZS1170.2	$V_{sit,\beta} = VM_d M_{(z,cat)}\ M_s\ M_t$	$q_z = (1/2)\ \rho_{air} V_{des,\theta}^2$	$p = q_z\ C_{fig}\ C_{dyn}$

[a] The subscript *f* denotes *D* (for walls) or *R* for roofs in the AIJ.

The formulas (in the right-hand column) appear to be quite different from each other, but they all contain quasi-steady or mean pressure coefficients (C_p, c_{pe}, C_f, C_{fig}) and factors to adjust the resulting pressures to approximate peak values. In the case of ISO 4354 and AIJ, there are gust factors on pressure (C_{dyn} and G_f); in the case of EN 1991-1-4.6, the gust factor is incorporated into the exposure coefficient, $c_e(z)$, as discussed previously.

In ASCE 7, the quantities G and C_p are usually combined together as (GC_p) in tables. In AS/NZS1170.2, the 'aerodynamic shape factor', C_{fig}, consists of pressure coefficients, multiplied by factors for area reduction, combination of roof and wall surfaces, local pressure effects, and porous cladding. The local pressure factor K_ℓ is always greater than 1, and the area reduction factor K_a, which allows for correlation effects over large areas in separated flow regions, is less than 1. AS/NZS1170.2 is the only one in having a factor (K_p) for porous cladding.

The tables of shape factors and pressure coefficients of exterior surfaces of buildings given in the various documents are also sources of significant differences. In all cases the nominal wind directions are normal to the walls of buildings of rectangular plan.

The ASCE-7 Standard and AS/NZS1170.2 require alternative positive roof pressure coefficients to be considered. These are important values for the design of frames, especially for those in colder climates where dead loads are often high, as pointed out by Kasperski (1993).

EN 1991-1-4.6 gives tables of external pressure coefficients c_{pe} which are comparable to those in ASCE-7 and AS/NZS1170.2, since they are effectively applied to gust dynamic pressure through the use of the exposure coefficient $c_e(z)$. The tables give two values: $c_{pe,1}$, intended for tributary areas less than 1 m^2, i.e. local cladding design, and $c_{pe,10}$ intended for major structural members. It appears that the numerical values for flat and gable ('duo-pitch') roofs in EN 1991-1-4.6 are comparable to those in ASCE-7 and AS/NZS1170.2, and alternative (positive or lower negative) values are given for most roof pitches. However, no variation with height/width ratio is given.

The factors incorporated into the shape factor in the Australia/New Zealand Standard AS/NZS1170.2 for flat and gable-roofed buildings have already been discussed. However, it should also be mentioned that the effect of the tributary area and correlation effects are dealt with by the use of the three factors: K_a (area reduction factor), K_c (action combination factor) and K_ℓ (local pressure factor). The action combination factor, K_c, in the Australian and New Zealand Standard, allows for a reduction when wind pressures from more than one building surface, for example walls and roof, contribute significantly to a load effect.

The AIJ Recommendations also separate the specification of loads on the structural frames and on the 'components and cladding' of buildings. The specification of pressure coefficients is separated from the specification of the gust factor. Unlike any of the other documents, the gust factor, G_R, for the loads on the roofs of low-rise buildings has a dependency on natural frequency. Buildings are classified as those with heights less than, or greater than 45 m, a somewhat greater height than that used in the other documents.

15.6 BUILDING INTERNAL PRESSURES

The treatment of internal pressures varies considerably from one document to another.

For buildings with uniformly distributed openings, EN 1991-1-4.6 gives a graph of c_{pi}, varying from +0.35 to −0.5, as a function of an opening ratio, μ. For a dominant opening, the internal pressure coefficient is expressed as a fraction of the external pressure coefficient on the face with the opening. This document also gives fairly detailed guidance on pressures on walls and roofs, with more than one skin.

ASCE 7 (in Chapter 26) specifies three different situations: open, partially enclosed and enclosed buildings, and specifies values of GC_{pi} between +0.55 and −0.55. A feature, not found in the other standards, is a reduction factor, R_i, for large building volumes.

AS/NZS1170.2 gives two tables with various positive and negative values of internal pressure coefficients, $C_{p,i}$. For one of these tables, the values depend on the ratio of dominant openings on the windward wall to the total open area on other walls and roof. ISO 4354 gives similar internal pressure coefficients to those in AS/NZS 1170.2.

The AIJ Recommendations does not specify a positive internal pressure, that is the possibility of dominant openings is not considered. For buildings without dominant openings, values of C_{pi} of 0 or −0.5 are specified.

15.7 OTHER SHAPES AND SECTIONAL FORCE COEFFICIENTS

Apart from the AIJ Recommendations, which is intended exclusively for buildings, all the surveyed documents contain shape, or force, coefficients for a variety of structure shapes and cross-sections. Table 15.3 summarises the data given.

The data from all these documents appear to be based on modern wind-tunnel measurements for the most part. EN 1991-1-4.6 clearly contains the most comprehensive set of data. ISO 4354 only provides a limited amount of data on shape factors in Annex D, but suggests that other sources may be used in conjunction with the Standard with appropriate adjustment for gust averaging time and exposure. These sources include other codes or standards.

15.8 DYNAMIC RESPONSE CALCULATIONS

The five standards contain procedures for the calculation of dynamic response for wind-sensitive structures, such as slender, flexible, lightly damped tall buildings. ISO 4354, in Clause E.2.2, has a relatively complex set of numerical criteria to determine whether a structure is 'dynamically sensitive' in the along-wind, cross-wind and torsional modes. EN 1991-1-4:2005, in Clause 6.2, also has a set of empirical criteria to determine whether a structure need not be treated as dynamic; however, generally buildings with height to along-wind depth (i.e. h/d) greater than 4, require calculation of the structural factor. ASCE 7 and AS/NZS1170.2 classify wind-sensitive structures as those with a first-mode natural frequency less than 1 Hz; ASCE 7 also requires a height to breadth (or depth) ratio greater than 4.

In ISO 4354, information for calculation of along-wind response is given in two alternative formats – one for use with a (10-min) mean dynamic pressure (Clause E3), and a second for use with the (3-s) peak dynamic pressure (Clause E4). These are respectively the *gust*

Table 15.3 Shape factors contained in five documents (excluding rectangular enclosed buildings)

Type	*ISO 4354*	*EN 1991*	*ASCE 7*	*AIJ*	*AS/NZS1170.2*
Stepped roofs	No	No	Yes	No	No
Free-standing walls, hoardings	Yes (Walls only)	Yes	Yes	No	Yes
Parapets	No	Yes[a]	Yes	No	No
Free-standing roofs (canopies)	Yes	Yes	Yes	Yes	Yes
Attached canopies	No	No	No	No	Yes
Multispan roofs (enclosed)	No	Yes	Yes	Yes[b]	Yes
Multispan canopies	No	Yes	No	No	No
Arched roofs	No	Yes	Yes	Yes[b]	Yes
Domes	No	Yes	Yes	Yes[b]	Yes[c]
Bins, silos, tanks	Yes	Yes	Yes	No	Yes
Circular sections	Yes	Yes	Yes	Yes	Yes
Polygonal sections	No	Yes	Yes	No	Yes
Structural angle sections	Yes	Yes	No	No	Yes
Bridge decks	No	Yes	No	No	No
Lattice sections	Yes	Yes	Yes	No	Yes
Flags	No	Yes	No	No	Yes
Sphere	No	Yes	No	No	Yes
Roof-mounted solar panels	No	No	No	No	Yes

[a] Treated as free-standing walls.
[b] Given in commentary section of Japanese language version.
[c] Treated as arched roofs.

response factor discussed in Section 5.3.2 of this book, and the *dynamic response factor* format of Section 5.3.4. However, the user should be aware that, in the latter formulation, the specified peak factor, g_v, of 3.4 in equation (E.4.3) of ISO 4354 is too high when used with gust pressures derived from a 3-s gust, and is incompatible with Table B.1 in Annex B of ISO 4354 (see also Table 3.4 of this book) – potentially leading to un-conservative values of C_{dyn}.

In Section E6.1 of ISO 4354, some recommended values of structural damping are given as a function of construction material and building height. These values are apparently for ultimate limit states design criteria, as 75% of the specified values is recommended for 'habitability to horizontal vibrations'.

ISO 4354 recognises the importance of vortex shedding in causing dynamic cross-wind effects in slender prismatic and cylindrical structures. For rectangular cross-sections, the critical wind speed at which large-amplitude motions may result can be calculated as a function of side ratio and Scruton number (see Section 11.5.1 of this book).

A 'structural factor', in EN 1991-1-4.6, denoted by $c_s c_d$, incorporates the combined effects of size (c_s) and resonant dynamic response (c_d) for overall structural loads, or loads on major structural elements. The structural factor is equivalent to the 'gust effect factor' for dynamic structures in ASCE 7, and the 'dynamic response factor' discussed in Section 5.3.4.

This factor is given in the Eurocode by

$$c_s c_d = \frac{1 + 2k_p I_v(z_s)\sqrt{B^2 + R^2}}{1 + 7 \cdot I_v(z_s)} \qquad (15.9)$$

where

z_s is a reference height. For buildings and towers, this is normally taken as 60% of the roof height
k_p is a peak factor for the response
$I_v(z_s)$ is the longitudinal turbulence intensity at the reference height
B^2 is a background factor
R^2 is a resonance response factor

As discussed in Clause 6.3.1 in EN 1991-1-4.6, the structural factor is actually specified as a combination of a size factor, c_s with a dynamic factor, c_d, to form the combined structural factor, c_sc_d. The size factor separately takes account of correlation effects.

Alternative methods of calculation of the factors, k_p, B^2 and R^2, are given in Annexes B and C, with a simplified graphical method for estimation of c_sc_d, given in Annex D. Alternative expressions for the standard deviation of along-wind acceleration, for serviceability limit states, are also given in Annexes B and C.

Comprehensive information, including working equations, regarding vortex excitation and other aeroelastic effects such as vortex-induced large-amplitude lock-in-type vibrations, galloping (Section 5.5.2 of this book), various types of interference excitations and flutter (Section 5.5.3), are included in Annex E of EN 1991-1-4.6. Recommended calculation procedures for dynamic structural properties including natural frequencies, mode shapes, equivalent masses and logarithmic decrement are given in Appendix F of EN 1991-1-4.6.

In ASCE 7-10, an analytical procedure for the determination of a 'gust effect factor', G_f, for the along-wind vibrations of flexible buildings and other structures, is presented, in the commentary section of that Standard. The development of this factor was described by Solari and Kareem (1998). The gust effect factor is, in fact, a *dynamic response factor* (Section 5.3.4), defined in the same way as the structural factor, c_sc_d, in EN 1991-1-4.6, that is it is based on a 3-s gust wind speed, as adopted by ASCE -7. The calculation procedure is nearly identical to that in EN 1991-1-4.6, making use of the closed-form equations of Solari (1983). Expressions for maximum along-wind displacement and standard deviation and maximum along-wind acceleration are also given. However, no analytical procedure for cross-wind response is given.

In the AIJ Recommendations, a detailed procedure is applied to estimate the dynamic response of wind-sensitive structures. For along-wind response, a standard gust response factor approach along the lines of Equation 15.3 is used to determine a gust effect factor G_D. A mode shape correction for prediction of peak base bending moments for buildings with nonlinear mode shapes is provided (Tamura et al. 2004).

Vortex-induced cross-wind vibration and wind loads can also be determined from the AIJ, based on r.m.s. cross-wind base moment data obtained from wind-tunnel tests. Expressions for effective cross-wind-load distributions, displacement and acceleration are given. However, the cross-wind-response calculations are restricted to prismatic cross-sections with a height-to-breadth ratio no greater than 6, and to wind directions normal to a face of the building. Expressions for torsional angular acceleration and torsional wind-load distribution are also given. Guidelines for assessing potential aeroelastic instabilities including lock-in-type vortex resonance and galloping instabilities are presented.

The dynamic along-wind and cross-wind responses of tall buildings and towers are dealt with through a dynamic response factor, C_{dyn}, in Section 6, of AS/NZS1170.2; this corresponds with the definition of dynamic response factor given in Section 5.3.4 of this book. The methodology is a greatly modified version of that originally described by Vickery (1971).

Cross-wind base overturning moment and acceleration can be determined from cross-wind force spectrum coefficients, derived from wind-tunnel test data for a series of square and rectangular section buildings, with the incident wind normal to a face. Suggested values of damping for a range of steel and concrete structures under different stress levels are given. The importance of aeroelastic instabilities such as lock-in, galloping, flutter and interference are discussed separately in the Handbook for AS/NZS1170.2. However, a 'diagnostic' method for the cross-wind response of chimneys, masts and poles of circular cross-section is provided in the Standard itself.

15.9 INTER-CODE COMPARISONS

As discussed in Section 15.1, several numerical comparisons have been made of wind-load calculations by various international codes and standards. The results from two of those comparisons are summarised here.

Holmes et al. (2009) described an extensive comparison of calculations of wind loads on three buildings from 15 codes and standards in the Asia-Pacific region. The three buildings comprising a generic low-rise, a medium-rise and a high-rise building are shown in Figure 15.2. The low-rise building was a typical steel-framed portal frame structure, located in a rural area; the medium-rise building was a 48-m high office building located in an urban area, and the high-rise building was 183 m tall, and also located in an urban area. In all three cases, design wind speeds at the top of each building were pre-specified; wind speeds with averaging times of 3-s, 10-min and 1-h were all specified, and participants selected an appropriate time, according to the stated averaging time in their own code or standard at the time of the comparison.

The comparisons for the low-rise building showed coefficients of variation of 20–31% in the *net* pressures across the building surfaces. These relatively high values were partly caused by differences in the treatment of *internal* pressures, since a large opening in the windward wall of the building was specified. Somewhat smaller coefficients of variation of 13–26% were obtained for the wind loads on a large door and a small window in the building.

For the medium-rise building, the coefficients of variation for the predictions of base shear and bending moments, and cladding pressures, were consistent at 22–23%. The coefficients of variation for along-wind and cross-wind base shears and moments on the 183-m high-rise building were relatively small at 14–17%. This was attributed to the common origin of many of the code provisions for dynamic response to wind.

The comparisons of Holmes (2014), although limited in scope, are interesting because of the direct comparison of code values with wind-tunnel data. For wind acting normal to the wide face of the tall building in question (similar in dimensions to the high-rise building shown in Figure 15.2), AS/NZS 1170.2:2011 produced along-wind base bending moments that are within −4 to +8% of the averages of the wind-tunnel data (for three different wind speeds). ASCE 7-10 produced values 8–17% below the average of the wind-tunnel data, and the Hong Kong Code of Practice at the time (Buildings Department 2004) produced values 27–33% below the average of the wind-tunnel data. The paper gave explanations for the discrepancies; in the case of the Hong Kong Code of Practice, the differences were directly associated with low drag coefficients specified in the Code. That discrepancy will be addressed in a subsequent revision of the Hong Kong CoP.

15.10 GENERAL COMMENTS AND FUTURE DEVELOPMENTS

This chapter has reviewed the provisions of five major and current (at the time of writing) standards for wind loading. Considerable differences exist in both format and the type of information presented in these documents.

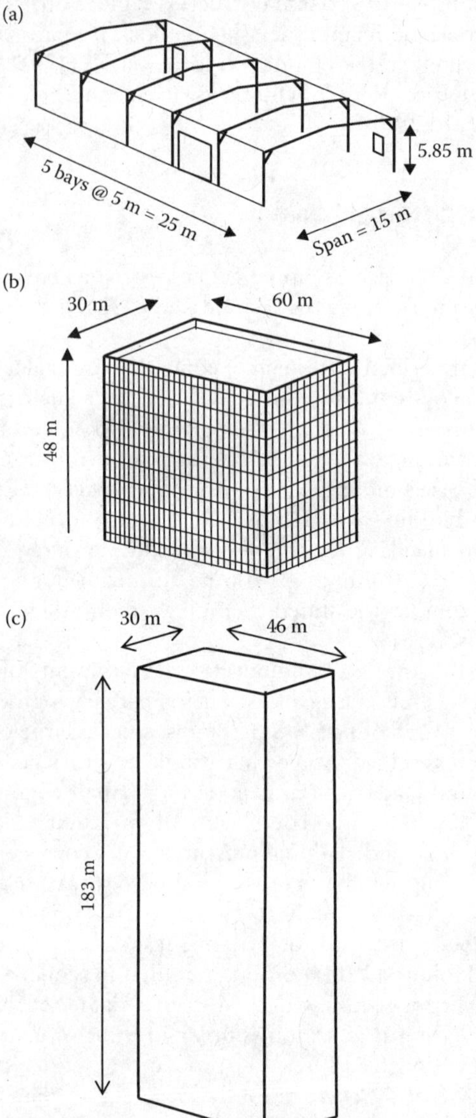

Figure 15.2 Three buildings used for an inter-code comparison in the Asia-Pacific region. (a) Low-rise industrial building; (b) medium-rise office building; (c) high-rise building. (From Holmes, J.D., Tamura, Y. and Krishna, P. 2009. Comparison of wind loads calculated by fifteen different codes and standards, for low, medium and high-rise buildings. *11th Americas Conference on Wind Engineering*, San Juan, Puerto Rico, 22–26 June.)

At present, the international standard on wind loading, ISO 4354:2009, is not generally used for structural design. It is difficult to use as an operating standard for design, without an accompanying set of wind climate data – that is the first link in the wind-loading chain of Figure 15.1. ISO 4354 has been adopted by Ethiopia, but without the necessary basic design wind speeds.

Eurocode 1 (EN 1991-1-4.6), however, has been adopted in nearly all of the European countries. It has also been adopted in South Africa, and in parts of Asia. ASCE 7 is also widely used outside of the United States, such as in the Middle East, and earlier versions have been adapted for certain Asian jurisdictions (such as the Philippines and Taiwan). AS/NZS 1170.2 is widely used in the South Pacific islands, and a version has been adapted for Malaysia.

It seems that a true international standard, although desirable for wind loads, remains, so far, an unachieved objective. The first requirement is a common format and notation. As indicated by Table 15.2, there are significant differences in the notation used across the codes, for what are essentially the same parameters.

For wide international acceptance in tropical and sub-tropical, as well as temperate climates, the special requirements of regions affected by typhoons (tropical cyclones or hurricanes) and thunderstorms will need to be incorporated. Some greater consistency in the treatment of effects such as directionality, topography and internal pressures could be achieved, as there is general agreement amongst researchers on the underlying physical and statistical principles behind these effects.

One problem appears to be that, in some jurisdictions, loading codes are the responsibility of government agencies, and regulators without the necessary wind engineering expertise, or practical experience that comes from using the documents for design. In other cases, the users of the document (i.e. practicing structural engineers) are the dominant 'players' in the development of a document. In further cases, academic wind engineers may be dominant, possibly leading to a document that is generally technically 'correct', but may contain ambiguities and is generally user-unfriendly. Ideally, all three groups should play a role in the development of a wind code or standard, to achieve a satisfactory user-friendly, non-ambiguous document that is also technically acceptable.

REFERENCES

American Society of Civil Engineers. 2010. Minimum design loads for buildings and other structures. ASCE/SEI 7-10, ASCE, Reston, VA, USA.

Architectural Institute of Japan. 2004. *AIJ Recommendations for Loads on Buildings*. AIJ, Tokyo.

Australian Building Codes Board. 2004. *Building Code of Australia*. ABCB, Canberra, Australia.

Australasian Wind Engineering Society. 2012. Wind loading handbook for Australia and New Zealand – Background to AS/NZS 1170.2 Wind actions. AWES-HB-001-2012.

British Standards Institution. 2005. Eurocode 1: Actions on structures – Parts 1–4: General actions – Wind actions. BS EN 1991-1-4.6, BSI, London, UK.

Buildings Department, Hong Kong. 2004. *Code of Practice on Wind Effects in Hong Kong*. Government of the Hong Kong Special Administrative Region, Buildings Department, Mongkok, Hong Kong, China.

CEN (European Committee for Standardization). 1994. Eurocode 1: Basis of design and actions on structures. Parts 2–4: Wind actions (draft). ENV-1991-2-4, CEN, Brussels.

Cook, N.J. 1990. *The Designer's Guide to Wind Loading of Building Structures. Part 2 Static Structures*. Building Research Establishment and Butterworths, London.

Davenport, A.G. 1977. The prediction of risk under wind loading. *Proceedings, 2nd International Conference on Structural Safety and Reliability*, Munich, Germany, 19–21 September, pp. 511–96.

Davenport, A.G. 1982. The interaction of wind and structures. In *Engineering Meteorology*, ed. E.J. Plate, pp. 527–572. Elsevier, Amsterdam, Netherlands.

Dorman, C.M.L. 1984. Tropical cyclone wind speeds in Australia. *Civil Engineering Transactions, Institution of Engineers, Australia*, CE26: 132–9.

ESDU International. 1983. Strong winds in the atmospheric boundary layer. Part 2: Discrete gust speeds, ESDU Data Item 83045, ESDU International, London, UK (revised 2002).

Holmes, J.D. 2002. A re-analysis of recorded wind speeds in Region A. *Australian Journal of Structural Engineering*, 4: 29–40.

Holmes, J.D. 2014. Along-wind and cross-wind response of a generic tall building – comparison wind-tunnel data with codes and standards. *Journal of Wind Engineering and Industrial Aerodynamics*, 132: 136–41.

Holmes, J.D., Allsop, A. and Ginger, J.D. 2014. Gust durations, gust factors and gust response factors in wind codes and standards. *Wind and Structures*, 19: 339–52.

Holmes, J.D., Baker, C.J., English, E.C. and Choi, E.C.C. 2005a. Wind structure and codification. *Wind and Structures*, 8: 235–50.

Holmes, J.D. and Ginger, J.D. 2012. The gust wind speed duration in AS/NZS 1170.2. *Australian Journal of Structural Engineering*, 13: 207–17.

Holmes, J.D., Kasperski, M., Miller, C.A., Zuranski, J.A. and Choi, E.C.C. 2005b. Extreme wind structure and zoning. *Wind and Structures*, 8: 269–81.

Holmes, J.D. and King, A.B. 2005. *A Guide to AS/NZS1170.2:2002 – Wind actions*. Warreen Publishing, Mentone, Victoria, Australia.

Holmes, J.D., Tamura, Y. and Krishna, P. 2009. Comparison of wind loads calculated by fifteen different codes and standards, for low, medium and high-rise buildings. *11th Americas Conference on Wind Engineering*, San Juan, Puerto Rico, 22–26 June.

International Standards Organization. 2009. Wind actions on structures. ISO International Standard. ISO 4354.

Kasperski, M. 1993. Aerodynamics of low-rise buildings and codification. *Journal of Wind Engineering and Industrial Aerodynamics*, 50: 253–63.

Kasperski, M. and Geurts, C. 2005. Reliability and code level. *Wind and Structures*, 8: 295–307.

Kijewski, T. and Kareem, A. 1998. Dynamic wind effects: A comparative study of provisions in codes and standards with wind tunnel data. *Wind and Structures*, 1: 77–109.

Letchford, C.W., Holmes, J.D., Hoxey, R.P. and Robertson, A.P. 2005. Wind pressure coefficients on low-rise structures and codification. *Wind and Structures*, 8: 283–94.

Mehta, K.C. 1998. Wind load standards. *Proceedings, Jubileum Conference on Wind Effects on Buildings and Structures*, Porto Alegre, Brazil, 25–29 May.

Ministry of Land, Infrastructure and Transport, Japan. 2000. Building Standard Law of Japan, Enforcement Orders Regulations and Notifications.

Peterka, J.A. and Shahid, S. 1998. Design gust wind speeds in the United States. *Journal of Structural Engineering (ASCE)*, 124: 207–14.

Solari, G. 1983. Gust buffeting II: Dynamic along-wind response. *Journal of Structural Engineering (ASCE)*, 119: 383–398.

Solari, G. and Kareem, A. 1998. On the formulation of ASCE-7-95 gust effect factor. *Journal of Wind Engineering and Industrial Aerodynamics*, 77–78: 673–84.

Standards Australia and Standards New Zealand. 2002. *Structural Design Actions. Part 0: General Principles*. Standards Australia, Sydney, and Standards New Zealand, Wellington. Australian/New Zealand Standard AS/NZS1170.0:2002 (amended 2003, 2005 and 2011).

Standards Australia and Standards New Zealand. 2011. *Structural Design Actions. Part 2: Wind Actions*. Standards Australia, Sydney, and Standards New Zealand, Wellington. Australian/New Zealand Standard AS/NZS1170.2:2011 (amended 2012 and 2013).

Tamura, Y., Kareem, A., Solari, G., Kwok, K.C.S., Holmes, J.D. and Melbourne, W.H. 2005. Aspects of the dynamic wind-induced response of structures and codification. *Wind and Structures*, 8: 251–68.

Tamura, Y., Kawai, H. Uematsu, Y., Okada, H. and Ohkuma, T. 2004. Documents for wind resistant design of buildings in Japan. *Workshop on Regional Harmonization of Wind Loading and Wind Environmental Specifications in Asia-Pacific Economies*, Atsugi, Japan, 19–20 November 2004.

Vickery, B.J. 1971. On the reliability of gust loading factors. *Civil Engineering Transactions, Institution of Engineers, Australia*, CE13: 1–9.

Appendix A: Terminology

Aerodynamic admittance The transfer function relating the gust spectral density to the spectral density of an aerodynamic force (Sections 5.3, 5.3.1 and 12.3.3).

Aerodynamic damping Aerodynamic forces proportional to the velocity of a structure, and additional to (or subtractional from) the structural damping (Section 5.5.1).

Background response That part of dynamic response to wind excluding the effects of resonant amplifications.

Bernoulli's Equation Equation describing irrotational and inviscid fluid flow (Section 4.2.1).

Blockage effect Distortion effect of wind-tunnel walls on measurements, particularly force and pressure measurements (Section 7.7).

Bluff body A body with a large frontal dimension, from which the airflow separates.

Body axes Axes defined by the body or structure (Section 4.2.2).

Boundary layer A region of reduced air velocities near the ground or the surface of a body (Section 3.1).

Cauchy Number The ratio of internal forces in a structure to inertial forces in the air (Chapter 7).

Coriolis force Apparent force on moving air due to the rotation of the earth.

Correlation The statistical relationship between two fluctuating random variables (Section 3.3.5).

Downburst Severe downdraft of air occurring in thunderstorms (Sections 1.3.5, 3.2.6 and 3.3.7).

Drag Along-wind force.

Dynamic response factor The ratio of the expected maximum structural response including resonant and correlation effects, to that ignoring both effects (Section 5.3.4).

Ekman spiral The turning effect of the wind vector with height in the atmospheric boundary layer (Chapter 3).

Flutter One-, or two-, degree-of-freedom aeroelastic instability, involving rotational motion (Section 5.5.3).

Friction velocity The non-dimensional measure of surface shear stress (Section 3.2.1).

Froude Number The ratio of inertial forces in the air to gravity forces (Chapter 7).

Galloping Single-degree-of-freedom translational aeroelastic instability (Section 5.5.2).

Geostrophic drag coefficient The ratio of friction velocity to geostrophic wind speed (Section 3.2.4).

Gradient wind Upper-level wind that can be calculated from the gradient wind equation (Section 1.2.4).

Gust factor The ratio of the expected maximum to mean value of wind speed, pressure or force.

Gust response factor The ratio of the expected maximum to mean structural response (Section 5.3.2).

Helmholtz resonance The resonance in internal pressure fluctuations associated with the compressibility of the air within a building, and the mass of air moving in and out of a dominant opening.

Inviscid Fluid flow in which the effects of viscosity are non-existent or negligible.

Isotach The contour of the constant basic design wind speed.

Jensen Number The ratio of the building dimension (usually height) to roughness length in the atmospheric boundary-layer flow (Section 4.4.5).

Lift Cross-wind force, usually but not necessarily, vertical.

Limit states design A method of structural design, which separately considers structural failure through collapse or overturning, from the functional (serviceability) aspects.

Lock in The enhancement of fluctuating forces produced by vortex shedding due to the motion of the vibrating body (Section 5.5.4).

Logarithmic law A mathematical representation of the profile of mean velocity with height in the lower part of the atmospheric boundary layer.

Manifold A device for averaging pressure measurements from several measurement positions (Section 7.5.2).

Mechanical admittance The transfer function relating the spectral density of aerodynamic forces to the spectral density of structural response (Section 5.3.1).

Peak factor The ratio of maximum minus mean value, to standard deviation, for wind velocity, pressure, force or response (Section 5.3.3).

Peak gust The maximum value of wind speed in a defined time period.

Pressure coefficient Surface pressure made non-dimensional by the dynamic pressure in the wind flow (Section 4.2.1).

Quasi-steady A model of wind loading that assumes that wind pressures on buildings fluctuate directly with the fluctuations in wind speed immediately upstream.

Return period Inverse of probability of exceedence of an extreme value (Chapter 2).

Reynolds Number The ratio of inertial forces to viscous forces in fluid flow (Section 4.2.4).

Roughness length A measure of the aerodynamic roughness of a surface, which affects the boundary-layer flow over it (Section 3.2.1).

Safety index A measure of the probability of failure of a structure. 'Reliability Index' is also used (Section 2.6.2).

Scruton number A non-dimensional parameter incorporating the ratio of structural mass to fluid mass, and structural damping, which is a measure of the propensity of a structure to resonant dynamic response (Section 11.5.1).

Shear stress (fluid flow) The force per unit area exerted by a layer of moving fluid on the adjacent layer.

Spectral density A measure of the contribution to a fluctuating quantity (e.g. wind velocity, wind pressure and deflection) within a defined frequency bandwidth.

Stagnation point The point on a body where the approaching flow is brought to rest.

Stationary The description of a random process whose statistical properties do not change with time.

Strouhal Number A non-dimensional vortex-shedding frequency (Section 4.6.3).

Synoptic winds Winds created by large-scale meteorological systems, especially gales produced by extra-tropical depressions.

Thunderstorm A thermally driven local storm capable of producing strong downdraft winds (Section 1.3.3).

Tornado A local intense storm formed from thunderclouds, with intense winds rotating around a vortex structure (Sections 1.3.4 and 3.2.7).

Tropical cyclone An intense tropical storm which can occur over warm tropical oceans. A generic name which incorporates 'hurricane' (used for Caribbean and northwest Atlantic storms) and 'typhoon' (used in the northwest Pacific) (Section 1.3.2).

Turbulence Fluctuations in fluid flow. In meteorology and wind engineering, the term 'gustiness' is also used.

von Karman's constant The dimensionless constant in the logarithmic law for the profile of mean velocity in a turbulent boundary layer.

Vortex shedding The periodic shedding of eddies formed from the rolling up of the boundary layer shed from a bluff body.

Vulnerability curve A graphical relationship between the damage produced by a severe wind event and a representative wind speed – usually a peak gust in the event (Section 1.6.2).

Wake The region of low velocity and turbulent flow in the region downstream of a body.

Wind axes Axes parallel and normal to the mean wind direction (Section 4.2.2).

Appendix B: List of symbols

Note: Symbols that are specific to particular wind-loading codes and standards described in Chapter 15 are not listed in this appendix.

a	i. Coriolis acceleration (Chapter 1)
	ii. Scale factor in probability distribution (Chapter 2)
	iii. Empirical constant (Equation 3.9)
	iv. Modal (generalised) coordinate
	v. Parameter in wideband fatigue (Equation 5.59)
a_s	Speed of sound
b	i. Dimensional parameter (Equation 1.22)
	ii. Cross-wind breadth of the bluff body
	iii. Parameter in wideband fatigue (Equation 5.59)
	iv. Ratio of bulk modulus of air to the bulk modulus of the building (K_A/K_B)
	v. Diameter of antenna dish (Chapter 14)
c	i. Translation speed of a hurricane (Chapter 3)
	ii. Ground clearance of an elevated hoarding (Figure 4.8)
	iii. Chord of a flat plate (Section 4.6.3)
	iv. Damping constant
	v. Scale factor in Weibull Distribution
	vi. Distance of the solar panel from the roof edge (Chapter 14)
d	i. Effective diameter of rod-type objects (Chapter 1)
	ii. Along-wind dimension of a building or bluff body, and chord of a bridge deck
	iii. Standoff of a solar panel from the roof surface (Chapter 14)
	iv. Depth of antenna dish
	v. Diameter of a pole (Appendix E)
$d()$	Drag force per unit length
e	i. Eccentricity
	ii. Eaves overhang
	iii. Wall thickness of a pole (Appendix E)
f	i. Coriolis parameter ($=2\Omega \sin \lambda$)
	ii. Force per unit length
	iii. Focal length of the parabolic antenna dish (Chapter 14)
$f()$	Probability density function
$f(\phi)$	Function of mode shape (Equation 11.17)
g	i. Gravitational constant
	ii. Peak factor
h	Height of the building or bluff body
$h(t)$	Unit impulse response function (Chapter 5)

h_c Height of a canopy (Chapter 14)
h_e Height to the eaves
h_i Depth of inflow in a tornado (Chapter 7)
h_p Height of a parapet (Chapter 8)
i, j Indices
k
i. $(\rho_a C_F)/(2 \rho_m \ell)$ (Chapter 1)
ii. Shape factor in generalised extreme value distribution (Chapter 2 and Appendix C)
iii. von Karman's constant (Chapter 3)
iv. Constant for the type of topographic feature
v. Decay constant for a hurricane (Equation 3.31)
vi. Spring stiffness
vii. Average surface roughness height (Chapter 4)
viii. Orifice constant (Chapter 6)
ix. Mode shape parameter (Equation 11.15)
x. Parameter (Equation 14.9)
xi. Parameter in the generalised Pareto distribution (Appendix C)

k_x Exponent (Equation 9.22)
k_y Exponent (Equation 9.23)
ℓ
i. Characteristic dimension for compact objects
ii. Correlation length
iii. Length of a tornado path (Chapter 13)
iv. Length of a solar panel (Chapter 14)

ℓ_e Effective length of air 'slug' (Chapter 6)
$\ell()$ Lift (cross-wind) force per unit length (Chapter 11)
m
i. Mass or mass per unit length
ii. Exponent in fatigue s–N relationship
iii. Mean value (Chapter 9)
iv. Exponent (Equation 9.18)

n
i. Frequency
ii. Stress cycle
iii. Number of events, for example, number of tornado occurrences in a region (Chapter 13)
iv. Exponent (Equations 5.64 and 9.17)

n_c Characteristic frequency for internal pressure fluctuations
n_s Vortex-shedding frequency
p
i. Pressure
ii. Probability (Chapter 2)
iii. Exponent (Equation 5.66)

p_o
i. Central pressure of a tropical cyclone (Chapter 1)
ii. Ambient (static) pressure

p_f Probability of failure (Chapter 2)
p_n
i. Atmospheric pressure at the edge of a storm (Chapter 1)
ii. Net pressure (Chapter 14)

p_L Leeward face pressure
p_w Windward face pressure
q Dynamic pressure
r
i. Radius of curvature – of isobars (Chapter 1), or square section (Chapter 4)
ii. Risk (Chapter 2)

iii. Radius in a downburst (Equation 3.13)
iv. General structural response, or load effect
v. Ratio A_L/A_W (Chapter 6)
vi. Roughness factor ($=2I_u$)
vii. Radius of gyration (Chapter 12)
viii. Rate of intersection of tornadoes with a transmission line

r_c Radius of the core in a tornado (Chapter 7)
r_u Radius of the updraft region in a tornado (Chapter 7)
s i. Position factor (Equation 3.33)
ii. Stress
iii. Height for the calculation of load effects
iv. Span length of a transmission line or free-standing wall (Chapters 13 and 14)
v. Clear space under a bridge (Chapter 14)
vi. Skewness (Appendix C)

t i. Time
ii. Thickness of sheet objects (Chapter 1)
iii. Parameter for interference factor (Equation 14.9)

u, v and w Orthogonal velocity components
u Location parameter of extreme value distribution (Chapter 2)
u_o Wind speed level (Chapter 2)
u_* Friction velocity (Chapter 3)
v Fractional change of internal volume (Chapter 6)
v_m Velocity of flying debris
w i. Shape factor in Weibull Distribution (Chapter 2)
ii. Shortest distance from the smoothed coastline (Chapter 3)
iii. Width of the tornado or downburst path (Chapter 13)
iv. Width of a building (Figure 14.12)

w_a Average width of a tower (Appendix E)
w_b Base width of a tower (Appendix E)
w_c Width of a canopy (Chapter 14)
w_o Assumed wind load per unit height (Equation 7.11)
x Horizontal distance travelled by a wind-borne missile (Chapter 1)
Distance travelled inland by a hurricane (Chapter 3)
$x(t)$ Random process, structural response
x_i Distance to the inner boundary layer
x, y and z Cartesian coordinate system (z is vertical)
z Variable of integration, or transformed random variable (Appendix C)
z_h Zero-plane displacement
z_o Roughness length
z^* Characteristic height (Equation 3.13)
A i. Scaling parameter (Equation 1.10)
ii. Reference or frontal area
iii. Parameter in cross-wind response (Equation 11.19)
iv. Area of a region (Chapter 13)

A_i Flutter derivative for rotational motion (Chapters 5 and 12)
A_i^* Normalised flutter derivative for rotational motion (Chapter 12)
A_L Area of openings on leeward wall
A_W Area of openings on windward wall

B i. Exponent (Equation 1.10)
ii. Background factor (also B_s)
iii. Bandwidth parameter (Equation 11.18)

C i. Decay constant (Equation 3.30)
ii. Modal damping

C_d Coefficient of drag force per unit length

C_D Drag coefficient

C_f Coefficient of aerodynamic force per unit length

C_F Aerodynamic force coefficient

C_I Inertial coefficient = $\ell_e/\sqrt{A}$ (Chapter 6)

C_k Equivalent glass design coefficient

C_M Moment coefficient

C_N Normal force coefficient

C_p Pressure coefficient

C_p^* Effective peak pressure coefficient (Equation 9.7)

C_{pn} Net pressure coefficient (Chapter 14)

C_{ps} Equivalent pressure coefficient for glass loading (Equation 9.20)

C_T Torque coefficient

C_X Coefficient of X force

C_Y Coefficient of Y force

C_Z Coefficient of Z force

$Co()$ Co-spectral density

D i. Damage index (Chapter 1)
ii. Drag
iii. Nominal dead load (Chapter 2)
iv. Accumulated damage (Chapters 5 and 9)

D_a Antenna drag (Chapter 14)

D_e Effective tower drag with antenna attached (Chapter 14)

D_t Tower drag (Chapter 14)

E i. Young's modulus
ii. Non-dimensional spectral density (Chapter 11)

F i. Force
ii. Non-dimensional parameter (Section 11.4)

$F()$ Cumulative probability distribution function

F_i Parameter in along-wind response (Chapter 11)

G i. Generalised mass
ii. Gust factor, gust response factor
iii. Shear modulus

$G()$ Complementary cumulative probability distribution (Appendix C)

H Height of the hill (Chapter 3)

H_i Flutter derivative for vertical motion (Chapters 5 and 12)

H_i^* Normalised flutter derivative for vertical motion (Chapter 12)

$H()$ Dynamic amplification factor; square root of mechanical admittance

I i. Fixing strength integrity parameter (Chapter 1)
ii. Influence coefficient
iii. Mass moment of inertia (Chapters 7 and 12)
iv. Second moment of area
I_u, I_v and I_w – Turbulence intensities

Je Jensen Number

K i. $\frac{1}{2}\frac{\rho_a}{\rho_m}\frac{U^2}{gl}\frac{\ell}{t}$ Tachikawa Number (Chapter 1)
ii. Modal stiffness
iii. Constant in fatigue s–N relationship
iv. Bulk modulus (Chapter 6)
v. Constant (Equation 7.1)
vi. Constant (Equation 9.18)
vii. Mode shape factor (Equation 11.16)

K_{ao} Parameter for negative aerodynamic damping (Equation 11.19)
K_i Interference factor
K_p Porosity factor
K_A Bulk modulus of air (Chapter 6)
K_B Bulk modulus of a building (Chapter 6)
K_W Correlation length factor (Equation 11.16)
K_θ Wind incidence factor (Equation 11.5)

L i. Lifetime of a structure (Chapter 2)
ii. Lift (cross-wind) force
iii. General length (Chapter 7)
iv. Length of a building (Figure 10.4)
v. Length of a transmission line (Chapter 13)

L_N Parameter to calculate frequency of the lattice tower (Appendix E)
L_S Span of a bridge (Appendix E)
M Moment
M_b Base bending moment
M_t Topographic multiplier

N i. Number of wind-direction sectors (Chapters 2 and 13)
ii. Cycles to failure by fatigue
iii. Number of patch loads for a guyed mast (Chapter 11)
iv. Number of samples of a random variable (Appendix C)

Q i. Generalised force
ii. Volume flow rate

R i. Return period (Chapter 2)
ii. Structural resistance
iii. Characteristic radius (Equation 3.13)
iv. Radius of maximum wind in a tornado (Figure 3.4)
v. Reduction factor for hurricane wind speeds (Section 3.6)
vi. Resonant response factor
vii. Radius of a liquid damper (Chapter 9)
viii. Rise of an arch (Figure 10.4)

R_c Combined return period for winds from more than one type of storm
R_j Structural response (load effect) due to unit modal coordinate, in mode, j
Re Reynolds Number
Re_b Reynolds Number based on diameter
Re_k Reynolds Number based on roughness height
Ro Rossby Number

S i. Structural load effect (Chapter 2)
ii. Size factor
iii. Span of arch

S^* Non-dimensional parameter for internal pressure (Chapter 6)

$S()$	Spectral density
Sc	Scruton Number
St	Strouhal Number
T	i. Time of flight of a missile ii. Sample time (Chapter 3) iii. Time period
U	Wind speed
U_0	Free-stream velocity (Chapter 4)
U_f	Wind speed for threshold of flight of debris
U_R	Wind gust speed corresponding to return period, R
V	Wind speed (in some code notations – Chapter 15)
V_o	Internal volume
W	i. Nominal wind load (Chapter 2) ii. Weighting factor
X	General random variable (Appendix C)
$X(t)$	Deflection of a structure

Superscripts

$\bar{}$	Mean (time-averaged) value
$'$	Fluctuating value
$\dot{}$	Differentiation with respect to time
$\hat{}$	Maximum value
$\check{}$	Minimum value
$*$	Characteristic value (Chapter 3)

Subscripts

$_0$	Initial value, free-stream value
$_1$	Natural frequency (Chapter 5)
$_{10}$	10-m height
$_a$	Air
$_{av}$	Average
$_b$	Base of the building, tower or pole, breadth or diameter and background
$_c$	Tornado core, canopy, cable and conductor
$_d$	Drag force per unit length, (antenna) dish
$_e$	i. External ii. Eaves iii. Effective (Chapter 6)
$_{\text{eff}}$	Effective
$_{\text{env}}$	Envelope
$_f$	i. Flight speed (Chapter 1) ii. Failure (Chapter 2)
$_g$	Geostrophic (Chapter 3)
$_h$	Zero-plane displacement (Chapter 3)
$_i$	i. Index of the position or wind direction ii. Internal iii. Inflow layer of the tornado (Chapter 7)
$_j$	Index of the mode of vibration
$_k$	Roughness height
$_\ell$	Lift (cross-wind) force per unit length
$_{\text{lat}}$	Lateral

m i. Missile (Chapter 1)
ii. Area of members (Chapter 4)
iii. Model (Chapter 7)
iv. Bending moment (Chapter 11)
max Maximum
n Net (pressure)
ni Normal to face *i* (Chapter 11)
p i. Peak
ii. Porosity (Chapter 4)
iii. Prototype (Chapter 7)
iv. Parapet (Chapter 8)
q Shearing force
r i. Radial wind velocity component in a tornado (Chapter 3)
ii. Ratio of the model to prototype value (Chapter 7)
ref Reference value
s i. Local gust speed (Equation 1.21)
ii. Sound (speed of)
iii. Structure
iv. Shedding
v. Sloshing
t Topographic, total area (Chapter 4), top of the building, tower or pole
u Updraft region in the tornado (Chapter 7)
v Vertical wind velocity component in a tornado (Chapter 3)
x *x*-Direction response
y *y*-Direction response
B i. Background response
ii. Barrier (Figure 7.4)
D Drag force
F Flutter (Chapter 12)
H Helmholtz resonance
L Leeward, lift (cross-wind) force
N Net force (Chapter 4)
Pl Patch load (Equation 11.27)
R Return period (Chapter 2), resonant response
T i. Top of a wind tunnel (Figure 7.3)
ii. Torsional (frequency) (Chapter 12)
TL Turbulence length scale (Equation 11.28)
V Vertical bending (frequency) (Chapter 12)
W Windward
1 First mode of vibration
θ Tangential wind velocity component in a tornado (Chapter 3)

Greek symbols

α i. Exponent in power law (Equation 3.7)
ii. Decay parameter for hurricanes (Equation 3.39)
iii. Angle of attack
iv. Span reduction factor for transmission lines or walls (Chapters 13 and 14)
v. Roof pitch angle
α_j Coefficient for influence of resonant mode, *j*, on a structural load effect
β i. Safety index (Chapter 2)

ii. Decay constant for hurricanes (Equation 3.40)
iii. Mode shape exponent (Chapter 7)
iv. Angle of intersection of tornado path width with the transmission line (Chapter 13)
v. Angle of a solar panel to the roof surface (Figure 14.12)

γ Ratio of specific heats of air
δ Solidity of a porous body
ε
i. Characteristic height (Equation 3.13)
ii. Spectral bandwidth parameter (Equation 5.62)

ϕ Upwind slope of a topographic feature
$\phi()$ Mode shape
γ Ratio of specific heats of air (Chapter 6)
γ_D Dead load factor
γ_W Wind load factor
η Ratio of damping to critical
φ
i. Resistance factor (Chapter 2)
ii. Wall porosity (Chapter 6)
iii. Phase angle (Chapter 11)
iv. Transmission line direction (Chapter 13)

κ Surface drag coefficient
λ
i. Angle of latitude (Chapter 1)
ii. Crossing rate (Chapter 2)
iii. Scaling factor (Equation 3.13)
iv. Parameter in wideband fatigue (Equation 5.59)
v. Factor for guyed mast response (Equation 11.28)
vi. Parameter in Poisson distribution (Appendix C)
vii. Parameter for pole frequency calculation (Appendix E)

μ
i. Dynamic viscosity
ii. Moment of spectral density (Chapter 5)

ν
i. Kinematic viscosity of air
ii. up-crossing rate and cycling rate

ξ Empirical factor in an approximate formula for the natural frequency of steel Chimneys (Appendix E)
π
i. Ratio of circumference to the diameter of a circle
ii. (With subscript) non-dimensional group

θ
i. Angular rotation
ii. Angle of incidence
iii. (As a subscript) tangential velocity component in a tornado (Chapter 3)
iv. Angle of downburst path to the transmission line (Chapter 13)

ρ, $\rho(y)$ Correlation coefficient
$\rho(n)$ Normalised co-spectral density
ρ_a Air density
ρ_b Average building density
σ
i. Scale factor (Chapter 2)
ii. Standard deviation
iii. Parameter in lognormal distribution (Appendix C)
iv. Parameter in the generalised Pareto distribution (Appendix C)

τ
i. Equilibrium, or response, time for internal pressure (Chapter 6)
ii. Dummy time variable for integration (Chapter 5)

iii. Gust duration (Chapter 3)
iv. Averaging time (Chapter 7)

τ_o Surface shear stress
υ Rate of occurrence of tornadoes per unit area (Chapter 13)
ω Circular frequency
Δ Mean deflection (Figure 9.1)
Φ Non-dimensional parameter (Chapter 6)
$\Phi()$ Cumulative distribution function of a normal (Gaussian) random variable (zero mean and unit standard deviation)
$\Gamma()$ Gamma function
Γ Imposed circulation on the tornado (Chapter 7)
Ω Angular velocity of rotation of the earth
Π Repeated multiplication
Σ Repeated summation
$X^2(n)$ Aerodynamic admittance

Appendix C: Probability distributions relevant to wind engineering

C.1 INTRODUCTION

Probability distributions are an essential part of wind engineering as they enable the random variables involved such as wind speeds, wind directions, surface pressures and structural response (e.g. deflections and stresses), to be modelled mathematically. Some of these variables are random *processes*, i.e. they have time-varying characteristics, as shown in Figure C.1. The probability density describes the distribution of the magnitude or amplitude of the process, without any regard to the time axis.

The appendix will cover: (1) some basic statistical definitions; (2) a selection of probability distributions for the complete population of a random variable—the normal (Gaussian), log-normal, Weibull and Poisson—will be considered; and (3) the three types of Extreme Value Distributions and the closely related Generalised Pareto Distributions will be discussed.

C.2 BASIC DEFINITIONS

C.2.1 Probability density function

The probability density function (PDF), $f_X(x)$ is the limiting probability that the value of a random variable, X, lies between x and $(x + \delta x)$. Thus, the probability that X lies between a and b is

$$\Pr\{a < X < b\} = \int_a^b f_X(x)\,dx \tag{C.1}$$

Since any value of X must lie between $-\infty$ and $+\infty$:

$$\int_{-\infty}^{\infty} f_X(x)dx = \Pr\{-\infty < X < \infty\} = 1$$

Thus, the area under the graph of $f_X(x)$ versus x must equal 1.0.

C.2.2 Cumulative distribution function

The cumulative distribution function (CDF) $F_X(x)$ is the integral between $-\infty$ and x of $f_X(x)$.

$$\text{That is } F_X(x) = \int_{-\infty}^{x} f_X(z)dz = \Pr\{-\infty < \mathrm{X} < \mathrm{x}\} = \Pr\{\mathrm{X} < \mathrm{x}\} \tag{C.2}$$

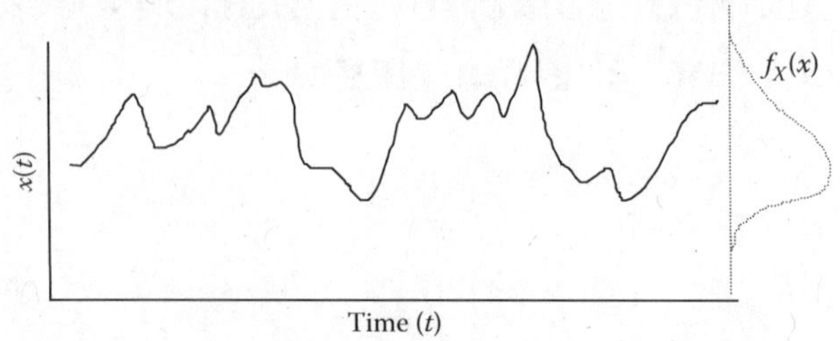

Figure C.1 A random process and amplitude probability density.

The complementary CDF, usually denoted by $G_X(x)$, is

$$G_X(x) = 1 - F_X(x) = \Pr\{X > x\} \quad \text{(C.3)}$$

$F_X(a)$ and $G_X(b)$ are equal to the areas indicated in Figure C.2.
Note that

$$f_X(x) = \frac{dF_X(x)}{dx} = -\frac{dG_X(x)}{dx} \quad \text{(C.4)}$$

The following basic statistical properties of a random variable are defined and their relationship to the underlying probability distribution is given.

Mean

$$X = (1/N)\,\Sigma_i x_i = \int_{-\infty}^{\infty} x\, f_X(x) dx \quad \text{(C.5)}$$

Thus, the mean value is the first moment of the PDF (i.e. the x coordinate of the centroid of the area under the graph of the PDF).

Variance

$$\sigma_X^2 = (1/N)\,\Sigma_i [x_i - X]^2 \quad \text{(C.6)}$$

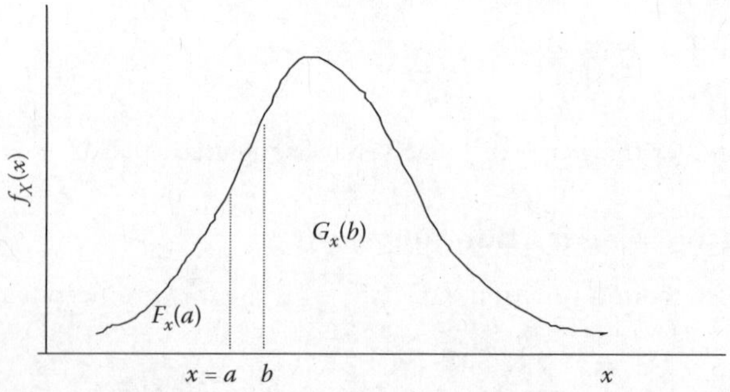

Figure C.2 Probability density function and cumulative distribution functions.

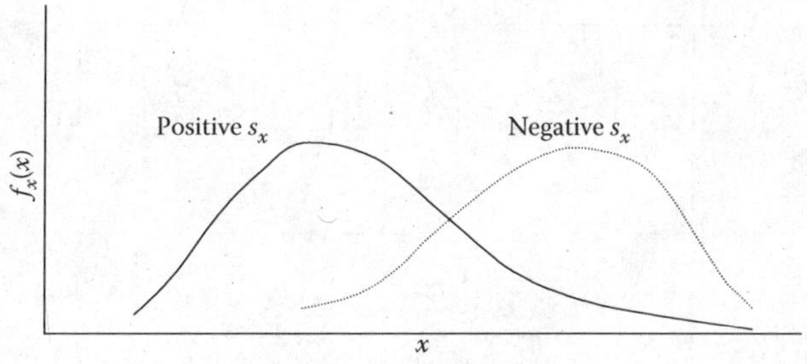

Figure C.3 Positive and negative skewness.

σ_X (the square root of the variance) is called the standard deviation

$$\sigma_X^2 = \int_{-\infty}^{\infty} (x - \bar{X})^2 f_x(x) dx \tag{C.7}$$

Thus, the variance is the second moment of the PDF about the mean value. It is analogous to the second moment of the area of a cross section about a centroid.

Skewness

$$s_X = [1/(N\sigma_X^3)]\Sigma_i[x_i - \bar{X}]^3$$
$$= (1/\sigma_X^3)\int_{-\infty}^{\infty} (x - \bar{X})^3 f_X(x) dx \tag{C.8}$$

The skewness is the normalised third moment of the PDF. Positive and negative skewness are illustrated in Figure C.3. A distribution that is symmetrical about the mean value has a zero skewness.

C.3 PARENT DISTRIBUTIONS

C.3.1 Normal or Gaussian distribution

For $-\infty < X < \infty$,

$$f_x(x) = \frac{1}{\sqrt{2\pi}\sigma_X} \exp\left[\frac{-(x - \bar{X})^2}{2\sigma_x^2}\right] \tag{C.9}$$

where $\bar{X}$, σ_X are the mean and standard deviation.

This is the most commonly used distribution. It is a symmetrical distribution (zero skewness) with the familiar bell shape (Figure C.4).

$$F_X(x) = \Phi\left(\frac{x - \bar{X}}{\sigma_X}\right) \tag{C.10}$$

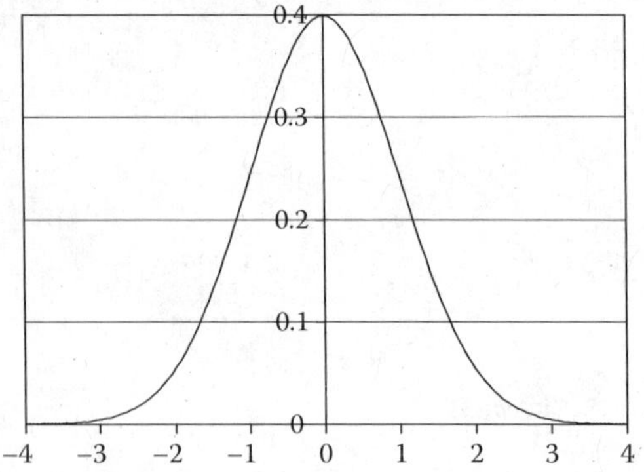

Figure C.4 Normal distribution.

where Φ() is the CDF of a normally distributed variable with a mean of zero and a unit standard deviation, that is

$$\Phi(u) = \left(\frac{1}{\sqrt{2\pi}}\right)\int_{-\infty}^{u} \exp\left(\frac{-z^2}{2}\right) dz \tag{C.11}$$

Tables of Φ (u) are readily available in statistics textbooks.

If $Y = X_1 + X_2 + X_3 + \cdots + X_N$,

where $X_1, X_2, X_3, \ldots, X_N$, are random variables with any distribution, the distribution of Y tends to become normal as N becomes large. If $X_1, X_2, \ldots$ themselves have normal distributions, then Y has a normal distribution for any value of N.

In wind engineering, the normal distribution is used for turbulent velocity components, and for response variables (e.g. deflection) of a structure undergoing random vibration. It should be used for variables that can take both negative and positive values; hence, it would not be suitable for scalar wind speeds that can only be positive.

C.3.2 Lognormal distribution

$$f_X(x) = \frac{1}{\sqrt{2\pi}\sigma x}\exp\left[\frac{-\{\log_e(x/m)\}^2}{2\sigma^2}\right] \tag{C.12}$$

where

the mean value $\bar{X}$ is equal to $m \exp(\sigma^2/2)$

the variance $\sigma_X{}^2$ is equal to $m^2 \exp(\sigma^2)\,[\exp(\sigma^2) - 1]$

$\log_e x$ in fact has a normal distribution with a mean value of $\log_e m$ and a variance of σ^2.

If a random variable $Y = X_1 \cdot X_2 \cdot X_3 \cdots X_N$, where $X_1, X_2, X_3, \ldots, X_N$ are random variables with any distribution, the distribution of Y tends to become lognormal as N becomes large. Thus, the lognormal distribution is often used for the distribution of a variable that is itself the product of a number of uncertain variables – for example, wind speed factored by multipliers for terrain, height, shielding, topography and so on.

The lognormal distribution has a positive skewness equal to $[\exp(\sigma^2) + 2][\exp(\sigma^2) - 1]^{1/2}$.

C.3.3 'Square-root-normal' distribution

Now, consider the distribution of $Z = X^2$, where X has the normal distribution.

$$f_Z(z) = \frac{1}{2(\sigma_X/\bar{X})\sqrt{2\pi z}}\left\{\exp\left[-\left(\frac{1}{2}\right)\left(\frac{\sqrt{z}-1}{(\sigma_X/\bar{X})}\right)^2\right] + \exp\left[-\left(\frac{1}{2}\right)\left(\frac{\sqrt{z}+1}{(\sigma_X/\bar{X})}\right)^2\right]\right\} \tag{C.13}$$

and the CDF is

$$F_Z(z) = \Phi\left(\frac{\sqrt{z}-1}{(\sigma_X/\bar{X})}\right) + \Phi\left(\frac{\sqrt{z}+1}{(\sigma_X/\bar{X})}\right) - 1 \tag{C.14}$$

This distribution is useful for modelling the pressure fluctuations on a building which are closely related to the square of the upwind velocity fluctuations, which can be assumed to have a normal distribution (e.g. Holmes, 1981).

C.3.4 Weibull distribution

$$f_X(x) = \left(\frac{wx^{w-1}}{c^w}\right)\exp\left[-\left(\frac{x}{c}\right)^w\right] \tag{C.15}$$

$$F_X(x) = 1 - \exp\left[-\left(\frac{x}{c}\right)^w\right] \tag{C.16}$$

where c (>0) is known as the scale parameter, with the same units as x, w(>0) is the shape parameter (dimensionless).

The shape of the PDF for the Weibull distribution is quite sensitive to the value of the shape factor, w, as shown in Figure C.5. The Weibull distribution can only be used for random variables that are always positive. It is often used as the parent distribution for wind

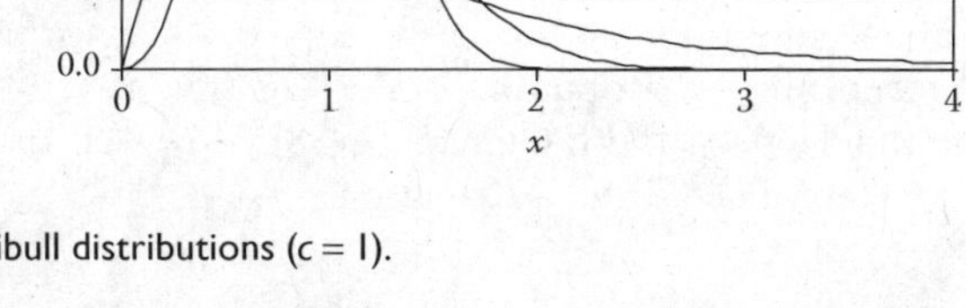

Figure C.5 PDFs for Weibull distributions (c = 1).

speeds, with k in the range of about 1.5–2.5. The Weibull distribution with $w = 2$ is a special case known as the Rayleigh distribution. When $w = 1$, it is known as the Exponential distribution.

C.3.5 Poisson distribution

The previous distributions are applicable to *continuous* random variables, that is x can take any value over the defined range. The Poisson distribution is applicable only to positive *integer* variables, for example, the number of cars arriving at an intersection in a given time, number of exceedences of a defined pressure level at a point on a building during a windstorm.

In this case, there is no PDF but instead a probability mass function:

$$p_X(x) = \lambda^x \frac{\exp(-\lambda)}{x!} \tag{C.17}$$

where λ is the mean value of X. The standard deviation is $\lambda^{1/2}$.

The Poisson distribution is used quite widely in wind engineering to model exceedences or up-crossings of a random process such as wind speed, pressure or structural response, or events such as number of storms occurring at a given location. It can also be written in the form:

$$p_X(x) = (\upsilon T)^x \frac{\exp(-\upsilon T)}{x!} \tag{C.18}$$

where υ is now the mean rate of occurrence per unit time, and T is the time period of interest.

C.4 EXTREME VALUE DISTRIBUTIONS

In wind engineering, as in other branches of engineering, we are often concerned with the largest values of a random variable (e.g. wind speed) rather than the bulk of the population.

If a variable Y is the maximum of n random variables, $X_1, X_2, \ldots, X_n$ and the X_i are all independent

$$F_Y(y) = F_{X1}(y) \cdot F_{X2}(y) \cdots F_{Xn}(y)$$

since $P[Y < y] = P[\text{all } n \text{ of the } X_i < y] = P[X_1 < y] \cdot P[X_2 < y] \cdots P[X_n < y]$

In the special case that all the X_i are identically distributed with CDF $F_X(x)$,

$$F_Y(y) = [F_X(x)]^n \tag{C.19}$$

If the assumptions of common distribution and independence of the X_i hold, the shape of the distribution of Y is insensitive to the exact shape of the distribution of the X_i. In this case, three limiting forms of the distributions of the largest value Y, as n becomes large may be identified (Fisher and Tippett, 1928; Gumbel, 1958). However, they are all special cases of the generalised extreme value (GEV) distribution.

C.4.1 Generalised extreme value (GEV) distribution

The CDF may be written as

$$F_Y(y) = \exp\left\{-\left[1 - \frac{k(y-u)}{a}\right]^{1/k}\right\} \tag{C.20}$$

In this distribution, k is a shape factor, a is a scale factor and u is a location parameter. There are thus three parameters in this generalised form.

The three special cases are:

Type I ($k = 0$). This is also known as the Gumbel distribution.
Type II ($k < 0$). This is also known as the Frechet distribution.
Type III ($k > 0$). This is a form of the Weibull distribution.

The Type I can also be written in the form:

$$F_Y(y) = \exp\{-\exp[-(y-u)/a]\} \tag{C.21}$$

The GEV is plotted in Figure 2.1 in Chapter 2, with k equal to −0.2, 0 and 0.2 such that the Type I appears as a straight line, with a reduced variate, z, given by

$$z = -\log_e\{-\log_e[F_Y(y)]\}$$

As can be seen, the Type III ($k = +0.2$) curves in a way to approach a limiting value at high values of the reduced variate (low probabilities of exceedence). Thus, the Type III distribution is appropriate for phenomena that are limited in magnitude for geophysical reasons, including many applications in wind engineering. The Type I can be assumed to be *a* conservative limiting case of the Type III, and it has only two parameters (a and u), since k is predetermined to be 0. For that reason, the Type I (Gumbel distribution) is easy to fit to the actual data, and is very commonly used as a model of extremes for wind speeds, wind pressures and structural response.

C.4.2 Generalised Pareto distribution

The complementary CDF is

$$G_X(x) = \left[1 - \left(\frac{kx}{\sigma}\right)\right]^{\frac{1}{k}} \tag{C.22}$$

The PDF is

$$f_X(x) = \left(\frac{1}{\sigma}\right)\left[1 - \left(\frac{kx}{\sigma}\right)\right]^{\left(\frac{1}{k}\right)-1} \tag{C.23}$$

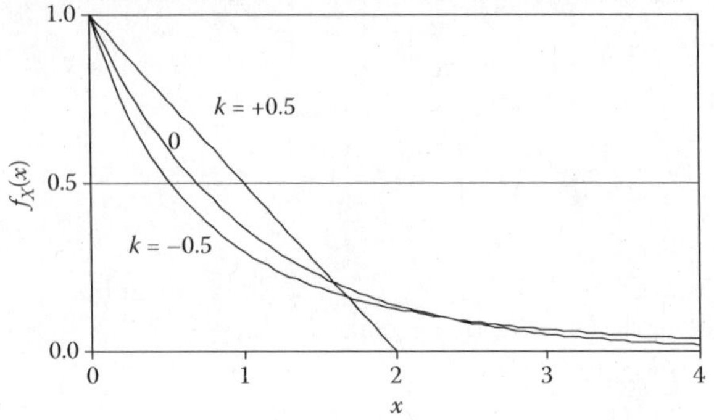

Figure C.6 PDFs for Generalised Pareto distributions ($\sigma = 1$).

where k is the shape parameter and σ is a scale parameter. The range of X is $0 < X < \infty$ when $k < 0$ or $k = 0$. When $k > 0$, $0 < X < (\sigma/k)$. Thus, positive values of k only apply when there is a physical upper limit to the variate, X. The mean value of X is as follows:

$$\bar{X} = \frac{\sigma}{k+1} \tag{C.24}$$

The PDFs for various values of k are shown in Figure C.6.

The Generalised Pareto has a close relationship with the GEV distribution (Hosking and Wallis, 1987), so that the three types of the GEV are the distributions for the largest of a group of N variables that have a Generalised Pareto parent distribution with the same shape factor, k. It also transpires that the Generalised Pareto distribution is the appropriate one for the excesses of independent observations above a defined threshold (Davison and Smith, 1990). This distribution is used for the excesses of maximum wind speeds in individual storms over defined thresholds (Holmes and Moriarty, 1999, and Section 2.4). From the mean rate of occurrence of these storms, which are assumed to occur with a Poisson distribution, predictions can be made of wind speeds with various annual exceedence probabilities.

C.5 LEVEL-CROSSING RATES AND EXPECTED MAXIMUM VALUES

C.5.1 Introduction

As previously discussed in Section C.1, the probability density, or distribution, gives no information on the rate of change of a random process, the rate of crossing of a given level or of the maximum value in a time T. For example, imagine that the sample of a random process previously shown in Figure C.1 represents a year's variation of hourly mean wind speeds at a measurement site such as an airport, as in Figure C.7.

Now, imagine that the process is 'sped-up' so that the year's variation is now compressed into 6 months (Figure C.8). The value that was the maximum value in 1 year is now the maximum value in 0.5 years. However, the probability density and distribution for the 'sped-up' process are exactly the same as that for the original process. Thus, it is clear that the probability density/distribution alone is not sufficient information to determine values

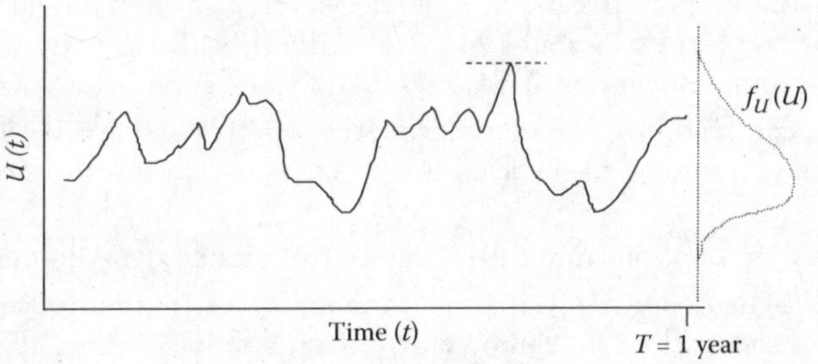

Figure C.7 Long-term wind variation at a site.

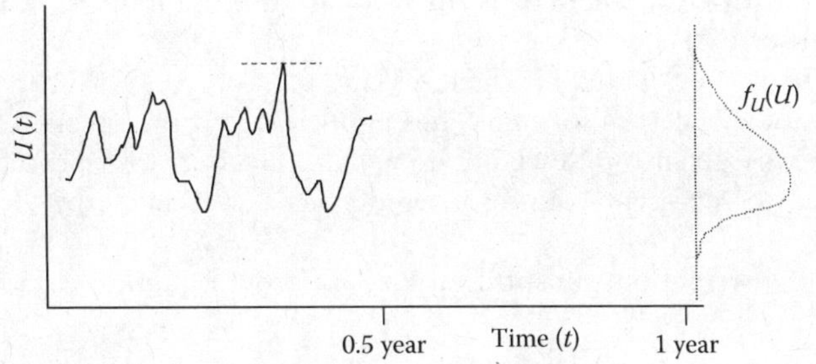

Figure C.8 'Compressed' random variation of wind speed.

such as 'expected annual maximum', or the 'value that is expected to be exceeded once per year'. Additional information is required on the time rate of change (derivative) of the process.

C.5.2 Level-crossing rates: Rice's formula

The expected rate of crossing of defined levels by a random process is itself useful, and it is also a necessary intermediate step in the determination of the expected maximum value in a defined period T. Figure C.9 shows five 'up-crossings' of the value $U = a$ by the random process $U(t)$ in a time period of T equal to 1 year.

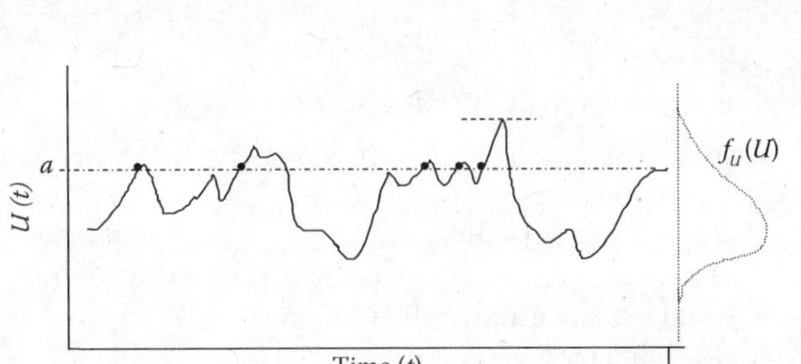

Figure C.9 Up-crossings of the level $U = a$ by the random process $U(t)$.

A formula for the rate of crossing of a defined level, a, by a stationary random process was published by Stephen O. Rice in 1944 (Rice, 1944). Although originally developed for application in telecommunications, it has also found many applications in wind engineering.

Rice's formula can be derived by considering the joint probability of $U(t)$ and $\dot{U}(t)$ under the conditions that a crossing of the level $U = a$ occurs with a positive slope within a short time interval.

Making further assumptions that the processes $U(t)$ and $\dot{U}(t)$ are un-correlated with each other, and that $\dot{U}(t)$ has a normal (Gaussian) probability distribution, Rice's formula, for the rate of crossing of the level $U = a$, can be written, in its simplest form, as

$$v_a^+ = \sqrt{2\pi} V_U \sigma_U f_U(a) \tag{C.25}$$

where σ_U is the standard deviation of the process $U(t)$, and $V_U = \sigma_{\dot{U}}/2\pi\sigma_u$ is known as the 'cycling rate' of the process, where $\sigma_{\dot{u}}$ is the standard deviation of the process, $\dot{U}(t)$ – that is the rate of change of $U(t)$.

$f_U(a)$ is the PDF of U evaluated for the level $U = a$.

The full derivation of Rice's formula, and application to the special case of a Gaussian random process was given by Newland (1993). Application to a process with a Weibull distribution, used to represent long-term wind speeds, was made by Gomes and Vickery (1977).

For the Weibull distribution, substituting for $f_U(a)$ from Equation C.15 in Equation C.25:

$$v_a^+ = \sqrt{2\pi} V_U \sigma_U \left(\frac{wU^{w-1}}{c^w} \right) \exp\left[-\left(\frac{U}{c} \right)^w \right] \tag{C.26}$$

where c is the Weibull scale parameter, and w is the shape parameter.

For a Gaussian random process, $x(t)$, with a zero mean, the up-crossing rate is

$$v_a^+ = \frac{1}{2\pi} \frac{\sigma_{\dot{x}}}{\sigma_x} \exp\left[-\frac{a^2}{2\sigma_x^2} \right] \tag{C.27}$$

Note that for this special case, $v_0^+ = v_x$. That is the crossing rate of the mean value (i.e. $a = 0$) is equal to the cycling rate for the process. Note that this relationship does not apply to processes with other probability distributions.

The expected largest value of a random process can be obtained by determining the level, a, which is expected to have *zero* crossings in a time T. This was derived for Gaussian random processes by Davenport (1964).

C.5.3 Example of application of Rice's formula

An analysis of 10-min average wind speeds for the Melbourne area recorded over a 40-year period gave the following values for all-direction wind speeds at standard conditions (flat, open-country terrain, 10-m height):

$\sigma_U = 3.23$ m/s (standard deviation of 10-min means)
$w = 1.91$ (Weibull shape factor)
$c = 6.25$ m/s (Weibull scale factor)

The value of the cycling rate, v_U, was found to be about 600 cycles per annum.

The wind speed crossed on average once per year, obtained from Equation C.26, can be obtained by trial and error, by finding the value of a that makes v_a^+ equal to 1. This is 18.9 m/s.

To check this, substitute the above values and $a = 18.9$ m/s into the right-hand side of Equation C.26:

$$v_a^+ = \sqrt{2\pi}(600)(3.23)\left(\frac{1.91 \times 18.9^{0.91}}{6.25^{1.91}}\right)\exp\left[-\left(\frac{18.9}{6.25}\right)^{1.91}\right] = 1.03 \cong 1.0$$

Similarly, the value of a exceeded, on average, once every 50 years can be obtained by finding the value of a that gives a value of v_a^+ of 0.02; this value is 23.4 m/s. This agrees quite well with a value obtained from a direct analysis of daily gust extremes (see Chapter 2), when thunderstorm gusts are excluded from the analysis.

C.5.4 Two-dimensional outcrossing rates

Some methods of prediction of the extreme wind speeds and responses are based on the 'out-crossing' formula, which gives the rate of crossing by a two-dimensional process of a defined boundary (see Section 9.11).

With certain assumptions, this rate was shown to be given by (Davenport, 1977)

$$N_R = \sqrt{2\pi}V_U\sigma_U \int_0^{2\pi} \sqrt{1 + \left(\frac{1}{r_R}\frac{dr_R}{d\theta}\right)^2} f_U(r_R, \theta)d\theta \tag{C.28}$$

where r, θ are the polar coordinates which define the two-dimensional process, and the response boundary, R. $f_U(U, \theta)$ is the joint probability density of wind speed and direction.

For the special case of a circular response boundary (i.e. r_R is constant with θ), the term $dr_R/d\theta$ becomes zero, the integral reduces to $f_U(r_R)$ and Equation C.28 reduces to Equation C.25.

C.6 OTHER PROBABILITY DISTRIBUTIONS

There are many other probability distributions apart from the ones discussed in this appendix. The properties of the most common ones are listed by Hastings and Peacock (1974).

The general application of probability and statistics in civil and structural engineering was discussed in specialised texts by Benjamin and Cornell (1970) and Ang and Tang (1975).

The theory and application of random processes are covered in detail by Bendat and Piersol (2010), and Newland (1993).

REFERENCES

Ang, A.H.-S. and Tang, W.H. 1975. *Probability Concepts in Engineering Planning and Design – Vol. I – Basic Principles*. John Wiley, New York.

Bendat, J.S. and Piersol, A.G. 2010. *Random Data: Analysis and Measurement Procedures*. 4th edition. Wiley, New York.

Benjamin, J.R. and Cornell, C.A. 1970. *Probability, Statistics and Decision for Civil Engineers.* McGraw-Hill, New York.

Davenport, A.G. 1964. Note on the distribution of the largest value of a random function with application to gust loading. *Proceedings, Institution of Civil Engineers*, 28: 187–96.

Davenport, A.G. 1977. The prediction of risk under wind loading. *Proceedings, 2nd International Conference on Structural Safety and Reliability,* Munich, Germany, 19–21 September, pp. 511–96.

Davison, A.C. and Smith, R.I. 1990. Models for exceedances over high thresholds. *Journal of the Royal Statistical Society, Series B*, 52: 393–442.

Fisher, R.A. and Tippett, L.H.C. 1928. Limiting forms of the frequency distribution of the largest or smallest member of a sample. *Proceedings, Cambridge Philosophical Society*, 24 (pt.2): 180–90.

Gomes, L. and Vickery, B.J. 1977. On the prediction of extreme winds from the parent distribution. *Journal of Industrial Aerodynamics*, 2: 21–36.

Gumbel, E.J. 1958. *Statistics of Extremes*. Columbia University Press, New York.

Hastings, N.A.J. and Peacock, J.B. 1974. *Statistical Distributions*. John Wiley, New York.

Holmes, J.D. 1981. Non-Gaussian characteristics of wind pressure fluctuations. *Journal of Wind Engineering and Industrial Aerodynamics*, 7: 103–8.

Holmes, J.D. and Moriarty, W.W. 1999. Application of the generalized Pareto distribution to extreme value analysis in wind engineering. *Journal of Wind Engineering and Industrial Aerodynamics*, 83: 1–10.

Hosking, J.R.M. and Wallis, J.R. 1987. Parameter and quantile estimation for the generalized Pareto distribution. *Technometrics*, 29: 339–49.

Newland, D.E. 1993. *An Introduction to Random Vibrations, Spectral and Wavelet Analysis*. 3rd edition. Longman Group Limited, London and New York.

Rice, S.O. 1944–1945. Mathematical analysis of random noise. *Bell System Technical Journal*, 23: 282–332 and 24: 46–156. Reprinted in Wax, N., *Selected Papers on Noise and Stochastic Processes.* Dover, New York, 1954.

Appendix D: Extreme wind climates – A world survey

In this appendix, an attempt has been made to describe the general type of extreme wind climate, and to catalogue the reliable design wind speed information available, from many countries in the world. Classification is done on a national basis, although, of course, extreme wind climates do not follow national boundaries. For small countries without wind-loading standards, or building codes with wind-loading information, it would be appropriate to use information from neighbouring countries.

It should be noted that wind-loading codes and standards are constantly under revision, and the values of design wind speed, zoning systems and so on given in this appendix may change periodically.

D.1 SEVERE WIND STRENGTH CLASSIFICATION SYSTEM

There have been cases where major errors have been made in the general level of design wind speeds used for a particular country or region, by engineers from other parts of the world. This is most likely to happen in the tropical and Equatorial regions, where the interface between very severe winds produced by tropical cyclones (typhoons, hurricanes) and the low extreme winds near the Equator where tropical cyclones do not occur may not be clearly defined. It is very useful to have a general idea of the level of design wind speeds in a country or parts of a large country. This information may in fact be sufficient for the design of small buildings, and less important structures such as signs or poles.

Table D.1 presents a simple classification system, which can be used to 'grade' any country or region in terms of its general level of wind speed. Nothing is stated in this table with regard to the type of wind storm that is dominant in a country. A dominant storm type in one country can produce similar extreme value statistics to another storm type in a different country. Note that for some storm types, such as downdrafts generated by thunderstorms, a peak gust may be a more relevant indicator than a 10-min mean.

D.2 COUNTRY-BY-COUNTRY SURVEY

Unless stated otherwise, all design wind data in the following paragraph are referred to a 10-m height, in a flat, open-country, terrain.

Table D.1 A classification system for design wind speeds

Level	*1-sec gust[a] (m/s)*	*10-min mean (m/s)*
I	<35	<22
II	35–45	22–30
III	45–55	30–35
IV	55–65	35–40
V	>65	>40

Note: 50-Year return period wind speeds at 10-m height.

[a] This is a nominal value, based on a moving-average definition of gust duration (see Section 3.3.3). Peak gusts with durations ranging from 0.2 to 3 s are used, or are in consideration, for national wind codes and standards.

D.2.1 Antigua (see Leeward Islands)

D.2.2 Argentina

Argentina is a large country and is affected by a range of different types of wind storms, although tropical cyclones do not occur. Large extra-tropical depressions are the dominant winds in the south (Patagonia and Tierra del Fuego). In the northeast (Cordoba region), the dominant winds are caused by severe thunderstorms; tornadoes and downbursts ('tormentas') have caused failures of several high-voltage transmission lines. Downslope and 'fonda' winds with severe gustiness occur in the Andes.

Early extreme value analyses (Riera and Reimundin, 1970; Viollaz et al., 1975) used the Frechet (Type II) Extreme Value Distribution to fit data from 63 stations in Argentina. This distribution (Section 2.2.1) is known to give excessively conservative predictions at high return periods. More recent extreme value analysis, based on the Gumbel Distribution, for six stations in the northeast of the country, gave 50-year return period gusts of 44–47 m/s. Thunderstorm winds were dominant in these records (de Schwarzkopf, 1995).

Extreme Wind Classifications: II and III

D.2.3 Australia

This large continental country has a variety of severe wind types, with large extra-tropical gales along the south coast and Tasmania, moving from the west, and 'East Coast lows' in the Tasman Sea affecting the eastern coastline. Thunderstorm-generated downbursts originating from local convection are the dominant windstorms in the interior. The strongest recorded winds, at 10-m height, in the four major capitals of Sydney, Melbourne, Adelaide and Brisbane are also caused by local downbursts from thunderstorms. Severe tropical cyclones can affect the coastline within about 100 km from the sea between 25° South and 10° South latitudes. The most common and most severe occur on the west coast between 20°S and 25°S.

Analyses of extreme wind speeds for Australia have been carried out by Whittingham (1964), Gomes and Vickery (1976a,b), Dorman (1983), Holmes (2002), Cechet and Sanabria (2011) and Wang et al. (2013), with the latter two carried out in the context of climate change effects. Wind speeds for structural design are given in the Australian Standard AS/NZS 1170.2 (Standards Australia, 2011). Four regions are defined; these are labelled from A to D with increasing the basic design wind speeds. Regions C to D are considered dominated

Table D.2 Australia

Region	*Description*	*Classification*
A	Thunderstorm downbursts and synoptic winds (gales)	II
B	Weakening tropical cyclones	III
C	Moderately severe-to-severe tropical cyclones	IV
D	Severe tropical cyclones	V

by severe-to-very-severe tropical cyclones. In Region B (covering Brisbane), and the tropical coastal strip between 50 and 100 km inland, weaker tropical cyclones can occur (Table D.2).

D.2.4 Austria

No values were specified in the original draft Eurocode (C.E.N., 1994). However, the wind climate should be similar to southern Germany and Switzerland.

Extreme Wind Classification: II

D.2.5 Bahamas

This island group is subjected to frequent Atlantic hurricanes. The U.S. Department of Defense gives 50-year return period gust speeds of 66–72 m/s. However, a more recent report by Applied Research Associates (Vickery and Wadhera, 2008) showed a lower value of about 54 m/s. The difference may be partially explained by the longer gust duration (3 s) used in the latter report.

Extreme Wind Classification: V

D.2.6 Bahrain

Analysis of combined gust data from Bahrain and Doha (Qatar) by the author gave a 50-year return period peak gust of 35 m/s.

Extreme Wind Classification: II

D.2.7 Barbados (see Windward Islands)

D.2.8 Belgium

The original draft Eurocode (C.E.N., 1994) specified a single value of 10-min mean wind speed with a 50-year return period, of 26.2 m/s.

Extreme Wind Classification: II

D.2.9 Belize

Belize in Central America experiences severe winds from hurricanes. Analysis of extreme wind speeds for the Commonwealth Caribbean was carried out by Shellard (1972). These results were used by the Caribbean Uniform Building Code (Caribbean Community Secretariat, 1986). This code specified a 50-year return period 10-min mean wind speed of 36 m/s for the north of the country, and 30.5 m/s for the south.

Extreme Wind Classifications: IV (North) and III (South)

D.2.10 Brazil

In Brazil, extreme winds are primarily produced by a mixture of large extra-tropical depression systems, and local thunderstorm downdrafts (Riera and Nanni, 1989; Loredo-Souza, 2012). The coastline of the South Atlantic normally does not experience tropical cyclones due to the low water temperature; however, a cyclonic system (named 'Catarina') with the characteristics of a weak hurricane formed off the coastline of southern Brazil in January 2004 (Loredo-Souza and Paluch, 2005).

Salgado Vieira Filho (1975) carried out extreme value analyses of wind speeds for 49 Brazilian stations, but used the conservative Frechet Distribution for predictions, and apparently the data were not separated by storm type. Later analyses by Riera and Nanni (1989) indicate that thunderstorm winds are dominant in most locations. Jeary (1997b) lists 10 years of recorded wind speeds (from 3-m height) for three stations in Rio de Janeiro. The Brazilian wind-loading code of 1987 (NBR-6123, 1987) gave isotachs of 3-s gust speeds with 50-year return period varying from 30 (north half of the country) to 50 m/s (extreme south).

Extreme Wind Classifications: I (North of 20°); II (South Central) and III (Extreme South)

D.2.11 Canada

Extreme winds in Canada are primarily generated by large-scale synoptic systems, and surface extreme winds can be quite well predicted from gradient wind observations (Davenport and Baynes, 1972). An appendix to the National Building Code of Canada (NRC, 2005) gives values of dynamic pressures. The equivalent hourly mean wind speeds, with a 30-year return period, range from 24 to 28 m/s in the main populated area around the Great Lakes (including Toronto, Montreal and Ottawa), to 30–35 m/s in Newfoundland, and in the Hudson's Bay area.

Extreme Wind Classifications: III (Newfoundland and North) and II (Rest of Canada)

D.2.12 China (Including Taiwan)

China is a large country with a range of extreme wind types ranging from severe gales arising from synoptic systems in Siberia in the northwest, to typhoons along the southern coastline. There is a region with downslope winds.

A combined loading code published in 1994 by the Department of Standards and Norms, Ministry of Construction includes a wind-loading section. This was also available as an English translation (GBJ-9 – Department of Standards and Norms, 1994). A map was included with this standard which gives contours of dynamic pressure in kN/m^2 (kPa). The standard stated that the 'wind reference pressure' is calculated from the 10-min mean wind speed at 10-m height by the formula, $w_o = v_o^2/1600$. These values had a 50-year return period. Values of dynamic pressure on these contours ranged from 0.30 to 0.90 kPa. For most of the country, the values were in the range of 0.30–0.50 kPa (Table D.3).

Table D.3 China

Region	*Description*	*Classification*
Central Mainland	Pressure contours from 0.30 to 0.50 kPa	II
Northwest and inner southern coast	Pressure contours from 0.60 to 0.70 kPa	III
Outer southern coast and islands, Hainan	Pressure contours from 0.80 to 0.90 kPa	IV
Taiwan	Severe typhoons	V

D.2.13 Croatia

The extreme wind climate in Croatia was described by Bajic and Peros (2005). Most of the country has been assigned a value of 25 m/s for the 10-min mean wind speed with a return period of 50 years. The exceptions are some valleys and mountain passes with accelerated *bora* (downslope) winds (see Section 1.3.6). In these regions, it is stated that values of 38–55 m/s were obtained.

Extreme Wind Classifications: II – Most of the country; IV–V – Downslope wind areas

D.2.14 Cuba

Cuba is affected by Caribbean hurricanes. A report by Applied Research Associates (Vickery and Wadhera, 2008) showed an effective 50-year return period peak gust speeds of 36 m/s in the east of the country increasing to 54 m/s in the west.

Extreme Wind Classifications: II (East side) and III (West side)

D.2.15 Denmark

Wind speed observations have been made in Denmark since the 1870s. The dominant source of extreme winds in Denmark is severe extra-tropical depressions moving in from the north Atlantic Ocean. Extreme value analyses of extreme wind speeds have been made by Jensen and Franck (1970) and several others.

The draft Eurocode (C.E.N., 1994) specified a single value of 10-min mean wind speed with a 50-year return period, for Denmark, of 27.0 m/s.

Extreme Wind Classification: II

D.2.16 Dominican Republic

The Dominican Republic, occupying the eastern part of the island of Hispaniola in the Caribbean, is affected by hurricanes. A report by Applied Research Associates (Vickery and Wadhera, 2008) showed an effective 50-year return period peak gust speeds of 36 m/s in the northwest of the country increasing to 54 m/s in the south.

Extreme Wind Classifications: II and III

D.2.17 Dubai

See United Arab Emirates.

D.2.18 Egypt

Locations on the coast have similar extreme winds to other Mediterranean locations – for example, the 50-year gust for Alexandria is given as 38 m/s by the U.S. Department of Defense. Inland values appear to be significantly lower.

Extreme Wind Classifications: I and II

D.2.19 Eritrea

Eritrea in East Africa has a mountain part and a coastal plain with different climates. The capital, Asmara, is an elevated terrain where extreme winds are produced by intermittent thunderstorms; for this location, the lowest classification (I) applies. At the port of Massawa on the Red Sea coastline, extreme winds are higher (II).

Extreme Wind Classifications: I and II

D.2.20 Ethiopia

Ethiopia is a land-locked country in the tropics with an elevated terrain between 1800 and 2400 m above sea level in the central part. Extreme winds in the capital Addis Ababa are generated by thunderstorms and are low.

Extreme Wind Classification: I

D.2.21 Fiji

The Fijian islands are subject to periodic visits from tropical cyclones (e.g. 'Kina' (1993), 'Ami' (2003) and 'Evan' (2012)) and consequent high winds.

Extreme Wind Classification: IV

D.2.22 Finland

The original draft Eurocode (C.E.N., 1994) specified a single value of 10-min mean wind speed with a 50-year return period, for Finland, of 23 m/s.

Extreme Wind Classification: II

D.2.23 France

Like other Western European countries, the extreme wind climate of France is dominated by synoptic gales from large depression systems moving in from the Atlantic Ocean.

The original draft Eurocode (C.E.N., 1994) specified four values of 10-min mean wind speed with a 50-year return period, for four zones in metropolitan France, ranging from 24 to 30.5 m/s. The highest values occur for Zone 4, which includes parts of Brittany and Normandy, the Mediterranean coastline and Eastern Corsica.

Extreme Wind Classifications: (Zones 1–3): II and Zone 4: III

D.2.24 French Caribbean (see also Leeward and Windward Islands)

The original draft Eurocode (C.E.N., 1994) specified a value of 10-min mean wind speed with a 50-year return period, of 34 m/s, for the French territories of Guadeloupe and Martinique in the southern Caribbean. These territories experience visits from hurricanes, although apparently less frequently than the islands of the northern Caribbean.

Extreme Wind Classifications: III (Martinique) and IV (Guadeloupe)

D.2.25 Germany

The original draft Eurocode ENV-1991-2-4 gave a map with a system of four zones. The highest wind speed zone, 3, is on the North Sea coast. The main source of strong winds is gales accompanying large-scale depressions moving into Germany from the west. The zone system given differed from an earlier zoning system for the Federal Republic (West Germany) by Caspar (1970).

Analyses by Schueller and Panggabean (1976) for stations in the former West Germany gave distributions for gust speeds, which give 50-year return period values between 35 and 50 m/s. An exception was Feldberg with 60 m/s; this is a mountain station, with topographic influences.

Kasperski (2002) carried out an analysis of hourly mean wind speeds for 183 stations in Germany. This analysis included separation of winds from different storm types – namely

Table D.4 Germany

Zone (Kasperski, 2002)	*Description*	*Classification*
I	Central	II
II	Northern Germany and southern alpine	II
III	Northwest and Saxony	II
IV	North Sea and Baltic coasts	III
V	North Sea offshore	III

frontal depressions, thunderstorms and gust fronts. He proposed five wind zones with 50-year return period hourly mean speeds between 22.5 and 32.5 m/s. His Zones I to III correspond to Classification II in this appendix; his Zones IV and V correspond to Classification III in Table D.4.

D.2.26 Greece

The draft Eurocode ENV-1991-2-4 specified a 50-year return period 10-min mean wind speed for Greece, of 36 m/s for the islands and coastal areas of the mainland within 10 km of the coastline. For the rest of the country, the value is 30 m/s.

Extreme Wind Classification: III

D.2.27 Greenland

The U.S. Department of Defense recommended 50-year gust wind speeds between 54 and 74 m/s for U.S. Air Force Bases in Greenland.

Extreme Wind Classification: IV

D.2.28 Guam (United States)

This Pacific Island has experienced some of the strongest recorded tropical cyclones. The ASCE-7 Loading Standard specified a 50-year return period gust speed of 76 m/s.

Extreme Wind Classification: V

D.2.29 Guyana

This country has an Equatorial climate with low wind speeds. Analysis of extreme wind speeds for the Commonwealth Caribbean was carried out by Shellard (1972). These results have been used by the Caribbean Uniform Building Code (Caribbean Community Secretariat, 1986) and by the Code of Practice of the Barbados Association of Professional Engineers (1981). The former specified a 50-year return period 10-min mean speed of 18 m/s, and the latter gave a 50-year return period 3-s gust speed of 22 m/s.

Extreme Wind Classification: I

D.2.30 Haiti

Haiti, occupying the western part of Hispaniola, is occasionally affected by Caribbean hurricanes. A report by Applied Research Associates (Vickery and Wadhera, 2008) showed an effective 50-year return period peak gust speeds (3 s duration) of 36 m/s in the north of Haiti increasing to 49 m/s in the south east.

Extreme Wind Classifications: II and III

D.2.31 Hong Kong and Macau

As for the rest of the south China coastline, Hong Kong and Macau are subjected to frequent visits from moderate-to-severe typhoons. Hong Kong has good-quality-recorded wind speed data extending more than a hundred years from 1884 to 1957, from the Hong Kong Observatory, and since 1957, from Waglan Island. Analyses of extreme winds from typhoons have been carried out by a number of authors including Faber and Bell (1967), Chen (1975), Davenport et al. (1984), Melbourne (1984), Jeary (1997a) and Holmes et al. (2009). Most of these studies have normalised the wind speeds to a height of 50 m, rather than 10 m. Design wind speeds in Hong Kong and Macau are set by the respective building departments.

The Hong Kong Code of Practice (Buildings Department, 2004) implies a 50-year return period 3-s gust wind speed at 50-m height of about 65 m/s.

Extreme Wind Classification: IV

D.2.32 Iceland

Iceland is subject to Atlantic gales. The original draft Eurocode ENV-1991-2-4 specified a 50-year return period 10-min mean at 10-m height, of 39 m/s for coastal areas within 10 km of the coastline. For inland areas, the value is 36 m/s. (ENV-1991-2-4 incorrectly described these values as 3-s gusts.)

Extreme Wind Classification: II

D.2.33 India

India, a large sub-continental tropical country, has a range of extreme wind zones, with extreme tropical cyclones being dominant on the east (Bay of Bengal) coast, and less frequent ones on the west coast. In inland areas, thunderstorms and monsoon winds are prevalent.

India has a good network of meteorological stations, and there have been a number of extreme value analyses of wind speeds for the country. The Indian Standard for Wind Loads IS875 Part 3 (Bureau of Indian Standards, 1987) divided the country into six zones, giving 50-year return period gust wind speeds ranging from 33 to 55 m/s (Table D.5).

Extreme Wind Classifications: I, II, III, IV and V

D.2.34 Indonesia

Like Malaysia and Singapore, Indonesia is entirely in the Equatorial zone, does not experience typhoons and design wind speeds from weak thunderstorms and monsoonal winds are low.

Extreme Wind Classification: I

Table D.5 India (refer to map in IS 875: part 3, for details of zones)

Zone	*Description*	*Classification*
1	Tripura, Mizoram and Ladakh	IV
2	Coastal strips of Tamil Nadu (including Madras), Andhra Pradesh, Orissa, Gujarat, West Bengal (including Calcutta) and Assam	III
3	Northern India including Delhi, central Tamil Nadu	III
4	Coastal strip on Arabian Sea, including Bombay, inland Madhya Pradesh and Orissa	II
5	Most of southern India	II
6	Inland Karnataka, including Bangalore	I

D.2.35 Ireland

Ireland is a small island nation exposed to severe Atlantic gales on its west coast. A map showing contours of extreme wind speeds for Ireland appears in the 1997 British Code of Practice (British Standards Institution, 1997). The map in the original draft Eurocode (C.E.N., 1994) had higher values, although nominally also 10-min mean wind speeds with a 50-year return period. The values in the British Standard are believed to be more reliable.

Extreme Wind Classification: II

D.2.36 Israel

Israel experiences several different wind storm types, including the 'sharqiya' easterly downslope winds, similar to the 'bora' of Croatia (Saaroni et al., 1998). The highest wind gust recorded in Israel was 44 m/s at Jerusalem on 20/1/1974.

Extreme Wind Classification: II

D.2.37 Italy

An extreme value analysis of 10-min mean wind speeds for Italy was carried out by Ballio et al. (1999). In the original draft Eurocode (C.E.N., 1994), Italy was divided into nine zones with five basic wind speeds. These are 10-min mean speeds with a 50-year return period, ranging from 25 to 31 m/s (Table D.6).

Extreme Wind Classifications: II and III

D.2.38 Jamaica

Jamaica is in a region of hurricane tracks in the Caribbean, and experiences severe winds from these events. Analysis of extreme wind speeds for the Commonwealth Caribbean was carried out by Shellard (1972). These results were used by the Caribbean Uniform Building Code (Caribbean Community Secretariat, 1986) and by the Code of Practice of the Barbados Association of Professional Engineers (1981). The former specified a 50-year return period 10-min mean wind speed of 36.5 m/s, and the latter specified a 56 m/s peak gust. A more recent report by Applied Research Associates (Vickery and Wadhera, 2008) showed an effective 50-year return period peak 3-s gust speed of about 45 m/s.

Extreme Wind Classification: III

D.2.39 Japan

Japan is subject to typhoons from the Pacific in Kyushu and Okinawa, and temperate synoptic systems in the north of the country. The Architectural Institute of Japan has a contour map of design wind speeds (10-min mean, 100-year return period) in its wind load recommendations. Values range from 26 to 44 m/s on the main islands, to 50 m/s on Okinawa which is subject to frequent severe typhoons.

Extreme Wind Classifications: II, III, IV and V

Table D.6 Italy (zones are those given in the draft Eurocode, ENV 1991-2-4)

Zones	*Description*	*Classification*
1, 2	Northern Italy (25 m/s)	II
3	Central and southern Italy (27 m/s)	II
4, 5, 6	Sardinia and Sicily (28 m/s)	II
7	Liguria (29 m/s)	II
8, 9	Trieste and islands (31 m/s)	III

D.2.40 Korea

The coastline of South Korea has some influence from typhoons on the south and east coasts and the island of Cheju, but these are relatively infrequent.

The Architectural Institute of Korea has a map of 10-min mean 100-year return period wind speeds varying from 25 m/s in the inland centre to 40 m/s at some points on the eastern and southern coastline. Seoul is specified as 30 m/s.

Extreme Wind Classifications: II, III and IV

D.2.41 Leeward Islands

This group of islands is affected by hurricanes in the Caribbean. Analysis of extreme wind speeds for Commonwealth countries in the Caribbean was carried out by Shellard (1972). These results were used by the Caribbean Uniform Building Code (Caribbean Community Secretariat, 1986) and by the Code of Practice of the Barbados Association of Professional Engineers (1981). The latter specified a 50-year return period 3-s gust speed of 64 m/s, based on studies for Antigua. This value was also applicable to St. Kitts-Nevis, Montserrat and the Virgin Islands. However, a more recent report by Applied Research Associates (Vickery and Wadhera, 2008) showed effective 50-year return period 3-sec peak gust speeds of 53–58 m/s for the latter islands.

Extreme Wind Classification: IV

D.2.42 Luxembourg

The draft Eurocode (C.E.N., 1994) specified a single value of 10-min mean wind speed with a 50-year return period, of 26 m/s.

Extreme Wind Classification: II

D.2.43 Madagascar

No direct information is available, but the eastern coast can be assumed to have a similar extreme wind climate to Reunion Island, and Mauritius.

Extreme Wind Classification: III

D.2.44 Malaysia

Malaysia is entirely in the Equatorial zone, does not experience typhoons and has very low extreme winds from weak thunderstorms and monsoonal winds. Monthly maximum wind data are available from more than 30 stations in the country, including Miri and Kuching in East Malaysia (Sarawak).

Analysis of these data for 50-year return period gust values for 20 stations by the Malaysian Meteorological Service gave values between 24 and 32 m/s. There is evidence of higher wind speeds in the highland stations away from the coastal plains.

Extreme Wind Classification: I

D.2.45 Mauritania

An investigation of design wind speeds, for offshore oil and gas platforms, indicated strong winds caused by severe thunderstorm events with a 50-year return period gust wind speed of around 55 m/s.

Extreme Wind Classification: III

D.2.46 Mauritius

Like neighbouring Reunion, Mauritius in the Indian Ocean experiences landfall from a tropical cyclone about once in every 5 years (Sites and Peterson, 1995).

Extreme Wind Classification: III

D.2.47 Mexico

Mexico experiences extreme winds from hurricanes on both its Pacific and Caribbean coasts. For inland areas, thunderstorms are dominant. Extreme value analyses were carried out by Vilar et al. (1991), (also Lopez and Vilar, 1995), for the Mexican Electrical Utility (CFE) using the GEV Distribution, for data from 57 stations. An isotach map resulting from this study shows 50-year return period 3-s gusts widely ranging from 28 m/s in the Mexico City area to 61 m/s on the Pacific coast.

Table D.7 is not official but describes zones based on the isotach map.

Extreme Wind Classifications: I, II, III and IV

D.2.48 Morocco

The 50-year gust for Casablanca on the Atlantic coast was given as 40 m/s by the U.S. Department of Defense. Inland mountainous regions may have higher values.

Extreme Wind Classifications: II and III

D.2.49 Netherlands

The Netherlands is exposed to gales from the North Sea on the coast. The original draft Eurocode ENV-1991-2-4 specified three 'areas' with 50-year return period 10-min wind speeds, of 25, 27.5 and 30 m/s. Amsterdam and Rotterdam are in Area 2 (27.5 m/s).

Extreme Wind Classification: II

D.2.50 New Caledonia

New Caledonia in the South Pacific Ocean is affected by tropical cyclones (Holmes and Weller, 2002).

Extreme Wind Classification: IV

D.2.51 New Zealand

Turner et al. (2012) described some damaging extreme wind events in New Zealand; these include localised tornadoes and downslope winds. However, the main extreme winds affecting most of New Zealand are temperate synoptic systems, although the north of the country can experience the effects of decaying tropical cyclones. The map of basic wind speeds in the Australian/New Zealand Standard (Standards Australia, 2011) shows three wind regions.

Table D.7 Mexico

Description	*Classification*
South of 24°S excluding coastline	I
North of 24°S excluding coastline	II
Within 50 km of Caribbean coast	
50–100 km from Pacific coast	III
Within 50 km of Pacific coast	IV

These have similar all-directional basic wind speeds (1000-year return period gusts) ranging between 46 and 53 m/s, but differ in their directional wind speeds. There are a number of mountain areas, especially in the South Island, where downslope winds occur – for these, the wind speed is increased by a 'Lee Multiplier' of up to 1.35.

Extreme Wind Classifications: II and III (Some mountain areas)

D.2.52 Norway

The draft Eurocode, ENV-1991-2-4, specified several wind velocity profiles of 3–5-s gust wind speeds, which incorporate terrain effects as well as height variation. Two of these curves, E and F, incorporate gust speeds at 10-m height in the open terrain of 45 and 50 m/s, respectively. Curves A and B correspond to 35 and 40-m/s, respectively. (Curves C and D apply to built-up areas, corresponding to open-country exposure for Curves A and B, respectively.)

Extreme Wind Classifications: II (A and B) and III (E and F)

D.2.53 Oman

Oman has a coastline facing the Arabian Sea and Gulf of Oman. In winter, it suffers from north-westerly 'shamal' winds. A study by Almaawali et al. (2008) determined a 50-year return period gust ranging from 36.5 m/s (Masirah – offshore island) to 52.6 m/s for Nizwa (inland city). The region around the capital, Muscat, has a value of about 40 m/s.

Extreme Wind Classification: II

D.2.54 Papua-New Guinea

The majority of Papua-New Guinea (including Port Moresby) is in the Equatorial zone, and the design winds, originating from thunderstorms produced by local convective activity, are quite low. An extreme value (Gumbel) analysis for Port Moresby by Whittingham (1964) using only 11 years of data gives a 50-year return period gust of 31 m/s. The addition of some extra years gives even lower values. The P-NG loading code gives a contour map with 50-year return period gust wind speeds ranging from 24 to 32 m/s. For the southwest tip, 40 m/s is specified. Values for major centres are: Port Moresby 28 m/s, Lae 23 m/s and Rabaul 26 m/s. A speed limit of 34 m/s is specified for Honiara (Solomons).

The islands on the southwest are occasionally exposed to developing Coral Sea cyclones, and should have higher design wind speeds.

Extreme Wind Classifications: I (Most of the Country) and II (Southwest Tip)

D.2.55 Philippines

The Philippines experiences typhoons from the southwest Pacific Ocean, which often cross the northern Philippines (Luzon) and re-form in the south-China sea. On the other hand, the southern island of Mindanao has little or no influence from typhoons, and effectively has an Equatorial extreme wind climate. An extreme value analysis of 1-min average extreme wind speeds in the Philippines was carried out in the early 1970s by Kintenar (1971). This gave widely ranging 50-year return period values, and was probably suffering from sampling errors due to short records.

The National Structural Code of the Philippines specifies three extreme wind zones with 1-min sustained wind speeds of 200 kph (55.5 m/s), 175 kph (48.6 m/s) and 150 kph (41.7 m/s), respectively. These values have been used for Table D.8, but are probably conservative for Mindanao.

Extreme Wind Classifications: III, IV and V

Table D.8 Philippines

Zone in structural code	*Description*	*Classification*
I	Eastern Luzon	V
II	Remainder of Philippines	IV
III	Western Mindanao, Palawan	III

D.2.56 Poland

The Polish wind-loading standard PN-77/B-02011 gives 'characteristic' wind speeds of 20, 24, 27 and 30 m/s for four zones. In the largest zone, 20 m/s is specified. These values are 10-min mean speeds, with a return period of 50 years.

Extreme Wind Classifications: II and III

D.2.57 Portugal

The draft Eurocode (C.E.N., 1994) specified a value of 10-min mean wind speed with a 50-year return period, of 31 m/s, for the Azores, Madeira, and the 5-km coastal strip of the mainland; 28 m/s is specified for the rest of the country.

Extreme Wind Classifications: II and III

D.2.58 Puerto Rico

As for other Caribbean islands, Puerto Rico is subjected to hurricane winds. A report by Applied Research Associates (Vickery and Wadhera, 2008) showed an effective 50-year return period peak 3-s gust speeds of 54–58 m/s.

Extreme Wind Classification: IV

D.2.59 Qatar

Analysis of combined gust data from Doha (Qatar) and Bahrain by the author gave a 50-year return period peak gust of 35 m/s.

Extreme Wind Classification: II

D.2.60 Reunion I

This small French island in the southern Indian Ocean has a design wind speed (10-min mean, 50-year return period) of 34.0 m/s specified in the draft Eurocode (C.E.N., 1994). According to Sites and Peterson (1995), Reunion experiences landfall of a tropical cyclone about once in every 5 years.

Extreme Wind Classification: III

D.2.61 Romania

The Romanian Standard STAS 10101/20-78 on Actions on Structures specified five zones for design wind pressures. These pressures correspond to peak gust wind speeds (10-year return period) ranging from 27 to 37 m/s.

Extreme Wind Classifications: I and II

D.2.62 Russia

Russia has a vast land area, with a range of extreme wind climates. The Russian loading code SniP 2.01.07.85 specified eight zones for design wind pressures. The specified values are 5-year return period pressures with a 10-min averaging time, and range from 240 Pa for the central part of the country to 1200 Pa on the coastal part of the Far East, and the islands of the Barents Sea (Popov, 2001).

Extreme Wind Classifications: II, III, IV and V

D.2.63 Samoa

Samoa in the South Pacific (latitude 13–14°S) is affected by tropical cyclones (Holmes and Weller, 2002). Cyclones 'Ofa' (1990) and 'Val' (1991) did a significant damage. The U.S. Department of Defense recommended a 50-year gust wind speed of 67 m/s for the capital, Apia.

Extreme Wind Classification: IV

D.2.64 Saudi Arabia

The national oil company has an Engineering Standard (Saudi Aramco, 2005) containing meteorological and seismic design data intended for use with the American Loading Standard ASCE-7; this contains 50-year return period gust data for many locations in the country. These range from 41 to 51 m/s, with a value of 46 m/s for the capital, Riyadh.

Extreme Wind Classifications: II and III

D.2.65 Singapore

Like Malaysia, Singapore in the Equatorial zone does not experience typhoons, and has very low extreme winds from weak thunderstorms and monsoonal winds (Choi, 1999). Good-quality-corrected monthly maximum extreme gust data are available from Tengah and Changi airfields. A Gumbel extreme value analysis for data up to 1997 from these data gives 50-year return period gusts of 33 and 25 m/s, respectively. (However, a gust of 40.1 m/s was recorded at Tengah in 1984.)

Extreme Wind Classification: I

D.2.66 South Africa

Kruger et al. (2010, 2012) described the various mechanisms producing strong wind gusts in South Africa. The country is subjected to severe thunderstorms on the inland high plains, and synoptic winds in the south, with the sources of the latter being subdivided into troughs, ridges and cold fronts. The 1989 Code of Practice for Loading of the South African Bureau of Standards (SABS 0160-1989) had a map showing design wind speeds for the country (50-year return period, 3-s gust). This map was based on the analysis of annual maximum wind speeds by Milford (1987). The value given for the majority of the country was 40 m/s. This value is specified for the main cities of Johannesburg, Pretoria, Cape Town and Durban. A small zone around Beaufort West had a value of 50 m/s.

Extreme Wind Classifications: II and III

D.2.67 Spain

No data were given in the draft Eurocode (C.E.N., 1994), but the map in the E.C.C.S. Recommendations for the Calculation of Wind Effects on Structural Steelwork (E.C.C.S,

1978) gives values of 10-min mean wind speeds (50-year return period) of 22–26 m/s for Spain. There are some downslope wind areas in the Pyrenees.

Extreme Wind Classification: II

D.2.68 Sri Lanka

The east coast of Sri Lanka is exposed to relatively weak tropical cyclones. A Building Code was prepared by an Australian consulting group in the 1970s. Three design wind zones were specified (Table D.9).

Extreme Wind Classifications: I, II and III

D.2.69 Sweden

The draft Eurocode (C.E.N., 1994) gave a contour map with values of 10-min mean wind speed with a 50-year return period, between 22 m/s (northeast) and 26 m/s south and west. For Stockholm, the value is 24 m/s.

Extreme Wind Classification: II

D.2.70 Switzerland

The draft Eurocode (C.E.N., 1994) specified values of 10-min mean wind speed with a 50-year return period, of 27.2 m/s over the vast majority of the country, including Zurich, Basel, Bern and Lausanne. There are a number of mountain areas where downslope wind occurs, and for which higher values of 30 and 33.3 m/s are specified.

Extreme Wind Classifications: II (Most of the country) and III (Some mountain areas)

D.2.71 Taiwan (see China)

D.2.72 Thailand

Thailand has a particularly mixed wind climate. Most of the country appears to be dominated by extreme winds from thunderstorms and monsoons. However, occasionally, typhoons have impacted on southern Thailand, as did Typhoon 'Gay' in 1989, inflicting considerable damage. Post-landfall typhoons can also affect northwest Thailand.

An analysis of historical gust data for 60 meteorological stations is described by Davenport et al. (1995), using Type I (Gumbel) Extreme Value Distributions. There were apparently siting problems for many of the anemometers, and although extreme winds caused by typhoons were separated, those from thunderstorms apparently were not.

The analysis by Davenport et al. (1995) proposed two design wind speeds based on a 50-year return period 10-min means of 26.5 and 30 m/s. The latter value applies to small zones on the east and northwest of Thailand. In recent proposals for a new Thailand loading code (Lukkunaprasit, 1997), these values were converted into nominal mean hourly speeds of 24.9 and 28.2 m/s, respectively.

Extreme Wind Classification: II

Table D.9 Sri Lanka

Zone in building code	*Description*	*Classification*
1	50 km from east coast	III
2	Inland strip	II
3	South and west (including Colombo)	I

D.2.73 Tonga

Tonga, an island group in the South Pacific, suffers regular visits by tropical cyclones (Holmes and Weller, 2002), including 'Isaac' (1982), 'Hina' (1997), 'Cora' (1998) and 'Waka' (2001), all of which did a significant damage.

Extreme Wind Classification: IV

D.2.74 Trinidad and Tobago

Analysis of extreme wind speeds for the Commonwealth Caribbean was carried out by Shellard (1972). These results were used by the Caribbean Uniform Building Code (Caribbean Community Secretariat, 1986) and by the Code of Practice of the Barbados Association of Professional Engineers (1981). The former specified a 50-year return period 10-min mean speeds of 20–28 m/s, and the latter gave a 50-year return period 3-s gust speed of 45 m/s for Trinidad, and 50 m/s for Tobago. The latter values were based on a Frechet (Type II) Extreme Value Distribution (Section 2.2.1), and may be conservative.

Extreme Wind Classification: II

D.2.75 Turkey

The extreme wind climate of coastal Turkey can be assumed to be similar to that of Greece. The U.S. Department of Defense recommends a 50-year return period gust of 44 m/s for Ankara.

Extreme Wind Classifications: II and III

D.2.76 Ukraine

Horokhov and Nazim (2001) give a zoning map of wind velocities for the Ukraine. The values have an averaging time of 2 min, with a 10-year recurrence interval. Values range between 20 and 45 m/s. The higher values are primarily in mountainous regions or adjacent to the Black or Azov Seas.

Extreme Wind Classifications: II, III and IV

D.2.77 United Arab Emirates

The U.A.E. along the southeastern tip of the Arabian Peninsula is a federation of seven emirates: Abu Dhabi, Dubai, Sharjah, Ajman, Umm al-Quwain, Ras al-Khaimah and Fujairah. Analyses by various wind-tunnel laboratories for building projects in Dubai give predictions of 50-year gust speed between 35 and 39 m/s. The dominant wind directions for the extreme winds are in the north-to-west quadrant.

Extreme Wind Classification: II

D.2.78 United Kingdom

The U.K. has a close network of meteorological stations, and high-quality data. The main strong wind source is severe gales moving in from the Atlantic on the west. Design winds are generally stronger on the west, reducing further east.

Analyses of extreme winds for the United Kingdom were carried out by Shellard (1958, 1962) and Cook and Prior (1987). The latter work was used for the design wind speed data in the British Standard BS6399:2. A further analysis of hourly extremes for the United

Kingdom was undertaken by Miller et al. (2001); however, this was only based on data from 1970 to 1980.

BS6399:2 contained a map of 1-h mean wind speeds (50-year return period) ranging from 20 to 30 m/s. The latter values occur only for the Shetland Islands in the north. The map also covers the whole of Ireland. The map in the draft Eurocode (C.E.N., 1994) was identical, with Ireland omitted.

Extreme Wind Classification: II

D.2.79 United States

The USA has a vast array of meteorological stations operated by the U.S. Weather Bureau, and other agencies, such as those involved in defence. Until 1995, the standard extreme wind was the 'fastest mile of wind', calculated from the time taken by a cup anemometer to rotate through 1 mile. The introduction of automatic weather stations has seen this measure replaced by a peak gust wind speed.

There have been many extreme value analyses for the United States, including those by Thom (1960, 1968), Simiu et al. (1979) and Peterka and Shahid (1998). The latter analysis resulted in the design wind map in the 2006 ASCE Loading Standard (ASCE, 2006). The latter contained two main zones, with 50-year return period gust speeds of 40, and 38 m/s. The lower value applies to the west-coast states. The Atlantic Ocean and Gulf of Mexico coastlines have isotachs ranging from 67 to 45 m/s. Alaska has contours from 40 to 58 m/s. Hawaii has a basic wind speed of 47 m/s (Table D.10).

Extreme Wind Classifications: II, III and IV

D.2.80 Uruguay

Uruguay in South America lies between the latitudes of 30°S and 35°S. The extreme wind climate of Uruguay has been discussed in detail by Duranona (2011, 2013). Extreme winds are produced by extra-tropical cyclones formed over the South Atlantic Ocean, and severe convective downdrafts from thunderstorms, with the latter affecting mainly the inland areas. The national wind code UNIT 50-84 (Instituto Uruguayo de Normas Tecnicas, 1984) gives a 3-s gust speed of 158 km/h (43.9 m/s) for a return period of 10 years; however, this may be an overestimate since it was based on measurements above a 43-m-high building in Montevideo (Duranona, 2011, 2013).

Extreme Wind Classification: II

D.2.81 Vanuatu

Vanuatu is an island group in the South Pacific, often affected by tropical cyclones (Holmes and Weller, 2002), including 'Uma' (1987), 'Dani' (1999) and 'Ivy' (2004), which did a significant damage.

Extreme Wind Classification: IV

Table D.10 United States

Description	*Classification*
Central and Western States	II
Atlantic and Texas coasts	III
Southern Florida and Louisiana, Alaska coasts	IV

Table D.11 Vietnam

Zone in loading code	*Description*	*Classification*
I	Inland north and south	II
II	Inland north and southern coast	II
III	Central and northern coastline	III
IV, V	Offshore islands in north	IV

D.2.82 Vietnam

Vietnam is influenced by typhoons over most of its coastline, although the influence is weaker on the southern provinces. For design wind speeds, Vietnam is divided into five zones with 20-year return period gust speeds ranging from 33 to 55 m/s, in the national loading code TCVN-2737 (values of dynamic pressure are given in the code). The zones of higher wind speeds occur close to the coast and reflect different degrees of influence from typhoons (Table D.11).

Extreme Wind Classifications: II, III and IV

D.2.83 Windward Islands

These islands in the Caribbean are visited by developing hurricanes, and weaker tropical storms. Analysis of extreme wind speeds for the former British colonies in the Caribbean was carried out by Shellard (1972). These results were used by the Caribbean Uniform Building Code (Caribbean Community Secretariat, 1986) and by the Code of Practice of the Barbados Association of Professional Engineers (1981). The latter specified a 50-year return period 3-s gust speed of 58 m/s, based on studies for Barbados. This value was also applicable to St. Vincent, St. Lucia, Grenada and Dominica. However, a more recent report by Applied Research Associates (Vickery and Wadhera, 2008) showed an effective 50-year return period peak 3-s gust speeds of 41 m/s for Barbados and St. Vincent, and 45–47 m/s for St. Lucia and Dominica. The same report gives 38 m/s for Grenada, but this seems too low, given that the island was struck by the very damaging Hurricane 'Ivan' in 2004.

Extreme Wind Classification: III

D.2.84 Zimbabwe

Zimbabwe is an elevated land-locked country with most land at 1000 m above sea level or greater. The country falls between 15° and 22° South in latitude. The expected dominant wind storm in this environment would be thunderstorm winds created by local convection. A code of practice for wind loads (Central African Standards Institution, 1977) specified a basic design wind speed (50-year return period gust) of 35 m/s, for the whole country. An analysis by Lewis (1983) for five different locations found higher and lower values than this.

Extreme Wind Classifications: I and II

REFERENCES

Almaawali, S.S.S., Majid, T.A. and Yahya, A.S. 2008. Determination of basic wind speed for building structures in Oman. *Proceedings, International Conference on Construction and Building Technology*, Kuala Lumpur, Malaysia, 16–20 June, 235–44.

American Society of Civil Engineers. 2006. *Minimum Design Loads for Buildings and Other Structures*. ANSI/ASCE 7-05, ASCE, New York.

Bajic, A. and Peros, B. 2005. Meteorological basis for wind loads calculation in Croatia. *Wind and Structures*, 8: 389–405.

Ballio, G., Lagomarsino, S., Piccardo, G. and Solari, G. 1999. Probabilistic analysis of Italian extreme winds: Reference velocity and return criterion. *Wind and Structures*, 2: 51–68.

Barbados Association of Professional Engineers. 1981. *Code of Practice for Wind Loads for Structural Design*. Consulting Engineers Partnership Ltd. and Caribbean Meteorological Institute, Barbados, West Indies.

British Standards Institution. 1997. *Loading for Buildings. Part 2. Code of Practice for Wind Loads*. BS 6399: Part 2, London, U.K.

Buildings Department, Hong Kong. 2004. *Code of Practice on Wind Effects – Hong Kong*. Hong Kong Special Administrative Region, China.

Bureau of Indian Standards. 1987. *Indian Standard IS: 875 (Part 3)*. Bureau of Indian Standards, New Delhi.

Caribbean Community Secretariat. 1986. *Caribbean Uniform Building Code. Part 2. Section 2. Wind Load*. Caribbean Community Secretariat, Georgetown, Guyana.

Caspar, W. 1970. Maximale windgeschwindigkeiten in der BRD. *Bautechnik*, 47: 335–40.

Cechet, R.P. and Sanabria, L.A. 2011. *Australian Extreme Wind Baseline Climate Investigation Project: Intercomparison of Time Series and Coincident Dines and Cup Anemometer Observations*. Geoscience Australia, GA Record 2011/2023, Australian Government, Canberra, Australia.

C.E.N. (European Committee for Standardization). 1994. *Eurocode 1: Basis of Design and Actions on Structures. Parts 2–4: Wind Actions*. ENV-1991-2-4 (draft), C.E.N., Brussels.

Central African Standards Institution. 1977. *Wind Loads on Buildings. CASI 160, Part 2*. Central African Standards Institution, Harare.

Chen, T.Y. 1975. *Comparison of Surface Winds in Hong Kong*. Hong Kong Royal Observatory, Technical Note 41, Hong Kong, China.

Choi, E.C.C. 1999. Extreme wind characteristics over Singapore – An area in the Equatorial belt. *Journal of Wind Engineering and Industrial Aerodynamics*, 83: 61–9.

Cook, N.J. and Prior, M.J. 1987. Extreme wind climate of the United Kingdom. *Journal of Wind Engineering and Industrial Aerodynamics*, 26: 371–89.

Davenport, A.G. and Baynes, C.J. 1972. An approach to the mapping of the statistical properties of gradient winds over Canada. *Atmosphere*, 10: 80–92.

Davenport, A.G., Georgiou, P.N., Mikitiuk, M., Surry, D. and Lythe, G. 1984. The wind climate of Hong Kong. *Proceedings, 3rd International Conference on Tall Buildings*, Hong Kong and Guangzhou.

Davenport, A.G., Lukkunaprasit, P., Ho, T.C.E., Mikitiuk, M. and Surry, D. 1995. The design of transmission line towers in Thailand. *Proceedings, 9th International Conference on Wind Engineering*, New Delhi, 9–13 January, 57–68.

Department of Standards and Norms (China). 1994. *Load Code for the Design of Building Structures*. (English Translation). GBJ-987. New World Press, Beijing.

Dorman, C.M.L. 1983. Extreme wind gust speeds in Australia, excluding tropical cyclones. *Civil Engineering Transactions, Institution of Engineers, Australia*, CE25: 96–106.

Duranona, V. 2011. Wind impact on Uruguay: Vulnerability to extreme winds and estimation of their risk. *Proceedings, 13th International Conference on Wind Engineering*, Amsterdam, Netherlands, 10–15 July.

Duranona, V. 2013. Highest wind gusts in Uruguay: Characteristics and associated meteorological events. *Proceedings, 12th Americas Conference on Wind Engineering*, Seattle, USA, 16–20 June.

European Committee for Structural Steelwork (E.C.C.S.). 1978. *Recommendations for the Calculation of Wind Effects on Buildings and Structures*. Technical Committee T12, E.C.C.S., Brussels.

European Convention for Standardization (C.E.N.). 1994. *Eurocode 1: Basis of Design and Actions on Structures. Parts 2–4: Wind Actions*. ENV-1991-2-4, C.E.N., Brussels.

Faber, S.E. and Bell, G.J. 1967. *Typhoons in Hong Kong and Building Design*. Hong Kong Royal Observatory, Reprint No. 37, Hong Kong, China.

Gomes, L. and Vickery, B.J. 1976a. On thunderstorm wind gusts in Australia. *Civil Engineering Transactions, Institution of Engineers, Australia*, CE18: 33–9.

Gomes, L. and Vickery, B.J. 1976b. Tropical cyclone gust speeds along the northern Australian coast. *Civil Engineering Transactions, Institution of Engineers, Australia*, CE18: 40–8.

Holmes, J.D. 2002. A re-analysis of recorded wind speeds in Region A. *Australian Journal of Structural Engineering*, 4: 29–40.

Holmes, J.D. and Weller, R. 2002. *Design Wind Speeds for the Asia-Pacific Region*. Handbook, HB212, Standards Australia, Sydney, NSW, Australia.

Holmes, J.D., Kwok, K.C.S. and Hitchcock, P. 2009. Extreme wind speeds and wind load factors for Hong Kong. *7th Asia-Pacific Conference on Wind Engineering*, Taipei, Taiwan, 8–12 November.

Horokhov, Y.V. and Nazim, Y.V. 2001. Wind and sleet loads on the aerial power lines in Ukraine. *Journal of Wind Engineering and Industrial Aerodynamics*, 89: 1409–19.

Instituto Uruguayo de Normas Tecnicas. 1984. Code wind actions on structures. (In Spanish). UNIT 50-84.

Jeary, A.P. 1997a. The wind climate of Hong Kong. *Journal of Wind Engineering and Industrial Aerodynamics*, 72: 433–44.

Jeary, A.P. 1997b. *Designer's Guide to the Dynamic Response of Structures*. E.F. and N. Spon, London.

Jensen, M. and Franck, N. 1970. *The Climate of Strong Winds in Denmark*. Danish Technical Press, Copenhagen.

Kasperski, M. 2002. A new wind zone map of Germany. *Journal of Wind Engineering and Industrial Aerodynamics*, 90: 1271–87.

Kintenar, R.L. 1971. *An Analysis of Annual Maximum Wind Speeds in the Philippines*. UNESCO, Manila.

Kruger, A.C., Goliger, A.M., Retief, J.B. and Sekele, S. 2010. Strong wind climatic zones in South Africa. *Wind and Structures*, 13: 37–55.

Kruger, A.C., Goliger, A.M., Retief, J.B. and Sekele, S. 2012. Clustering of extreme winds in the mixed wind climate of South Africa. *Wind and Structures*, 15: 87–109.

Lewis, G. 1983. Probabilistic estimation of extreme climatological parameters over Zimbabwe. *Proceedings, Institution of Engineers, (UK), Part 2*, 75: 551–5.

Lopez, A. and Vilar, J.I. 1995. Basis of the *Mexican Wind Handbook* for the evaluation of the dynamic response of slender structures. *Proceedings, 9th International Conference on Wind Engineering*, New Delhi, 9–13 January, 1890–1900.

Loredo-Souza, A.M. 2012. Meteorological events causing extreme winds in Brazil. *Wind and Structures*, 15: 177–88.

Loredo-Souza, A.M. and Paluch, M.J. 2005. Brazil storm Catarina: Hurricane or extratropical cyclone? *Proceedings, 10th Americas Conference on Wind Engineering*, Baton Rouge, 31 May–4 June.

Lukkunaprasit, P. 1997. Seismic and wind loading codes in Thailand. *International Workshop on Harmonization in Performance-Based Building Structural Design*, Tsukuba, Japan, 1–3 December.

Melbourne, W.H. 1984. Design wind data for Hong Kong and surrounding coastline. *Proceedings, 3rd International Conference on Tall Buildings*, Hong Kong and Guangzhou.

Milford, R.V. 1987. Annual maximum wind speeds for South Africa. *The Civil Engineer in South Africa*, January: 15–19.

Miller, C.A., Cook, N.J. and Barnard, R.H. 2001. Towards a revised base wind speed map for the United Kingdom. *Wind and Structures*, 4: 197–212.

National Research Council (Canada). 2005. *Climatic Information for Building Design in Canada*. Appendix C to the National Building Code of Canada. N.R.C., Ottawa.

Peterka, J.A. and Shahid, S. 1998. Design gust wind speeds in the United States. *Journal of Structural Engineering (ASCE)*, 124: 207–14.

Popov, N.A. 2001. Wind load codification in Russia and some estimates of a gust load accuracy provided by various codes. *Journal of Wind Engineering and Industrial Aerodynamics*, 88: 171–81.

Riera, J.D. and Nanni, L.F. 1989. Pilot study of extreme wind velocities in a mixed climate considering wind orientation. *Journal of Wind Engineering and Industrial Aerodynamics*, 32: 11–20.

Riera, J.D. and Reimundin, J.C. 1970. Sobre la distribucion de velocidades maximas de viento en la Republica Argentina. *Simposio sobre Acciones en Estructuras*, National University of Tucuman, Argentina.

Saaroni, H., Ziv, B., Bitan, A. and Alpert, P. 1998. Easterly wind storms over Israel. *Theoretical and Applied Climatology*, 59: 61–77.

Salgado Vieira Filho, J.M. 1975. Velocidades maximas do vento no Brasil. Master's thesis, Federal University of Rio Grande do Sul, Brazil.

Saudi Aramco. 2005. Meteorological and seismic data. Engineering Standard SAES-A-112. Environmental Protection Department, Saudi Aramco, Dharan, Saudi Arabia, December 2005.

Schueller, G.I. and Panggabean, H. 1976. Probabilistic determination of design wind velocity in Germany. *Proceedings, Institution of Engineers, (UK), Part* 2, 61: 673–83.

de Schwarzkopf, M.L.A. 1995. Meteorological weather patterns and wind storm types. Course notes. *Design Loadings on Transmission Lines*. Brisbane, Queensland, Australia, 5–7 July.

Shellard, H.C. 1958. Extreme wind speeds over Great Britain and Northern Ireland. *Meteorological Magazine*, 87: 257–65.

Shellard, H.C. 1962. Extreme wind speeds over the United Kingdom for periods ending 1959. *Meteorological Magazine*, 91: 39–47.

Shellard, H.C. 1972. Extreme wind speeds in the Commonwealth Caribbean. *Journal of the Barbados Association of Professional Engineers*, December: 1–8.

Simiu, E., Changery, M.J. and Filliben, J.J. 1979. *Extreme Wind Speeds at 129 Stations in the Contiguous United States. NBS Building Science Series 118*, National Bureau of Standards, Washington, DC.

Sites, J.S. and Peterson, R.E. 1995. Climatology of Southwest Indian Ocean tropical cyclones – (1962–1987). *Proceedings, 9th International Conference on Wind Engineering*, New Delhi, 9–13 January.

South African Bureau of Standards. 1989. *The General Procedures and Loadings To Be Adopted in the Design of Buildings*. South African Standard SABS 0160-1989.

Standards Australia. 2011. *Structural Design Actions. Part 2: Wind Actions*. Standards Australia, Sydney. Australian/New Zealand Standard AS/NZS1170.2:2011.

Thom, H. 1960. Distributions of extreme winds in the United States. *ASCE Journal of the Structural Division*, 85: 11–24.

Thom, H. 1968. New distributions of extreme winds in the United States. *ASCE Journal of the Structural Division*, 94: 787–1801.

Turner, R., Revell, M., Reese, S., Moore, S. and Reid, S. 2012. Some recent extreme wind events in New Zealand. *Wind and Structures*, 15: 163–176.

Vickery, P.J. and Wadhera, D. 2008. Wind speed maps for the Caribbean for application with the wind load provisions of ASCE 7. ARA Report 18108-1, Applied Research Associates, Raleigh, NC, USA, prepared for the Pan American Health Organization.

Vilar, J.I. et al. 1991. Analisis estadistico de datos de vientos maximos. Reporte Interno 42/2929/I/02/P. Departamento de Ingenieria Civil. Instituto de Investigaciones Electricas, Mexico.

Viollaz, A., Riera, J.D. and Reimundin, J.C. 1975. Estudio de la distribucion de velocidades maximas de viento en la Republica Argentina, Informe I-75-1, Structures Laboratory. National University of Tucuman, Argentina.

Wang, C.-H., Wang, X. and Khoo, Y.B. 2013. Extreme wind gust hazard in Australia and its sensitivity to climate change. *Natural Hazards*, 67: 549–67.

Whittingham, H.E. 1964. *Extreme Wind Gusts in Australia*. Bureau of Meteorology. Bulletin No. 46, Melbourne, Australia.

Appendix E: Some approximate formulas for structural natural frequencies

A necessary pre-requisite for dynamic response estimation is knowledge of the natural frequencies in the lowest sway modes of the structure. It is also useful to know these values to determine whether or not dynamic response calculations to wind are, in fact, necessary.

Most modern frame-analysis or finite-element computer programmes will of course give this information. However, if the structure is still in the early design stage, application of simple empirical formulae may be useful. Some of these are given here.

- For multi-storey office buildings, which are approximately uniform in plan (Jeary and Ellis, 1983):

  $$n_1 \approx 46/h \qquad \text{(E.1)}$$

  where h is the height of the building in metres.

  An alternative approximate formula for tall buildings is (Hirsch and Ruscheweyh, 1971)

  $$n_1 \approx 0.4\,(100/h)^{1.58} \qquad \text{(E.2)}$$

- For cantilevered masts or poles of uniform cross-section (in which bending action dominates):

  $$n_1 = \frac{0.56}{h^2}\sqrt{\frac{EI}{m}} \qquad \text{(E.3)}$$

  where
  EI is the bending stiffness of the section
  m is the mass/unit height

  (This of course is an exact formula for uniform masts or towers; it can be used for those with a slight taper, with average values of EI and m.)

- An approximate formula for cantilevered, *tapered*, circular poles (European Convention for Structural Steelwork, 1978)

  $$n_1 \approx \frac{\lambda}{2\pi \cdot h^2}\sqrt{\frac{EI}{m}} \qquad \text{(E.4)}$$

  where h is the height, and E, I and m are calculated for the cross-section *at the base*.

λ depends on the wall thicknesses at the tip and base, e_t and e_b, and external diameter at the tip and base, d_t and d_b, according to the following formula:

$$\lambda = \left[1.9 \exp\left(\frac{-4d_t}{d_b}\right)\right] + \left[\frac{6.65}{0.9 + (e_t/e_b)^{0.666}}\right] \quad \text{(E.5)}$$

Note that for $(d_t/d_b) = (e_t/e_b) = 1.0$, i.e. a uniform cylindrical tube, $\lambda = 3.52$, and Equation E.3 results.

- For steel, tubular chimneys, Hirsch and Bachmann (1997a) give the following approximate formula:

$$n_1 = \xi \cdot 1010 \cdot \left(\frac{b}{h}\right)^2 \quad \text{(E.6)}$$

where b is the external diameter in metres, h is the height in metres and ξ is an empirical factor, approximately equal to 1.15 times (mass of steel part/total mass of the chimney, including insulation, etc.)$^{1/2}$.

- For free-standing lattice towers (without added ancillaries such as antennas, lighting frames, etc.) (Standards Australia, 1994):

$$n_1 \approx 1500 w_a / h^2 \quad \text{(E.7)}$$

where
w_a is the average width of the structure in metres
h is the tower height

An alternative formula for lattice towers (with added ancillaries) is (Wyatt, 1984)

$$n_1 \approx \left(\frac{L_N}{h}\right)^{2/3} \left(\frac{w_b}{h}\right)^{1/2} \quad \text{(E.8)}$$

where
w_b = tower base width
L_N = 270 m for square base towers, or
230 m for triangular base towers

- A formula which seems to fit data on bridges with spans between 20 and 1000 m (Jeary, 1997) is

$$n_1 \approx 40(L_s)^{-3/4} \quad \text{(E.9)}$$

where L_s is the span in metres (main span in the case of a multi-span structure). An alternative formula for cable-stayed bridges is (Hirsch and Bachmann, 1997b)

$$n_1 \cong 110/L_s \quad \text{(E.10)}$$

where L_s is the length of the main span in metres.

REFERENCES

European Convention for Structural Steelwork (E.C.C.S.). 1978. *Recommendations for the Calculation of Wind Effects on Buildings and Structures*. Technical Committee T12, E.C.C.S., Brussels.

Hirsch, G. and Bachmann, H. 1997a. Wind-induced vibrations: Chimneys and masts. In *Vibration Problems in Structures*. 2nd edition. Birkhäuser Verlag, Basel, Switzerland.

Hirsch, G. and Bachmann, H. 1997b. Wind-induced vibrations: Suspension and cable-stayed bridges. In *Vibration Problems in Structures*. 2nd edition. Birkhäuser Verlag, Basel, Switzerland.

Hirsch, G. and Ruscheweyh, H. 1971. Newer investigations of non-steady wind loadings and the dynamic response of tall buildings and other constructions. *Proceedings, 3rd International Conference on Wind Effects on Buildings and Structures*, Tokyo, Japan, 6–9 September, pp. 811–23, Saikon Co. Ltd.

Jeary, A.P. 1997. *Designer's Guide to the Dynamic Response of Structures*. E.F. and N. Spon, London.

Jeary, A.P. and Ellis, B.R. 1983. On predicting the response of tall buildings to wind excitation. *Journal of Wind Engineering and Industrial Aerodynamics*, 13: 173–82.

Standards Australia. 1994. Design of steel lattice towers and masts. Standards Australia, North Sydney, Australian Standard AS3995-1994.

Wyatt, T.A. 1984. Sensitivity of lattice towers to fatigue induced by wind gusts. *Engineering Structures*, 6: 262–7.

Appendix F: Example of application of the LRC method for the effective static wind loads on a simple structure

F.1 INTRODUCTION

In this appendix, the LRC formula of Kasperski (1992) is applied to a simple structure – a pitched free roof – to illustrate the method of determining the effective static wind pressures. Data were obtained from wind-tunnel tests carried out by Ginger and Letchford (1991).

F.2 WIND PRESSURE DATA

A model of a pitched free roof (i.e. no walls), with a roof pitch of 22.5°, at a geometric scaling ratio of 1/100, was tested in a boundary-layer wind tunnel by Ginger and Letchford (1991). Net area-averaged pressures across the windward and leeward roof slopes were measured. Three panels per roof half were used, but the data used here apply to the central panels, that is the central third of the roof.

Figure F.1b shows the mean and standard deviation pressure coefficients for a wind direction normal to the ridge as shown; the latter values are in brackets. Maximum and minimum panel pressure coefficients were also recorded, and are shown in Figure F.1c. The directions for positive net panel pressures are shown in the figure.

F.3 EFFECTIVE STATIC LOADS FOR TOTAL LIFT AND DRAG

At first, one might assume that the maximum total lift force should be obtained from the two recorded minimum pressures on the two roof panels. Similarly, the maximum drag could be obtained from the maximum on panel 1 and the minimum on panel 2. However, this would be incorrect, and conservative, as these values do not occur simultaneously. The *expected* pressure coefficients coinciding with the maximum and minimum lift and drag are derived in the following section.

F.3.1 Mean lift and drag

The mean lift force (positive upwards) is obtained as follows:

$$\bar{L} = (-1)\,(0.46)\,q_h(d/2) + (-1)\,(-0.60)\,q_h(d/2) = 0.14q_h(d/2)$$

where

q_h is the reference mean dynamic pressure at roof height $\left(=(1/2)\rho_a \bar{U}_h^2\right)$

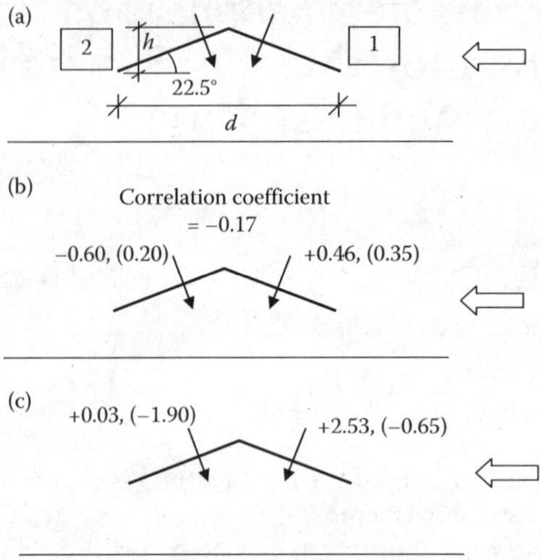

Figure F.1 Pressure coefficients for a pitched free roof. (a) Roof geometry; (b) mean (standard deviation), pressure coefficients and correlation coefficient; (c) maximum and (minimum) pressure coefficients.

d is the along-wind length of the roof

In this case, the *influence coefficients* for the *lift force* are both equal to $-(d/2)$.
The mean drag force is given by

$$\bar{D} = (+1)\,(0.46)\,q_h(h) + (-1)\,(-0.60)\,q_h(h) = 1.06q_h(h) = 0.44q_h(d/2)$$

since $h/(d/2) = \tan 22.5° = 0.414$.

The *influence coefficients* for the *drag force* are equal to $+h = (d/2)\tan 22.5°$ for panel 1, and $-h = -(d/2)\tan 22.5°$, for panel 2.

F.3.2 Standard deviations of lift and drag

The root mean square (rms) fluctuating, or standard deviation, lift and drag forces can be obtained by *covariance integration* (Holmes and Best, 1981; Ginger and Letchford, 1991, 1994).

The standard deviation of the lift force, σ_L, is obtained as follows:

$$\sigma_L = q_h\,(d/2)\,[(0.35)^2 + (0.20)^2 + 2(-0.17).(0.35)\,(0.20)]^{1/2}$$

$$= 0.372\,q_h\,(d/2)$$

The standard deviation of the drag force, σ_D, is

$$\sigma_D = q_h\,(d/2)\tan 22.5°[(0.35)^2 + (0.20)^2 - 2(-0.17)\cdot(0.35)\,(0.20)]^{1/2}$$

$$= 0.432\,q_h\,(d/2)\tan 22.5° = 0.179\,q_h\,(d/2)$$

F.3.3 Effective pressures for peak lift force

The expected pressure on panel 1 when the *lift* is a maximum is given by (Kasperski, 1992)

$$(p_1)_{\hat{L}} = q_h[\bar{C}_{p1} + g\rho_{p1,L}\sigma_{Cp1}]$$

where g is a peak factor for the lift (it will be taken as 4), and $\rho_{p1,L}$ is the correlation coefficient between the pressure $p_1(t)$ and the lift $L(t)$.

The *covariance* between the pressure $p_1(t)$ and the lift $L(t)$ is given by

$$-q_h^2(d/2)[\sigma_{Cp1}^2 + \overline{p_1'p_2'}] = -q_h^2(d/2)[(0.35)^2 + (-0.17)(0.35)(0.20)]$$

$$= -(0.111)\ q_h^2(d/2)$$

Then,

$$\rho_{p1,L} = \frac{-0.111}{(0.35)(0.372)} = -0.853$$

Hence,

$$(p_1)_{\hat{L}} = q_h[\bar{C}_{p1} + g\rho_{p1,L}\sigma_{Cp1}] = q_h[(0.46) + 4\,(-0.853)\,(0.35)] = -0.73\ q_h$$

Similarly, the *covariance* between the pressure $p_2(t)$ and the lift $L(t)$ is given by

$$-q_h^2(d/2)[\sigma_{Cp2}^2 + \overline{p_1'p_2'}] = -q_h^2(d/2)[(0.20)^2 + (-0.17)(0.35)(0.20)]$$

$$= -(0.028)\ q_h^2(d/2)$$

Then,

$$\rho_{p2,L} = \frac{-0.028}{(0.20)(0.372)} = -0.376$$

Hence,

$$(p_2)_{\hat{L}} = q_h[\bar{C}_{p2} + g\rho_{p2,L}\sigma_{Cp2}] = q_h[(-0.60) + 4\,(-0.376)\,(0.20)] = -0.90\ q_h$$

Thus, the expected pressure coefficients corresponding to the maximum lift (acting upwards) are

$$(C_{p1})_{\hat{L}} = -0.73 \qquad (C_{p2})_{\hat{L}} = -0.90$$

The pressures corresponding to the *minimum* lift force (downwards) are also of interest. In this case,

$$(p_1)_{\check{L}} = q_b[\bar{C}_{p1} - g\rho_{p1,L}\sigma_{Cp1}] = q_b[(0.46) - 4\ (-0.853)\ (0.35)] = +1.65\ q_b$$

and,

$$(p_2)_{\check{L}} = q_b[\bar{C}_{p2} - g\rho_{p2,L}\sigma_{Cp2}] = q_b[(-0.60) - 4\ (-0.376)\ (0.20)] = -0.30\ q_b$$

Hence,

$$(C_{p1})_{\check{L}} = +1.65 \qquad (C_{p2})_{\check{L}} = -0.30$$

These pressure coefficients are shown in Figure F.2a and b.

F.3.4 Effective pressures for maximum drag force

The expected pressures for the maximum *drag* force can be determined in a similar way as the lift force, as follows.

The *covariance* between the pressure $p_1(t)$ and the drag $D(t)$ is given by

$$q_b^2\,(d/2)\ \tan\ 22.5°[\sigma_{Cp1}^2 - \overline{p_1'p_2'}] = q_b^2\,(d/2)\ \tan\ 22.5°[(0.35)^2 - (-0.17)(0.35)(0.20)]$$

$$= (0.134)\ q_b^2(d/2)\ \tan 22.5°$$

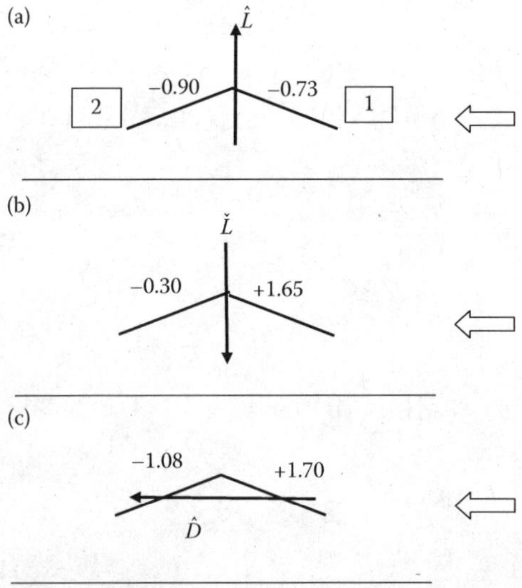

Figure F.2 Pressure coefficients for a pitched free roof. (a) Effective pressures for maximum lift force; (b) effective pressures for minimum lift force; (c) effective pressures for maximum drag force.

Then,

$$\rho_{p1,D} = \frac{0.134}{(0.35)(0.432)} = 0.886$$

Hence,

$$(p_1)_{\hat{D}} = q_b[\bar{C}_{p1} + g\rho_{p1,D}\sigma_{Cp1}] = q_b[(0.46) + 4\ (0.886)\ (0.35)] = 1.70\ q_b$$

(again taking a peak factor of 4).

Similarly, the *covariance* between the pressure $p_2(t)$ and the drag $D(t)$ is given by

$$-q_b^2(d/2)\ \tan\ 22.5°[\sigma_{Cp2}^2 - \overline{p_1'p_2'}] = -q_b^2(d/2)\ \tan\ 22.5°[(0.20)^2 - (-0.17)(0.35)(0.20)]$$

$$= -(0.052)\ q_b^2(d/2)$$

Then,

$$\rho_{p2,D} = \frac{-0.052}{(0.20)(0.432)} = -0.602$$

Hence,

$$(p_2)_{\hat{D}} = q_b[\bar{C}_{p2} + g\rho_{p2,D}\sigma_{Cp2}] = q_b[(-0.60) + 4\ (-0.602)(0.20)] = -1.08\ q_b$$

Thus, the expected pressure coefficients corresponding to the maximum drag are

$$(C_{p1})_{\hat{D}} = +1.70 \qquad (C_{p2})_{\hat{D}} = -1.08$$

These pressure coefficients are shown in Figure F.2c.

F.4 DISCUSSION

The effective pressure coefficients for maximum and minimum lift, and maximum drag, as summarised in Figure F.2, are clearly quite different to each other, and indicate the difficulty in specifying a single set of pressure coefficients in a code or standard, for a structure such as this.

It can be checked that the values obtained in the previous section will in fact give the correct values of the peak load effects. For example, the maximum lift can be obtained in two ways as follows.

From the effective static pressure coefficients:

$$\hat{L} = (-1).(-0.73)\ q_b(d/2) + (-1)\cdot(-0.90)\ q_b(d/2) = 1.63q_b(d/2)$$

Directly from the mean and standard deviation:

$$\hat{L} = L + 4\sigma_L = 0.14q_h(d/2) + 4 \times 0.372q_h(d/2) = 1.63q_h(d/2)$$

The effective static pressure coefficients for each panel should lie between the limits set by the maximum and minimum pressure coefficients for each panel. This is the case here (see Figures F.1 and F.2), except the value on panel 1 for $\hat{L}$, −0.73, is slightly more negative than the measured minimum value of −0.65. This could result from a sampling error in the measured peak, or the choice of a slightly conservative peak factor of 4 for the lift force.

F.5 CONCLUSIONS

This example has explained, using a simple two-panel case, the LRC methodology for determining the expected pressure distributions corresponding to peak load effects, based on correlations. More complex cases, such as large roofs, require a large number of panels, and a matrix of correlation coefficients, but the principles of the calculation are the same.

REFERENCES

Ginger, J.D. and Letchford, C.W. 1991. Wind loads on canopy roofs. University of Queensland, Department of Civil Engineering, Research Report, CE132, June.

Ginger, J.D. and Letchford, C.W. 1994. Wind loads on planar canopy roofs—Part 2: Fluctuating pressure distributions and correlations. *Journal of Wind Engineering and Industrial Aerodynamics*, 51: 353–70.

Holmes, J.D. and Best, R.J. 1981. An approach to the determination of wind load effects on low-rise buildings. *Journal of Wind Engineering and Industrial Aerodynamics*, 7: 273–87.

Kasperski, M. 1992. Extreme wind load distributions for linear and nonlinear design. *Engineering Structures*, 14: 27–34.

Index

A

B

C

D

E

T

U

V

Z